Congress Proceedings

Volume II

XIX International Coal Preparation Congress
13 – 15 November, 2019,

New Delhi – India

Editor

Raj K. Sachdev

Coal Preparation Society of India

WOODHEAD PUBLISHING INDIA PVT LTD

New Delhi

Published by Woodhead Publishing India Pvt. Ltd.
Woodhead Publishing India Pvt. Ltd.,
303, Vardaan House, 7/28, Ansari Road,
Daryaganj, New Delhi - 110002, India
www.woodheadpublishingindia.com
Phone: 91-11-23266107, 011-43612145

First published 2019, Woodhead Publishing India Pvt. Ltd.
© Woodhead Publishing India Pvt. Ltd., 2019

Woodhead Publishing India Pvt. Ltd. ISBN: 978-93-88320-19-1
Woodhead Publishing India Pvt. Ltd. e-ISBN: 978-93-88320-20-7

Typeset by Bhumi Graphics, New Delhi
Printed and bound by Replika Press Pvt. Ltd.

Acknowledgement

I sincerely thank all the 300 plus authors and co-authors who have contributed articles in these two volumes of proceedings. I also thank the Members of the International Organising Committee (IOC) of the XIX International Coal Preparation Congress (ICPC) who took pains in reviewing all 170 plus 'ABSTRACTS' received and shortlisted some 115 abstracts for inclusion in the congress agenda. On behalf of the Coal Preparation Society of India (CPSI), I express deep gratitude to the Ministries of Coal, Power, New and Renewable Energy, Environment, Forest & Climate Change, Steel, Mines, Science & Technology, Earth Sciences and Economic Diplomacy & States Division, External Affairs for extending their full support to this prestigious global event on COAL.

I also thank the editorial team of the publishers Woodhead Publishing India Pvt. Ltd., for their efforts in bringing out this two volume set of 'Congress Proceedings'.

My special thanks to the Saint Petersburg Mining University and the International Competence Centre for Mining-Engineering Education, Saint Petersburg (Russia) under the auspices of UNESCO for their appreciation and recognition of this work.

Raj K. Sachdev

Foreword

India is a country, which has been fortunate to be endowed with very large reserves of coal. As of date, the reserve is over 300 billion tonnes of coal. While the abundance of this fuel is acknowledged,most of this coal has a very high ash content, which makes beneficiation a necessary process to improve its quality and to render its usage as a fuel, which is more environment friendly. The need to wash and improve the quality of the coal has been endorsed by the government of India repeatedly, but the progress in establishing washeries by Coal India Ltd., the main producer of coal, is abysmally poor. Coal Preparation Society of India is the front-runner in promoting the use of washed coal. There are some very critical issues involved in promoting this activity despite clear benefits in respect of preservation of the environment. Any enterprise would welcome any sort of move to get and use good quality coal to improve its business. We in CPSI are constantly engaged in creating forums and encouraging industries to recognize both environmental and commercial benefits of washing.

CPSI is a member of the Organising Committee of the International Congress on Coal Preparation (ICPC) and in that capacity has become the host for the 19thcongress, which is to be held in New Delhi from 13th to 15th November 2019. The first ICPC was held in 1950 in Paris, France. The main objective was to make the coal available and usable from the damaged mines in Europe after the World War. ICPC is coming back to India after 37 years. The 9th ICPC was held in New Delhi in 1982. ICPC is a forum where intensive deliberations take place on varied technical aspects of Coal Preparation reflecting the worldwidestatus with regard to the latest state-of-the-art technologies being developed or deployed in the industry in respect of coal preparation. This congress will provide all stakeholders, which may include coal miners, steel and cement manufacturers, thermal power generating companies, manufacturers of equipment used in coal handling and washing and many others an excellent opportunity to gather information on the latest technologies and also be able to interact with the technology providers. CPSI finds itself in an exceedingly advantageous situation in furthering its objectives as the organization, which is primarily responsible for holding the 19th ICPC.

ICPC is a forum where only technical matters are presented. We in CPSI initiated the process of seeking papers for presentation in the congress more than a year ago. Initially, abstracts were obtained which were scrutinized by the International Organising Committee of ICPC and thereafter the authors of the articles were requested to send us the full text of the articles. We were fortunate enough to receive more than 110 articles for presentationin the Congress. These proceedings contain the 100 articles, which are being presented in the Congress. All the proceedings of ICPC since its initiation have become an invaluable record of scientific advancements that have taken place in the area of coal preparation over the late 70 years. We are sure these proceedings will be of immense value to the industry and the scientists.

Alok Perti
Chairman, Coal Preparation Society of India (CPSI)
Chairman, National Organising Committee (NOC)

Editorial

Coal has been and will continue as the main source of energy for India for many more years. India's hunger for coal is more than justified because of its meagre known hydrocarbon resources and the country imports nearly 85% of its consumption of oil and natural gas. Furthermore, out of its 1.35 billion population about 25% is yet to be served with clean cooking fuel and electricity. Therefore, continuing use of coal which provides about 75% of India's electricity generation becomes a critical necessity. However, notwithstanding the fact that per capita electricity is only about 1100 Kwhr, country's economy has to grow at a reasonable rate of 8 to 9%, that makes it imperative that India continues to depend on coal based electricity, despite heavy investment on creating non-fossil based generation capacity of 40% of total electricity generation capacity of 800 – 900 GW required by 2030. With country's aim of becoming a Five Trillion Dollar economy by 2025, the requirement of coal for generation of electricity, for making iron & steel, cement for meeting requirement of growing infrastructure like ports, highways, housing, education, health facilities and other basic needs to provide required quality of life to its growing population will also grow commensurately.

With poor generic quality of domestic coal, it is imperative that all coals must be washed to reduce the ash content and improve the heat value so that it burns more efficiently with lesser emissions. With this, as its main objective, the Coal Preparation Society of India (CPSI) has been dedicatedly promoting washing of coal. Fortunately, with persistent efforts, CPSI has been able to convince the government as well all stakeholders about positives of coal washing and the benefits that accrue to the Nation as a whole.

Coal Preparation Society (CPSI) with strong industry support has made a niche for itself as a credible professional body representing the Indian coal industry. It was support of the member companies that prompted CPSI to undertake the onus of hosting the XIX International Coal Preparation Congress & Expo (ICPC).

The 9th ICPC was held in India in 1982 when the world was faced with sudden doubling of oil prices after Iranian revolution and beginning of Iran-Iraq war. However, coal continued to be the King. The XIX ICPC is now being held under the shadow of climate change, fears of melting glaciers, fast changing weather patterns and concerns about floods, droughts and crop losses. Naysayers are predicting death of coal. However, China, India and many other countries having limited choice between renewable and coal, will continue to use coal for power, iron and steel, cement etc for many more years to come. International Energy Agency (IEA) has predicted that in 2045 or so, around 26% (from present 38%) of world electricity will still be coming from coal. But we cannot afford to adopt an ostrich approach. India has accepted the need to wash coal, adopt high efficiency low emission (HELE) technologies for power generation and in coming years CCUS projects are sure to become reality.

In the above background, we are fortunate to have the XIX ICPC at this juncture. With expected participation of around 500 delegates including some 120 subject experts from overseas, this prestigious international event on coal offers a great opportunity to all participants including government policy makers to share experience and exchange knowledge on various topical issues on coal. At the same time, this is a unique opportunity for business community, investors, project developers and technology suppliers to show case their strengths and team up with local entities to establish business in India and elsewhere.

Coming to the subject of this publication, we have over 110 high quality papers from some 15 countries, covering all aspects of coal processing and also touching upon gasification, coal to chemicals and emerging uses of coal through its conversion into high value, low polluting energy and petro substitutes.

While every care is taken to edit these two volumes of 'Congress Proceedings', we will be open to receive suggestions for any omissions or errors. In case any reader desires to get in touch with the author(s) for any clarification, he or she may approach the concerned author directly. CPSI will provide the e mail contact of the concerned author, if requested.

I must record my sincere thanks to all 300 plus authors and co authors, my own CPSI team and the editorial team of Woodhead Publishing (India) who have put their heads together to bring out a two volume set of 'Congress Proceedings' running into over thousand pages.

Raj K. Sachdev

President, Coal Preparation Society of India (CPSI),
Chairman, International Organising Committee (IOC),
XIX International Coal Preparation Congress & Expo (ICPC),
New Delhi, India

Messages

भारत के उपराष्ट्रपति
VICE-PRESIDENT OF INDIA
<u>MESSAGE</u>

The most crucial element required for rapid socio-economic development is energy. In India, despite a serious effort to increase the dependence on renewable energy, **COAL** will remain the most important source for electricity generation for several years. Electricity, which in India is largely coal based is crucial for building modern infrastructure, supporting urbanization and improving the quality of life. The objective is to produce 1.5 billion tonne of coal by 2020 and increase it to 2.5 billion by 2032. With such increase in coal production and consumption, adoption of clean coal technologies for improved efficiency in power generation should be part of our energy security plan.

With every passing day, global climate change concerns are becoming more pronounced. India is committed to fulfill its declared targets in reduction of CO_2 emissions. In this effort the government has taken significant policy initiatives and I am hopeful that these efforts will helps in reducing overall CO_2 emissions, so that, India can meet its climate priorities.

I extend my heartiest felicitations to the **Coal Preparation Society of India** for spearheading the cause and concept of washing of coal in India. I am happy to note that **19th International Coal Preparation Congress** will be held in 2019 in New Delhi, India after a gap of 37 years. I extend heartiest welcome to the subject experts and delegates both from within the country and overseas who will be participating in this important event on coal. I wish the **Coal Preparation Society of India** all success in their undaunted efforts in making this international conference a great success.

(M. Venkaiah Naidu)

New Delhi
04th May, 2018.

Shri Narendra Modi
Hon'ble Prime Minister of India

"Due to the use of washed coal, the energy consumed in transportation, handling and milling, is optimized as the inert material from coal is eliminated. This helps in reducing the auxiliary consumption of equipment involved in coal processing because the use of improved quality coal ultimately results in reduction of emissionof GHG as compared to conventional coal."

'CONVENIENT ACTION – continuity for
Change' by Narendra Modi.

प्रल्हाद जोशी
PRALHAD JOSHI
ಪ್ರಲ್ಹಾದ ಜೋಶಿ

संसदीय कार्य, कोयला तथा खान मंत्री
भारत सरकार
नई दिल्ली
MINISTER OF PARLIAMENTARY AFFAIRS,
COAL AND MINES
GOVERNMENT OF INDIA
NEW DELHI

No. M(C/M&PA/Message/2019/   June, 2019

MESSAGE

Coal is valued as a vital energy and strategic resource essential to the world's sustainable development and energy security objectives.

With 75% of electricity generation, 65% steel production and 90% cement production being dependent of coal as a critical fuel; Coal is and will continue to be a major source of India's commercial needs for many more years to come. Therefore, it is essential that coal must be washed so that it is burnt efficiently and power plant emissions are also reduced. Therefore, washing of coal is an essential subset of coal production-supply chain in India.

It is heartening to note that the Coal Preparation Society of India (CPSI) has been dedicatedly working towards creating awareness about the benefits of using washed coal, among the coal producers as well as the consumers. I convey my best wishes to the Coal Preparation Society of India (CPSI) for its stupendous efforts in promoting coal washing in India.

I am happy to know that the XIX International Coal Preparation Congress (ICPC) will be held at New Delhi during 13-15 November, 2019 New Delhi and Coal Preparation Society of India (CPSI) is the organiser of this International event on Coal. I wish the XIX International Coal Preparation Congress (ICPC) a grand success and I am confident that this International Conference will be an excellent opportunity for exchange of experience, knowledge and expertise amongst subject experts from various coal producing countries attending this important congress on coal.

(PRALHAD JOSHI)

Office : Room No. 15, Parliament House, New Delhi-110001, Tel : 011-23017780, 23017798, 23018729, Fax : 011-23792341
Office : Room No. 504, C Wing, 5th Floor, Shastri Bhawan, Tel : 23070522, 23070524, Fax : 23070529
Residence : 5, G.R.G. Road, New Delhi-110001, Tel : 011-23094650, 23093497
H. No. 122-D, 'Kamitartha' Mayuri Estate, Keshwapur, Hubli-580023 (Karnataka)
Tel. No. : 0836-2251055, 2258955 E-mail : pralhadvjoshi@gmail.com

प्रकाश जावडेकर
Prakash Javadekar

मंत्री
पर्यावरण, वन एवं जलवायु परिवर्तन मंत्रालय,
सूचना एवं प्रसारण मंत्रालय
भारत सरकार
Minister
Ministry of Environment, Forest & Climate Change
Ministry of Information and Broadcasting
Government of India

MESSAGE

Coal is the backbone of India's commercial energy supply and long term energy security and sustainable development. Nearly 75% of India's electricity is generated using coal. The Iron & steel, cement and many other industries are also heavily coal dependent. Various projections made by credible institutions including NITI Ayog, indicate that COAL will continue to be a major source of our country's energy needs for few decades.

India is endowed with abundant coal resources but its oil and gas resources are very limited. Due to its very generic nature, the quality of our domestic coal is poor due to its high ash content. Since washing of coal reduces its ash content and its heat value also improves, it is essential that coal must be washed so that it is burnt efficiently and power plant emissions are also reduced, therefore, coal has to be washed for its efficient burning and with less environmental pollution. We must ensure that our coal is washed as it will help to a significant extent in meeting India's commitments made at the Paris Climate Treaty.

It is gratifying to note that the **Coal Preparation Society of India (CPSI)** has been dedicatedly spearheading the concept and cause coal washing by creating awareness about the benefits of using washed coal, among all stake holders.

I am delighted to learn that the **XIX International Coal Preparation Congress (ICPC)** will be held during 13-15 November 2019 at New Delhi and Coal Preparation Society of India(CPSI) is the organiser of this international event on COAL. It is further heartening to know that this prestigious international congress on coal is being held in India after 37 years. Last one the IX ICPC was held in 1982.

I convey my best wishes to **Coal Preparation Society of India (CPSI)** for its stupendous efforts in promoting coal washing and clean coal technologies in India.

I wish the **XIX International Coal Preparation Congress (ICPC)** a grand success. I am sure that this international conference will afford an excellent opportunity for exchange of knowledge and sharing amongst experts from abroad who will be participating in this prestigious international event on coal.

Date: 22.08.2019

(Prakash Javadekar)

Paryavaran Bhawan, Jor Bagh Road, New Delhi-110 003
Tel. : 011-24695136, 24695132, Fax : 011-24695329
E-mail : minister-efcc@gov.in

5th Floor, Shastri Bhawan, New Delhi-110 003
Tel. : 011-23384340, Fax : 011-23782118
E-mail : mib.inb@nic.in

अनिल कुमार जैन, भा॰प्र॰से॰
सचिव
ANIL KUMAR JAIN, IAS
SECRETARY
Tel.: 23384884 Fax : 23381678
E-mail : secy.moc@nic.in

भारत सरकार
GOVERNMENT OF INDIA
कोयला मंत्रालय
MINISTRY OF COAL
शास्त्री भवन, नई दिल्ली–110 001
SHASTRI BHAWAN, NEW DELHI-110 001
www.coal.gov.in

MESSAGE

Coal is the key stone of India's energy supply. Apart from nearly 75% of India's electricity being generated using coal, 65% of iron & Steel and 90% of cement production is also coal dependent. Besides, there are many other industries using coal either as their feedstock or for generating heat and electricity.

India's coal resources being of 'drift' origin, the ash content is high. This is true both for coking coal as well as thermal variety of coal. Therefore, coal must be washed to reduce the ash content and improve its heat value.

The **Coal Preparation Society of India (CPSI)** is well known for its efforts in spreading awareness among the stakeholders about the techno-economic benefits that accrue to the power plants of using washed coal.

I am very happy that the **Coal Preparation Society of India (CPSI)** is organizing **XIX International Coal Preparation Congress (ICPC)** from 13th to 15th November, 2019 at New Delhi. It is further heartening to know that this prestigious international congress on coal is being held in India after 37 years.

I convey my best wishes to **Coal Preparation Society of India (CPSI)** for its dedicated efforts in promoting coal washing among all stakeholders.

I wish the **XIX International Coal Preparation Congress (ICPC)** a grand success. I am sure that this international conference will afford an excellent opportunity to Indian coal professionals and experts from overseas for exchange of knowledge and sharing experiences.

(Anil Kumar Jain)

Date: 03.10.2019

विनय कुमार
सचिव
Binoy Kumar
Secretary

भारत सरकार
इस्पात मंत्रालय
GOVERNMENT OF INDIA
MINISTRY OF STEEL

28th August, 2019

MESSAGE

The Indian Steel Industry has entered into a new development stage envisaging 250 million tonnes of steel production in coming decade. In 2018 Indian Crude Steel production was 2nd highest in world and this position is poised to be retained..

While Indian Steel Sector is on a growth trajectory, it is required that our steel companies are globally competitive. For achieving this, both availability and the cost of major inputs like iron ore and coking coal are key factors. As the global suppliers of metallurgical grade coking coal are limited in number, our endeavour must be to maximize the use of domestic coking coal. Since around 12% of total coal resources are known be of coking variety but of higher ash content, washing of such coals must be undertaken to obtain clean coal of desired quality to suit the requirement of steel mills. This would naturally require appropriate designing and setting up of coal washeries based on coal specific washing technologies. More importantly, additional sourcing of coking coal from domestic sources will reduce our import dependence to a significant extent.

It is really commendable that the **Coal Preparation Society of India (CPSI)** has been dedicated to promoting awareness among the stakeholders about the techno-economic benefits of utilizing domestic coking coals after proper washing. Towards this endeavour I am happy that the CPSI is organizing **XIX International Coal Preparation Congress (ICPC)** from 13th to 15th November, 2019 at New Delhi. I wish this prestigious international event on **COAL** all success.

Binoy Kumar
(Binoy Kumar)

Room No. 291, Udyog Bhawan, New Delhi-110 011, Tel. : +91 11 23063912, Fax : +91 11 23063489
E-mail : secy-steel@nic.in

सचिव
भारत सरकार
पर्यावरण, वन एवं जलवायु परिवर्तन मंत्रालय
SECRETARY
GOVERNMENT OF INDIA
MINISTRY OF ENVIRONMENT, FOREST AND CLIMATE CHANGE

सी.के.मिश्रा
C.K.Mishra

MESSAGE

Energy is an essential requirement for any economy and coal is the major supplier of India's commercial energy supply and a critical fuel for our country's long term energy security and sustainable development.

Nearly 75% of India's electricity is generated using coal. The iron & steel, cement and many other industries are also heavily coal dependent.

While India has abundant coal resources, its known hydrocarbon oil and gas resources are rather limited. However, the quality of our domestic coal is poor due to its high ash content. Since washing of coal reduces its ash content and its heat value also improves, it is essential that coal must be washed so that it is burnt efficiently with lower emissions.

Therefore, we must ensure that our coal is washed as it will help to a significant extent in meeting India's commitments made at the Paris Climate Treaty.

It is well known that the **Coal Preparation Society of India (CPSI)** has been dedicatedly spreading awareness about the benefits of using washed coal among coal producers as well as coal consumers.

I am very happy that the **XIX International Coal Preparation Congress (ICPC)** will be held during 13 -15 November, 2019 at New Delhi and Coal Preparation Society of India (CPSI) is the organiser of this international event on COAL. It is further heartening to know that this prestigious international congress on coal is being held in India after 37 years. Last one, the IX ICPC was held in 1982.

I convey my best wishes to **Coal Preparation Society of India (CPSI)** for its stupendous efforts in promoting coal washing among all stakeholders.

I wish the **XIX International Coal Preparation Congress (ICPC)** a grand success. I am sure that this international conference will be a good opportunity for exchange and sharing of knowledge amongst Indian coal professionals and experts from overseas participating in this prestigious international event on coal.

[C. K. Mishra]

Dated: 5th August, 2019
Place: New Delhi

इंदिरा पर्यावरण भवन, जोर बाग रोड, नई दिल्ली-110 003 फोन : (011) 24695262, 24695265, फैक्स : (011) 24695270

INDIRA PARYAVARAN BHAWAN, JOR BAGH ROAD, NEW DELHI-110 003 Ph. : (011) 24695262, 2465265, Fax : (011) 24695270
E-mail : secy-moef@nic.in, Website : moef.gov.in

United Nations • International
Educational, Scientific and • Competence Centre for
Cultural Organization • Mining-Engineering Education

The Saint Petersburg Mining University and the International Competence Centre for Mining-Engineering Education in Saint Petersburg (Russia) under the auspices of UNESCO, recognise excellent efforts put in by Mr. Raj K Sachdev, President Coal Preparation Society of India in editing these **Congress Proceedings**, and appreciate his contribution to the coal preparation industry.

Vice-Rector for Scientific and Innovative Activities  V. Bazhin
of Saint Petersburg Mining University

Contents

Additional Articles

About International Coal Preparation Congress (ICPC)

The **International Coal Preparation Congress (ICPC)** was an offshoot of the Allied Coal Commission which was constituted as part of the Marshall Plan post World War II to initiate the process of reconstruction of Europe. The first ICPC was held in 1950 in France and subsequent Congresses were held in Germany, Belgium, UK, USA, Poland, Australia, Ukraine, Russia, India, Canada, Japan, South Africa, Turkey and China. ICPC is now held every three years in a country selected by the IOC through ballot.

The International Organizing Committee (IOC) of the ICPC is a body which has representatives from 15 countries. Representation on IOC is by a non-government organization which deals in their respective country with the issues relating to coal preparation.

The goals of the ICPC inter alia, are:

- to ensure the growth and development of the coal preparation industry, support and promotion of scientific and technical cooperation among all countries;
- to facilitate the exchange of information about the state-of-art technologies in the field of coal mining, preparation and transportation of coal. Also promote new and emerging technologies in the area of environment friendly and low emission coal utilisation.
- to bring together experts from different countries in order to form professional contacts, exchange experimental results, and promote international cooperation.

History of International Coal Preparation Congress

No.	Year & Place	No.	Year & Place
I.	1950 - France, Paris	XI	1990 - Japan, Tokyo
II	1954 - Germany, Essen	XII	1994 - Poland, Krakow
III	1958 - Belgium, Liege	XIII	1998 - Australia, Brisbane
IV	1962 - UK, Harrogate	XIV	2002 - Republic of South Africa, Johannesburg
V	1966 - USA, Pittsburgh	XV	2006 - Peoples Republic of China, Beijing
VI	1973 - France, Paris	XVI	2010 - USA, Lexington
VII	1976 - Australia, Sydney	XVII	2013 - Turkey, Istanbul
VIII	1979 - Ukraine, Donetsk	XVIII	2016 - Russian Federation, St. Petersburg
IX	1982 - India, New Delhi	XIX	2019 - India, New Delhi (13-15 November 2019)
X	1986 - Canada, Edmonton		

About Coal Preparation Society of India (CPSI)

Representing India's commitment to the use of Clean Coal to the world, Coal Preparation Society of India (CPSI) is a not - for - profit, non-government professional body of coal, power, iron and steel sectors and their allied industries. CPSI's mission is to promote washing of high ash domestic coal to improve its quality and enhance the calorific value, making it more environment friendly and suitable for use in High Efficiency Low Emission (HELE) power generating systems as well as for gasification and conversion into chemicals, CTL and other high value petroleum substitutes. It will, therefore, be a step, which will facilitate fulfilling the country's commitment made at the Paris Climate Treaty.

Main Objectives of CPSI inter alia are:
- To facilitate policy formulation in coal beneficiation and preparation;
- To provide an effective network amongst coal producers, consumers, R & D organisations etc;
- To provide an independent platform for deliberating technological, operational, financial, commercial and policy aspects of the Indian Coal Preparation Industry.
- To promote and encourage the exchange of technical information relevant to the Indian coal industry.
- Meeting India's commitment to the Climate Change Treaty

CPSI is registered under the Societies Registration Act, XXI of 1860 and its head office is located in New Delhi. CPSI is a member of the International Organizing Committee (IOC) of the International Coal Preparation Congress (ICPC). CPSI is a member of IOC representing India.

Domestic and Global Network:

CPSI is an Associate Member of the World Coal Association - a global industry association formed of major international coal producers and stakeholders and has a bilateral relationship with IEA Clean Coal Centre, UK for promoting clean coal technologies for use in High Efficiency Low Emission (HELE) power generating systems.

CPSI has bilateral arrangements with China National Coal Association (CNCA), Coal Preparation Society of Ukraine and similar organisations of other IOC member countries for technical knowledge sharing and exchange of experts and trainees.

CPSI is a Member of ASSOCHAM and also an Associate Member of the PHD Chamber of Commerce and Industry.

Major industry bodies like Federation of Indian Mineral Industries (FIMI), Sponge Iron Manufacturers Association (SIMA), Association of Power Producers (APP), ASSOCHAM of India, World Coal Association (WCA), IEA Clean Coal Centre (UK) are associating and supporting CPSI's efforts towards cleaning of coal and introduction of clean coal technologies so that India can meet its emissions reduction target committed at Paris Climate Treaty.

List of IOC Members

1. Raj K. Sachdev, India, Chairman, IOC
2. Alok Perti, India, Chairman, NOC
3. Andrew Swanson, Australia
4. Zhang Shaoqiang, P. R. China
5. Xie Wenbo, P. R. China
6. Zhou Shaolei, P. R. China
7. Maria Holuzko, Canada
8. Dieter Ziaja, Germany
9. Ireneusz Baic, Poland
10. Leonid A. Vaisberg, Russian Federation
11. Vladimir Yu Bazhin, Russian Federation
12. Quentin Campbell, South Africa
13. Neil Jenkinson, United Kingdom
14. Olexandr Yegernov, Ukraine
15. Ihor Shemelov, Ukraine
16. Barbara J. Arnold, United States of America

Corresponding Members

1. George Anastasakis, Greece
2. Bokanyi Ljudmilla, Hungary
3. ...., Kazakhstan

To improve flotation of low-grade coal with flotation modifier

B.V. Sudhir Kumar[1], Ashutosh Kumar Pandey[1], Manish Kumar[1], Vineet Kumar[2]

[1]Tata Steel Ltd., Jamshedpur, India
[2]Nalco, an Ecolab company, Pune, India

Abstract: With the possibility of enhancing yield keeping the concentrate ash% in specified limits by usage of modifier 88150, and as a developmental initiative Nalco has worked to develop the flotation modifier. Preliminary testing was conducted at Nalco lab and the results were quite encouraging in terms of ash reduction as well as yield improvement. The results obtained from the flotation test clearly indicate higher yield but at reduced concentrate ash%. Based on the encouraging results another round of lab test was conducted at Tata Steel NRD laboratory to verify and assess the performance.

Modifier 88150 is basically a surface-active agent which used to enhance flotation efficiency by detaching the adhered clay from the coal particle so that the selective adsorption of collector takes place in order to separate the hydrophobic coal and hydrophilic gangue. It enhances the probability of attachment and thus increases the selectivity in terms of grade enrichment. The metallurgical results may be enhanced in terms of low concentrate ash and high tailings ash. These results in improved separation efficiency and selectivity thus increased coal recovery at lower ash.

- Separation efficiency increased by 1.67(5%) points.
- Overall plant productivity has gone up by 0.5% that translates 5000 MT of extra production/yr.
- More energy is produced as overall plant yield has gone up by 0.5% percentage point. Energy produced per annum 1.25 × 10¹¹ kJ/Yr (Calorific value: 25 kJ/kg).

Keywords: Separation efficiency, modifier, hydrophobic, flotation, selectivity, recovery

1. Background

Tata Steel West Bokaro always encouraged improvement initiatives. With depletion of high grade coal seams and subsequent processing of low grade ores having poor floatability, possibility of adherence of clay on the coal surface that detoriated the flotation performance always loomed large.

These seams have wider variation in ash percentage with poor floatability characteristics resulting into lower fine clean coal recovery.

The availability of indigenous coal has been drastically increased to cater to the increased requirement after expansion project at Tata Steel Jamshedpur. The supply of indigenous coal at lower ash level has always been a priority for Tata Steel West Bokaro in order to reduce the import requirement of clean coal from overseas considering the economics associated with it.

Nalco has pioneered the coal collector across all the Washeries of Tata Steel last year with an objective to improve the recovery as well as replace the HSD with an alternate synthetically developed product which helped increase the yield by 1–1.5%.

With the possibility of enhancing yield keeping the concentrate ash% in specified limits by usage of modifier, and as a developmental initiative Nalco has worked to develop the flotation modifier. Preliminary testing was conducted at Nalco lab and the results were quite encouraging in terms of ash reduction as well as yield improvement. The results of lab tests are summarized below.

Table 1: Summary of the lab flotation test at Nalco lab

Data	Collector - 88001 (ml)	Frother - 9840	Modifier - A	Feed		Froth		Tailings		Yield%	
				Wt %	Ash %	Wt %	Ash %	Wt %	Ash %	Weight basis	Ash basis
15-June	0.11	0.03	Blank	100	23.22	56.67	13.13	43.33	36.17	56.67	56.21
	0.11	0.03	0.015	100	23.22	60.67	12.02	39.33	40.58	60.67	60.78
	0.11	0.03	0.03	100	23.22	63.33	12.88	36.67	41.31	63.33	63.63
18-May	0.11	0.03	Blank	100	23.34	53.33	13.24	46.67	34.89	53.33	53.35
	0.11	0.03	0.015	100	23.34	68.00	11.31	32.00	49.84	68.00	68.78
	0.11	0.03	0.03	100	23.34	76.67	13.79	23.33	53.44	76.67	75.91

The results obtained from the flotation test clearly indicate higher yield but at reduced concentrate ash%. Based on the encouraging results another round of lab test was conducted at Tata Steel NRD laboratory to verify and assess the performance.

The flotation modifier was tested in NRD for VIII SE. The results are quite similar to the earlier tests and were very promising in terms of improved separation (20–25% better) efficiency and yield (4–5% overall yield increase) at the same ash level. Based on the encouraging lab results plant scale trail was recommended for comparative study in different seams.

Table 2: Summary of the lab scale flotation test at NRD, Tata Steel

Data	Seam	Batch	Sample	Details	Feed		Froth		Tailings		Yield %	SE
					Wt %	Ash %	Wt %	Ash %	Wt %	Ash %	Ash basis	
07.08.2013	8SE	70	Test-1	No modifier	100	21.83	77.35	13.94	22.65	48.79	77.36	30.78
			Test-2	Half	100	20.38	80	9.72	20	63.02	80.00	47.45
			Test-3	Equal	100	20.78	81.67	11	18.33	64.33	81.66	43.18
11.08.2013	8SE	70	Test-4	No modifier	100	26.89	61.67	13.83	38.33	47.89	61.66	35.29
			Test-5	Half	100	26.89	65	10.87	35	56.64	65.00	47.21
			Test-6	Equal	100	27.36	69.39	12.87	30.61	60.19	69.38	44.07

Modifier is basically a surface active agent which is used to enhance the flotation efficiency by detaching the adhered clay from the coal particle so that the selective adsorption of collector takes place in order to separate the hydrophobic coal and hydrophilic gangue. It enhances the probability of attachment and thus increases the selectivity in terms of grade enrichment. The metallurgical results may be enhanced in terms of low concentrate ash and high tailings ash. These results in improved separation efficiency and selectivity thus increased coal recovery at lower ash level.

This new application has the potential to reduce the overall ash by 0.5–1% without compromising on yield. This will help to increase the yield by more than 1%. The anticipated benefits would be:

- The possibility of 1% ash reduction without compromising on yield.
- Lowering of ash will lead to more clean coal recovery (anticipated more than 1% yield gain).
- Floatation modifier is particularly very effective for inferior coal, resulting in increased coal recovery beyond the level that can be achieved synthetic collector alone.

2. First phase trial

Based on the encouraging result a plant scale trail for a short period was recommended for comparative study. First phase of plant trail was started on 25th Nov at 9:00 PM for the Seam VII of Q-SE and comparative performance of the modifier + collector vis-à-vis collector only was evaluated. As soon as the modifier dosing was started, the same was reflected in the process. Improvement in floatation was clearly visual, yield increased and was reflected in the clean coal belt readings/physically too. The results generated were very convincing to carry out the extended trial and the improvements observed are summarized below:

- Absolute 1–1.5% reduction in fines ash was observed.
- Separation efficiency improved by more than 20–25%.
- There was significant increase in overall yield (more than 1.5%). Lowering of fines ash and increase in contribution of fines in composite clean coal lead the overall yield gain.

Reagent	Feed	Concentrate	Tailings	Yield% ash basis	Separation efficiency
Without modifier	22.10	14.10	38.70	67.48	26.94
With modifier	21.80	12.37	49.40	74.53	36.13
Difference	−0.30	−1.73	10.70	7.05	9.19

3. Second phase trial

Based on the successful first phase small scale trial an extended one-month trial was released to accesses the performance in various seam. The second phase of modifier trail was started on 2nd January at 6:00 PM. The results obtained till 8th Jan'14 shift C are very promising and encouraging.

The salient features and of the plant trial observations are as follows:

- Flotation modifier trial started from 02-January on C-Shift in running plant. During the switchover VIII SE seam was running. Switch over data clearly indicate the difference in the performance of synthetic modifier.
- Qualitative data collected from control log sheet for the analysis purpose.
- Results were compared in terms of yield% on ash basis and separation efficiency.
- Histogram analysis of feed, forth, tailings ash, fines% in comp clean coal and yield% on ash basis has been carried out; results were compared with collector alone.
- Dosage of modifier optimized to assess the performance in clean coal ash reduction and yield increase of clean coal.
- Dosage was closely monitored and altered during the entire trial period with the help of the in house lab facilities.

3.1 Seam wise summary of the preliminary trial

Based on the data generated by NRD, analysis was carried out. The seam wise summary of the preliminary trial analysis indicate that,

- there is a substantial difference in tailings ash using modifier,
- the concentrate ash is more or less same using modifier,
- yield% on ash basis of modifier trial is much superior than blank,
- separation efficiency is also better in case of modifier trial.

Overall preliminary summary of the qualitative analysis with Nalco flotation modifier trial					
Reagent	Feed	Fines	Tailings	Yield % on ash basis	Separation efficiency
With modifier	21.22	11.37	50.28	70.61	34.91
Without modifier	22.87	11.4	39.14	56.37	31.43
Difference	−1.65	−0.02	11.14	14.24	3.48

- The average feed ash during modifier trial is 21.22 compared to 22.87 pretrial periods. The feed ash was almost consistent during trail and pretrial period.
- The average froth ash during modifier trial is 11.37 compared to 11.4 pretrial period.
- At the same froth ash level with modifier trial and pretrial data there is substantial increase in tailings ash. The average tailings ash with modifier is 50.28 compared to 39.14 without modifier.
- Yield% during modifier trial is 70.61 compared to 56.37 without modifier.
- The average separation efficiency during modifier is 34.91 compared to pretrial data without modifier is 31.43.
- There is substantial increase in yield% on ash basis at same ash level.

Reagent	Seam	Feed	Concentrate	Tailings	Yield % on ash basis	Separation efficiency
Without modifier	VIII SE	25.31	10.74	36.07	40.68	27.28
With modifier	VIII SE	22.32	11.54	52.27	70.90	36.26
Difference		−2.99	0.79	16.20	30.22	8.98
Without modifier	X SE	20.63	10.10	41.33	71.71	34.86
With modifier	X SE	18.26	10.50	54.53	81.81	37.98
Difference		−2.38	0.40	13.19	10.10	3.12
Without modifier	VI SE	22.60	12.41	36.29	56.75	29.02
With modifier	VI SE	22.40	11.75	42.52	60.21	31.49
Difference		−0.20	−0.66	6.23	3.46	2.47

4. Histogram analysis

4.1 Feed ash

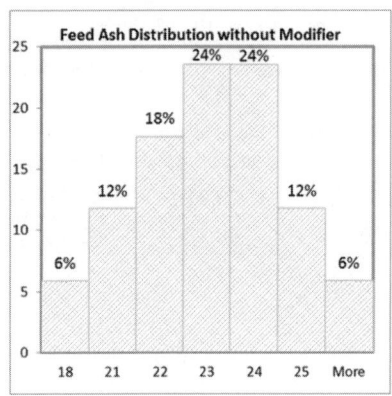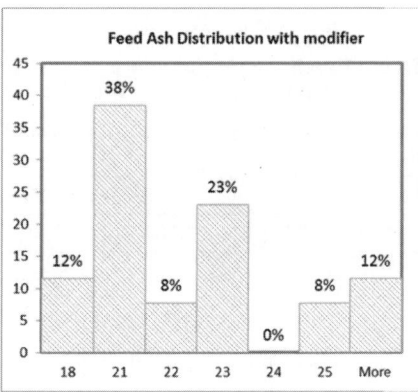

- The average feed ash during the trial is 21.22 compared to pretrail data 22.87.

4.2 Froth ash

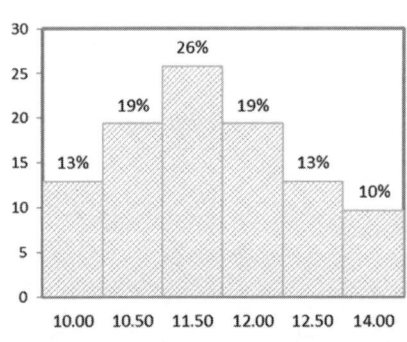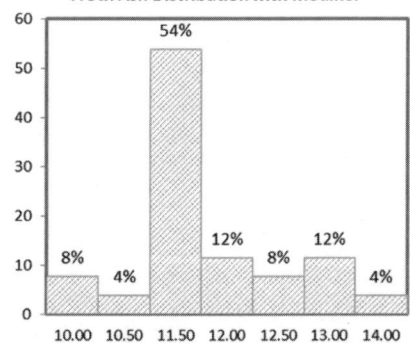

- The average froth ash during trial is 11.37 compared to pretrial data 11.40.
- The froth ash with modifier is very consistent. Almost 54% ash is b/w 11–11.5. Without modifier only 26% data is b/w 11–11.5.
- Most of the data with modifier (around 66%) is b/w 11–12% ash, which shows the consistency of the modifier.

4.3 Tailings ash

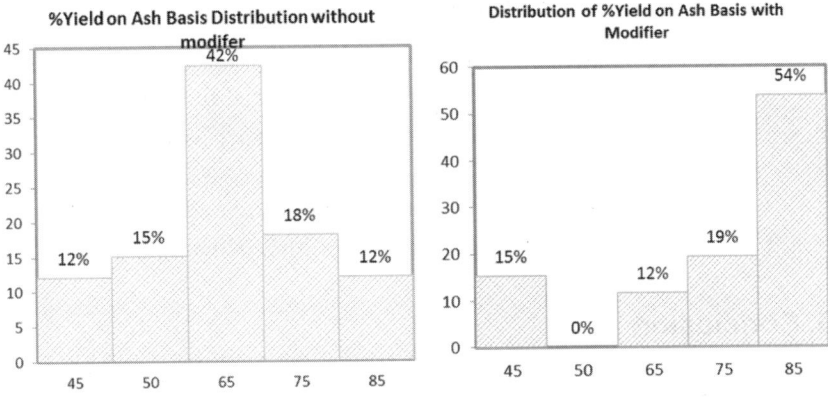

- Average tailings ash with modifier is 50.28 compared to 39.14 pretrial data.
- Only 14% times tailings ash without modifier is greater than 45 compared to 74% times tailings ash is greater than 45 with modifier.
- With modifier 27% times tailings ash is greater than 60% whereas without modifier none of times tailings ash is greater than 60%.
- Without modifier 40% times tailings Ash is lower than 35 with compared to only 8% times tailings ash is less than 35 with Modifier.

4.4 Yield % on ash basis

- The average yield% on ash basis with modifier is 70.61 compared to 56.41 without modifier.

- Only 30% times yield% is greater than 60 without modifier compared to 73% with modifier.

4.5 Separation efficiency

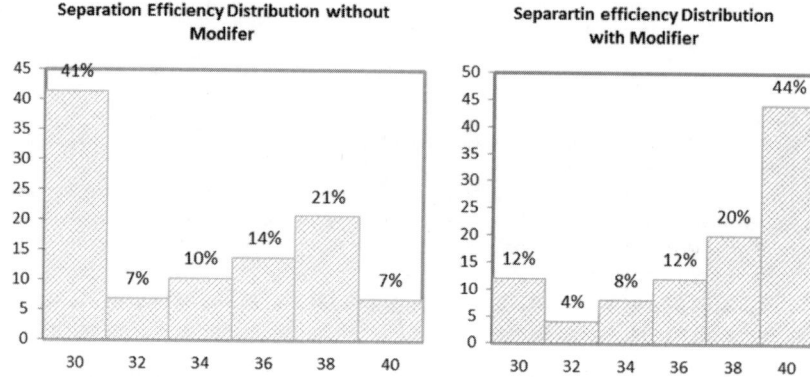

Separation efficiency is the most important parameter for evaluation of floatation performance.

- With modifier maximum times (76%) the SE is greater than 34 compared to only 42% SE is greater than 34 without modifier.
- Only 12% SE is below 30 with modifier but without modifier 41% data is coming below 30.

5. Results and discussion

- The comparative histogram summary indicates that the raw coal quality is almost similar during the modifier trial and without modifier.
- Tailings ash of with modifier is 9–10% higher as compared to the synthetic collector alone.
- The yield% on ash basis of fines coal has increased by 10–12%.
- Separation efficiency has improved by 20–25% with modifier.

6. Conclusion

With the experimental studies and subsequent plant scale trails it is evident that with the use of modifier there is significant improvement in separation efficiency of flotation cell and hence there is a potential saving in coal production.

2

Beneficiation of high ash coking coal fines using column flotation with new external sparger

Prasad Kopparthi, D. Sachinraj, Dr. A.K. Mukherjee

R&D, Tata Steel Ltd, Jamshedpur, India

Abstract: Mechanical flotation cell is the most widely used machine for the flotation of coal fines (-0.5 mm) in India. Though popular, flotation cells cannot recover low ash coal and significant loss in yield is observed owing to the inherent design disadvantages such as high turbulence, short circuiting of feed, and poor control on bubble size. To enhance the recovery of clean coal, including that of the low ash components, flotation column technology was used, along with an in-house designed external sparger, at laboratory column flotation. The new external sparger is a multi-layered static mixer containing (X) cross elements arranged one after another with alternate elements given a twist in orientation. The desired bubble size can be produced by varying the air, slurry flow rates, and frother dosage. In the current research work, the effect of operating variables such as air flow rate (AFR), collector dosage (CD), and frother dosage (FD) were optimized to produce high recoveries. With the new system, the recovery of fines increased and 5 units improvement in the coal yield is observed.

Keywords: Coal, column flotation, external sparger, fine bubbles, contact angle, zeta potential

1. Introduction (Jones et al., 1989)

In coal preparation plants, froth flotation is widely used process for beneficiation of fines (-0.5 mm). The clean coal particles are selectively separated from ash minerals using air bubbles due to difference in their hydrophobicity. Typically, mechanical flotation cells were popular however, the popularity of column flotation has been increasing due its features such as high residence time, wash water system, quiescent zones, and bubble generator system. The main advantages of column flotation compared to conventional cells are (1) a better product without sacrificing recovery, (2) a reduction in the number of stages of operation, (3) ability to handle a finer feed, (4) savings in collector requirements, (5) simplicity in design for construction without any moving parts, and (6) less floor space requirements. However, compared to conventional cells the water requirement per tonne of feed processed and consequently the frother requirement may be more (Jena et al., 2008).

The major disadvantages of flotation columns are the mixing in the axis of the column, the blockage of spargers (diffusers) and problems posed by the column height in installations. In the last few decades, many alternative column designs have been developed to eliminate these disadvantages and to increase the recovery. Microcel column flotation with external sparger is one of the alternative development to the conventional column flotation (Yoon, 1993; Fig. 1).

Figure 1: Lab scale flotation column

2. Literature review

Column flotation cell provide better grade than the mechanical flotation cells due to a deeper froth zone and employment of wash water to reduce the amount of hydrophilic material entrained in the rising froth that reports to the overhead product stream(Patil et al., 2010).Both conventional (Denver) and the column flotation provided a high (~90%) combustible recovery from the fine coal, however, the column provided a low ash clean coal (3.8%) compared to that provided by conventional flotation (7.0%)(Parekh et al., 1990). The performance of the column flotation strongly depends on the stability of froth and the effect of different column flotation parameters such as gas flow rate, wash water flow rate, froth height, wash water addition point, and feed solid concentration on froth stability(Tao et al., 2000).Pilot scale column flotation was used to separate vitrinite rich fractions from the feed vitrinite content of 76.9–96.8% to 95.2–99.8% for Guachinte and Yolanda coal samples from south-western Colombia respectively(Barraza and Piñeres, 2005).Harris et al. (1992) evaluate three flotation technologies such as batch Leeds type flotation cell, flotation column, and hybrid agitated column cell to beneficiate coal fines. It was observed that the selectivity in the flotation cell is less compared to the column and the recovery of coarse fraction coal increased with hybrid agitated column compared to conventional column(Harris et al., 1992, 1994). The effect of location of the wash water addition point and its flow rate on the performance of the column flotation was investigated using the microbubble flotation column. The increased wash water velocity improved the column product quality and reduced recovery. A value of 0.5 cm/s was considered as a maximum applicable superficial wash water velocity for the column however the optimum value was the 0.33 cm/s based on the grade and recovery relationship. The recovery also varied with the adjustment of wash water addition point in the cleaning zone of the column, and the variations were strongly governed by the wash water flow rate. The recommended wash water addition point of between 10 cmand 15 cm below the overflow level(Choung et al., 1993).

The coal particles were separated from bottom ash samples of Tuncbilek power station using column flotation and it was reported that the ash content reduced from 77.6% to 42.6% and it leads to increase in the calorific value from 1270 kcal/kg to 3840 kcal/kg (Demir et al., 2008). The performance of mechanical flotation cell and column flotation was compared for the coking coal fines and it was reported that column produced high selective clean coal with ash content of 10.1% compared to 14.4% of mechanical flotation cell from the feed content of 24.4%. The corresponding clean coal yield was 72% and 78%, respectively (Jena et al., 2008). Microcel column flotation with external sparger was evaluated against the Canadian Process Technology (CPT) Slamjet Column to beneficiate zinc at Red Dog Mine. The jetting-type system of CPT column flotation produced the averaged sauter mean bubble diameter d_{32} of 3.4 mm while the Microcel system averaged a d_{32} of 1.9 mm. Microcel column produced a higher average sphalerite concentrate grade (0.6% zinc absolute) at a higher average sphalerite recovery (2.8% sphalerite absolute) over the existing jetting-type column (Pyecha et al., 2006). Fine hard coal (<150 μm) from Hwa-Sun coal mine, South Korea was beneficiated using column flotation. It was reported that CPT Coalpro column could produce a clean coal concentrate of 85% combustible recovery with about 81% ash rejection at maximum separation efficiency of 62%, compared to conventional flotation which produces a relatively low combustible recovery of 70% with 70% ash rejection at a maximum efficiency of 42%. The results demonstrate that the CPT column is superior to conventional flotation for fine coal beneficiation (Han et al., 2014). The recovery of coarse coal particles (>250 μm) improved by 9% in modified column flotation with the self-aspirated bubble generator in the feed line compared to conventional flotation column at the same ash content of clean coal (Ni et al., 2017).

3. Experimentation

3.1 Materials

In this research work the coal sample namely VIIA seam has been collected from the coal Washery #2 in West Bokaro, Tata Steel Ltd, India. High speed diesel oil (HSD) and commercial frother (supplied by NLC Nalco) were used as collector and frother, respectively. The representative sample of the coal flotation feed (−0.5 mm) was screened into different particle size fractions. Total weight and mineral matter (ash) distribution in the particle size fractions were analyzed as shown in Fig. 2. The nature of the coal is medium coking coal. The size wise analysis shows that the feed sample contains 17% of coarse size (>500 microns), 30% of intermediate size (250 microns), and 22% of ultrafine size (<53 microns). The ash content of coarse size fraction, intermediate size and ultrafine size is 29%, 24%, and 33%, respectively. The overall ash percentage of coal sample is 27%.

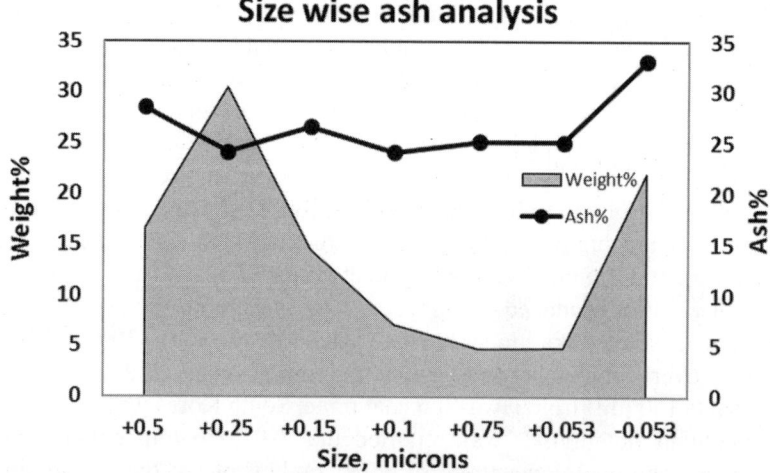

Figure 2: Size wise ash analysis of feed sample

3.2 Experimentation plan

To evaluate flotation column performance, experiments were carried out to find out the maximum yield. In the current investigation, the effect of variables such as air flow rate (AFR), collector dosage (CD), and frother dosage (FD) have been considered. The experimental plan of the flotation column is shown in Table 1.

Table1: Experimental parameters and their range

Sl. no	Parameter	Range
1	Air flow rate (AF), lpm	3, 4, 5, 6
2	Collector dosage (CD), kg/t	1, 1.5, 2.0
3	Frother dosage (FD), kg/t	0.05, 0.1, 0.15, 0.2

4. Results and discussions

4.1 Effect of air flow rate

In the flotation column separation occurs coal particles affinity towards air bubbles. There is direct relationship between air flow rate and gas bubbles. Bubble columns can be operated in four regimes; they are chain bubbling, fully developed bubbly, chum turbulent, and slug flow regimes. For effective separation flotation columns should be operated in bubbly flow. The flotation experiments were carried out with 3 lpm, 4 lpm, 5 lpm, and 6 lpm by keeping the wash water rate at 3 lpm, collector dosage 1 kg/t and frother dosage 0.1 kg/t. The flotation results are shown in Fig. 3. The yield of clean coal at 11% ash was 45%, 47%, 40%, and 34% at 3 lpm, 4 lpm, 5 lpm, and 6 lpm, respectively. The separation efficiency was 29%, 33.27%, 29.7%, and 25% at 3 lpm, 4 lpm, 5 lpm, and 6 lpm. The maximum yield and separation efficiency 47% and 33.27% observed respectively at 4 lpm.

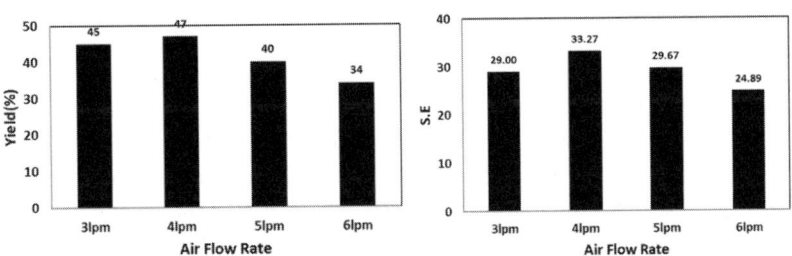

Figure 3: Effect of air flow rate on clean coal yield and separation efficiency

With increasing the air flow rate from 3 lpm to 4 lpm the yield of clean coal increases from 45% to 47%. This is due to that with increasing the gas flow rate the number of bubbles forming increase, eventually the yield of clean coal increases. From the flotation rate constant equation, the rate constant is directly proportional gas flow rate, so increase in air flow rate the flotation rate also increases (flotation rate constant, $k = 1.5PV_g/D_b$). With further rise in gas flow rate the yield is decreased from 47% to 40% and then 34%. The drop-in yield from 47% to 40% is due to with raising the air flow rate the size of the bubbles increases. The bigger air bubbles have high rise velocity, so the residence time is decreases and bubble collection efficiency decreases. Further rise in air flow rate from 5 lpm to 6 lpm, it creates high turbulence with in the column, resulting high backward flow of bubbles. In the turbulent conditions, there is no plug flow condition prevails, and air bubbles move in zigzag paths. In this regime, coal particles will get less contact time with air bubbles compared to laminar flow conditions. Even the attached coal particles to the air bubbles will detach due to high axial back mixing. So, the flotation column operating regime shifts from bubbly flow to the churn turbulent flow.

The separation efficiency was 29%, 33.27%, 29.7%, and 25% at 3 lpm, 4 lpm, 5 lpm, and 6 lpm. The maximum separation efficiency 33.27% observed at 4 lpm this is due to increasing the air flow rate the generation of air bubbles increases consequently selectivity increases. With further rise in the air flow rate the entrapment of ash particles increases there by the separation efficiency decreases.

4.1.1 Effect of collector dosage

Even though coal is naturally hydrophobic, collectors have been used to enhance the hydrophobicity of coal in coal flotation. Collectors are selectively adsorbed on the surface of coal consequently improves the hydrophobicity. Collectors can be divided into two classes one is non-ionic collectors and ionic (anionic and cationic) collectors. The mechanism of non-ionic collector is, it adsorb/spread over the coal surface due to high hydrophobic interaction between non-ionic collector and coal surface. The adsorbed collector increases the contact angle of coal consequently enhance the hydrophobicity of coal. The non-ionic collector has two branches one is polar and other is non-polar. The polar group reacts with polar sites over the mineral surface and non-polar group orients towards bulk solution. Thus, non- polar branch enhances the hydrophobicity of mineral surface. When air interface is provided in the form of an air phase and if the bond strength between the monolayer of the collector and the coal surface is strong enough the particle will be lifted by the buoyancy of the air bubble.

In the present investigation, high speed diesel oil was used as collector to increase hydrophobicity of coal. Diesel oil comes under the non-ionic collector. Diesel oil was added to the coal slurry prior to the flotation, in the conditioning tank. Diesel oil adsorbs or spreads as mono layer over the coal surface due to strong hydrophobic interaction between diesel oil and coal surface, consequently increases the hydrophobicity.

Column flotation experiments were carried out by varying the collector dosage from 1 kg/t to 2.5 kg/t with increasing 0.5 kg/t of diesel oil per experiment keeping other variables constant (wash water rate 3 lpm, air flow rate 4 lpm and frother dosage 0.1 kg/t). From the experimental results, as shown in Fig. 4, it was observed that clean coal yields were 47%, 48%, 54%, and 46% for 1 kg/t, 1.5 kg/t, 2 kg/t, and 2.5 kg/t, respectively. The clean coal yield was increased from 47% to 54% by increasing collector dosage from 1 kg/t to 2 kg/t. This is due the adsorption of collector over the number of coal particles increases with increasing the collector dosage, consequently the yield of clean coal increases. With further increase in collector dosage from 2 kg/t to 2.5 kg/t, the clean coal yield decreased from 54% to 46%. From the experimental studies reveals that the starvation limits of collector dosage for the column experiments was

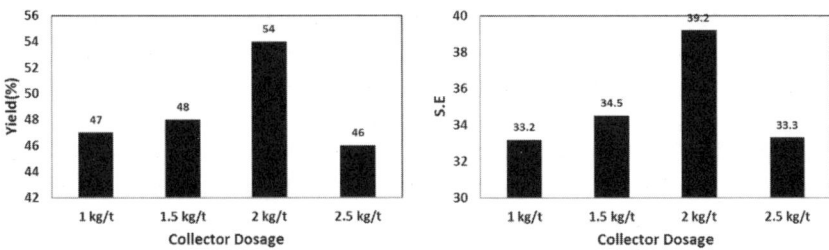

Figure 4: Effect of collector dosage on clean coal yield and separation efficiency

2 kg/t. This is due to after the starvation level further rises in collector concentration shows adverse effect on coal flotation. After the starvation limit of collector dosage, it adsorbs in multilayer over the coal surface. Multilayer collector has strong hydrophobic interactions, and due to these interactions among the coal particles, inter particle attraction will increase and tends to form coal agglomerations. The agglomerations of coal particles do not lift by air bubbles due to high density, so the yield of clean coal decreases.

The separation efficiency was 33.2%, 34.5%, 39.2%, and 33.3% for 1 kg/t, 1.5 kg/t, 2 kg/t, and 2.5 kg/t, respectively. The separation efficiency also increases from 33.2% to 39.2% by increasing the collector dosage from 1 kg/t to 2 kg/t, respectively. The increase in separation efficiency is due to collector adsorption over the coal surface increases the selectivity of coal. With further rise in collector dosage from 2 kg/t to 2.5 kg/t, the separation efficiency decreases from 39.2% to 33.3%, respectively. Beyond 2 kg/t collector concentrate, it may adsorb on the mineral particles thereby reduces the selectivity of coal decreases consequently separation efficiency decreases.

4.1.2 Effect of frother

One of the prerequisites for a successful flotation operation is the stability of the bubble-particle aggregate. A stable bubble is produced by using a frother, the function of which is to decrease the surface tension at the air–liquid interface. A frother has a number of functions in flotation – first, it reduces the surface tension of the air–liquid interface in order that a stable bubble is produced in the system; secondly, it influences the kinetics of bubble–particle adhesion; thirdly, it thins the liquid layer by interacting with collector molecules; and finally, it stabilizes the bubble-particle aggregate (Schulman and Leja, 1954; Leja, 1956–57).

Frothers are neutral molecules consisting of a medium chain length hydrocarbon entity and a polar group(s) entity. This gives the molecule dual affinity to water and air. The hydrophobic contribution is not as strong as in collectors. The polar groups of frothers are always hydroxyl – usually in the form of alcohol or glycol. A good frother should have less or negligible collecting properties like collectors. If frother has collecting property, then it has advantage and disadvantage simultaneously. The advantage is amount of required collector may reduce in flotation and disadvantage is it becomes impossible to alter the frothing characteristics and the collecting characteristics of the flotation operation independently.

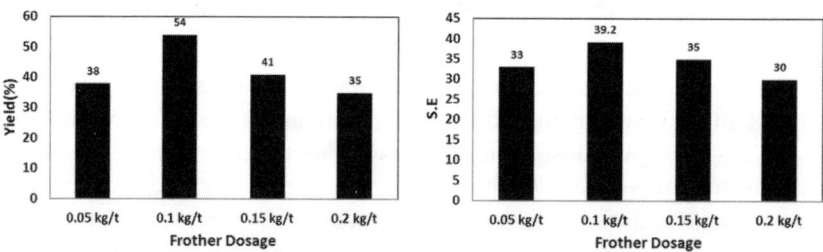

Figure 5: Effect of frother dosage on clean coal yield and separation efficiency

Column flotation experiments were carried out by varying the frother dosage (commercial frother supplied by Nalco Ltd) from 0.05 kg/t to 0.2 kg/t with increasing 0.05 kg/t of frother per experiment keeping other variables constant (wash water rate 3 lpm, air flow rate 4 lpm, and collector dosage 2 kg/t). From Fig. 5, it was observed that clean coal yield was 38%, 54%, 41%, and 35% for 0.05 kg/t, 0.1 kg/t, 0.15 kg/t, and 0.2 kg/t, respectively. The clean coal yield was increased from 38% to 54% by increasing frother dosage from 0.05 kg/t to 0.1 kg/t. The increase in the clean coal yield due to the addition of frother, increase in bubble formation, reduce breakage of bubble, capture of particles reaching to the top of the column.

With further increase in frother dosage from 0.1 kg/t to 0.2 kg/t, the clean coal yield decreased from 54% to 35%. The decrease in yield with further increasing the frother concentration is attributed to hydrogen bonding between frother and hydrated ash mineral matter which results in a greater recovery of entertainable ash bearing particles.

The separation efficiency was 33%, 39.2%, 35%, and 30% for 0.05 kg/t, 0.1 kg/t, 0.15 kg/t, and 0.2 kg/t, respectively. The separation efficiency increases from 33% to 39.2% by increasing the frother dosage from 0.5 kg/t to 0.1 kg/t, respectively. The increase in separation efficiency is due to increase in number of air bubbles and thereby increases the selectivity of coal. With further increase in the frother concentration the separation efficiency decreases due to collection of ash particles.

5. Conclusion

The flotation behavior of coking coal fines was investigated using column flotation with external sparger and observed that material responds to column flotation very well. The effect of operating parameters such as air flow rate, collector dosage, and frother dosage on column flotation performance in-terms of clean coal yield and separation efficiency evaluated. The clean coal yield and separation efficiency increased with air flow rate. However, with further rise in air flow rate more than 4 lpm the clean coal yield reduced due to bubble coalescence and high back mixing. The clean coal yield increased with increase in the collector and frother dosages however, the yield reduced with further rise in reagent dosages. The maximum clean coal yield 54% with 11% ash content observed at wash water rate 3 lpm, air flow rate 4 lpm, collector dosage 2 kg/t, and frother dosage of 0.1 kg/t.

6. References

[1] Barraza, J., Piñeres, J., 2005. A pilot-scale flotation column to produce beneficiated coal fractions having high concentration of vitrinite maceral. Fuel 84, 1879–1883. https://doi.org/10.1016/j.fuel.2005.03.021.

[2] Choung, J.W., Luttrell, G.H., Yoon, R.H., 1993. Characterization of operating parameters in the cleaning zone of microbubble column flotation. Int. J. Miner. Process. 39, 31–40. https://doi.org/10.1016/0301-7516(93)90050-K.

[3] Demir, U., Yamik, A., Kelebek, S., Oteyaka, B., Ucar, A., Sahbaz, O., 2008. Characterization and column flotation of bottom ashes from Tuncbilek power plant. Fuel 87, 666–672. https://doi.org/10.1016/j.fuel.2007.05.040.

[4] Han, O.H., Kim, M.K., Kim, B.G., Subasinghe, N., Park, C.H., 2014. Fine coal beneficiation by column flotation. Fuel Process. Technol. 126, 49–59. https://doi.org/10.1016/j.fuproc.2014.04.014.

[5] Harris, M.C., Franzidis, J.P., O'Connor, C.T., Stonestreet, P., 1992. An evaluation of the role of particle size in the flotation of coal using different cell technologies. Miner. Eng. 5, 1225–1238. https://doi.org/10.1016/0892-6875(92)90161-2.

[6] Harris, M.C., Franzidis, J.P., Breed, A.W., Deglon, D.A., 1994. An on-site evaluation of different flotation technologies for fine coal beneficiation. Miner. Eng. 7, 699–714. https://doi.org/10.1016/0892-6875(94)90101-5.

[7] Jena, M.S., Biswal, S.K., Das, S.P., Reddy, P.S.R., 2008. Comparative study of the performance of conventional and column flotation when treating coking coal fines. Fuel Process. Technol. 89, 1409–1415. https://doi.org/10.1016/j.fuproc.2008.06.012.

[8] Ni, C., Jin, M., Chen, Y., Xie, G., Peng, Y., Xia, W., 2017. Improving the recovery of coarse-coal particles by adding premineralization prior to column flotation. Int. J. Coal Prep. Util. 37, 87–99. https://doi.org/10.1080/19392699.2016.1140151.

[9] Parekh, B.K., Bland, A.E., Groppo, J.G., Yingling, J., 1990. A parametric study of column flotation for fine coal cleaning. Coal Prep. 8, 49–60. https://doi.org/10.1080/07349349008905172.

[10] Patil, D.P., Parekh, B.K., Klunder, E.B., 2010. A novel approach for improving column flotation of fine and coarse coal. Int. J. Coal Prep. Util. 30, 173–188. https://doi.org/10.1080/19392699.2010.497106.

[11] Pyecha, J., Lacouture, B., Sims, S., Hope, G., Stradling, A., 2006. Evaluation of a microcel TM sparger in the Red Dog column flotation cells. Miner. Eng. 19, 748–757. https://doi.org/10.1016/j.mineng.2005.09.044.

[12] Tao, D., Luttrell, G.H., Yoon, R.H., 2000. A Parametic study of froth stability and its effect on column flotation of fine particlse. Int. J. Miner. Process. 59, 25–43. https://doi.org/10.1016/S0301-7516(99)00033-2.

[13] Yoon, R.H., 1993. Microbubble flotation. Miner. Eng. 6, 619–630.

3

A case study on effect of water quality on froth flotation performance in coal washery

B.S. Tiwari, M.P. Sinha, Rajeev Kr. Ranjan, Aman Gupta, J.K. Singh, D.P. Chakraborty

Tata Steel Ltd, Jamshedpur, India

Abstract: As water resources are becoming scarcer and also due to strict rules and regulations of Pollutions control board, coal wash plants have made strategies for increasing water re-use and accessing multiple water sources for coal processing to save freshwater. Recently, a study was carried out by the authors to analyse water quality parameters and to determine the effect of water quality on flotation performance in Coal Washery BCPP, in Jharia. Main sources of processing water in BCPP (coal washery) are fresh water, underground water, re-circulation water (tailings pond, thickener) and the water from different sources differed in pH, total dissolved solids, total suspended solids and conductivity.

Main findings of the experiment are:

1. pH of water was observed as the main significant criteria effecting froth flotation cell operation.

2. Metal ion species and dissolved inorganics ($CaCO_3$) concentration was also observed as one of the most important factor effecting coal flotation cell performance and both of them are dependent on pH.

3. Typical contaminants in water sources like F^-, Mg^{2+}, Ca^{2+}, sodium nitrates, dissolved oxygen turbidity in thickener overflow water, were found as most significant constituents of water effecting froth flotation performance.

Keywords: Froth flotation cell, BCPP (Bhelatand Coal Preparation Plant), CC (clean coal)

1. Introduction

Presence of constituents (water composition) in water has a significant effect on coal flotation process. Froth flotation process utilises surface properties of coal and gangue material. Based on various studies, it is claimed that inorganic and organic substances, dissolved ions, salts in flotation are very important for process quality and performance. Based on source of process water, the concentrations of the in water affect the hydrophobicity of coal and performance of froth flotation cell. In general, flotation is most effectively undertaken with quality of water composition which has a significant effect on flotation. To understand the role of water composition in the process water a study was conducted in BCPP coal washery.

Need of water quality tests:

(i) To study impact of various dissolved ions and metal ion species in water on froth flotation process.

(ii) To reduce water consumption.

(iii) To enhance FF (froth flotation) cell separation efficiency and to enhance clean coal yield from FF cell circuit.

Sources of process water in BCPP:

BCPP uses various sources of water as process water for processing coal, are as follows:

1. Fresh water.

2. Underground water.

3. Re-circulation water:

    (a) Tailings pond and

    (b) Thickener overflow water.

2. Experiment and methodology

1. Different water parameters and dissolved ions were identified for test work as shown in Table 1.

Table 1: Key constituents of water identified for test work

Sl. no.	Parameter	Sl. no.	Parameter
1	Colour	18	Manganese as Mn
2	pH	19	Mercury as Hg
3	Turbidity	20	Lead as Pb
4	Electrical conductivity	21	Cadmium as Cd
5	Total suspended solids	22	Total chromium as Cr
6	Colloidal silica as SiO_2	23	Zinc as Zn
7	Reactive silica as SiO_2	24	Selenium as Se
8	Total dissolved solids	25	Sulphate as SO_4
9	Total hardness as $CaCO_3$	26	Total phosphates as PO_4
10	Total alkanity as $CaCO_3$	27	Nitrate as NO_3
11	Residual free chlorine	28	Cyanide as Cn
12	Chloride as Cl	29	Phenolic compounds as (C_6H_5OH)
13	Fluoride as F	30	Dissolved oxygen
14	Calcium as Ca	31	BOD at 27° for 3 days
15	Sodium as Na	32	COD
16	Magnesium as Mg	33	Oil and grease
17	Iron as Fe	34	Coliforms

2. Collection of water from different water sources were carried out as mentioned above.
3. Analysis of water sample was carried out as shown in Table 1 above.
4. Sample of raw coal fines feed to FF cell was collected from plant for flotation test with different sources of water.
5. Flotation test work was conducted at laboratory scale, synthetic collector and frother were used for test work.

Reagents dosing rates, agitator RPM and air pressures were maintained same for all tests as shown in Table 2.

Table 2: Process parameters maintained in froth flotation cell test with all water

Dossing (kg/ton)		RPM	Air pressure
Sy. collector	Frother		
0.25	0.014	800–850	0.18 kg/cm²

6. Laboratory flotation cell results were tabulated as shown in Table 3.
Raw feed ash to FF cell: 22.51%

Table 3: Flotation test results

Results	Fresh water	Mines water	Recirculation water	Thickener over flow water
Yield (%)	73.38	76.98	79.68	68.81
Ash (%)	16.2	15.26	15.65	15.14

7. Statistical tool (scatter diagram) was used to identify significant parameters as (R^2) out of total 34 water constituents identified earlier (see Table 1), against fines clean coal yield.
8. Further significant relationships R^2 was classified in three categories as shown in Table 4.

Table 4: Criteria selected for selection of key constituents of water

Categories	R^2 range	Action
Very critical parameter	>0.5	Selected
Critical parameter	0.20–0.49	Selected
Not critical parameter	<0.20	Not selected

Based on above selection procedure as discussed above, six main water constituents out of 34 parameters (see Table 1), were finally selected as shown in the Table 5.

Table 5: Relationship table with CC yield and various water parameters

Sl. no.	Very critical/critical parameter	R^2 value	Nature of relationship positive (+) or negative (−) observed
1.	pH	0.64	(+)ve
2.	Dissolve oxygen	0.6415	(+)ve
3.	Fe	0.5326	(−)ve
4.	Total hardness $CaCO_3$	0.4572	(−)ve
5.	Turbidity	0.4924	(−)ve
6.	Nitrate as NO_3	0.571	(−)ve

Note: Out of final six water constituents, except pH and dissolved oxygen all other constituents have negative impact on fines clean coal recovery.

3. Discussion

3.1 pH

pH value 8.16 for recirculation water was observed highest as shown in Table 5, below which indicates water from recirculation source was slightly alkaline in nature. An alkaline solution is caustic in nature and will help to clean the surface of coal. This strengthens the bond between the collector and each particle of coal. This may be the reason for higher clean coal yield with recirculation of water from flotation cell and also it was observed that as the pH value increased from 7.64 to 8.16, fines clean coal yield also increased as shown in Table 5, pH of water is source dependent. The floatability/ hydrophobicity of bituminous coal has been shown to be dependent on pH, generally exhibiting a maximum in the neutral pH region.

Table 6: Showing pH, CC yield and CC ash of different water sources

Parameters/water sources	Fresh water	Under ground water	Recirculation water	Thickener over flow water
pH	7.64	7.80	8.16	7.69
Clean coal yield (%)	73.38	76.98	79.68	68.81
Clean coal ash (%)	16.20	15.26	15.65	15.14

3.2 Effect of $CaCO_3$ on froth flotation cell performance

Water test analysis shows that total hardness as $CaCO_3$ is less than maximum permissible value of 600 mg/L in all water sources used as process water in plant. But total alkanity as $CaCO_3$ was observed lowest in recirculation water than other water sources which has helped to gain highest clean coal yield as shown in Table 3. As the $CaCO_3$ concentration decreases clean coal yield was observed increasing as shown in relationship Graph 1. The floatability of coal is also affected by the presence of dissolved inorganics in the system (Somasundaran and Liu, 1998; Somasundaran et al., 2000).

Table 7: Showing value of $CaCO_3$, CC yield and CC ash

Parameters/water sources	Fresh water	Under ground water	Recirculation water	Thickener over flow water
$CaCO_3$	227.2	326	153.1	192.7
Clean coal yield (%)	73.38	76.98	79.68	68.81
Clean coal ash (%)	16.20	15.26	15.65	15.14

Graph 1 Relationship graph showing as hardness $CaCO_3$, decreases fines clean coal increases

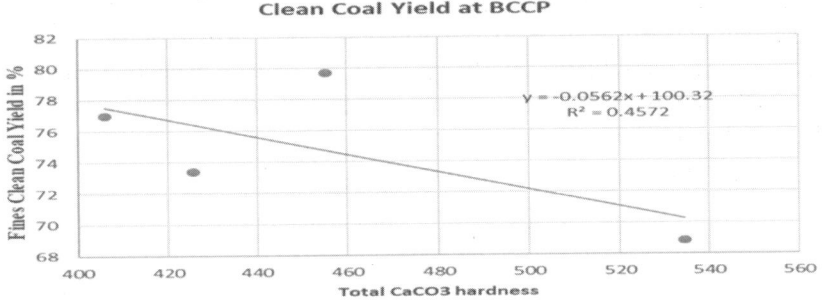

3.3 Effect of Fe ions on coal flotation cell

Relationship Graph 2a shows that as the concentration of Fe ions in water decreases, fines clean coal increases. Also, Graph 2b shows lower concentration of Fe at higher pulp density of water. Based on different studies it has been observed that the coal slurry as flotation feed contains a considerable amount of dissolved Fe and Ca ions and also mineral ions was found as a function of pH (Liu et al., 1994). The concentrations of dissolved Fe, Al, Ca and Mg decrease as the pH increases. In experiment work it was observed the lowest at pH value of 8.16. This result suggests that if the pH increases during coal processing, there will be precipitation of metal ion species whereas if the pH decreases, there will be dissolution of mineral species.

Graph 2(a and b) Relationship between Fe and fines clean coal yield and between Fe and pH of water of different sources

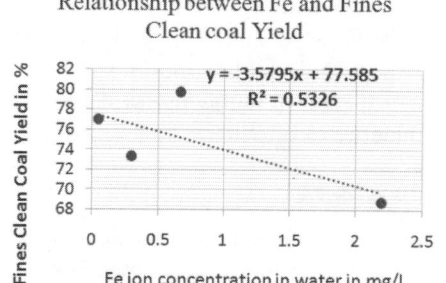

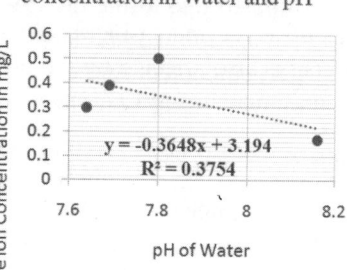

3.4 Effect of nitrate as NO_3

Nitrate is a polyatomic ion with the molecular formula NO_3. Nitrate is the compound predominantly found in groundwater and surface waters. The determination of nitrate in water is difficult because of interferences and much more difficult in wastewaters because of higher concentrations of numerous interfering substances. In BCPP concentration of nitrate in thickener overflow water was observed highest as shown in Graph 3 below. Based on statistical tool analysis, adverse impact of nitrate concentration was observed on fines clean coal yield.

Graph 3 Showing relationships between $NaNO_3$

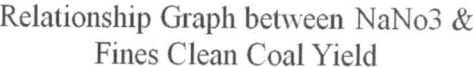

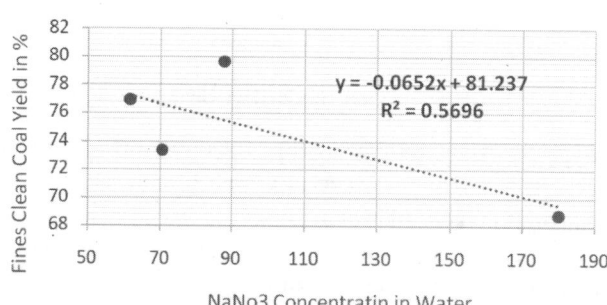

3.5 Effect of dissolved oxygen on froth flotation cell performance

Dissolved oxygen refers to the level of free, on-compound oxygen present in water or other liquid. Non-compound oxygen, or free oxygen (O_2), is oxygen that is not bonded to any other element. As the dissolved oxygen becomes oversaturated due to the decrease in ethanol–water ratio, oxygen nanoscale gas state are formed and stabilised on the hydrophobic surfaces so that the total oxygen content in the suspension is increased compared to the control solution without the particles.

In case of BCPP, in underground mines water and recirculation water, dissolved oxygen was observed 7.8% each with maximum yield recovery as shown in graph below in comparison to river water and thickener overflow water with 6.5 and 6.8 mg/L and have less recovery of clean coal from FF cell. It is obvious that as the oxygen content increased, fines clean coal yield also increased as shown in Graphs 4 and 5 which shows that dissolved oxygen in water is pH dependent and increases as pH increases. It was observed higher near pH value 8.16 of re-circulation water.

Graph 4 Relationship between pH and dissolved oxygen in water

Graph 5 Relationship of dissolved oxygen in water and clean coal field

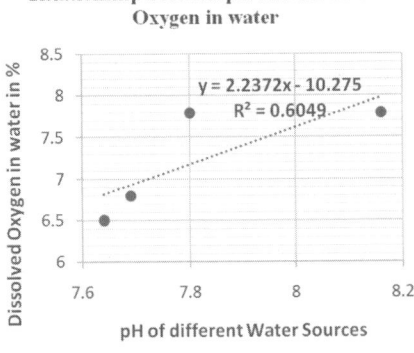
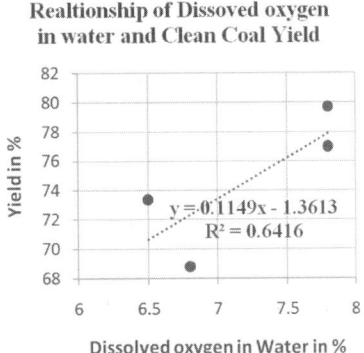

3.6 Effect of turbidity

Turbidity is caused by particles suspended or dissolved in water. Factors that influences turbidity. Suspended solids can be comprised of organic and inorganic materials such as sediment, algae and other contaminants. However, there are specific factors that can affect turbidity levels in a body of water. These are water flow, point source pollution, land use and resuspension.

In case of BCPP, refer to relationship Graph 7 there is inverse relationship between turbidity and fines clean coal yield. Turbidity was observed very high in high rate thickener overflow. It may be due to low settling rate of particles in thickener.

Graph 7 Relationship between turbidity and fines coal yield at BCCP

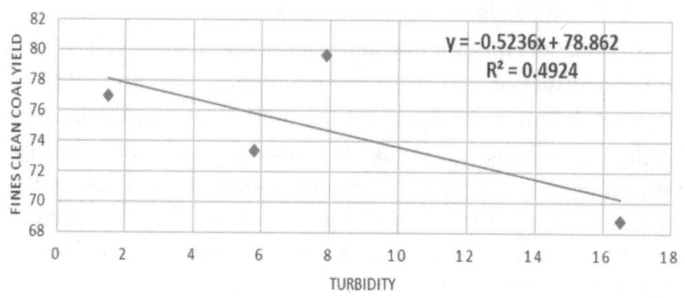

4. Conclusion and discussion

- Variations in different constituents of water of different sources were observed.
- Variation in fines clean coal yield was observed with different water sources.
- Scatter diagram (relationship graph) was plotted to find the relationship of different constituents present in water of all different water sources with fines clean coal yield.
- Based on R^2 value (very critical > 0.5, critical: range in 0.20–0.49), six main constituents out of 34 parameters (analysed during test) were selected.
- pH of water was observed the main significant criteria effecting froth flotation cell operation and yield.
- Metal ion species and dissolved inorganics ($CaCO_3$) concentration was also observed one of the most important factors effecting coal flotation cell performance and they were observed pH dependent.
- Typical contaminants in water sources were observed present as F^-, Mg^{2+}, Ca^{2+} and sodium nitrates.
- Dissolved oxygen in water sources was observed favourable for enhanced performance of froth flotation cell and also it was observed that as pH value increases, oxygen also increases.
- Turbidity in thickener overflow water was observed abnormally very high (1650 NTU), indicating low settling rate of solid particles in thickener.

- Negative effect of nitrate NO_3 on coal flotation cell performance was observed.

5. Way forward

1. To explore pH modifiers and its application in FF cell to improve FF cell performance.
2. Exploration of on line pH real time-based measurement system.
3. Automation in high rate thickener to enhance settling rate of ultra fines coal.

6. References

[1] Saeed Farrokhpay, Massimiliano Zanin: An investigation into the effect of water quality on froth stability. Advanced Powder Technology, 23(4):493–497, 2012.

[2] A.M. Gaudin, Robert H. Richards: The role of oxygen in flotation. Professor Emeritus of Mineral Engineering, Cambridge, Massachusetts, 1973.

[3] Marc A. Hampton, Anh V. Nguyen: Accumulation of dissolved gases at hydrophobic surfaces in water and sodium chloride solutions: Implications for coal flotation. Minerals Engineering, 22(9–10): 786–792, 2009.

[4] D. Liu, P. Somasundaran, T.V. Vasudevan, C.C. Harris: Role of pH and dissolved mineral species in Pittsburgh no. 8 coal flotation system – I. Floatability of coal. International Journal of Mineral processing 41(3–4):201–214, 1994.

[5] Edison Muzenda: An Investigation into the effect of water quality on flotation performance. World Academy of Science, Engineering and Technology, 2010.

[6] Hai Ou Song, Yang Zhou, Ai Min Li, Sandra Mueller: Selective removal of nitrate from water by a macroporous strong basic anion exchange resin. Chinese Chemical Letters, 23(5):603–606, 2012. DOI: 10.1016/j.cclet.2012.03.004.

[7] T.V. Subrahmanyam, Eric Forssberg: Froth stability, particle entrainment and drainage in flotation: A review. International Journal of Mineral processing 23(1–2):33–53, 1988.

4

Study on bubble formation mechanism and flotation experiment of jet-mixing flotation unit

Jinbo Zhu, Chao Wang, Wei Zhou, Chenguang Yang, Youli Han

Anhui University of Science & Technology, Department of Materials Science and Engineering, Huainan, China

Abstract: Combined with the advantages of available flotation devices in suction, energy saving, and stable mineralization environment, a jet-mixing flotation unit is proposed. The gas is injected by the jet device, and the impeller impinge on the active impeller to drive the driven impeller to stir the pulp, so as to ensure the uniform dispersion of pulp, chemicals and bubbles, and provide a stable mineralization environment in the flotation cell. In this paper, high-speed dynamic camera was used to test the mechanism of bubble formation, fracture and merger, and to explore the relationship between the size and number of bubbles with the structural parameters and working parameters of the device, so as to achieve the controllability of bubble formation. The flotation experiments show that the device has higher flotation efficiency, higher flotation yield and lower flotation cleaned coal ash, and has a better flotation effect of slime.

Keywords: Jet-mixing flotation, mechanism of bubble formation, controllability of bubble formation, flotation efficiency

1. Introduction

Flotation is to use the collision and adhesion of minerals in the flotation machine and a large number of fine bubbles, and the bubble rise to the liquid level of the foam layer to achieve mineral separation, so the suction performance and bubble formation mechanism in the flotation machine is the key factor affecting the performance of the flotation machine (Shen et al., 2013). The traditional flotation device is divided into mechanical mixing flotation machine, jet flotation machine, and flotation column, of which, the mechanical mixing flotation machine has the stable distribution of bubble particle size, larger processing capacity, lower energy consumption and simple system (Ma et al., 2014; Yang et al., 2015). The jet flotation machine has larger single processing capacity, stronger adaptability to flotation feed, lower energy consumption, and lower maintenance costs without any running parts (Sun and Tao, 2017b; Zhu et al., 2018). Bubbles and particles in flotation column collide and mineralize in highly turbulent slurry environment and narrow pipeline space, so the separation speed is fast and efficiency is high (Zhou et al., 2007; Sun and Tao, 2017a; Luo et al., 2009).

Combined with the advantages of the existing flotation device in the suction, energy saving, and stable mineralization environment. Zhu Jinbo's team proposed a jet-mixing flotation unit (Fei et al., 2017). The principle is shown in Fig. 1(a). When the gas-slurry jet is on the driving wheel, the impact force forces the driving wheel to rotate at a high speed, making the impeller in a driven state rotate at a constant speed. After the high-speed gas-bearing pulp flow in the feeding tank impacts on the driving wheel, the pulp is thrown towards the inner wall of the feeding tank. Under the double action of gravity and centrifugal force, the pulp moves downward in a spiral way in the discharging cylinder and diverted to the mixing wheel by the conical cover. The bubbles were cut into micro-bubbles again by high speed rotating mixing wheel and driving wheel, thus promoting the process of bubble mineralization. In this paper, the flotation mechanism of the device is studied and optimized from the aspects of suction performance, bubble generation mechanism, and flotation effect of mineralization.

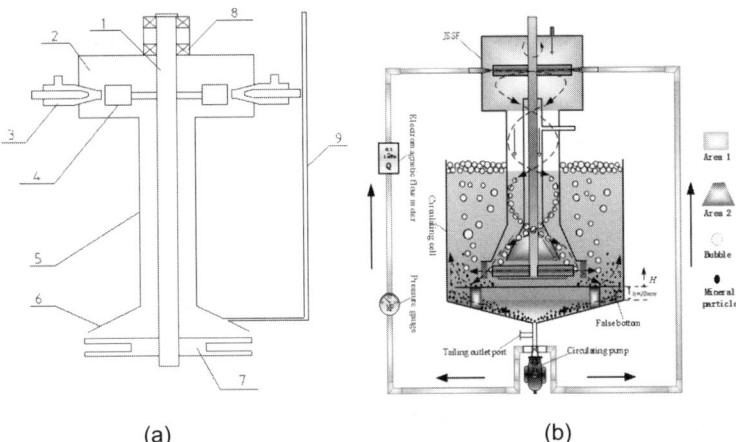

(a)   (b)

Figure 1: (a) Structural diagram of the jet-mixing flotation unit. (b) Separation test system for the jet-mixing flotation unit. 1, main shaft; 2, feed tank; 3, annular nozzle; 4, driving wheel; 5, discharging cylinder; 6, conical cover; 7, stirring wheel; 8, bearings; 9, aspirating tube

2. Bubble formation mechanism

In order to study the bubble diameter produced by the device, the method of controlling the speed of the impeller and the amount of air inhalation was adopted under the condition of different concentration of foaming agent (MIBC) to explore the scale distribution and influence law of the bubble produced by the device.

$C_6H_{14}O$ was used as the foaming agent in the experiment. The molecular weight was 102.18 and the concentration was denoted as c. The impeller speed n (r/min) and the inspiratory capacity q (L/min) were set to factor variables. Experimental results are described below.

2.1 Effect of foaming agent concentration on bubble size distribution

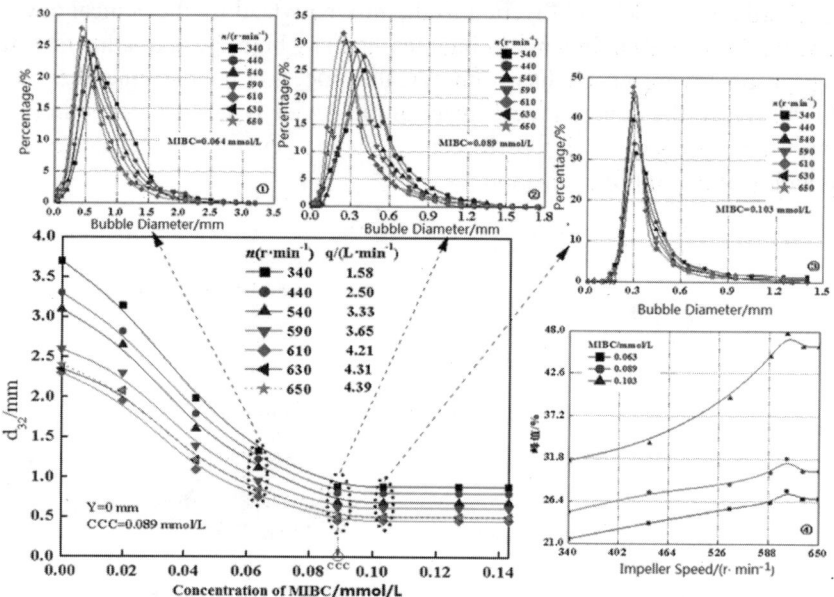

Figure 2: Effect of foaming agent concentration on bubble Sauter diameter and particle size distribution

Foaming agent MIBC can reduce the surface tension of the solution, prolong or prevent the process of bubble merging, which is the main factor to determine the bubble size (Li, 2013; Lau et al., 2013).

As shown in Fig. 2, with the increase of MIBC concentration, the number of MIBC molecules adsorbed on the gas–liquid interface increases, the solution surface tension further decreases, and the bubble diameter further decreases. When foaming agent concentration $c < 0.089$ mmol/L, Sauter diameter decreases with the increase of foaming agent concentration. But due to the low reagent concentration, Sauter diameter is bigger and the probability of a bubble merger is higher. Figure 3 shows the merger process of bubbles A and B. After 0.5 s, Bubble A "gobbles up" bubble B to form "conical" bubble C. With the passage of time, its tail gradually shrinking until 9.7 s, then stable bubbles larger than bubbles A and B are formed relatively. Therefore, when the concentration of foaming agent is less than the critical merger concentration, the foaming agent cannot completely prevent the merger of bubbles, but only prolong the time of merger process (Zhang et al., 2015).

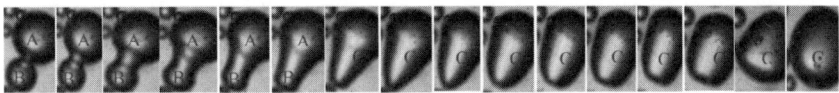

0s 0.1s0.2s 0.3s0.4s0.5s   0.6s   0.7s   0.8s 0.9s   1.0s 1.1s 1.2s   1.3s   2.3s9.7s

Figure 3: Process of bubble merger

When $c = 0.089$ mmol/L, the solution surface tension was further reduced, the bubble Sauter diameter was further reduced and then remained unchanged with the increase of the concentration of the agent, At this time, the pressure inside the bubble was stable, there is basically no merger between bubbles, and the minimum Sauter diameter was 0.45 mm. As can be seen from the illustration in Fig. 2, when the concentration of MIBC increases from 0.064 mmol/L to 0.103 mmol/L, the peak value of each curve gradually increases, indicating that the number of bubbles increases. At the same time, each curve is "offset" to the side where the bubble diameter decreases. In particular, when the concentration increased to 0.103 mmol/L, except the bubble diameter decreased, each peak value increased significantly, and the range of bubble diameter changed from wide to narrow. This phenomenon indicates that when the concentration of reagent exceeds the critical concentration of merger, merger is prevented, and the bubble diameter decreases due to the decrease of surface tension. Although Sauter diameter remained constant with increasing concentration, the number of bubbles with different grain sizes increased and the proportion increased, and all the bubbles changed at the diameter of 0.3 ± 0.01 mm, which indicates that the bubbles were more evenly distributed with the increase of concentration.

2.2 Influence of impeller speed on bubble size distribution

As shown in Fig. 4, Sauter diameter decreases more when the foaming agent concentration increases from 0.064 mmol/L to 0.089 mmol/L than when the foaming agent concentration increases from 0.089 mmol/L to 0.103 mmol/L. It can be seen that the critical annexation concentration is the key point for determining the variation range of bubble diameter. At the same time, the bubble Sauter diameter first decreases and then increases with the increase of the impeller speed under the same concentration of foaming agent. This is because the strong turbulent flow field and cutting effect generated by the impeller, which cause the bubble Sauter diameter to decrease. When $n \geq 610$ r/min, the bubble Sauter diameter first increased and then tended to be flat, and the comprehensive performance was first increased and then remained unchanged.

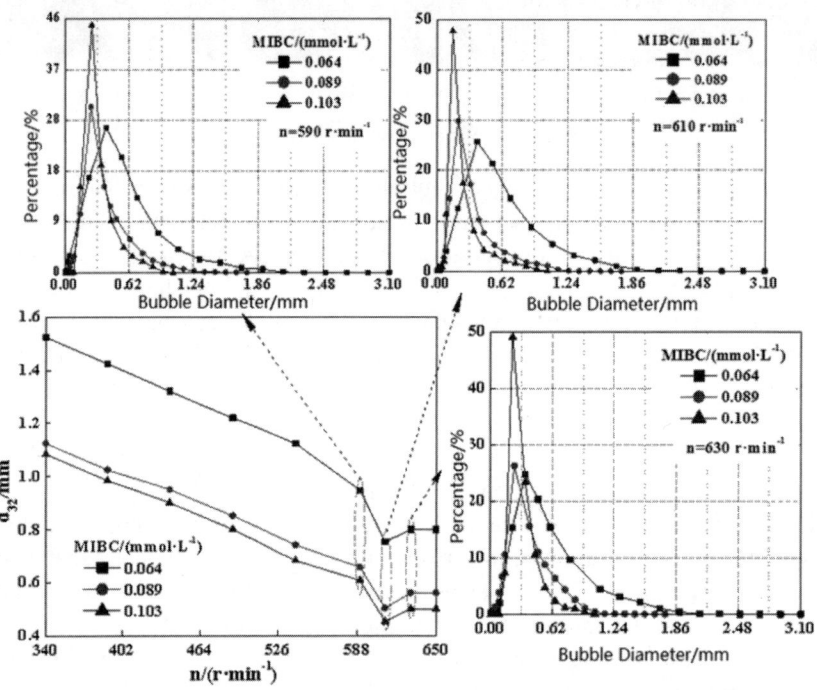

Figure 4: Effect of impeller speed on bubble Sauter diameter and size distribution

2.3 Influence of inspiratory capacity on bubble size distribution

Figure 5 shows the influence curve of the inspiratory capacity on the bubble Sauter diameter when the foaming agent concentrations are 0.064 mmol/L, 0.089 mmol/L, and 0.103 mmol/L, respectively. The dotted line in Fig. 5 represents the change curve of bubble Sauter diameter with the impeller speed, $n-d_{32}$ and $q-d_{32}$ represent the bubble Sauter diameter values corresponding to the impeller speed and inspiratory capacity, respectively, and the upper and lower abscissas represent inspiratory capacity and impeller speed, respectively. It can be seen from Fig. 5 that the bubble Sauter diameter increases with the increase of inspiratory capacity, and in the process that the bubble Sauter diameter is affected by both inspiratory capacity and impeller speed, there is an "intersection point" between the two curves. The intersection speed and inspiratory capacity are $n = 610$ r/min, $q = 4.08$ L/min, corresponding to this, Sauter diameters of bubbles with different foaming agent concentrations are respectively 0.75 mm, 0.5 mm, and 0.45 mm. It can be seen that the Sauter diameter at the intersection of the two curves is the smallest under the concentration of each foaming agent.

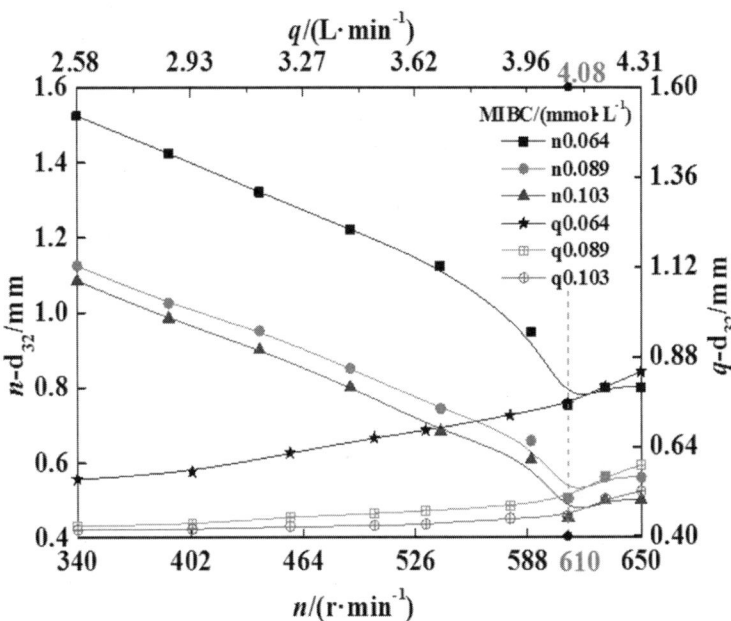

Figure 5: Effect of inspiratory volume on Sauter diameter of bubbles

3. Evaluation of flotation effect

3.1 Establishment of test system

Figure 1(b) shows the schematic diagram of system configuration. The black solid line and dotted line arrows in the figure indicate the flow direction of slurry, and the hollow arrow indicates the flow direction of air. In order to stabilize the flow field at the bottom of the stirring wheel so that the bubbles can rise and float in the tank body stably, a false bottom is set 15 mm below the sealing bottom plate at the bottom of the stirring wheel. The vertical upward indicating coordinate origin is set in the same plane with the false bottom, and the upward direction is positive. The separation liquid level is set at 230 mm.

3.2 Test conditions

The test coal sample is less than 0.5 mm grain raw coal, which was collected from Huainan pan I coal preparation plant. The concentration of test pulp was 20 g/L and the volume of flotation tank was 20 L. The foaming agent was MIBC, and the collector was $C_{12}H_{26}$. The feeding pressure was 0.14 MPa and the air filling capacity was 2.5 L/min. The separation test was conducted in accordance with GB/T4757 – 2001 "method for flotation test of pulverized coal (coal slime) laboratory unit".

According to GB/T 477-2008 "coal screening test method", particle size composition analysis of feed coal slime is carried out, and the results are shown in Table 1.

Table 1: Size composition of raw coal

Graded (mm)	Yield (%)	Ash (%)	Cumulative yield (%)	Cumulative ash (%)
0.5~0.25	17.56	15.85	17.56	15.85
0.25~0.125	24.19	20.17	41.75	18.35
0.125~0.075	20.12	22.09	61.87	19.57
0.075~0.045	12.96	19.64	74.83	19.58
<0.045	25.17	21.06	100.00	19.95

As can be seen from Table 1, the 0.25~0.075 mm grain size is the dominant grain size of the coal sample, and its yield is 44.31%, which is not significantly different from other grain sizes. The yield of <0.075 mm grain rank is 38.13%, and the weighted average ash content is 20.58%. The yield of this grain rank is relatively high, and the ash content was 19.95%.

3.3 Evaluation index

The recovery of combustible product and the perfect index of flotation used in this test as the evaluation index (National Standardization Administration of China, 2017), and its calculation formula is as follows:

$$E_j = \frac{\gamma_j(100 - A_{dj})}{(100 - A_{dy})} \times 100\% \tag{1}$$

$$\eta_{wf} = \frac{\gamma_j}{100 - A_{dy}} \times \frac{A_{dy} - A_{dj}}{A_{dy}} \times 100\% \qquad (2)$$

where E_j

100.00	66.51	7.38	100.00	33.50	44.92

In summary, the device has a better sorting effect on <0.25 mm micro-sized slime and has a strong ability to capture and mineralize low-ash granules in fine coal slime, and the recovery rate of clean coal reaches 70–80%. The surface modification of the slime particles is realized and the influence of the clay mineral cover on the fine coal slurry sorting is eliminated. According to the calculation, the recoverable rate of the flammable body of the flotation device is 76.95%, the perfect index of flotation is 52.36%, the flotation efficiency of the device is high, and the recovery effect of the coal sample is better.

4. Conclusion

(1) Foaming agent is the key factor affecting bubble diameter. Under the condition of controlling the rotating speed of the impeller and the suction amount, the Sauter diameter of the bubble decreased with the increase of the concentration of the foaming agent. The experimental results showed that the critical merger concentration of MIBC was 0.089 mmol/L. At the same concentration, Sauter diameter increased linearly with inhalation. There is a "critical intersection point" when the impeller speed and inspiratory capacity jointly influence the change of bubble size, that is, $n = 610$ r/min, $q = 4.08$ L/min, and the bubble Sauter diameter is at least 0.45 mm at this case. When the critical speed and inspiratory capacity were exceeded, the bubble Sauter diameter remained unchanged and the bubble distribution was more even.

(2) With the increase of the suction amount, the production of small bubbles will also increase. When the bubble dispersion uniformity is taken into account, the bubble size distribution can be more uniform by selecting the appropriate concentration of foaming agent and the impeller speed, which can not only ensure the ability of the bubble to separate minerals, but also take into account the transport capacity of the bubble after mineralization, and can improve the flotation efficiency of the device to a certain extent.

(3) The experimental research results show that the clean coal yield of the device is 66.51%, and the ash of clean coal is 7.38%, and the flotation effect is better for <0.25 mm fine coal slime. The recovery rate of clean coal of each grade is 70–80%, and that of combustibles is 76.95%, the flotation efficiency is 52.36%, which indicates that and the flotation effect is better.

5. Acknowledgement

This work was supported by Key University Science Research Project of AnHui Province (KJ2018A0093) and Natural science Foundation of AnHui province (1708085QE128) and National Natural Science Foundation of China (51374015).

6. References

[1] Fei Zhikui, Zhu Jinbo, Zhu Hongzheng, et al. Study on the inspiratory mechanism of the coupled flotation device with jet and agitation. Journal of China Coal Industry, 2017, 42(S2):472–478.

[2] Lau Y M, Deen N G, Kuipers J A M. Development of an image measurement technique for size distribution in dense bubbly flows. Chemical Engineering Science, 2013, 94:2019.

[3] Li Guosheng. Study on stability regulation of flotation foam and decarbonization of fly ash. Beijing: China University of Mining and Technology, 2013:13–18.

[4] Luo Shijun, Xie Jiemin, Xia Jingyuan. Analysis of mineral separation process characteristics of two flotation columns. Chromatid gold (ore separation part), 2009(6):41–45, 35.

[5] Ma Shan, Wang Zhan, Wang Bo, et al. Numerical study of gas–liquid–solid three-phase flow field in a mechanically stirred self-suction flotation machine. Journal of Engineering Design, 2014(1):62–67.

[6] National Standardization Administration of China. Evaluation method for flotation engineering effect of coal preparation plant: GB/T 34164 – 2017. Beijing: China Standard Press, 2017.

[7] Shen Zhengchang, Lu Shijie, Shi Shuaixing, et al. Analysis and discussion of bubble characteristic parameters in KYF flotation machine – KYF flotation machine flow field test and simulation research (5). Nonferrous Metals (Mineral Processing Section), 2013(5):44–49.

[8] Sun Fengjie, Tao Xiuxiang. Influence of "cyclone" on gas holdup and bubble size of flotation column. Metal Mines, 2017a(12):115–118.

[9] Sun Fengjie, Tao Xiuxiang. Study on gas-filled mixing performance of jet flotation machine. Coal Technology, 2017b(12):260–263.

[10] Yang Yingjiang, Chen Jianhua, Shen Zhengchang, et al. Simulation study on flow field characteristics of gas-filled agitator flotation machine. Mineral Protection and Utilization, 2015(2):22–26.

[11] Zhang Shijie, Liu Wenli, Zhao Shukai, et al. The effect of foaming agent and air volume on the particle size of flotation bubbles. Coal Engineering, 2015, 47(3):119–121.

[12] Zhou Lingfeng, Fu Lianhai, Zhang Qiang. High efficiency fine particle flotation column. Non-Ferrous Metals, 2007(2):55–58.

[13] Zhu Jinbo, Han Youyi, Fei Zhikui, et al. Research on jet-agitation-coupled slime flotation device. Clean Coal Technology, 2018(1):69–73.

5

Improving coking coal availability using hydrophobic hydrophilic separation technology for Indian coals

R.B. Mathur and Kalpit K. Dubey

Minerals Refining India Pvt. Ltd., Gurgaon, India

Abstract: Minerals Refining Company's patented, university-developed technology, called the hydrophobic–hydrophilic separation (HHS) Process, uses surface chemistry to simultaneously clean and dewater fine (−100 mesh) and ultrafine (−325 mesh) coal. The result is more tons of saleable coal, reduced waste, and improved plant economics. HHS uses the concept of chemically displacing the coal from both hydrophilic mineral matter and moisture. Conceived in the laboratories of Virginia Tech and commercialized in partnership with industry veterans, MRC's technology heralds a new future for coal and mineral processing. Extensive tests were conducted on the various samples of Indian coking coals in 2017 and 2018, which showed impressive results. Considering these results, Indian subsidiary was formed Minerals Refining India Pvt. Ltd. to implement commercial plants using HHS process to the existing coking coal washeries. In the Indian context, the fine coal slurry and rejects which are visible all around the washeries in BCCL present an excellent opportunity to re-treat and recover coking coal.

Keywords: Pilot plant, Hydrophylic, Slurry, Economics, Hydrophobic

1. Introduction

For nearly four decades, Dr. Roe-Hoan Yoon has focused his research on methods to improve separation and recovery of fine coal and minerals. As Director of the Center of Advanced Separation Technologies (CAST) at Virginia Tech, Dr. Yoon has invented and licensed equipment, additives, and processes that have been employed by the mining industry to produce billions of dollars' worth of raw materials. To date, he has 57 U.S. and international patents to his credit, many of which have been commercialized.

Dr. Yoon and research partner Dr. Gerald Luttrell made a laboratory breakthrough in recovering the finest fractions of minerals and coal, a process which showed great promise of both cleaning and dewatering the particles. Upon sharing batch lab scale results with Dr. Stanley Suboleski, former Professor and Department Head of Mining and Minerals Engineering at Virginia Tech, the group approached industry veteran and long-standing supporter of the university, Mr. E. Morgan Massey. Immediately recognizing the potential of the technology, Mr. Massey's Evan Energy Investments funded continuous laboratory-scale testing of the process, this proved successful. With the support of industry leaders Dr. Peter Bethell, Carl Bauer, and Bob Gentile, the group launched Minerals Refining Company (MRC) in 2012 to further prove and ultimately commercialize the HHS Process for coal processing plants around the world.

Minerals Refining India Pvt. Ltd. (MRIPL), a subsidiary of Minerals Refining Company was incorporated in 2017 in India to implement HHS based commercial plants in India.

2. HHS process

Figure 1 represents the concept of the HHS process. In Step I, a hydrophobic particle, *e.g.*, bituminous coal, placed in water phase is transferred to a hydrophobic liquid phase above. The process is spontaneous, with its free energy of transfer being negative, *i.e.*, $\Delta G_t < 0$. In Step 2, the hydrophobic particle is removed to a vapor phase. The residual hydrophobic liquid adhering to the surface is evaporated and condensed for recycling. The free energy change associated with the Step II is positive, *i.e.*, $\Delta G_e > 0$. However, the energy required for the vaporization is a fraction (14–16%) of what is required for vaporizing water in thermal drying.

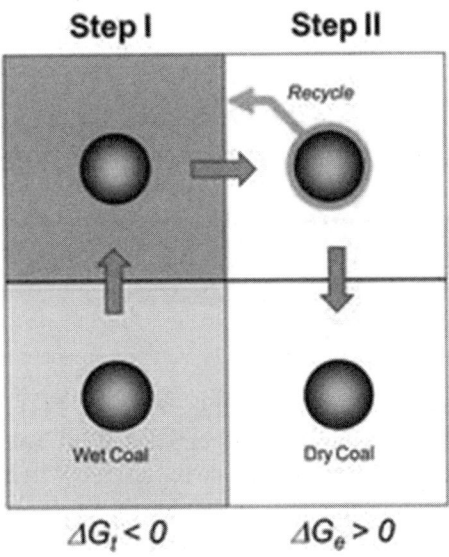

Figure 1: Steps involved in the HHS process

Step I of the HHS process is similar to the two-liquid flotation process, which has been shown to be superior to flotation for the recovery of ultrafine particles (Mellgren and Shergold, 1966; Lai and Fuerstenau, 1968; Shergold, 1976). In two-liquid flotation, hydrophobic particles are collected by oil, while in flotation hydrophobic particles are collected by air. For an air bubble to collect hydrophobic particles suspended in water with a finite water contact angle (q_w), it is necessary that wetting tension ($g_{SV}-g_{SW}$) be lower than the surface tension of water (g_{WV}), where the subscripts S, W, and V represent solid, water, and vapor phases, respectively. For an oil droplet to collect a hydrophobic particle with an oil contact angle q_o, it is likewise necessary that the wetting tension ($g_{SO}-g_{SW}$) be lower than the surface tension of water (g_{WV}). In general, $g_{SO} < g_{SV}$. It follows, therefore, that $q_w < q_o$.

Figure 2 shows the contact angles (q_o) of n-alkanes with $n = 4-10$ on a bituminous coal in water.

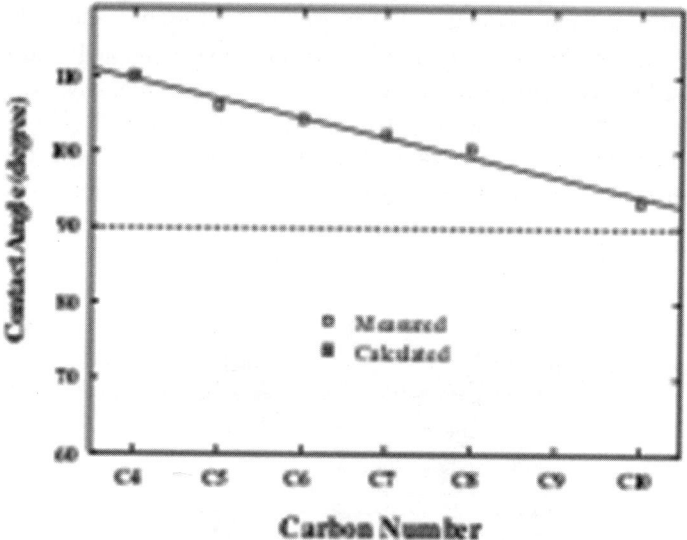

Figure 2: Contact angles of n-alkanes in water on a bituminous coal in water

As shown, the contact angles are in the range of 94–110°, which are well beyond the values obtainable with air bubbles. The captive bubble contact angles (q_w) of the U.S. bituminous coals are ~65° (Gutierrez-Rodriguez et al., 1984). Thermodynamically, it would, therefore, be easier to collect hydrophobic particles with oil droplets than with air bubbles. Kinetically, oil–particle attachment should be faster than bubble–particle attachment. For an air bubble to attach on a surface, it must overcome the repulsive van der Waals force, which should slowdown the process. On the contrary, the van der Waals forces in the wetting films formed between oil droplets and hydrophobic surfaces are attractive, which should lead to faster collection efficiencies.

The hydrophobic particles transferred to the hydrophobic liquid phase should be free of surface moisture. If hydrophilic particles are suspended in the same aqueous phase, they will stay behind and will not enter the organic phase. Therefore, the two liquid flotation process described above can achieve both recovery and dewatering of hydrophobic particles. For cleaning coal fines, the process can be used to separate hydrophobic particles from hydrophilic minerals and at the same time dewater the clean coal product. For obvious reasons, the process is referred to as hydrophobic–hydrophilic separation (HHS).

3. Application in coking coal

Customers pay premium prices for coal with coking characteristics, used to produce iron and steel. Coal sold for coke production requires higher purity levels (3–6% ash) compared to steam coal (7–12% ash). Advanced dewatering processes, such as plate-and-frame filter presses, have been used to dry high-value ultrafine coal in this market, although high equipment cost and limited moisture removal have severely restricted industrial applications. The HHS process offers an economic solution to recovering this high-value product.

4. Advantages

(1) **Increase revenue with little to no capital:** The HHS process offers a low-cost, highly profitable method of extracting greater value from your existing product stream. Our plants and operational agreements are tailor-made to each processing facility to ensure cost-effectiveness and a new revenue stream for our customers, with minimal capital.

(2) **Improve plant economics:** Despite the small fraction of fine and ultrafine coal in a processing facility's product stream, recovering this additional product could have a remarkable effect on the economics of a preparation plant. In addition, due to the quality specifications of product from the HHS process, it may be possible to raise the gravity in the other processing plant circuits and improve recovery of the coarser coals.

(3) **Extend slurry pond life:** The potentially recoverable coal in fine slurry typically represents 35–65% of the solids being pumped from the thickener to the waste impoundment. Recovering this coal as a

saleable product will extend the life of the slurry pond and reduce the potential for a black water spill.

5. Development of HHS based plants

(1) **POC plant:** Proof-of-concept unit was commissioned at Virginia Tech in June of 2013. Here, Minerals Refining Company tested coal-waste slurry from a number of plants in Virginia, West Virginia, Pennsylvania, and Kentucky. The HHS process consistently produced a salable coal product with ash content of 3–6% by weight and single-digit moisture values.

(2) **Pilot plant:** With a proven technology in hand, MRC entered into a commercial contract with the one of the largest U.S. coal companies to install a 1 clean-ton per hour Pilot Plant at an operation in Southwest Virginia. The plant became operational in late 2015 and completed testing in 2017. It successfully upgraded a split of the <40 µm (<325 mesh) deslime-cyclone overflow and recovered the coal as a high-quality product. The 5%-solids slurry feed averaged 58% ash, while the final product contained 4–1/2% ash and 5–9% moisture. The moisture is controllable while the ash is a function of the seam liberation characteristics. The ash achieved can be compared to the overall quality produced by plant, which is 6–1/2%.

(3) **Commercial plants:** On the strength of the pilot plant results, MRC is currently designing the first commercial plant, with construction scheduled to start in 2019. The first plant will be located in the coalfields of Central Appalachia, with an expected output of 20 tons/h of clean coal.

6. Indian operations

For India, MRIPL is working on two-pronged approach of working with the major consumers of coking coal for implementing HHS plant in their existing and upcoming coking coal preparation plants for their captive usage and also setting up a merchant standalone HHS plant for sale of clean coking coal to domestic users.

During the feasibility study for application of HHS process in India, it was found that we can create capacities of tune of 2–3 mtpa for HHS product with the currently available feed for the next 20 years. This will reduce India's dependence on imported coal and will reduce the FOREX outgo by approx. USD 600 million.

Initial test results of the samples of coking coal are very promising, with consistently showing less than 5% moisture and 10% ash, including excellent yield, which clearly suggest that HHS technology will have wide application across all the existing coking coal washeries in the country.

7. Test results of various Indian coal samples

(1) Following samples of Coal India's coking coal (of Central Coalfields Ltd.) were tested for application of HHS process at the labs at Virginia Tech.

    (a) Kathara: Fine coal slurry (0.5 mm × 0)

    (b) Kathara: Raw coal (80 mm × 0)

    (c) Kedla: Raw coal (80 mm × 0)

    (d) Rajrappa: Raw coal (80 mm × 0)

(2) Kathara and Kedla are low-volatile medium coking coals, while Rajrappa is low-volatile poor coking coal (Gautam et al., 2015).

(3) All three washeries have fine coal flotation circuit, but they are not in operation.

(4) Table 1 below depicts the summary lab test results of Kathara Fine Coal for HHS implementation.

Table 1: Summary of Kathara fine coal

Sample #	Feed (%wt.)	HHS product (%wt.)		Refuse ash (%wt.)	Mass yield (%wt.)	Recovery (%wt.)	
	Ash	Ash	Moisture			Combustible	Organic
As received D_{95} = 500 µm	33.3	4.0	5.8	56.3	43.9	63.2	66.3
		3.5	2.8	55.9	43.1	62.4	65.4
Pre-screened 300 µm × 0	36.5	4.2	8.0	72.6	52.8	79.6	84.1
		4.0	6.3	73.2	53.0	80.1	84.7
Pre-screened 150 µm × 0	40.4	5.1	9.3	78.5	51.8	82.6	88.1
		5.0	2.5	77.8	51.4	81.9	87.4
Pre-screened 75 µm × 0	46.4	5.0	1.9	78.8	44.0	77.9	84.8
		5.0	4.7	78.6	43.8	77.6	84.4

(5) Table 2 below depicts the summary lab test results of Kathara raw coal for HHS implementation.

Table 2: Summary of Kathara raw coal

Sample #	Feed (%wt.)	HHS product (%wt.)		Refuse ash (%wt.)	Mass yield (%wt.)	Recovery (%wt.)	
	Ash	Ash	Moisture			Combustible	Organic
Ground feed D80 = 220 µm	28.5	8.4	1.7	74.7	69.7	89.3	91.6
		8.6	1.9	74.3	69.8	89.2	91.4
Ground feed D80 = 160 µm	28.4	7.2	3.8	76.4	69.3	89.9	92.3
		6.8	2.7	76.5	69.0	89.8	92.2
Ground feed D80 = 65 µm	28.8	9.3	3.0	79.4	72.2	91.9	94.2
		9.9	2.7	79.2	72.7	92.1	94.3

(6)   Table 3 below depicts the summary lab test results of Kedla raw coal for HHS implementation.

Table 3: Summary of Kedla raw coal

Sample #	Feed (%wt.)	HHS product (%wt.)		Refuse ash (%wt.)	Mass yield (%wt.)	Recovery (%wt.)	
	Ash	Ash	Moisture			Combustible	Organic
Ground feed D80 = 240 µm	30.5	12.8	2.0	75.7	71.8	90.1	92.3
		13.0	1.8	75.4	71.9	90.0	92.2
Ground feed D80 = 140 µm	29.7	11.1	3.0	80.1	73.1	92.4	94.7
		9.5	3.2	79.9	71.4	91.8	94.2
Ground feed D80 = 55 µm	30.4	9.6	3.6	79.6	70.2	91.3	93.8
		11.0	5.2	80.0	71.8	91.9	94.3

(6)   Table 4 below depicts the summary lab test results of Rajrappa raw coal for HHS implementation.

Table 4: Summary of Rajrappa raw coal

Sample #	Feed (%wt.)	HHS product (%wt.)		Refuse ash (%wt.)	Mass yield (%wt.)	Recovery (%wt.)	
	Ash	Ash	Moisture			Combustible	Organic
Ground feed D80 = 400 µm	43.1	7.4	3.8	75.2	47.4	77.1	81.5
		7.1	2.2	75.4	47.4	77.3	81.8
Ground feed D80 = 150 µm	43.2	5.2	3.4	77.4	47.3	79.1	83.8
		5.6	5.6	77.1	47.4	78.9	83.6
Ground feed D80 = 80 µm	45.8	8.2	3.8	81.7	48.8	82.7	88.0
		7.8	2.4	81.1	48.2	82.0	87.3

8. Conclusion

Hydrophobic–hydrophilic separation (HHS) process has shown excellent results for various tested coal samples from India. The test results show that the HHS process can produce clean coal containing less than 5% ash and moisture. In general, the product quality improves with finer coal due to improved mineral liberation. Also, the product moisture is independent of particle size as the HHS process is designed to remove surface moisture by displacement.

Additionally, the sample results of various Indian coking coals have helped MRIPL is making business case for establishing HHS based plants in India for both the captive usage of the steel making companies and also standalone merchant plants.

9. Acknowledgement

The authors are thankful to Evan Energy Investments (USA), Minerals Refining Company (USA), Virginia Mining Resources Pvt. Ltd. (India), and Minerals Refining India Pvt. Ltd. (India) for their financial support. Authors also wish to thank Central Coalfields Ltd. (CCL) for their support in providing coal samples.

10. References

[1] 2016 SME Annual Conference, "Hydrophobic–hydrophilic separation (HHS) process for the recovery & dewatering of ultrafine coal", Nikhil Gupta, Biao Li, Gerald Luttrell, Roe-Hoan Yoon, Robert Bratton & James Reyher.

[2] Bethell, P., Luttrell, G., (2005), "Effect of Ultrafine Desliming on Coal Flotation Circuit", Ed. by Fuerstenau, M, Jameson, G., Yoon, R., Proceedings of the Centenary of Flotation Symposium, Brisbane, Australia, pp. 719–728.

[3] Gutierrez-Rodriguez, J., Purcell R., Aplan, F., (1984), "Estimating the hydrophobicity of coal", Colloids and Surfaces, Vol. 12(1):1–25.

[4] Honaker, R., Kohmuench, J., Luttrell, G., (2013), "Cleaning of fine and ultrafine coal", The Coal Handbook: Towards Cleaner Production, Ed. by Osborne, D., Woodhead Publishing Ltd., pp. 301–346.

[5] Hucko, R., Gala, H., Jacobsen, P., (1988), "Status of DOE-sponsored advanced coal cleaning project", Industrial Practice of Fine Coal Cleaning, Ed. by Klimpel, R., Luckie, P., SME, pp. 159–210.

[6] Lai, R., Fuerstenau, D., (1968), "Liquid–liquid extraction of ultrafine particles", Transactions AIME, Vol. 241:549–556.

[7] Luttrell, G., Bethell, P., Honaker, R., (2014), "Designing and operating fine coal processing circuits to meet market specifications", International Journal of Coal Preparation and Utilization, Vol. 34(3–4):172–183.

[8] Mellgren, O., Shergold, H.L., (1966), "Method for recovering ultrafine mineral particles by extraction with an organic phase", Transactions of the Institution of Mining and Metallurgy, Vol. 75:C267–268.

[9] Orr, F., (2002), "Coal waste impoundments: risks, responses, and alternatives", National Research Council report, Washington D.C.

[10] Shergold, H.L., (1976), "Two-liquid flotation for the treatment of mineral slimes, Ind. Minera. (S.-Etiennem Fr.) Mineralurgie, Nov. 1976, pp. 192–205.

[11] Wills, B., Napier-Munn, T., (2006), Mineral Processing Technology, 7th Edition, Elsevier.

[12] Yoon, R.-H., Aksoy, B.S., (1999), "Hydrophobic forces in thin water films stabilized by dedecylammonium chloride", Journal of Colloid Science, Vol. 211:1–10.

[13] HHS Process Laboratory Scale Tests for Coal India Limited, Stan Suboleski (MRC), Steve Sears (MRC), Trey Jones, Nikhil Gupta, Roe-Hoan Yoon, Jerry Luttrell Center for Advanced Separation Technologies, Virginia Tech.

6

Enrichment of small grain classes: Laboratory scale

Waldemar Mijał[1] and Barbara Tora[1]

AGH University of Science and Technology, Department of Environmental Engineering and Mineral Processing, Kraków, Poland

Abstract: Dry coal separation is currently used widely in many countries like China, USA, India, Turkey, South Africa, Australia, and last year it was tested in Poland. From theory and practical usage of dry separation, the best results can be obtained during the enrichment process of narrow grain class. Polish experience shows that grain class smaller than 6(8) mm influence the separation process. Before the separation process, some amount of grain class 0-6(8) mm is removed on the vibrating screen but some amount of small grains is still required to create a fluidized bed.

The article deals with problems of beneficiation small grain class 0–6(8) mm. This grain class is affecting the accuracy of the raw separation of the feed-in air vibrating table FGX. Next chapters will describe the methodology of research, effects of enrichment grain class 1–6(8) mm, and the maximum share of grain class 0–1 mm without affecting the separation process. All tests will be performed in laboratory air jig.

Keywords: Dry beneficiation, coal deshaling, dry coal separation, small coal particles, air jig, dry air separator

1. Introduction

The dry method of coal enrichment was popular at the beginning of the 20th century. In this method, a material consisting of a set of grains with different densities is separated by ascending or pulsating air stream. Dry enrichment is usually applied in places where there are water shortages to wet processes and in a harsh climate due to the possibility of freezing separation products after separation in a water medium. The raw materials that can be enriched with this method are mainly hard coals with a large proportion of coal-fired or waste fractions and for brown coals (hard types).

With the development of new equipment for coal enrichment in China, a new vibrating air separator type FGX was developed, which can be an independent processing plant or it can be used in processing plants for preliminary deshaling of coal output before being directed to wet enrichment processes. As a success of this construction can be a great interest from the United States, Turkey, India, South Africa, and a dozen other countries. After analyzing the experiences published by these countries in 2012 the Institute of

Mechanized Construction and Rock Mining, through the company WARKOP Sp. z o.o. (representative of Tangshan Shenzou Machinery Co. Ltd. on the Polish market) purchased a vibrating air separator type FGX-1. The Institute is currently conducting its own research related to among others with the removal of ecotoxic elements from energy hard coal and tests for Polish coal companies to check the applicability of the above devices in Polish mining (Baic et al., 2014a,b; Dawaasuren et al., 2016; Sobko et al., 2016).

For better separation results grain class 0–6 mm is removed by vibrating screen before air vibrating table. Pilot-scale tests show that some amount of this grain class is required to create fluidized bed (10–12% of grain class 0–6 mm in the feed). This situation creates an idea to check the possibility for beneficiation of small particles in a laboratory scale by using air jig.

2. Results of separation of small grains by different equipment used in world coal separation processes

In the last twenty years, dry coal separation methods become more popular. Nowadays appear more equipment suitable for this method which can be divided for separators intended for separation of bigger coal grains and smaller coal grains. In order to approximate the possibilities of fine grain enrichment, three currently used technologies will be described. First good example for separation of small particles in industrial scale is TFX air jig. The dry cleaning jig operates on a principle which is similar to that of a wet cleaning jigging machine. The fine size material on the deck tends to become gradually loosened and stratified under the combined effect of vibration force and the pulsation produced by the upward air-stream coming from deck bottom. The heavier material gradually sinks down to bed bottom while the lighter material gradually floats up to the top of the bed. During the proper jigging process, a table material bed is finally formed. The heavier material at bed bottom is discharged through the discharge unit while the other materials move forward to go through further separating process under the effect of vibration force. The TFX air jig unit is shown in Fig. 1 (Tangshan promotional materials 2016, Mijał et al., 2018) (Table 1).

KAT (Korean Advanced Technology) table is a type of pneumatic table separator. In this separation process is used differences in density between coal grains. The separation scheme is showed in Fig. 2. The denser material falls down on the deck to form a lower layer and moves to the oscillating direction of the table deck by the action of eccentric motion. The upper layer of light particles roll down by the means of airflow and table shaking to the lower part of the deck and accumulates along the blocking wall, and when the level of accumulated light particles exceed the height of blocking wall it will start overflow. This specific design of the KAT table generates an autogenetic medium of light particles along the lower end of the deck, which increases product quality. KAT table separates coal into three different fractions. The light fraction-coal, the heavy fraction-reject, and the middling (Dawaasuren et al., 2016, Mijał et al., 2018) (Table 2).

Figure 1: Air jig TFX unit (Tangshan promotional materials 2016, Mijał et al., 2018)

Table 1: Parameters Results of separation by using TFX air jig for coal in grain class 13–0 mm (Tangshan promotional materials 2016, Mijał et al., 2018)

Coal	Raw coal ash (%)	Clean coal		Middlings		Reject	
		Yield (%)	Ash (%)	Yield (%)	Ash (%)	Yield (%)	Ash (%)
Coking coal	19.75	77.49	9.10	7.50	28.43	15.01	70.41
Power coal	42.53	58.73	27.26	21.93	54.88	19.39	73.47
Power coal	35.85	60.97	20.71	20.90	46.27	18.13	74.73

Table 2: Results of preparation by using KAT process technology for coal in grain class 1–5 mm (Dawaasuren et al., 2016 , Mijał et al., 2018)

Test	Clean coal		Middlings		Reject	
	Yield (%)	Ash (%)	Yield (%)	Ash (%)	Yield (%)	Ash (%)
1	57.20	16.00	20.40	44.50	22.40	86.10
2	53.40	9.40	22.90	37.90	23.70	88.70
3	58.30	10.70	14.00	37.30	27.60	84.00

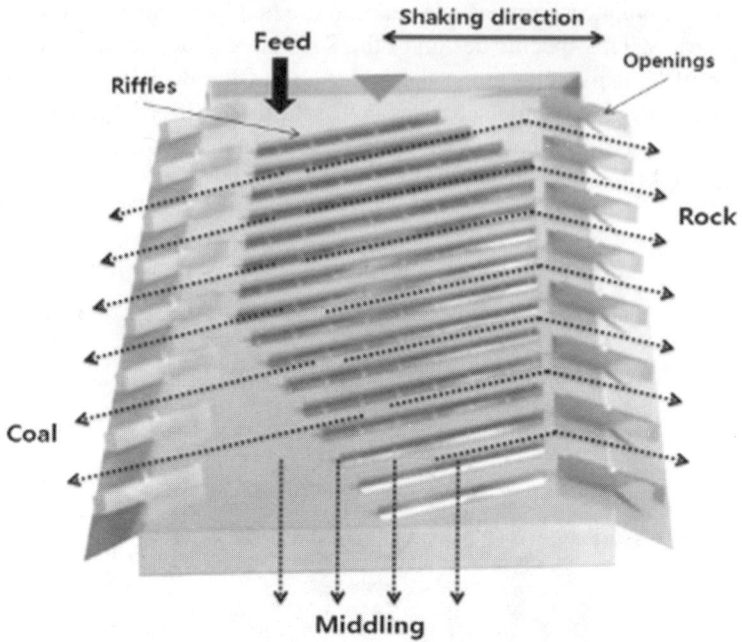

Figure 2: KAT air table deck – trajectory of separated products
(Dawaasuren et al., 2016 , Mijał et al., 2018)

3. Tests on laboratory scale air jig

Results mentioned in the previous chapter show that it is possible to separate small particles in air separators. KAT processing technology for small particles is very effective and guarantee rejects with the amount of ash at an average level of 86%, concentrates from this separator have ash content less than 20%. These facts create a chance to try separate small coal particles by air jig created in one Polish University.

Air jig contains an air chamber created from 5 to 6 different removable rings, receiving tank of separation products, fan, expansion tank, and dust bag system. In Fig. 3 we can see how the separation chamber looks from laboratory air jig.

For tests grain class (0–6 mm) separated from concentrate from one of the Polish coal mines was used. Tests were separated for two main parts:

- Separation with different amount of grain class < 1 mm (for tests was chosen 2% and 4% of particles less than 1 mm),
- Separation without grain class < 1 mm.

Figure 3: Working chamber of laboratory air jig

The main target in this research was to find the best possible way to reach ash content in concentrate < 4% with a high yield of the final product (Tables 3 and 4).

Table 3: Results of separation of grain class 0–6 mm

Product	Test 1 (2% content 0–1 mm)		Test 2 (4% content 0–1 mm)	
	Yield (%)	Ash (%)	Yield (%)	Ash (%)
Concentrate	74.7	4.4	9.2	4.8
Reject	24.0	10.0	27,9	9,4
Dust	1,4	18,7	2,9	16,9

Table 4: Results of separation of grain class 1–6 mm

Product	Test 1		Test 2	
	Yield (%)	Ash (%)	Yield (%)	Ash (%)
Concentrate	72.4	4.6	67.5	3.7
Reject	27.6	8.9	32.5	6.6

Figure 4: Pyrite particles separated during tests

4. Conclusion

Foreign experience shows that beneficiation of fine particles is possible. Concentrates with ash content between 9% and 16% and wastes with ash content higher than 85% shows that separation KAT processing can be applied as a supplement for FGX air vibrating table to separate clean coal from grains <6 mm. Quality of wastes separated by TGX air jig could eliminate this separator. Wastes with ash content <80% shows that they still include coal particles (mostly middlings).

Investigation prepared on laboratory air jig shows that it is possible to get concentrates with ash content <4%. During the tests, it was observed that the pulsation process in this air jig it has a positive effect on the release of pyrite. Few main conclusions will be described below:

- Dust removing system was not effective enough to handle a higher amount of particles smaller than 1 mm. The maximum limit adopted for the original design of the air jig is 4% of the grain class content <1 mm,
- Construction based on the jigging process is not effective enough to handle beneficiation process for separate ultra clean concentrate for special purposes,
- Beginning of tests aimed at separation of clean coal from raw feed not from already separated products,

- During these tests in rejects was found many pyrite particles or concrescence between coal and sulfur Fig. 4. Which shows that this method of dry separation can be used for removing pyrite and sulfur from concentrate.

5. References

[1] Baic I., Blaschke W., Szafarczyk J. 2014a. Dry coal cleaning technology, Inżynieria Mineralna – Journal of the Polish Mineral Engineering Society. No. 34, p 257–62.

[2] Baic I., Blaschke W., Szafarczyk J. 2014b. The first FGX unit in the European Union. CPSI Journal – a magazine by the coal preparation society of India. Vol. VI, Issue 16, p. 5–12.

[3] Dawaasuren J., Byoung-gon K., Ju-hyoung L., Davaatseren G., Bazarragchaa M. 2016. Dry Coal Preparation of Fine Particles by KAT Process, XVIII International Coal Preparation Congress (Saint Pettersburg) ISBN 978-3-319-40943-6, pp 1171–6.

[4] Tangshan Kaiyuan Technology Co., Ltd. – promotional materials 2016.

[5] Mijal W., Blaschke W., Baic I. 2018. Dry coal beneficiation methods in Poland. 25th World Mining Congress. Astana, Kazakhstan (Not published).

[6] Sobko W., Blaschke W., Baic I. 2016. Constructional improvements of FGX-1 air concentrating table aiming at optimization of operation, Inżynieria Mineralna – Journal of the Polish Mineral Engineering Society No. 37, p 37–46

7

Economic efficiency of the dry coal separation process: Polish experience

Gabriel Buchalik, Sebastian Motyczka, Józef Szafarczyk, Ireneusz Baic, Wiesław Blaschke

Warkop Sp. z o. o., Świerklany, Poland
Institute of Mechanised Construction & Rock Mining, Warsaw, Poland

Abstract: The dry separation technology based on the use of a new generation of FGX air-vibrating separators is currently used on an industrial scale in many countries where hard coal is mined. Installations based on dry separation technology are used as independent plants or as initial separation module before processes of coal wet cleaning. In recent years, also in Poland, thanks to the work undertaken by the industrial and scientific consortium Warkop Sp. z o. o. (sole representative of Tangshan Shenzhou Manufacturing Co., Ltd. China for the Central European market) and also by the Institute of Mechanized Construction and Rock Mining, the interest in this technology both among hard coal producers (preparation plants) and distributors has increased.

The paper presents the results of a simplified economic analysis of the implementation of dry separation technology in Polish conditions including the identification of investment outlays and operating costs in a generic manner and the assessment of project effectiveness in a generic system using static and discount methods for assessing investment effectiveness, i.e.: payback period (PP), simple rate of return (ROI), net present value (NPV) and internal rate of return (IRR).

Keywords: Economic analysis, economic efficiency, dry separation, air-vibrating separator

1. Introduction

The hard coal output from Polish mines is sold either in the form of raw coal or it is subjected to the processes of beneficiation in preparation plants. In the first case, this applies to the coal output whose parameters (calorific value, ash and sulphur content, etc.) allow for its direct use in the processes of electricity or heat production. If the aforementioned parameters are not sufficient for the users, coal output is subjected to the enrichment processes. The coal output is divided into the narrower grain fractions, crushing large coal grains to the dimensions below 250 (200) mm. In Polish conditions, particular grain classes undergo only the wet enrichment processes. In practice, grains larger than 20 (50) mm are enriched in dense liquid suspension (a mixture of water and magnetite). Grains below 50 (20) mm and larger than 1.0 (0.5) mm are enriched in pulsating grain jiggers or fine coal jiggers and in hydrocyclones. Whereas coal grains measuring 1.0 (0.5)–0 mm (mainly coking coals) are enriched in flotation machines or in Reichert spirals equipped with the set of filter presses (Baic, 2013).

An alternative to wet enrichment methods is the possibility of conducting the separation processes of gangue grains from coal grains through the dry separation method (Li Gongamin, 2006; Honaker, 2007B; Ghost et al., 2013). In Poland, the possibility of implementing dry separation methods of enrichment based on the FGX type air-vibrating separators has been considered for several years (Baic et al., 2014A). Installations for the dry coal separation may be independent coal enrichment plants for the grain classes bellow 100 (80), (75), (50) mm, etc. In the technological system of the preparation plant they can replace the wet enrichment process in jigs. Dry coal separation installations may also be a kind of a bypass in existing preparation plants that allow dedusting of the finest coal grains and removal of a certain amount of rock before providing of the partially enriched coal to the wet enrichment process (when the low ash content in the product is required), at the same time reducing the amount of the finest coal grains going to the water and sludge circulation (Baic and Blaschke, 2013; Baic et al., 2015A). In order to implement this technology a number of tests were performed in the several coal mines with positive results (Baic et al., 2014A, 2014B, 2015B; Blaschke et al., 2014, 2016; Mijał et al. 2018A, 2018B).

Currently, the concept of application the FGX type air-vibrating separators in the coking plants in front of the jiggers is being analysed in order to average the feed size grains that come from various exploitation fronts, with the amount of gangue that is changing in time (Blaschke and Baic, 2018B). This paper presents the results of economic analyses of the dry coal separation technology used in Poland for purification of the hard coal products in units that are independent enrichment plants.

2. Dry coal separation installations in Poland

Also thanks to the actions undertaken by the industrial and scientific consortium of Warkop Ltd. and the Institute of Mechanized Construction and Rock Mining, non-resident branch in Katowice, (possessing an experimental installation equipped with air-vibrating separator of FGX 1 type) (Baic et al., 2013, 2014C) the interest in dry coal separation technology, mainly among the hard coal distributors, in recent years has also increased in Poland. Currently two plants are operating in Poland for dry deshaling process of steam coal using air-vibrating separators of FGX type. The capacity of those installations is 30 Mg/h. Figure 1 presents the technological scheme of the installation with a capacity of 30 Mg/h which was started in August 2018. Figure 2 shows the pictures of the installation.

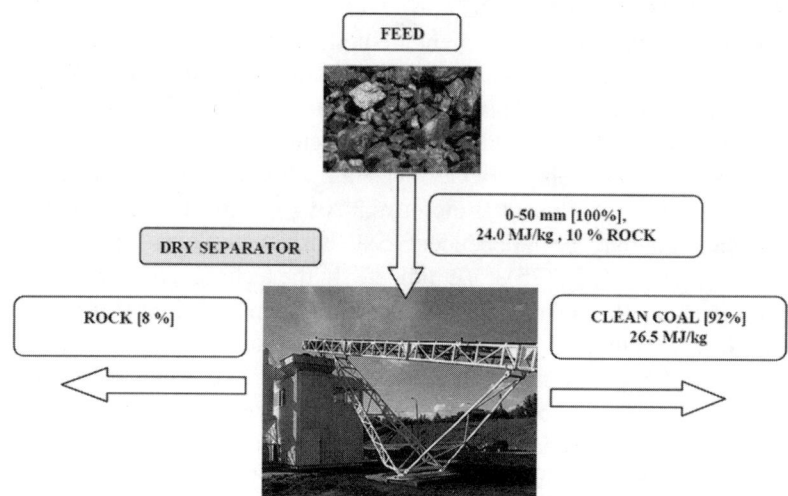

Figure 1: Technological system for coal enrichment with the use of air vibrating separator (Blaschke and Baic, 2018A)

Figure 2: Pictures of installations for dry separation (Blaschke and Baic, 2018A)

3. Economic efficiency of the dry separation process

Economic efficiency is an ambiguous concept that has been captured from various perspectives in both economic and management sciences. In economic sciences, economic efficiency is defined as the achievement of technological efficiency using the resources in the most effective way. The economic efficiency assessment of the investment is called all the assessments related to the comparison of the effects obtained during the lifetime of the completed investment with the expenditures necessary to achieve it. The basic elements of the economic efficiency assessment are: payback period (PP), return on interest (ROI), net present value (NPV) and internal rate of return (IRR) (Brigham and Daves 2016; Felis, 2016; Michalski, 2009).

3.1 World experience

Economic efficiency analysis of dry coal separation technology was a subject of many reports developed in USA and Republic of South Africa. The simplified results of the economic analysis obtained for a single dry separation installation operating in the USA are shown in Table 1.

Table 1: Approximate investment outlays and operating costs – USA (Honaker, 2007A)

Type of process	Investment expenditures ($/tone/h)	Operating costs ($/Mg)
Dry separation (FGX air-vibrating separator)	6200.00	0.50
Wet enrichment methods	13,000.00	1.95

The summary presented in Table 1 shows that investment outlays of dry coal separation technology account for 48% of investment outlays of wet enrichment methods (two times lower). Whereas operating costs of dry coal separation technology account for 25% of operating costs of wet enrichment methods (four times lower) (Honaker, 2007A). However, it should be emphasised here that the data presented in Table 1 reflect legal and environmental conditions as well as the location of coal mining plants and coal enrichment, quality parameters of coal, technological infrastructure as well as requirements set by the power sector in USA.

While the report developed in Republic of South Africa (de Korte, 2014) presents, among others, approximate investment outlays and operating costs of dry separation technology in comparison to wet enrichment methods using cyclones with heavy liquid – Table 2.

Table 2: Approximate investment expenditures and operating costs – Republic of South Africa (de Korte, 2014)

Type of process	Investment expenditures ($a)	Operating costs ($a/year)
Dry separation (FGX air-vibrating separator)	1,250,000.00	825,000.00
Wet enrichment methods (cyclones with heavy liquid)	5,000,000.00	2,480,500.00

a1$ = 14 Rand.

The summary in Table 2 shows that the investment outlays of dry separation technology account for 25% of investment outlays of wet enrichment methods (four times lower). While operating costs of dry separation technology account for 32% of operating costs of wet enrichment methods (three times lower). The economic efficiency measures – PP (payback period) and ROI (return on investment) as well as NPV (net present value) of the aforementioned technology were also compared – Tables 3 and 4.

Table 3: Approximate values (PP) and (ROI) (de Korte, 2014)

Type of process	Payback period (years)	Return on investment (%)
Dry separation (FGX air-vibrating separator)	0.72	39
Wet enrichment methods (cyclones with heavy liquid)	0.92	9

Table 4: Net present value (NPV) versus time (de Korte, 2014)

Years	Dry separation (FGX air-vibrating separator) ($a)	Wet enrichment methods (cyclones with heavy liquid) ($a)
1	563,377.00	−182,269.00
2	3,139,096.00	5,075,908.00
5	9,325,539.00	17,705,164.00
10	15,934,341.00	31,196,642.00
15	19,684,353.00	38,852,068.00
20	21,812,210.00	43,195,963.00

[a]1$ = 14 Rand.

The results presented in Table 3 show that dry separation systems using air-vibrating separators of the FGX type have a shorter payback period (PP) than those using for separation the cyclones with dense liquid. Moreover, they have a higher return on investment ratio (ROI).

Low investment outlays and operating costs of installations with FGX separators allows to obtain a positive NPV value already at the end of the 1st year of operation, whereas for installations using cyclones with dense liquid the NPV value at the end of the 1st year of operation is negative. After the second year of operation, the value of NPV index for installations using cyclones with dense liquid takes a higher value than NPV value for systems with FGX separators. As can be seen from the data presented in Table 4, the value of the NPV index also increases in the following years, whereas the increase of NPV for installations using cyclones with dense liquid is faster than increase of NPV for installations with FGX separators, due to the higher value of annual outlays.

The analysis shows that dry coal separation installations could be a good solution for short-term or low-capacity investments in South Africa. In case of long-term or high-capacity investments, installations based on wet separation technologies could be more economical, despite higher investment outlays and operating costs.

The author of the report (de Korte, 2014) emphasises that the selection of a given technology, apart from investment outlays and operating costs, is above all influenced by many other factors such as: availability of water for enrichment processes, location of enrichment plant and recipients of commercial products, characteristics of raw coal dispersibility, quality requirements of the clients and the availability of the necessary infrastructure.

3.2 Polish experience

In order to conduct a simplified analysis of economic efficiency for a dry separation installation equipped with an air-vibrating separator FGX3 operating in Poland, the following assumptions were made (Blaschke and Baic, 2018A):

- purchase price of installation for dry separation: air-vibrating separator FGX 3 – 381,579$, feed conveyor (25 m) – 36,842$, intake hopper – 7,105$, conveyor for concentrate collection – 11,842$, conveyor for rock collection – 11,842$, qualitative tests (for calibration) – 7.895$, installation and calibration – 34.211$; total price: 491,316$;
- service and maintenance price – 13.158 $/year;
- depreciation rate – 10% on an annual basis;
- operation time – 16h/day (2 production shifts, 3 shift for maintenance and repair), 22 days a month, 264 days a year;
- nominal capacity of the installation is 30 Mg/h – for the assessment the average capacity obtained under actual operation conditions at 25 Mg/h was assumed;
- amount of raw coal subjected to the separation process: 8,800 Mg/month, 105.600 Mg/year;
- energy consumption – 120 kWh, price: 1 kWh – 0.16$;

- number of employees operating the installation during one shift – 2;
- cost of man-hours of employees (gross) – 13.16$;
- purchase price of raw coal – 92.11 $/Mg, calorific value −24 MJ/kg; rock content −10%;
- price of car transport (25 Mg), installation location (one-way transport) – 0.05 $/km/Mg;
- average transport distance – 300 km (location in the centre of Poland);
- calorific value of the concentrate obtained after the separation process – 26.5 MJ/kg (2 MJ/kg increase);
- price for 1 MJ of coal with grain size 0–50 mm – 4.97$;
- efficiency of the separation process – 75%;
- amount of rock separated in the process – 8%, the rock is fully used for economic purposes (lack of incomes and costs).

The analysis does not take into account costs related to: preparation of the area for investment, obtaining the necessary decisions and permits, loading, transhipment and storage of coal, maintenance of the storage yard, administrative and lease costs of buildings and land. Table 5 presents the cost and income components of separation process per 1 Mg of raw coal.

Table 5: Cost and income components of dry separation process (Blaschke and Baic, 2018A)

No.	Components	($/Mg)
Costs		
1.	Depreciation	0.05
2.	Energy consumption	0.75
3.	Service and maintenance	0.12
4.	Labour	1.05
5.	Coal purchase including transport	106.32
Total		108.29
Incomes		
1.	Sales of coal after the dry separation process	113.94
Profit		5.65

Table 6 presents the results of the economic efficiency assessment.

Table 6: Results of the economic efficiency assessment

No.	Indicators	Unit	Value
1.	Cost of installation purchase	$	491,316.00
2.	Profit for 1 month	$	40,337.00
3.	Number of months needed to return the investment (PP – payback period)	–	12.2
4.	Period of capital return (in years)	–	1.02
5.	ROI – return on investment after 2 years	%	48.52
6.	ROE – return of equity after 2 years	%	197.04
7.	Net present value (NPV) at a discount rate of 12% per annum (1% on a monthly basis) and cash flows generated once a month		
7.1.	After 12 months	$	−54,754.10
7.2.	After 13 months	$	4,976.32
7.3.	After 24 months	$	335,027.54
8.	IRR – internal rate of return		
8.1.	After 12 months	%	−0.48
8.2.	After 13 months	%	1.09
8.3.	After 24 months	%	61.24

4. Conclusion

Simplified analysis of economic efficiency for a dry separation installation equipped with an air-vibrating separator FGX3 operating in Poland carried out with the adopted assumptions showed the possibility of achieving a profit of 5.65 $/Mg, which on a monthly basis allows to get income in the amount of approx. 40,000$. The number of months needed to return the investment (payback period) is 12.2 (the payback period of the capital invested expressed in years equals 1.02).

The identification of investment outlays and operating costs in a generic manner made it possible to calculate a return of investment index (ROI) and return on equity index (ROE) during the period of 2 years. They equalled respectively: ROI – 48.52% and ROE – 197.04%. The high positive values of ROI and ROE indexes indicate that the implementation of a dry separation installation is an investment of a high profitability.

Whereas the following indexes: net present value (NPV) and internal rate of return (IRR) indicate that with a discount rate of 12% on a yearly basis and cash flow equalled to the amount of the net operating profit index (EBIT), the planned investment after the period 13 months will be profitable.

To sum up, it should be stated that installations for dry separation of steam coal are characterized by high profitability and constitute an interesting business offer for hard coal distributors in Poland.

5. References

[1] Baic I. (2013) "Analiza parametrów chemicznych, fizycznych i energetycznych depozytów mułów węglowych zinwentaryzowanych na terenie woj. śląskiego", *Rocznik Ochrona Środowiska – Annual Set The Environment Protection*, Vol. 15, p. 1525–1548 [in Polish].

[2] Baic I., Blaschke W. (2013), "Analysis of the possibility of using air concentrating tables in order to obtain clean coal fuels and substitute natural aggregates", *Polityka Energetyczna – Energy Policy Journal*, Vol. 16, Issue 3, p. 247–260 [in Polish].

[3] Baic I., Blaschke W., Szafarczyk J. (2014A) "The first FGX unit in the European Union", *CPSI Journal a Magazine by the Coal Preparation Society of India*, No. 16, p. 5–12.

[4] Baic I., Blaschke W., Sobko W., Szafarczyk J. (2014B) "Badania nad wzbogacaniem węgli kamiennych na powietrznych stołach koncentracyjnych", *Wiadomości Górnicze – Mining News*, No. 7–8, p. 417–421 [in Polish].

[5] Baic I., Blaschke W., Szafarczyk J. (2014C) "Dry Coal Cleaning Technology", *Inżynieria Mineralna – Journal of the Polish Mineral Engineering Society*, No. 34, p. 257–62.

[6] Baic I., Blaschke W., Góralczyk S., Szafarczyk J., Buchalik G. (2015A) "Nowa ekologiczna metoda usuwania zanieczyszczeń skałą płonną z urobku węgla kamiennego". *Rocznik Ochrona Środowiska – Annual Set The Environment Protection*, Vol. 17, p. 1274–1285 [in Polish].

[7] Baic I., Blaschke W., Sobko W. (2015B) "Badania nad odkamienianiem energetycznego węgla kamiennego na powietrznych stołach koncentracyjnych". *Rocznik Ochrona Środowiska – Annual Set The Environment Protection*, Vol. 17, p. 958–972 [in Polish].

[8] Blaschke W., Baic I., Witkowska-Kita B. (2014) "Badania podatności węgli kamiennych na proces rozdziału metodą suchej separacji". *Polityka Energetyczna – Energy Policy Journal*, Vol. 17 Issue 4, p. 117–126 [in Polish].

[9] Blaschke W., Szafarczyk J., Baic I., Sobko W. (2016) "A Study of the Deshaling of Polish Hard Coal Using an FGX Unit Type of Air Concentrating Table". *Proceedings of the 18th International Coal Preparation Congress*. Saint Petersburg, Russia. Vol. 2. p. 1143–1148. Ed.Springer.

[10] Blaschke W., Baic I. (2018A) "Ocena efektywności odkamieniania próby węgla importowanego klasy ziarnowej 50–25 mm na urządzeniu typu FGX-3", Raport IMBiGS, Katowice [in Polish].

[11] Blaschke W., Baic I. (2018B) "Poprawa parametrów rozdziału węgla w osadzarkach poprzez wstępne uśrednianie nadawy metodą odkamieniania na sucho". *Zeszyty Naukowe Instytutu GSMiE PAN – The Bulletin of the Mineral and Energy Economy*

Research Institute of Polish Academy of Science, Cracow, No. 104, p. 163–172 [in Polish].

[12] Brigham E.F., Daves P.R. (2016) "Intermediate Financial Management", *Cengage Learning, Boston*, p. 469

[13] Felis P. (2016) "Finansowa ocena inwestycji rzeczowych", *Difin, Warszawa* [in Polish].

[14] Ghost T., Path D., Parekh B.K., Honaker R.Q. (2013) "Upgrading low Rank Coal Using a Dry Density – Based Separator Technology". *Proceedings of the 17th International Coal Preparation Congress*. Istanbul. p. 295–308.

[15] de KORTE GJ. (2014) Dry Processing of Coal – Status Update, *Report CSIR/Nre/ Mmr/Er/2014/0040/B*, Coaltech, South Africa.

[16] Honaker R.Q. (2007A) "Dry Coal Cleaning Technologies for India Coal", *Workshop on Coal Beneficiation and Utilization of Rejects: Initiatives, Policies and Best Practices Ranchi, India*.

[17] Honaker R.Q. (2007B) "Development of an Advanced Deshaling Technology to Improve the Energy Efficiency of Coal Handling, Processing, and Utilization Operations". *U.S. Department of Energy, Industrial Technologies Program, Mining of the Future, ID Number: DE-FC26-05NT42501*.

[18] Li Gongamin (2006) "Coal Compound Dry Cleaning Technique-study and Practice", *Proceedings of 15th International Coal Preparation Congress. Beijing*. Vol. II, p. 439–447.

[19] Michalski M.Ł. (2009) "Analiza metody oceny efektywności inwestycji rzeczowych", *Ekonomia Menadżerska*, p. 119–128 [in Polish].

[20] Mijał W., Blaschke W., Baic I. (2018A) "Dry coal beneficiation methods in Poland". *Proceedings of the 25th World Mining Congress. Astana, Kazakhstan*.

[21] Mijał W., Blaschke W., Baic I. (2018B) "Sucha metoda wzbogacania węgla w Polsce". *Przegląd Górniczy – Polish Mining Review*, Issue 11, p. 9–18 [in Polish].

8

Dry beneficiation of small South African coal using an air dense medium fluidization bed

ES Peters, M Le Roux, QP Campbell, N Hughes

School of Chemical and Minerals Engineering, North-West University, Potchefstroom, South Africa

Abstract: This study investigated the air dense medium fluidized bed (ADMFB) as a suitable dry beneficiation technique to upgrade small (+0.5 mm–13.2 mm) South African coal. The separation efficiency of the bed was assessed by determining the influence, if any, of particle size range, dense medium (magnetite) to coal feed ratios, and the addition of vibration to the bed. During batch operation, most of the high ash coal collected in the bottom layer of the bed, yielding coal with lower ash values towards the top layers. Product yields of 75% and 82% were found to contain ash of 13% and 22%, respectively. The addition of +300 µm magnetite (d50 = 490 µm) did not have a significant influence on the separation ability of +2.8 mm coal particles, nor did added vibration to the system (apart from lowering the minimum fluidization velocity). The feed ash values obtained for the −2.8 mm ranges were lower than for the +2.8 mm ranges, leading to the conclusion that most of the valuable macerals have been liberated and hence carrying less mineral matter. Nevertheless, the −2.8 mm beneficiated product still adhered to the local thermal coal market demands.

Keywords: Small coal, dry coal beneficiation, magnetite, air dense medium fluidization

1. Introduction

Despite the fact that several countries are progressively investigating and developing clean energy alternatives, these are unlikely to replace coal as an energy resource in the near future because of its abundance and affordability (WEC, 2018). South Africa's national energy resource base is currently dominated by coal, with coal-fired power stations meeting more than 75% of the country's electricity demands (DOE, 2017). The South African Chamber of Mines (2018) recently noted that roughly 30% of the coal produced domestically is exported leaving coal as the largest contributor to the country's GDP. This highlights the importance of coal export in terms of foreign revenues. South Africa has recently been labelled a water-stressed country with a recorded average annual rainfall of 492 mm, which is half of the global average (Rand Water, 2017), putting strain on washing coal, decreasing in quality. This calls for efficient beneficiation techniques that do not require copious amounts of water or high operating and plant costs (de Korte, 2017). As such, coal processing industries worldwide are investigating dry processing instead of conventional wet processing as a possible solution. FGX separators, X-ray sorters, air jigging, pneumatic oscillating tables and air dense medium fluidized beds (ADMFB) are some of the emerging dry beneficiation technologies that have been investigated in recent years and the performance was acceptable (Chen and Wei, 2003; Zhao et al., 2015). ADMFB holds the advantages of reducing the risks associated with water pollution and costly dewatering processes by eliminating the use of water (Chen and Yang, 2003). The principle of air dense medium fluidization is that coal particles are stratified according to differences in density, with less dense coal particles moving to the top, and more dense coal particles (gangue) sinking (He et al., 2016). For the past two decades, numerous small-scale successes for beneficiating coal, in a number of countries, using ADMFB's technology have been recorded. Table 1 shows that the separation efficiency (E_p) recorded for numerous countries. From the information in Table 1, it is apparent that the separation efficiency becomes worse as the particle size decreases (PSD), which indicates an increased degree in difficulty when beneficiating smaller coal sizes.

Table 1: Separation results of various coal size fractions

Country	PSD	E_p (g/cm³)	Medium	Reference
China	−50+6 mm	0.05–0.07	Magnetite	Zhao et al. (2011)
India	−25+13 mm	0.06	Magnetite	Mohanta et al. (2011)
India	−13+4.75 mm	0.19	Magnetite	Mohanta et al. (2011)
Canada	−1+0.42 mm	0.10	Magnetite	Choung et al. (2006)

Air dense medium fluidization has been implemented successfully in China (Mohanta et al., 2011); however, South Africa's coal seams differ from other coal-producing countries and are notorious for being difficult to beneficiate. Therefore, thorough evaluation of the effectiveness of this coal processing technology on South African coal seams is necessary (de Korte, 2017).

2. Experimental

2.1 Materials used

The coal used for this study was sourced from the Witbank area in South Africa. The feed to the ADMFB was prepared by drying, crushing, and screening the coal into various particle size fractions. Research revealed that adequate fluidization of coal particles is attained at surface moisture levels below 5%wt (Sahu et al., 2013; Terblanche, 2013). The coal (as received) was air-dried by laying it out on a flat surface, at ambient air conditions, for 24 h. Hereafter it was crushed and screened into the various particle size ranges including PSD-11 (+11.2–13.2 mm), PSD-10 (+9.5–11.2 mm), PSD-09 (+8.0–9.5 mm), PSD-08 (+6.7–8.0 mm), PSD-07 (+5.6–6.7 mm), PSD-06 (+4.75–5.6 mm), PSD-05 (+4.0–4.75 mm), PSD-04 (+2.8–4.0 mm), PSD-03 (+2.0–2.8 mm), PSD-02 (+1.0–2.0 mm), and PSD-01 (+0.5–1.0 mm). A representative sample of the material in each PSD was taken by the cone-and-quarter method. The proximate analysis and calorific values for each PSD fraction were determined and the results fell within the ranges tabulated below.

Table 2: Range of average initial proximate and calorific values for feed coal

Description	Range	Standards
Moisture content (%wt) air dried	0.6–3.4	ACT-TPM-010 based on SANS 5925: 2007
Volatiles (%wt) air dried	21.4–25.4	ACT-TPM-012 based on ISO 562: 2010
Ash yield (%wt) air dried	27.0–33.2	ACT-TPM-011 based on ISO 1171: 2010
Fixed carbon (%wt) by difference	41.3–49.5	By difference
Gross calorific value (MJ/kg)	21.7–25.5	ACT-TPM-014 based on ISO 1928: 2009

The magnetite dense medium used for this study was prepared by air-drying it to <5%wt and screening to +300 μm (d_{50} of 490 μm). The true density of the magnetite was 4.8 g/cm^3 and was determined by helium pycnometry (SANS 1014: 1985).

2.2. Equipment used

An air dense medium fluidized bed (ADMFB) unit, shown in Fig. 1, was used to conduct batchwise experimental runs. The unit consisted of three conjoined sections that are identified as: (1) an airflow system, (2) fluidized bed, and (3) dust control section. The airflow system consists of an air blower, airflow sensor as well as an air-distribution compartment. Air from the blower is distributed by a distribution device at the bottom of the bed, which consists of a packed bed of ceramic balls sandwiched between two distribution plates and a 100 μm wire mesh. Eight identical rectangular layers (0.3 m × 0.3 m × 0.05

m) were constructed from clear blue PVC and stacked on top of each other to form the fluidized bed section of 0.4 m high. Each layer could be detached and removed from the fluidized bed section, individually, to reveal a cut-point height of the material. The fluidized bed section was extended with a 0.3 m layer, also constructed of PVC, to allow enough space for the particles to fluidize. The dust control section of the ADMFB unit is composed of a tapered trapezoidal extension (20° hopper angle), covered by a 100 µm stainless steel wire mesh. This effectively limits ultra-fine particles from escaping the unit as the pressure drop is increased and the particles fall back into the bed. A 0.18 kW oscillating vibration motor was mounted onto the outside structure of the unit, which could induce vibration to the bed. The frequency and amplitude range provided by the vibration motor is 15 Hz and 0.7–1.0 mm, respectively.

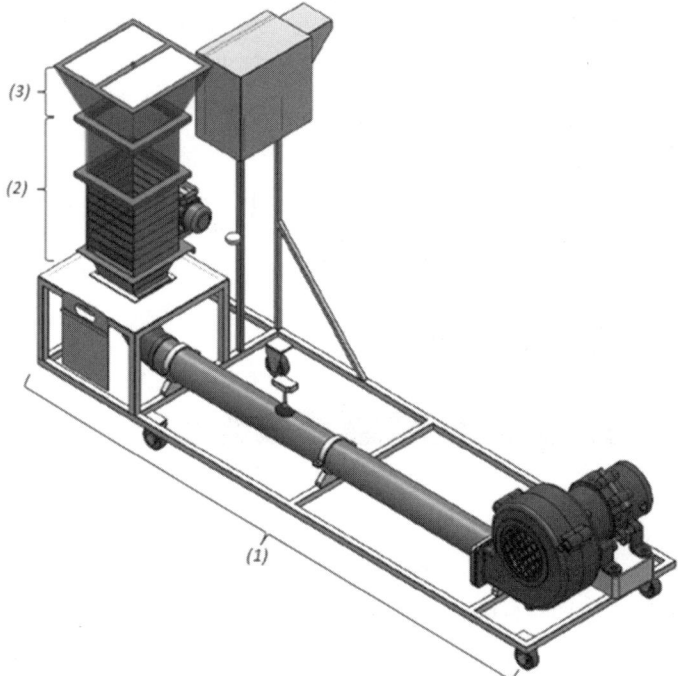

Figure 1: Three-dimensional illustration of the ADMFB with (1) airflow system, (2) fluidized bed system, and (3) dust control section

The fluidized bed section was loaded with a predetermined mass ratio of magnetite to coal. The load was fluidized at an operating velocity (U^*) of 1.05–1.30 times the minimum fluidization velocity (U_{mf}). After 10 min of fluidization, the ADMFB was switched off and the bed sections were disassembled from the top to the bottom. The material in each layer was retained in the individual frames by inserting a cutting plate (flat metal plate) between the layers and lifting the layer off the stack with the contents intact. The contents of each layer was coned and quartered until a representative sample for ash, calorific value and density analysis was obtained.

3. Results and discussion

To study the influence of PSD, dense medium to coal mass ratio (DM) as well as the effect of vibration (vibration activated, VA, or vibration deactivated, VD), the ash yield and calorific value (CV) for each layer along the height (bottom to top) of the bed was measured and recorded. PSD-06, without vibration (VD) and dense medium (DM 0:1) was selected for demonstrative purposes. The ash yield (%wt) of the four bed layers, along the bed height, together with the feed ash value is shown in the figure below.

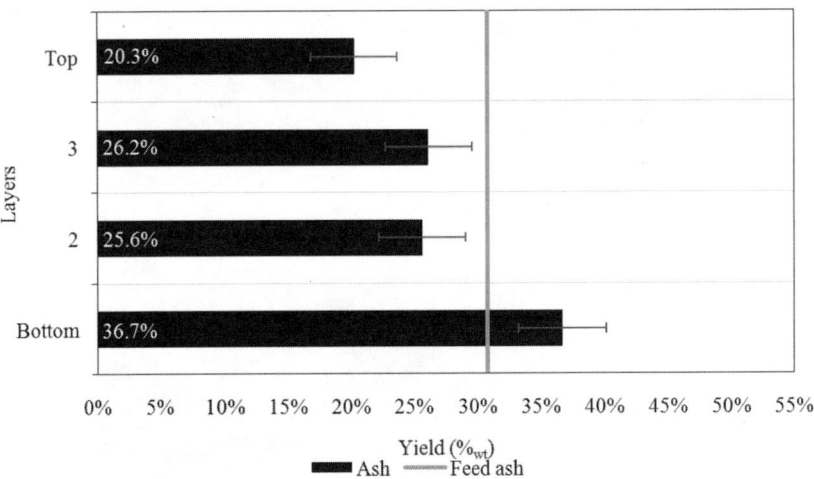

Figure 2: Ash values of control run according to bed layers (PSD-06; DM 0:1; VD)

The feed coal (the red line in Fig. 2) had an ash yield and CV value of 30.8%wt and 22.51 MJ/kg, respectively, producing a top layer, which measured a 20.3%wt ash value after fluidization and CV of 25.22 MJ/kg. The ash yield recorded in the bottom layer, after fluidization, increased by about 16.5%wt, which is indicative of de-stoning. The ash yields in the second and third layer of the bed (middlings portion) differs marginally, indicating that individual near-dense coal particles are constantly moving between the layers during operation and only settle once the airflow is switched off.

3.1. Effect of particle size range

The effect of particle size range on the minimum (U_{mf}) and operating (U) fluidization velocities of the bed is briefly discussed in this section. Several PSD's (DM 0:1), without vibration, along with the respective bed masses, U_{mf} and U values are summarized in Table 3.

Table 3: Minimum (U_{mf}) and operating (U) fluidizing conditions for various PSD's

PSD (mm)	Bed mass (kg)	U_{mf} (m/s)	U (m/s)
PSD-11 (+11.2–13.2)	15.06	15.1	16.1
PSD-09 (+8.0–9.5)	15.35	14.7	14.8
PSD-07 (+5.6–6.7)	15.20	9.2	10.4
PSD-05 (+4.0–4.75)	11.78	7.1	7.9
PSD-03 (+2.0–2.8)	12.67	5.7	6.3
PSD-01 (+0.5–1.0)	12.43	0.9	1.0

It is apparent from the values in Table 3 that an increase in U_{mf} is prompted by an increase in PSD. The reason for this is that a greater upward force is required due to the increased surface area of the larger particles, and in order to ensure the particles are fully suspended; a larger minimum fluidization velocity is therefore required. For pure coal beds, an operating velocity in close range with U_{mf} delivered a stable suspended bed with minimal turbulence and back mixing. The results obtained for PSD-11, PSD-09, and PSD-07 feed to the bed, operating it without dense medium (DM 0:1) or addition of vibration (VD), are summarized in Table 4.

Table 4: Ash, CV, and density value comparison of PSD-11, PSD-09, and PSD-07

PSD (mm)	Layer	Ash (%wt)	CV (MJ/kg)	Density (g/cm3)
PSD-11 (+11.2–13.2)	Feed	28.3	21.8	1.61
	Top	21.6	25.0	1.56
	Bottom	36.2	19.3	1.72
PSD-09 (+8.0–9.5)	Feed	30.6	21.3	1.66
	Top	24.0	24.0	1.54
	Bottom	41.6	17.1	1.84
PSD-07 (+5.6–6.7)	Feed	33.4	20.4	1.73
	Top	23.1	24.3	1.59
	Bottom	43.2	16.5	1.89

Comparable results were obtained for the bottom and top layers for PSD-11, PSD-09, and PSD-07 in Table 4. The degree of separation for PSD-11 to PSD-07 from top to bottom remained similar, however it seems the bottom layers were somewhat dependent on the feed coal characteristics. The ash values in the bottom layer increases in accordance to the feed ash values, which may be indicative of an increased fraction of dense particles present in the feed relocating to the bottom layer.

The ash yield obtained for the cumulative ash percentage in the product is represented in the form of performance curves like the one shown in Fig. 3. The results of PSD-11, PSD-04, and PSD-01 are collectively compared to the float-sink test results (solid line in Fig. 3) of PSD-11. The float-sink results for PSD-11 was selected for comparisons, since the float-sink curves of the remaining particle size ranges generated similar results (standard deviation of 0.14). Moreover, the yield versus the cumulative ash percentage is shown by this curve and it sets the basic standard for the best performance obtainable for the given coal samples.

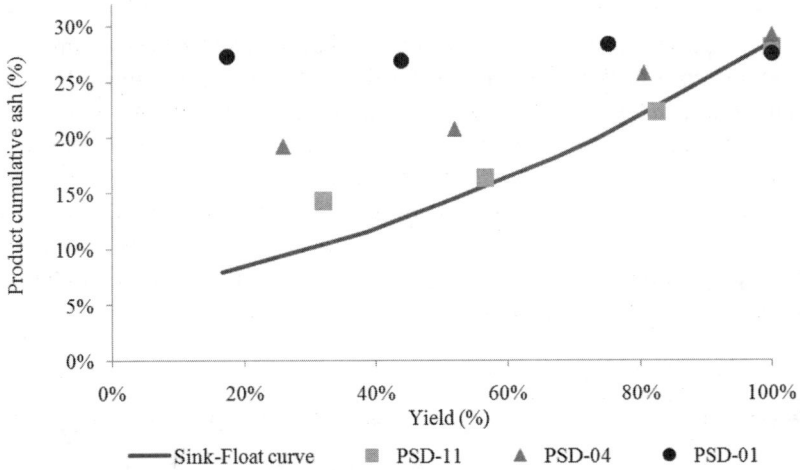

Figure 3: Comparison of bed performance curves for various PSD's (DM 0:1; VD)

The curves in Fig. 3 highlight the ease of separation for the larger particle size ranges as compared to the smaller ones. Both the PSD-11 and PSD-04 produced acceptable performance to the float-sink curve at high yields. However, a limit in ash close to 15% and 20% was reached for PSD-11 and PSD-04, at coal yields of 30% and 25%, respectively. In contrast, at the same yield, the ash value of PSD-01 was found to be close to that of the feed coal, showing little to no beneficiation in the bed for these sized particles. This indicates that the stratification effect of the pseudo fluid bed is stronger for the larger particle sizes. Yang et al. (2015) found that a decrease in PSD yielded substandard results, which is confirmed by the performance of PSD-01 observed in Fig. 3.

3.2. Effect of dense medium

The effect of dense medium on the separation performance of the ADMFB was investigated by varying the coal to dense medium ratio in the feed and comparing the results in the top and bottom layers of the bed after fluidization. For the experimental runs employing dense medium, the ratios were determined by weighing and mixing the dense medium and coal masses before placing it in the bed. About 7.5 kg magnetite and 7.5 kg coal was mixed for DM 1:1 experiments while 10 kg magnetite and 5 kg coal was mixed for DM 2:1 experiments. In order to ensure diverse, uniform distribution of coal and magnetite particles across the bed, it is essential to mix the magnetite and coal prior to loading. Table 5 compares the ash yields of the top and the bottom layers of the experimental runs for PSD-11 with DM 0:1, 1:1, and 2:1.

Table 5: Comparison of dense medium ratio of PSD-11 coal (VD)

Ash yield (%wt)			
Layer	DM 0:1	DM 1:1	DM 2:1
Feed	28.3	22.9	23.6
Top	21.2	20.3	21.2
Bottom	53.8	33.7	34.8

Minute differences are visible in the ash values of the top layers of the different beds, shown in Table 5, which indicates that the results obtained are close to best performance for each of the three beds. On the other hand, a difference in the ash values of the bottom layers of the dense medium beds in comparison to the pure coal bed was noticed. Both the dense medium beds behaved similarly to yield ash values in the proximity of 34%, while the ash values of the pure coal bed was 20% points higher, close to 54%. A larger feed ash for the pure coal bed did contribute to this discrepancy, however it was evident that better destoning occurred within the pure coal bed. This was observed for all the dense medium beds at each PSD interval, with and without the addition of vibration to the bed.

In addition, it was also established from this study that the minimum fluidization velocity (U_{mf}) of a bed is influenced by the dense medium to coal feed ratios. In principle, a fluidized bed is fed with dense medium to create a fluid-like suspension of magnetite particles in which coal particles sink or float according to density. An experimental run using a pure magnetite bed (DM 1:0) of 15 kg was tested and the minimum fluidization velocity (U_{mf}) and bed density thereof was found to be 8.0 m/s and 1.31 g/m³, respectively. This was done to compare the minimum fluidization of the DM1:1 and DM 2:1 beds to the pure magnetite (DM 1:0) bed. The minimum fluidization velocity (U_{mf}) values of PSD-11, PSD-07, and PSD-03 coal (total material mass per experimental run was recorded as ±15 kg) as well as the operating velocities (U), are shown in Table 6, both without the addition of vibration.

Table 6: U_{mf} (m/s) and U (m/s) of PSD-11 and PSD-07 coal

DM ratio	Bed height (cm)	U_{mf} (m/s)	U (m/s)
PSD-11			
0:1	20	15.1	16.1
1:1	15	8.6	9.8
2:1	15	9.2	10.5
PSD-07			
0:1	20	9.2	10.4
1:1	15	7.2	8.3
2:1	15	5.2	6.9
PSD-03			
0:1	20	5.7	6.3
1:1	15	4.2	5.0
2:1	15	2.6	2.9

When dense medium is introduced to the system, the bed height as well as the minimum fluidization velocity is effectively reduced for all the PSD's. Once dense medium is added to the bed, a very fine particle with high density is added to the system. These high-density, fine particles collect in the pores between the large particles (blocking them), essentially exposing a solid surface to the air and thus is lifted with more ease from its resting position. It is apparent that the minimum fluidization velocity (U_{mf}) of the PSD's in Table 6 falls within a narrow range from the 8.0 m/s of the magnetite bed, with the exception of PSD-11 (DM 0:1) and PSD-03 (DM 1:1), (DM 2:1). PSD-03 (2:1) has a minimum fluidization velocity (U_{mf}) of 2.6 m/s, which is considerably low compared to the minimum fluidization velocity of the pure magnetite bed (DM 1:0). It was found that when dense medium is added to the

−2.8 mm coal particle size ranges, the pseudo fluid effect of the bed dwindles. This is because the coal and magnetite particles form a tightly packed bed in the ADMFB, making it difficult for the air to penetrate the particles, causing particle segregation and bed stratification to suffer.

3.3. Effect of vibration

Vibration as a parameter was investigated by adding a vibration motor to the bed that was able to induce a horizontal vibration motion. The addition of vibration to the bed lowered the minimum fluidization velocity (U_{mf}) while improving fluidization stability due to sufficient particle movement and bubble formation (He et al., 2015). The fluidization velocities of PSD-11, PSD-09, and PSD-07 in Table 7 show the influence of adding vibration to the system.

Table7: Comparison of minimum and operating fluidizing conditions for varying vibration (DM 0:1)

Vibration	U_{mf} (m/s)	U^*	U (m/s)
PSD-11			
No	15.1	1.07	16.1
Yes	14.7	1.04	15.3
PSD-09			
No	14.7	1.01	14.8
Yes	11.1	1.15	12.8
PSD-07			
No	9.2	1.13	10.4
Yes	8.0	1.15	9.2

A decrease in minimum fluidization velocity (U_{mf}) is evident in Table 7 for PSD-11, PSD-09, and PSD-07, when introducing vibration to the system. The same pattern was observed for all the other particle sizes tested during this study.

4. Conclusion

Upgrading of the coal quality was achieved to varying degrees according to the set variables. During batch operation, most of the high ash coal collected in the bottom part of the bed, yielding sequentially lower ash yield values towards the top layer where ash ranging between 13.0% and 22.4% were found for corresponding yields of 82.2% and 74.8%. Negligible influences on the separation efficiency for all the +2.8 mm fractions were observed when applying vibration to the bed or feeding dense medium, +300 μm magnetite (d_{50} = 490 μm) with the coal to the batch setup. The addition of the vibration to the bed did however lower the minimum fluidization velocity and aided in particle striation and bed stability. Very little upgrading was observed for all particle size fraction less than 2.8 mm. In general the feed ash to the bed for these fractions were lower than for the +2.8 mm, leading to the conclusion that most of the valuable macerals have been liberated at these sizes and hence carrying less mineral matter. In addition, the separation of coal and magnetite at these size ranges were extremely difficult due to adherence of magnetite to the additional exposed coal surfaces. But even with these less than desirable results, it was still possible to produce a product that adhered to the local thermal coal market demands. It can be concluded from this project that operating the ADMFB without the addition of magnetite or vibration will yield a coal product that comply to thermal standards, and in some cases to metallurgical standards. This operation philosophy aids in downstream processing by eliminating magnetic separation and recycling of dense media.

5. Acknowledgement

- SACPS
- DST – SARChI

The authors wish to acknowledge the financial and technical contribution made by The Coaltech Research Association NPC Ltd., and the Centre of Excellence in Carbon Based Fuels, North-West University, South Africa.

This work is based on the research supported in part by the National Research Foundation of South Africa (Coal Research Chair Grant Numbers: 86880; and UID85643, UID85632). Any opinions, findings and conclusions or recommendations expressed in any publication generated by the NRF supported research is that of the author(s) alone, and that the NRF accepts no liability whatsoever in this regard.

6. References

[1] Chen, Q. & Wei, L. 2003. Coal dry beneficiation technology in China: The state-of-the art. *China Particuology*, 1(2):52–56.

[2] Chen, Q. & Yang, Y. 2003. Development of dry beneficiation of coal in China. *Coal Preparation*, 23:3–12.

[3] Choung, J., Mak, C. & Xu, Z. 2006. Fine coal beneficiation using an air dense medium fluidized bed. *Coal Preparation*, 26(1):1–15.

[4] Department of energy (DOE), Republic of South Africa. 2017. Coal resources: Overview. http://www.energy.gov.za/files/coal_frame.html. Date of access: 16 June 2017.

[5] de Korte, G.J. 2017. New dry beneficiation tech has multiple benefits. Mining Weekly. http://www.miningweekly.com/article/new-dry-beneficiation-tech-has-multiple-benefits. Date of access: 17 March 2017.

[6] He, J., Zhao, Y., Zhao, J., Luo, Z., Duan, C. & He, Y. 2015. Separation performance of fine low-rank coal by vibrated gas–solid fluidized bed for dry coal beneficiation. *Particuology*, 23: 100–108.

[7] He, J., Tan, M., Zhao, Y., Duan, C., He, Y. & Luo, Z. 2016. Fluidization characteristics and density-based separation of dense-medium gas–solid fluidized bed: An experimental and simulation study. *Journal of Taiwan Institute of Chemical Engineers*, 61(2016):223–233.

[8] Mohanta, S., Chakraborty, S. & Meikap, B.C. 2011. Influence of coal feed size on the performance of air dense medium fluidized bed separator used for coal beneficiation. *Industrial and Engineering Chemistry Research*, (50):10865–10871.

[9] Rand water. 2017. Water Wise. http://www.waterwise.co.za/site/water/environment/situation.html. Date of access: 11 May 2017.

[10] Sahu, A.K., Tripathy, A. & Parida, A. 2013. Study on particle dynamics in different cross sectional shapes of air dense medium fluidized bed separator. *Fuel*, 111:472–477.

[11] South African Chamber of Mines. 2018. Coal Strategy 2018: National Coal Strategy for South Africa. www.chamberofmines.org.za. Date of access: 10 January 2019.

[12] Terblanche, A., 2013. Dry beneficiation of fine coal using a fluidized dense medium bed, Potchefstroom: North West University, School of Chemical and Minerals Engineering.

[13] WEC (World Energy Council). 2018. World Energy: Issues Monitor. www.worldenergy.org. Date of access: 12 January 2019.

[14] Zhao, Y., Li, G., Luo, Z., Liang, C., Tang, L., Chen, Z. & Xing, H. 2011. Modularized dry coal beneficiation technique based on gas–solid fluidized bed. *National Natural Science Foundation of China*, 18:374–380.

[15] Zhao, Y.M., Zhang, B., Luo, Z.F., He, J.F., Dong, L., Peng, L. & Cai, L. 2015. Effect of lump coal shape on separation efficiency of gas–solid fluidized bed for dry coal beneficiation. *Procedia Engineering*, 102:1123–1132.

9

Advantages of utilising dry jigging and X-ray sorting technology for coal beneficiation in India

E Orena, G Van Wykb, K S Ashvanic

Allmineral Aufbereitungstechnik GmbH & Co, Düsseldorf, Germany
STEINERT Australia Pty. Ltd, Bayswater, Australia
Allmineral Asia Pvt. Ltd, Kolkata, India

Abstract: Traditional practice in coal beneficiation includes wet processes such as heavy media cyclones wet jigging and spirals. The most obvious drawback for these processes is the need for water, dewatering of the fines produced and slurry handling. With no requirement for water and magnetite, dry jigging of coal has emerged as an effective means of coal beneficiation throughout the world. Dry processing eliminates the clean coal moisture penalty. It is relatively easy to obtain the permits required for a dry wash plant as compared to a wet coal wash plant. The coal industry in countries such as USA, India, Columbia, Spain, Turkey, South Africa and Ukraine has responded positively to allmineral's allair® dry coal solution. More than 70 units have been installed worldwide for beneficiation of all type of coals including coking, non-coking, anthracite, lignite, etc., in different feed sizes ranging from 0 mm to 50 mm. India's electricity sector is dominated by fossil fuels, and in particular coal, which in 2017–18 produced about three fourths of all electricity. Dry beneficiation has the potential to optimise coal processing and enhance profitability through a reduction in capital and operational costs which is the need of hour for countries like India.

Apart from dry de-shaling technologies such as air jigs, a relatively new dry coal beneficiation process is the dual energy X-ray transmission sorting technology. STEINERT has developed a XSS®T X-ray transmission sorting system with working widths to 2 m. At size ranges from 50 mm to 150 mm the sorter can achieve feed rates up to 200 t/h. This technology measures the rate of X-ray absorption and thereby determine the average atomic density of each particle. The particle-by-particle sorting decisions employed by its processor are based on high-resolution X-ray images, similar to those used in airport luggage scanners. One of the main benefits of this process is that, unlike optical sorters, it is not sensitive to surface conditions (e.g. dust). Therefore, it can be used in completely dry environments without the added operating costs of wet processing. X-ray sorting technology is well proven for particle sizes above 50 mm. Both X-ray sorting and allmineral's allair® can improve the quality of coal by reducing ash and sulphur content.

In this paper we describe a unique way of dry coal beneficiation combining air jigs for 0–50 mm sizes and X-ray sorting for above 50 mm. Also we shall focus on the benefits of dry coal beneficiation and its modular operation.

Keywords: Dry jigging, Modular, Gravity separation, Sorting, Deshaling, Mechanisation

1. Introduction

Popular methods of coal beneficiation are heavy media cyclones, wet jigging and spirals, etc. Generally the efficiency of the wet processes is better than conventional dry processes, but limitations and disadvantage are water usage, fines dewatering, slurry storage and ultra-slimes disposal. This is the reason why continuous research is on-going in the field of dry beneficiation. Dry jigging and X-ray sensor sorting of coal has emerged as an effective means of coal beneficiation and is widely being accepted throughout the world. Pneumatic separation was introduced to the U.S. coal industry prior to 1924 in the form of the air table, a device with many similarities to a Deister table (Mitchell, 1942). Commercial use of the air jig commenced in the early 1930s in the U.S., and became the dominant technology for dry cleaning coal. The original design of the air jig changed only modestly over the decades, and the last commercial plants in the USA were decommissioned in the early 1990s. In the late 1990s, a new air jig design was introduced to the coal industry by allmineral. The new air jig employed a different air fluidisation system, a different air distribution system, and an instrumentation and control system for controlling the separating gravity. Since 2001, new air jig plants have been installed in the USA, Colombia, Spain, Ukraine and India for both thermal coal and coking coal. Dry processing eliminates the clean coal moisture penalty.

2. Why dry beneficiation?

Heavy media may be viewed as the right solution for cleaning high ash coal with typically high NGM (near gravity material). However, especially for power utilities with a 35% ash requirement, the required separating gravities will usually be in the range of 1.9–2.0 t/m³. Certain heavy media installations recently constructed in India are experiencing problems maintaining such a high gravity separation. The lack of availability of high quality magnetite further adds up to the problem for heavy media operations. Therefore, coal preparation with jigging technology, both wet and dry, is a superior choice for Indian coals. The separating gravity achieved by jigging processes is only limited by the relative density of the feed material. The fundamentals of jigging technology are stratification and separation. In comparing beneficiation of coal through jigging involving wet or dry techniques, dry jigging seems to be more promising due to several advantages as mentioned below:

(1) A dry product, resulting in a higher calorific value per tonne.
(2) The water source problem can be acute in arid and semi-arid areas. Wet processing of coal can require large quantities of water. Water is consumed as product moisture, tailings disposal and evaporation.

(3) The fines and rejects generated in dry processing are ideal fuel for fluidised bed combustors.

(4) It is relatively easier to get environmental clearance for dry coal wash plants as compared to wet wash plants, resulting in a shorter lead time for installation of the plant.

(5) The dry process for de-shaling in pit reduces transportation cost to carry ROM (run-of-mine) to wash.

(6) Less skilled manpower is required to operate the wash plant thus reducing the cost of manpower.

(7) Modular and mobile plant can be installed and shifted mine to mine. Capacity can be increased and decreased in modules.

3. Dry beneficiation technology

The allmineral allair® utilises the basic principles of jigging; it stratifies the feed material by specific gravity and subsequently measures and discharges the high density (and high ash) strata. To stratify the material, the allair® uses pulsating and constant air flow through a perforated jig bed. Vibrating mechanisms assist the transport of material across the bed. The feed star gate provides an even feed distribution over the width of the bed, and the discharge star gate provides an even removal of the heavy particles (rock) from the jig, thereby maintaining a residual layer of refuse below the light particles (coal). Continuous sealing at both ends for allair® decreases the consumption of air.

To measure the coal–rock interface, a nuclear density measuring device is located towards the discharge end of the jigging bed. This device automatically controls the discharge mechanism (the star gate) of the jig. If more dilution enters the jig, the star gate will speed up and discharge more rock. If more coal enters the jig, the star gate will slow down thereby discharging less rock and minimising the loss of clean coal in the refuse. To complete the allair® jig process, dust particles are removed via a baghouse filter (Fig. 1).

The allair® is designed to handle material up to a maximum particle size of 50 mm. The maximum feed capacity of the jig depends on the particle size distribution of the feed material. A throughput for coarse material (50 mm top size) of 40 t/h per meter of width, and 30 t/h per metre of width for fine feeds (6 mm top size) will ensure optimum performance.

After the great success of 50 t/h dry allair® and keeping in mind the growing demand of washed coal in the industry, allmineral has come up with a 100 t/h model of allair. It can operate at double capacity and an improved efficiency as compared to the older model with same operating parameters and feed sizes.

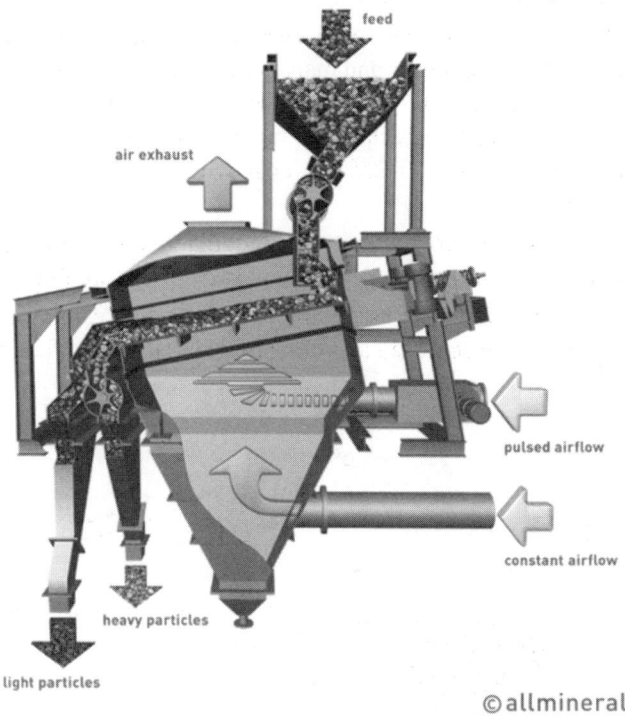

Figure 1: Schematic of allair® dry jig

4. Mobile or semi mobile modular plant for mine sites

Mobile or semi mobile allair® plants are pre-assembled and can be transported to the mine site for immediate operations (Fig. 2). The advantages could be briefly described:

- No need for erecting and installation set up with associated risks and costs.
- The plant can be installed at the mine itself, eliminating transport cost of material to the wash plant and providing backfill options as well.
- Capacity can be increased or decreased in modules.
- Shorter duration of project. Normally in any project, civil work is done first before any fabrication erection and installation at site. In our modular design we can shorten that period because civil work

can go parallel while fabrication and assembly takes place at the workshop. Once the equipment is ready along with its structure, it can directly be fitted at site.

- Reduced capital costs. A shorter timeline for project execution will utilise less manpower and management, and with effective, organised and efficient use of resources at one place, cost of the overall project can be reduced appreciably in modular and skid-mounted plants. We can save around 20–25% in cost with modular and skid-mounted designs.

- Lower safety risks. Since the modular plant is prefabricated and assembled, the onsite construction safety risks are reduced.

- High quality. Onsite assembly and fabrication cannot match quality which can be achieved at workshop using high end facilities.

Figure 2: Allair® mobile plant

5. Continuous improvement of allair®

allmineral strongly believes the key to success is continuous research and development work to benefit the mineral industry. A team of engineers work on the development and troubleshooting of allmineral equipment. The parameters measured from different operations all over the world are applied to computerised simulations for improving the design. Figure 3 below shows the expected air speed distribution of an allair® before approval for construction. The simulation considers possible changes during the natural life of mine.

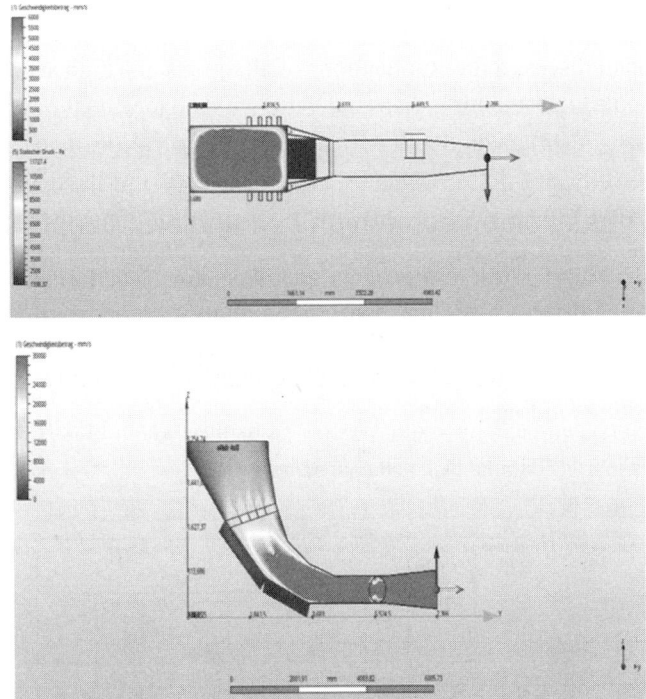

Figure 3: Allair® air speed distribution graphs

An industrial size of allair® can be tested on the user's site for evaluating real process parameters and results. A photo of the unit can be seen below (Fig. 4).

Figure 4: Allair® dry jig

6. Results and discussion

Here we shall discuss a few applications of allair on some high ash Australian coal and some high sulphur Indian coal. Also we shall look on the performance of X-ray sensor sorters on South African Coal – as per deposit near Witbank.

6.1 Ash reduction on some high ash and NGM coal

Here we shall discuss test results of a high ash coal received from some Indian client. Jigging of coal was done at size 5–20 mm. At the same time sink float analysis of coal was also done to compute the efficiency of separation. Table 1 shows the sink float analysis of coal. Characteristic washability curves were drawn as shown in Fig. 5.

Table 1: Washability data for high ash coal sample

[1]		[2]	[3]	[4]	[5]	[6]	[7]	[8]	[9]	[10]	[11]	[12]
Relative density fractions Mass		Ash (air dry basis)	Proportion of ash	Relative density (ρ) [2]×[3]	Size fraction basis						Percentage mass of NGM (ρ±0.1) %(m/m)	
					Cumulative floats			Cumulative sinks				
					Mass Σ[2]	Proportion of ash Σ[4]	Ash [7]/ [6]	Mass Σ[2]	Proportion of ash Σ[4]	Ash [10]/[9]		
Sinks	Floats	%	%(m/m)			%		%(m/m)	%			
									100	5407.49	54.07	
	1.4	0.60	20.52	12.39	1.4	0.60	12.39	20.52	99.40	5395.11	54.28	
1.4	1.5	8.59	23.37	200.69	1.5	9.19	213.07	23.18	90.81	5194.42	57.20	24.66
1.5	1.6	16.07	29.31	471.12	1.5	25.26	684.19	27.08	74.74	4723.30	63.20	21.71
1.6	1.7	5.63	39.29	221.33	1.6	30.90	905.52	29.31	69.10	4501.97	65.15	15.12
1.7	1.8	9.49	44.3	420.30	1.7	40.39	1325.82	32.83	59.61	4081.67	68.47	16.12
1.8	1.9	6.63	53.98	357.81	1.8	47.01	1683.63	39.96	52.99	3723.87	70.28	15.84
1.9	2	9.21	61.11	562.96	1.9	56.23	2246.59	54.07	43.77	3160.91	72.21	
2	–	43.77	72.21	3160.91	2	100.00	5407.49					
Total		100.00			2.2							

Note: NGM (near gravity material) = percentage mass for ρ + 0.1 in [6] minus percentage mass for (ρ – 0.1) in [6].

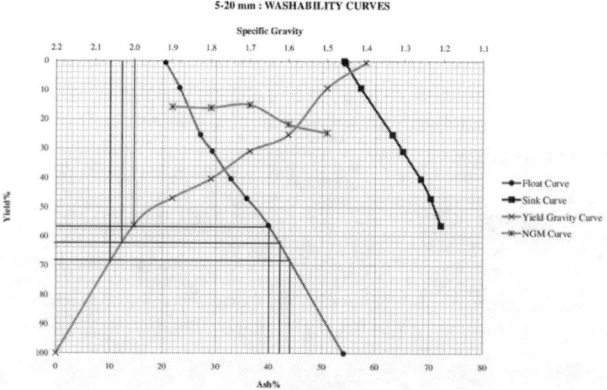

Figure 5: Washability curves for sink float analysis of coal

Jig yields were drawn with respect to ash% and FC (fixed carbon) % to determine yields at different target ash% and different target FC%. Figure 6 shows the jig performance curve.

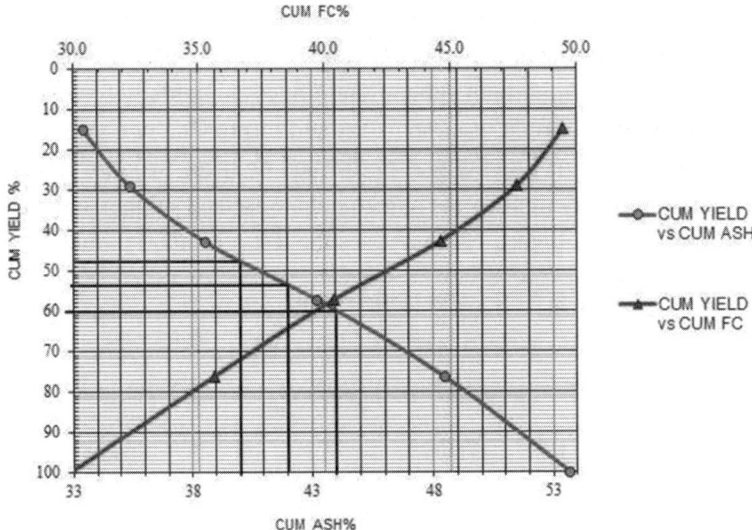

Figure 6: Jig performance curve

The predicted yields were calculated from both washability curve and jig performance curve at same target ash percentage to calculate organic efficiency of separation. Table 2 shows the result at various target ash percentages.

Table 2: Jig performance and comparison with sink float results at various target ash%

Target ash %	Jig yield %	Ideal yield from sink-float data	Organic efficiency	Specific gravity of separation	Reject ash %	Ash reduction
44	60	68	88.24	2.06	68.18	−9.67
42	54	62	87.10	2.03	67.37	−11.67
40	48	56	85.71	2.00	66.29	−13.67

It is noticeable here that the NGM was high (15–20%) at selected specific gravity of separation but due to proper liberation at the selected size fraction the efficiency of separation was high.

6.2 High ash and high sulphur Indian coal

Test work was done on a high ash, high sulphur coal received from north eastern region of India. The objective of the test work was to beneficiate coal through the cheapest possible route and reduces its ash and sulphur content.

The as-received sample was 0–100 mm material. It was decided to beneficiate the coal in two sizes 0–6 mm and 6–30 mm. The sample was crushed in such a way so that there is minimum generation of fines and initially screened at 30 mm. The 0–30 mm sample was then screened to produce two size fractions 0–6 mm and 6–30 mm. Jigging was performed on these two size fraction in allmineral's allair® dry jigging machine to reduce ash and sulphur and increase fixed carbon content. Figure 7 summarises the test procedure (Table 3).

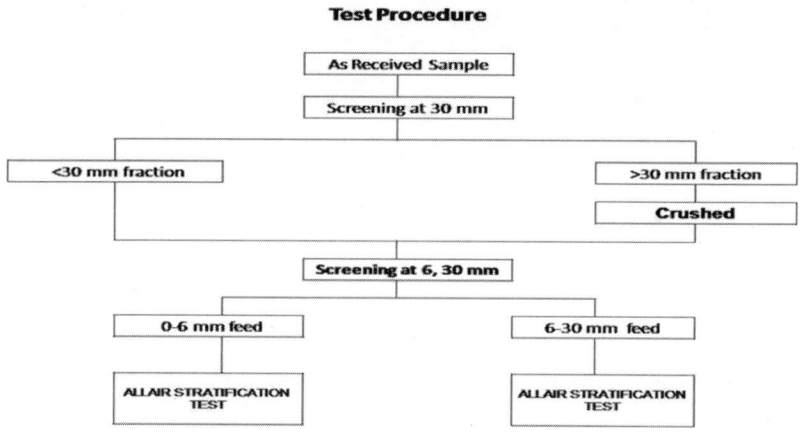

Figure 7: Allair® dry jig test procedure

Table 3: Size analysis of feed

Analysis of feed						
Size	**Wt %**	**Ash %**	**Moisture %**	**VM %**	**FC %**	**S %**
0–6 mm	22.15	32.49	5.48	28.7	33.33	3.71
6–30 mm	77.85	41.25	4.85	23.25	30.66	4.19
Feed	100	39.31	4.99	24.46	31.25	4.08

For the 6–30 mm size fraction the jig yield was plotted against target ash% and FC%, the yield curves are shown in Fig. 8. Curves were also plotted showing yield and corresponding sulphur in clean coal (Fig. 9).

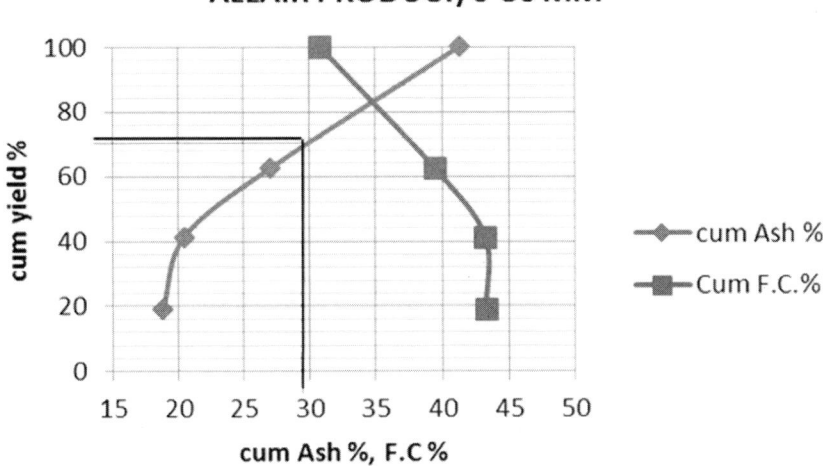

Figure 8: 6–30 mm yield versus ash% and FC%

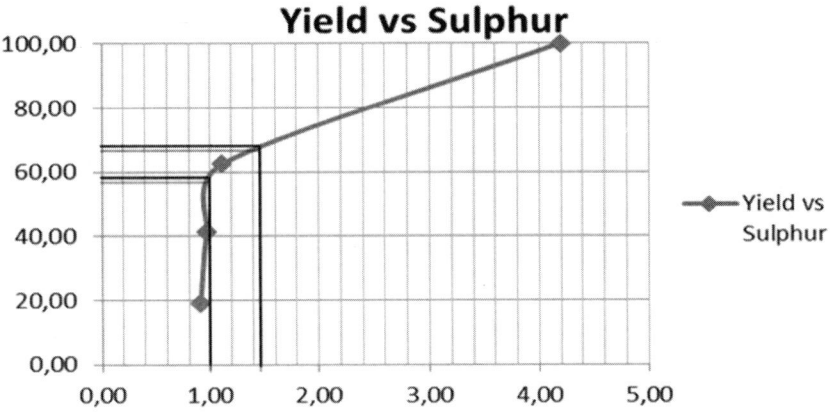

Figure 9: 6–30 mm yield versus sulphur%

For the 0–6 mm size fraction the same yield curves have been plotted – see Figs. 10 and 11 below.

Both fractions 0–6 mm and 6–30 mm, give promising results in dry stratification with the allair® jig.

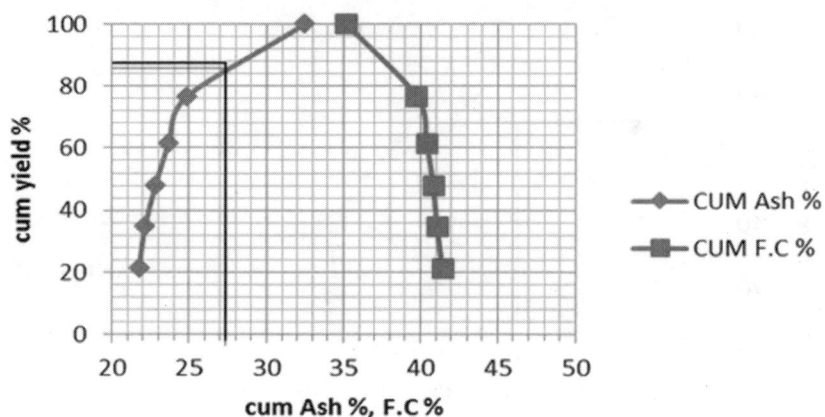

Figure 10: 0–6 mm yield versus ash% and FC%

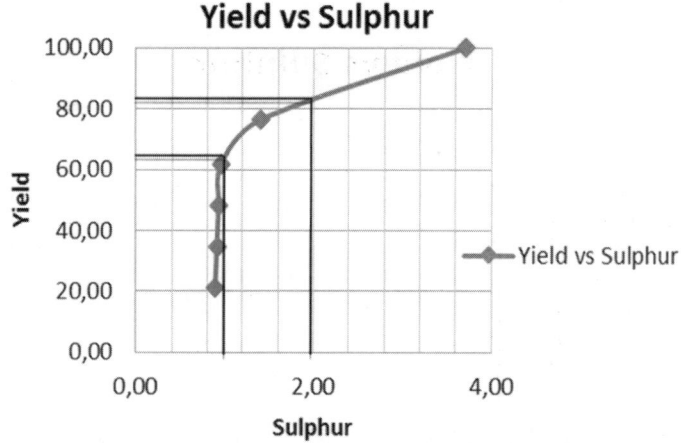

Figure 11: 0–6 mm yield versus sulphur%

For the 0–6 mm size fraction the predicted yield is 84% corresponding to a target ash of 27%. Predicted sulphur content in the product is 1.2% in this case. For the 6–30 mm size the predicted yield is 68% corresponding to a target ash of 27%. Sulphur content in product will be 2.0% in this case. All together (combined product 0–30 mm) we can achieve a predicted yield of 71.54% at 27% target ash with 1.79% sulphur. The final reject (28.46% of feed) contains 70.26% ash and 9.85% sulphur.

All findings of the test work are summarised in a flow diagram – see Fig. 12.

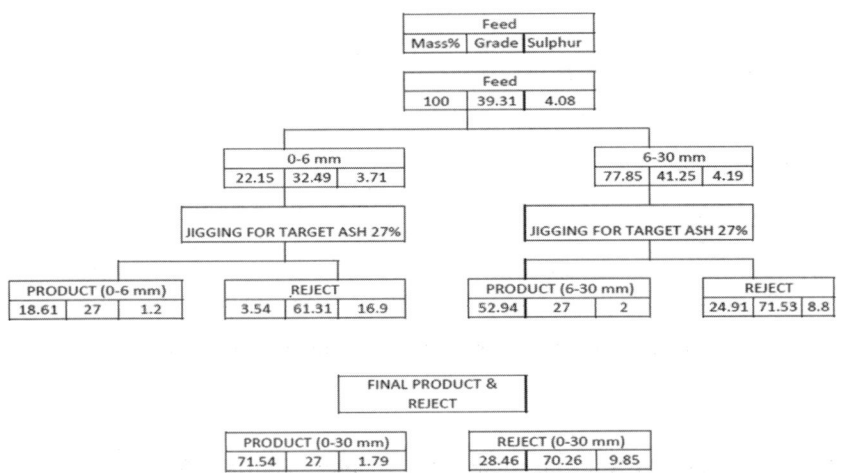

Figure 12: Predicted dry jig performance for a target ash content of 27%

6.3 De-stoning and de-shaling of South African coal from a deposit near Witbank with X-ray sorting

The material was taken from a coal deposit with a relatively high ash content in some cases exceeding 50%. The <50 mm material was screened off and the remaining >50 mm sample was subjected to X-ray de-stoning technology (XSS®T). The objective of the test was to differentiate and eject all non-coal material. Therefore, rock, shale and pyrite rich material would be targeted and removed to reduce ash and sulphur content. Two tests were done with two different samples having feed ash of 40.69% and 58.91%, respectively (Sample 1 and Sample 2). Results are shown in Table 4, Fig. 13 and Table 5, Fig. 14

Table 4

	Yield Wt %	Ash content %	Sulphur %	Calorific value (MJ/kg)
Feed	100.00	40.69	1.19	16.85
Product	67.17	32.11	0.18	20.22
Reject	32.83	58.25	3.25	9.97

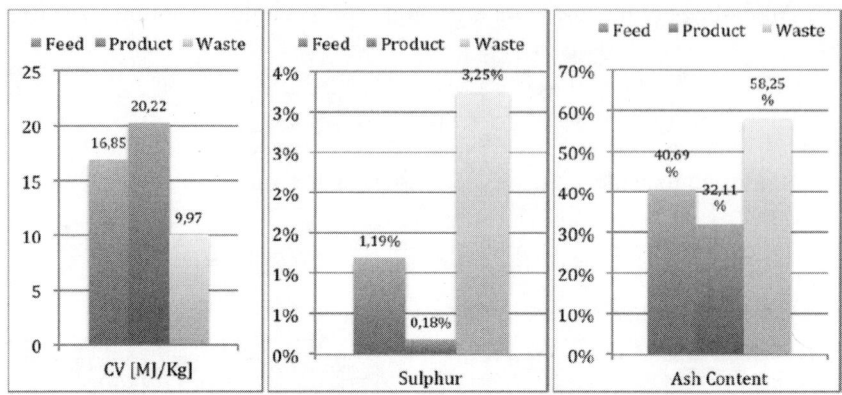

Figure 13: Results for Sample 1 – 40.69% ash content

Table 5

	Yield Wt %	Ash content %	Sulphur %	Calorific value (MJ/kg)
Feed	100.00	58.91	0.93	9.98
Product	25.42	31.17	0.21	20.88
Reject	74.58	68.36	1.17	6.27

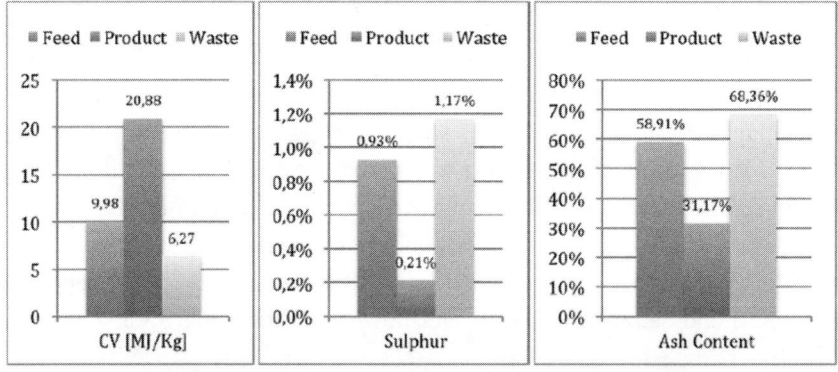

Figure 14: Results for Sample 2 – 58.91% ash content

In Sample 1 the ash reduction was about 27% while in Sample 2 the ash was reduced by almost 50%. Sulphur reduction in Sample 1 was by a factor of 6.6 times and in Sample 2 it was 4.4 times lower. The CV value was increased by 20% in Sample 1 and by almost 50% in Sample 2.

These results were achieved using a fixed sorting algorithm (accept/reject) for each test. It is possible to adjust the sorting algorithm to reduce the rejection of particles containing a modestly elevated ash content so as to maximise the coal mass yield as possible. Similarly, a setting (sorting algorithm) can be chosen to reject particles with even minimal ash inclusions so to achieve a better quality coal but at same time concede higher coal losses. The high variability of mass yields and ash and sulphur reduction shown with Sample 1 versus Sample 2 strongly suggests that the sorting algorithm can be finely adjusted to achieve the project's specific coal quality objectives.

6.4 Complete dry beneficiation plant utilising both jigging and sorting

It is known that dry jigging technology gives its best results in the size range of 0–50 mm whereas X-ray sorting technology provides better results for the coarser particles preferably above 40 mm. We can thus use both the technologies to design a complete beneficiation plant in which the 0–50 mm size fraction will be handled with allair® dry jigging and the >50 mm size fraction will be treated with X-ray sorter XSS®T. Figure 14 shows a conceptual flowsheet of the plant.

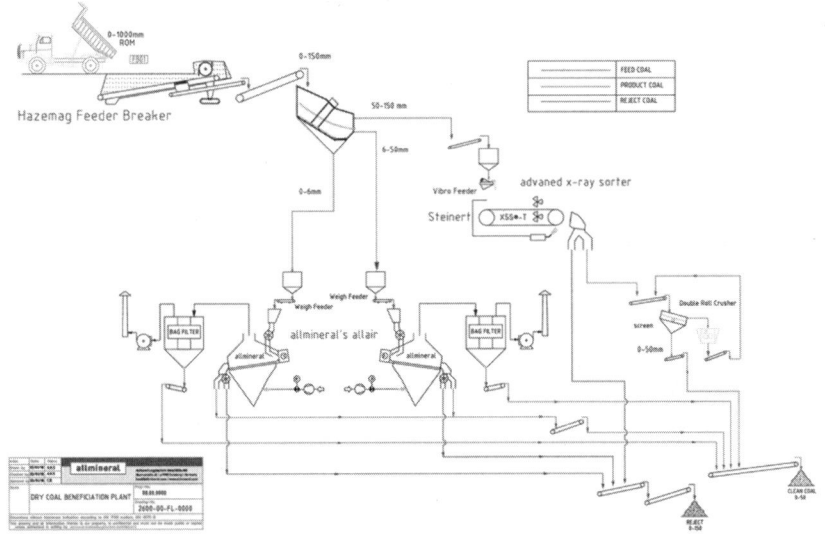

Figure 15: Conceptual flowsheet combining allair® dry jig and XSS®T X-ray sorter

ROM coal having size 0–1000 mm is fed to plant with dumper in a HAZEMAG feeder breaker which reduces the material size to 150 mm. 150 mm material is dry screened into three size fractions 0–6 mm, 6–50 mm and 50–150 mm. The 0–6 mm material is fed to a fines allair® whereas 6–50 mm material is fed to coarse allair®. The 50–150 mm material is fed to the X-ray sorter XSS®T for de-stoning. All the rejects from three machines are collected on one conveyor and stocked. The 50–150 mm product from the X-ray sorter is fed to a double roll crusher for size reduction to 0–50 mm and then combined with the allair® clean coal product to produce an overall 0–50 mm final clean coal product along with fines from the bag filter.

7. Conclusion

(1) A primary benefit of the dry beneficiation process for coal is to utilise the given resources whilst providing a more reliable and consistent quality clean coal fuel with lower operating costs.

(2) Water is an increasingly scarce and expensive global commodity, increasing the attractiveness of dry versus wet processing technology in industrial applications. This is a major advantage of the dry beneficiation of coal.

(3) Use of AFBC/CFBC boilers is becoming a growing trend in world especially in captive power plants. allair® dry rejects along with dust collector fines can be directly fed to these boilers for power generation.

(4) Dry beneficiation with both jigging and sorting can effectively reduce sulphur along with ash in the clean coal, which is an added advantage for mine operators.

(5) Dry beneficiation processes are more amenable to modular and mobile plant designs than wet processes. Mobile and modular plants provide greater flexibility and operational advantages, as well as potentially reduced capital and time overrun risks for coal projects

(6) The concept of utilising both jigging and sorting for different size fractions provides another dimension to dry beneficiation and provides flexibility with respect to treating a wide range of particle size distribution.

(7) Dry beneficiation involves less time for installation (7–8 months) and payback time is also less (6–7 months).

8. References

[1] Mineral Council of Australia, 2nd Edition, www.minerals.org.au/files/publications, Coal Hard Facts page 6–10.

[2] Department of Industry innovation and Science, Australian Government, Coal resources, https://www.industry.gov.au/resource/Mining/AustralianMineralCommodities/Pages/ Coal.

[3] Mitchell, D.R., 1942, Progress in Air Cleaning of Coal, Transactions of the American Institution of Mining and Metallurgical Engineers.

[4] Heribert Breuer, Andreas Hees and Hakan Oezdemir, South Africa Coal Preparation Society (SACPS) 2015, Dry jigging of coal.

[5] Kai Bartram, Johan van Zyl and Ezio Viti, International Coal Preparation Congress 2013, Dual Energy X-ray Transmission Sorting of Coal for De-shaling and Ash Content Reduction.

10

Latest developments and experiences in the beneficiation of coal using TOMRA X-ray transmission sorting machines

Jens-Michael Bergmann

TOMRA Sorting GmbH, Wedel, Germany

Abstract: Around the world the processing of coal, ores and industrial minerals is combined with the use of a huge quantity of water. This water must be obtained and – after use – recycled, discharged, or it leaves the process with the product or the tailings. This produces a wet coal and a large environmental challenge.

To overcome this TOMRA developed its own, unique X-ray transmission (XRT) sorting technology already in 2003, using the patented Dual Energy XRT approach. The Dual-Energy-XRT-technology enables materials to be recognized and separated based on their Planar projection atomic density. This creates a high purity level in the sorted materials irrespectively of size, moisture or contamination. In the recent years a lot of experience were gained while applying this technology on various minerals including coal. Also sorting performance increased considerably in the recent years: nowadays one sorter can (depending on grain size) handle up to 500 tph feed capacity and accurately remove the various contaminants.

Above all stands the human health. XRT-sorters are designed to meet all applicable health and safety regulations. The radiation shielding guarantees no human will be exposed to radiation beyond the legal limits. Safety devices monitor inappropriate access during operation and must shut down the X-ray source immediately. Fail-safe warning lamps permanently inform about the machine status.

Keywords: TOMRA, Dual Energy X-ray transmission, coal sorting, lignite sorting, XRT, health and Safety

1. Introduction

Sensor-based sorting is widely applied in the upgrading and beneficiation of industrial minerals and the recovery of diamonds for decades. The fast development of sensor-based sorting within the recycling industry, where thousands of machines are installed, has pushed the ability to unknown limits.

TOMRA Sorting is the worldwide leader in supplying sensor-based sorting solutions. Besides equipment for mining applications, sorters for the recycling and the food processing industry are the focus. The development of various sensors, electronic hard- and software and the controlled mechanical or pneumatic ejection systems is happening in a centralized competence center. This setup allows it to utilize the experience of various applications and cross-benefit from this to all applications.

New sensors using laser scattering effects, near-infrared-spectroscopy, and X-ray transmission (XRT) show a huge potential for various applications. The proven detection systems have been integrated into platforms which are designed and optimized for mining environments and mining applications.

2. Coal sorting

2.1 Coal preparation technologies

There are various technologies of coal preparation currently available and proven on the market. The main technologies are shown here in respect to the size range they are applied in (Fig. 1):

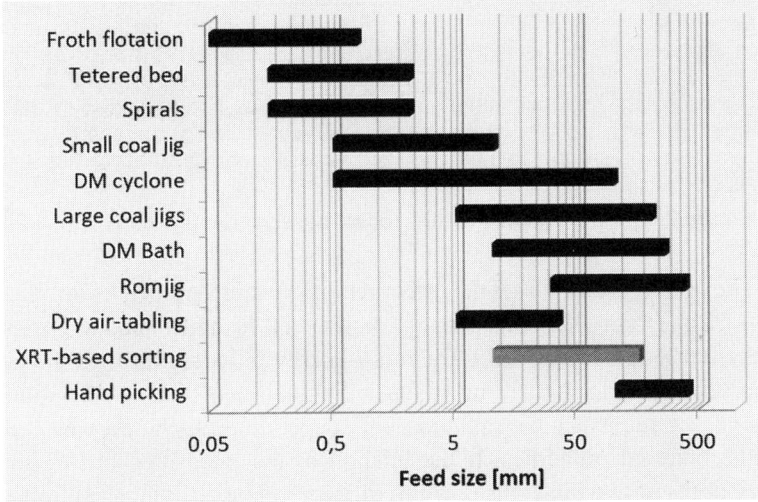

Figure 1: Overview on separation technologies and the achievable separation cut-points, modified from Sanders (2007)

Only the last three technologies work dry, that means with the strict absence of water and all its impacts.

2.2 Principle of an XRT-sorter

The principles of sensor-based ore sorting should be well-known among the industry meanwhile. A short summary is provided below.

The sorting machine must be fed with properly screened material to ensure an effective separation through the air blasts. In practice, a screen directly feeding onto the sensor-based sorter is mandatory. The screened feed material then enters the machine and is placed in a mono-layer onto a fast-running conveyor belt (Fig. 2).

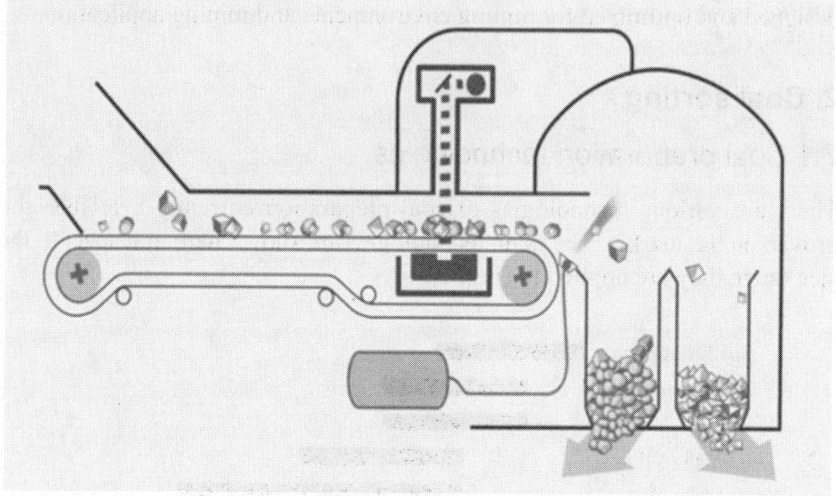

Figure 2: Working principle of XRT sensor-based sorter (Kleine et al., 2102)

The transport belt passes the detection system and the particles are scanned individually (throughout this paper particles shall be understood as 8 mm and larger). An X-ray source sends the multi-energy radiation through the object to a sensor module. According to the individual elementary composition of the object it attenuates the radiation only depending on chosen wavelengths and the material properties. If the relation of the intensities of two chosen wavelengths of two materials is different, they can be distinguished from each other (Jong et al., 2004).

A specialized computer then takes a decision for each particle based on predefined separation criteria. According to this decision, an array of highspeed air-valves is controlled. There are usually around 160 nozzles placed per meter working width (Robben et al., 2013). Single particles are ejected, thus deflected from their flight parabola. As the logic can be switched around, both fractions can be set as discard and product fractions. The lower amount fraction is ejected to save compressed air and therefore energy and operating costs.

2.3 Limitations of sensor-based sorting

2.3.1 Capacity

Since sensor-based sorting – independent of the applied sensing technology – needs to inspect every rock one by one, the rocks must be exposed to the detection system side by side in a monolayer. It must be avoided to have the rocks lying in double layers and covering each other. This fact has two physical consequences as follows:

(a) The particle size ratio, which can be processed in one run, needs to be limited, otherwise large pieces can cover their small neighbors. A size ratio of 1:3 (smallest to largest piece) has proven to deliver the best performance, under certain circumstance a ratio of 1:5 is still working reasonably. If this ratio is not maintained well, the ejectors will shoot more material than actually detected. This effects recovery of wanted material when ejecting the unwanted or increases product dilution in case the sorter is programmed to eject the product.

(b) The capacity of a sorter is simply depending on the volume which can be processed per time. The volume is determined, besides width, speed and area coverage, by the layer height. This layer height is directly depending on the average particle size. That means that the same mechanical equipment can process 20 or 100 tons per hour, only depending on the feed size. Consequently, the particle size directly influences the cost effectiveness of a sorter.

2.3.2 Radiation safety

Coal sorting with sensor sorters is mainly done using X-ray transmission. Here, any human being starts feeling uncomfortable: a radiation which is invisible and not tangible. It must be the highest priority for any manufacturer to protect the health of the people dealing with the equipment. This can be done by suitable radiation protection shielding as well as intelligent design and safety switches which shut down the energy source immediately one of the safety circuits is opened.

To penetrate coal or stones by X-rays it needs a certain amount of energy. The higher the applied energy, the higher is the thickness of the rocks which can be penetrated. Vice-versa, the higher the applied energy the more effort must be taken to provide the correct radiation shielding, which can make a sorting machine unnecessarily heavy and expensive. Therefore typically the power of an X-ray source for industrial applications is found between 1 and 3 kW.

2.3.3 Education of operation personnel

The operation of X-ray-based sorting machines is no rocket science. Operators and maintenance staff need to be trained on the potential dangers caused by the radiation and on safe operation and maintenance procedures. If the procedures are followed – in combination with the safe machine design – any educated electrician or mechanic offers the skills to operate these units.

3. Why does the next generation of XRT-sorters offer highest potential for the future?

3.1 Separation accuracy

The accuracy of separation depends on several factors, many of them hidden deeply inside the computer software, improving the data processing.

To increase separation accuracy the power of the X-ray source is a key factor. It allows to penetrate thicker rocks, increase the signal to noise ratio, improves the identification of small high-density inclusions (e.g. pyrites in coal) inside the particles.

Mechanically, having the nozzles as close as possible to the particles when shooting them reduces the amount of necessary compressed air tremendously, and compressed air is the main energy consumer of a sensor sorter. Having a discrete connection between one valve and one nozzle also cuts air consumption. At the same time this setup provides more ejection accuracy, less misplacement and experiences less blockages of valves.

In other words: accuracy.

3.2 Capacity

Efficient and high-volume separation is a lever to reduce the total capital and operational expenditure of a sorting plant. Higher volumes can be achieved by these factors:

(a) Belt speed increased from 2.7 to 3.5 m/s (+30%).

(b) Area occupancy from 25 to around 35% (+40%).

(c) More powerful X-ray system which allows sorting of larger particle sizes due to better X-ray penetration (+10%).

All these factors together allow it to double the specific capacity. To carry it to the extremes, by doubling the sorter width from 1.2 to 2.4 m, capacities of up to 500 t/h (coal) or even 700 t/h (iron ore) can be realized capacity and creating a new industry benchmark (MRA, 2018).

3.3 Operability, serviceability and availability

Over recent years, TOMRA sorters have notched up more than half a million operating hours under harsh conditions ranging from arid deserts to arctic environments, with both wet and dry feed. "We learned early on, for instance, that sorters made for the recycling sector do not necessarily fit the purposes of mining processes. The higher tonnages and abrasive materials running through the sorter mean that the design needs to be more robust and must include increased wear protection."

The mechanical platform of the latest TOMRA COM XRT 2.0 was redesigned for safer, simpler, quicker and more ergonomic maintenance, by providing easier access to all relevant components. The separation chamber, in-feed chute, belt and air system components are all easy and quick to access, making the maintenance stops more efficient and giving the unit high availability. Modular wear components also enable safe and ergonomic manual exchange in the shortest possible time. Increased wear resistance and component lifetime improve the mean time between failures (MRA, 2018).

3.4 Safety features

Safety is the ultimate priority on any equipment. In case of X-rays the radiation safety becomes a key factor. The X-ray source inside the sorters emits radiation. The energy level of this radiation is directly depending on the power of the X-ray source. Appropriate shielding is applied to ensure that at no point above, beneath or around the machine the maximum permissible value is exceeded. Even more, at TOMRA we reach by far better values than any legal requirements.

Before the units leave the workshop, radiation measurements are made to ensure everything is manufactured correctly. Later, measurements are made at site, after installation and during commissioning. Repetitive measurements during operation ensure the highest possible safety level.

Special design features prevent to introduce parts of the body into the encapsulated part of the sorter. Doors are equipped with monitored safety switches, which shut down the energy immediately if one of the circuits is broken. Monitoring windows are made of lead glass and protective covers cannot be removed without special tools.

Coal dust explosions are another threat in any coal application. The dust itself must be aspirated as good as possible to avoid agglomerations, but additionally, TOMRA designed the whole sorting machine in such a way that no sparks are generated by a rock hitting any surface of the sorter.

3.5 Economics

All the technical features do not convince anyone to invest in a sorting machine. The economics of such an investment must match company objectives. Simply, it must pay off.

Two main points that help here are the absence of water and latest sorting technology.

3.5.1 No processing water necessary

At first there is the absence of process water. Water is one of the most essential resources in our live. Conventional coal preparation methods are mainly based on water (see Fig. 1). Sensor-based sorting cannot replace all the wet techniques, but it is a proven and reliable complementary technique to these wet processes. The advantages are mainly the reduced consumption of water itself, the reduced installation of water-pumping, transport and treatment technologies, lower consumption of dense medium powders and flotation reagents. Finally the costs for all equipment and procedures to erect and maintain the necessary tailings dams can be seriously reduced. Not to forget, reducing the use of land, the environmental impact and the permitting efforts for the tailings dams is also a countable benefit. The product is a dry coal where the usable CV is higher than that of the same coal after a wet beneficiation process.

That means, for all particle sizes larger than 8 mm and up to 150 mm, sensor-based sorting is a technically viable alternative. Economically, this size must be shifted to larger particles to increase the capacity, certainly all depending on each individual application.

3.5.2 The sorter

The higher capacity makes the latest TOMRA XRT-sorter a particularly valuable proposition for larger mines, as it significantly cuts down the number of machines required (Fig. 3).

The result is also a small plant footprint and comes down to low capital costs for mineral treatment projects, which translates into the lowest specific operating costs per ton. An unrivaled separation efficiency which shows a low misplacement of discard to product and product to discard. The system's energy efficient design requires only about half of the energy for compressed air when compared to conventional sensor-based ore sorting systems.

Feeding as much material as possible, recovering close to 100% of the valuable mineral, pure removal of the impurities, lowest consumption of compressed air and energy economizes on both the capital and operating costs (Fig. 4).

Figure 3: Part of the 1800 tph-installation at MWSPC's phosphate mine

Figure 4: Two XRT sorters dry-processing coal side-by-side at a remote site @ 200 tph

4. Conclusion

Sensor-based sorting and especially XRT-sorting has found its way into the mining industry. It is no longer considered as too sensitive for this "rough" business. It has proven its rigidity and viability in a mining environment under worst conditions in many cases.

The latest technological developments have lifted the technology to another level. Doubled capacities, reduced energy consumption, low wear and simple maintainability reflect directly into an increased availably and a considerable cost reduction per ton.

Depending on each individual application, coal sorting with XRT-sorting machines is a realistic alternative to conventional, water-based, processing methods and should always be considered when planning a greenfield or brownfield installation.

Economically, caused by the lower operation and capital expenditures compared to a wet processing plant, sensor-sorting has become a very interesting alternative.

5. References

[1] Jong, T. de, Houwelingen, J.A. van, Kuilman, W., 2004. Automatic sorting and control in solid fuel processing: Opportunities in Europe and perspective (Geologica Belgica (2004) 7/3–4: 325–333).

[2] Kleine, C., de Korte, J. de, Ketelhodt, L. von, Robben, M., 2012. Recent developments in dry coal sorting using X-ray transmission (Proceedings of the 26th International Mineral Processing Congress, New Delhi, India).

[3] MRA Mining Review Africa, TOMRA X-ray sorter a game changer for minerals processing (Issue 11, 2018, pages 53–56).

[4] Robben, C., Korte, J. de, Wotruba, H., Robben, M., 2013. Experiences in dry coarse coal separation using X-ray-transmission-based sorting (17th International Coal Preparation Congress, Turkey).

[5] Sanders, G. J., 2007. Principles of Coal Preparation (Australian Coal Preparation Society).

11

Release analysis of low volatile coking coal fines

Divya Jyoti[1], Sumantra Bhattacharya[2], Soni Jaiswal[3], Sanchita Chakravarty[4]

[1]*Chemical Engineering Department, Indian Institute of Technology, New Delhi, India*
[2]*Department of Fuel and Mineral Engineering, Indian Institute of Technology (ISM), Dhanbad, Jharkhand, India*
[3]*Research and Development, Tata Steel, Burma Mines, Jamshedpur, Jharkhand, India*
[4]*National Metallurgical Laboratory, Burma Mines, Jamshedpur, Jharkhand, India*

Abstract: Release analysis of three low volatile coking coal fines obtained from two different coalfields of India was carried out at three different size levels, −0.5 mm, −0.5+0.1 mm and −0.1 mm. The collectors used were n-dodecane and commercial diesel oil. Incomplete liberation of coal fines even at a size of −100 μm, corroborated by photomicrographs of run-of-mine (ROM) and clean coal macerals, in a great measure reduced the flotation-based cleaning potential of the fines. Field emission scanning electron microscopy (FESEM) images of fines revealed the adherence of ultra-fines onto the surface of relatively coarser particles and that reduced the clean coal recovery. The results obtained also appear to be coal, reagent combination and particle size and particle surface specific. Only one coal indicated the liberation as is usually expected for −0.1 mm particles with a 65% yield at 18% ash.

Keywords: Coking coal, release analysis, particle size

1. Introduction

Coal flotation is considered to be difficult, if the percentage of near gravity material (NGM), i.e. essentially locked particles, in flotation feed is >10% (Aplan, 1993), which is usually the case with all the coking (metallurgical) coal fines in India though the fines are rich in vitrinite. Release analysis proposed by Dell (1953) aimed to determine the ultimate floatability of coal. Several other techniques had been proposed but release analysis technique has been considered to be a vast improvement over the batch flotation process (Forrest et al., 1994). In most of the coal producing countries current practice is to treat up to 0.25 mm or 0.1 mm by density separation processes and to treat by froth flotation only −0.25 mm or −0.1 mm size fractions. Feed size to coal flotation in India remains however, −0.5 mm. Release analysis is believed to be independent of collector and frother dosage and type, but it is influenced by the "state of release of the sample." The shape of release curve is an indication of the degree of liberation. For poorly liberated coal the release analysis curves are much closer to 45° angles, rather sharp inflection points (elbow) signify easy cleaning potential. Release curves with shallow elbows indicate majority of the mineral matter is in locked form and cannot be removed readily by froth flotation (Forrest et al., 1994). From the comparison with tree analysis it was found that the release analysis provides the best recovery-grade relationship for fines containing middling particles (Mohanty et al., 1998) which is a typical feature of coal flotation feeds in India. The finely disseminated mineral matter causes the non-coal matter to float at high flow rates, and thereby increases the yield and ash in the concentrate. However, with the decrease in size from −0.5 mm to −0.1 mm the release analysis shows better results.

By release analysis of a medium coking coal with 26% ash content, approximately 32% coal could be floated at 15% ash (Das et al., 2008). The fines from Bhojudih wash plant located in Jharia coalfield, having proximate analysis as 24.4% ash, 19.8% volatile matter and 53.8% fixed carbon on dry basis provided 75% clean coal yield at an ash level of 13.7% through release analysis (Jena et al., 2008). Release analysis of a low volatile coking (LVC) coal carried out by Singh et al. (2011) indicated that for the size fraction 0.5–0.1 mm, n-dodecane could hardly achieve any noteworthy yield at ≤16% ash whereas with a synthetic collector the clean coal ash got reduced to 14.5% with 29% yield. For the +0.075 mm size fraction the yield increased (48–62%) at 16–19% ash content by using the same synthetic collector. The yield reduction factor per 1% ash drop however was high, in the range of 7–25%. The corresponding tailing ash in one case was as high as 63%. For the ultra-fine size fractions, −0.075 mm and −0.05 mm, yields of 48–71% could be obtained with n-dodecane at 19–20% ash with practically no recovery at a lower target ash. At sizes finer than 0.1 mm, release curves of the LVC coal began acquiring a distinct similarity with the corresponding curves for typical −0.5 mm *in-situ* coal. Curves obtained for size fractions −0.5 mm and +0.1 mm were much closer to 45° angles (Singh et al., 2011).

2. Material and methods

Three coal sources, Kathara from East Bokaro and Joyrampur and Patherdih from Jharia coalfield were selected for this study. Representative samples of the coal were crushed to −13 mm by jaw crusher followed by double roll crusher. Fines, −0.5 mm, were one of the major contributors to the size consist of the crushed coal. In case of Joyrampur and Patherdih, the share of this fraction was about 12%, whereas for Kathara, it was about 6.5%. These fines have been considered in this study for release analysis. Release analysis was performed for each coal at three levels of sizes, −0.5 mm,

−0.5+0.1 mm and −0.1 mm following the procedure laid down by BS 7530-2:1995, using collectors n-dodecane and commercial diesel oil. FESEM (field emission scanning electron microscopy) set up, model no MONOCL 4 was used to obtain SEM images. Petrographic and optical studies were carried out in polished sections under the microscope using reflected light with oil immersion objective for run-of-mine (ROM) and clean coal.

3. Results and discussion

Size by size proximate analysis of fines (Table 1) indicates that −0.1 mm size fraction has a significant presence in the fines and it is this fraction, which is high ash and low volatile in nature. The fines of all the three coals belong to same LVC coal type, yet their proximate analysis differs.

Table 1: Size by size proximate analysis of fines (on air dried basis) (VM, volatile matter and M, matter)

Size (mm)	Joyrampur (%)				Patherdih (%)				Kathara (%)			
	Weight	Ash	VM	M	Weight	Ash	VM	M	Weight	Ash	VM	M
−0.5	100	22.6	22.5	0.7	100	21.9	19.0	2.8	100	27.9	23.7	2.4
−0.5+0.1	80	23.5	22.9	0.9	60	21.7	19.7	2.5	80	27.7	23.8	2.3
−0.1	20	26.3	20.5	1.2	40	32.0	18.7	2.4	20	36.5	20.5	2.4

3.1 Release analysis by n-dodecane

Figures 2–4 shows size by size release analysis for all three coals by n-dodecane and MIBC (methylisobutylcarbinol) combination. For Joyrampur coal all three size fractions have similar yield of 50–52% at 18% ash. At 17% ash however there appears to be practically no yield for the size fraction −0.1 mm, though size fractions −0.5 mm and −0.5+0.1 mm have 44% and 49% yield, respectively. The release curves of −0.5 mm and −0.5+0.1 mm size fractions appear to have an overlap (Figs. 2 and 3) possibly because of dominance of particles coarser than 0.1 mm, nearly 80% of total mass of fines. Vertical nature of curve (Fig 4) for about 50% of mass in −0.1 mm size fraction indicates significant liberation of "coal" particles at this size. Patherdih shows notable increase in clean coal yield to 30% at 18% ash (Table 2) only for the size fraction −0.5+0.1 mm (Fig. 3). No significant difference could be observed in the pattern of the three release curves and the curves of two coarser size fractions for about 80% of the masses appear to be parallel. Kathara coal shows certain similarity in pattern (Fig. 4) to the release curves obtained for Patherdih coal. Substantial clean coal yield has been obtained only for the −0.5+0.1 mm size fraction.

3.2 Release analysis by diesel oil

Release curves obtained with diesel oil and MIBC combination for the composite fines appears to show considerable difference between the three coals (Fig. 2) with insignificant yields (Table 2). Considerable improvement in yield could be observed for the size fraction, −0.5+0.1 mm, possibly arising from the absence of −0.1 mm size (Fig. 3). Similar, results were reported by Huang et al. (2018). Release curves of −0.1 mm size fraction appear to be distinctly different for all the three coals (Fig. 4). Except for Joyrampur coal, the curves do not indicate the liberation as is usually expected for −0.1 mm particles. Highest and a reasonable yield is obtained with Joyrampur coal for the size fraction −0.1 mm (Table 2).

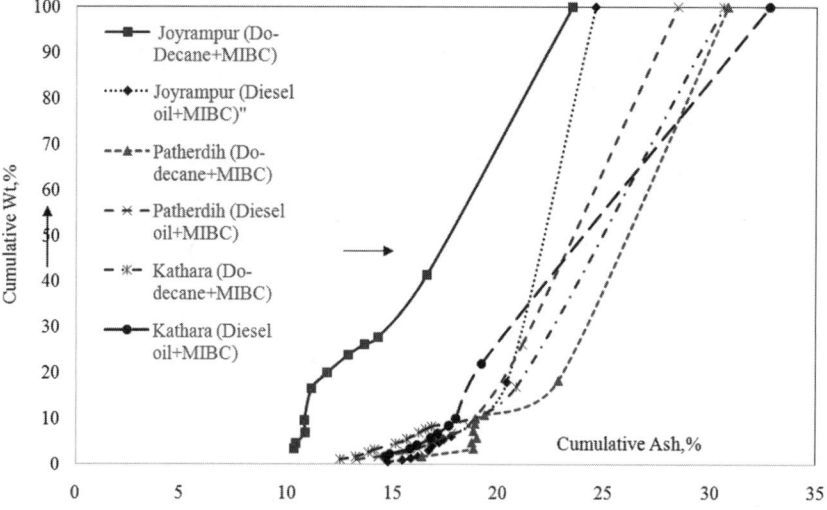

Figure 2: Release analysis curves for the size fraction −0.5 mm

Figures 2–4 and Table 2 indicate no specific patterns in release curves and on the effect of collector on release analysis. Since all three coals were of same type and belonged to same seam, at least at −0.1 mm size level some overlap between the curves (Fig 4) were expected, which could not be observed. Some similarity between the curves appears to be there in the ash range of <20%. Yield variation between the two collectors appears to be coal, reagent combination and particle size specific. For Joyrampur and Kathara coal, at the size level of −0.5+0.1 mm both the collectors deliver approximately the same yield, whereas significant difference is recorded for Patherdih coal (Table 3). Almost 50% plus yields are obtained for all the three coals depending on particle size and collector used, that included high purity n-dodecane and routine commercial diesel oil procured from fuel supply station. Best yields typically vary between 21% and 65%, depending on particle size and collector used. With both the reagent combinations, yield differences between 17% and 18% cut-off ash shows quite significant yield reduction factor (YRF). Tables 2 and 3 indicate that for all larger yield values, YRF are also high. Large YRF values indicate that liberation of "coal" from associated mineral matter remains incomplete, in particular for Kathara and Patherdih coals, even at the size of −0.1 mm. That appears to be the reason why the release curves obtained are different from the curve re-produced in Fig. 1 for *in-situ* coal after BS 7530-2:1995.

3.3 Release analysis results and feed characteristics

Figure 5 showing photomicrographs of ROM and clean coal of the three coals appear to be quite indicative of the moderate and somewhat inconsistent results of release analysis. Occurrence of fine to extremely fine dissemination of clay, mineral matter and inertinite on vitrinite surface appears to be a major determining factor of the results. Photomicrographs of coal macerals shown on 100 μm scale indicate that complete or say significant liberation possibly is not achievable even at less than 100 μm size, though 100 μm has become a kind of limiting size for technology availability, process selection and plant operation for fines in coal preparation plants worldwide. For all practical purposes flotation is the only process available for cleaning the −0.1 mm coal particles.

Table 2: Clean coal yields at 18% ash level obtained by release analysis

Collector	Joyrampur			Patherdih			Kathara		
Size (mm)	−0.5	−0.5+0.1	−0.1	−0.5	−0.5+0.1	−0.1	−0.5	−0.5+0.1	−0.1
n-Dodecane	52	57	52	2	30	4.5	9	54	26
Diesel oil	7	54	65	7	51	1	15	54	21

Table 3: YRF between 17% and 18% ash level recorded by release analysis

Collector	Joyrampur			Patherdih			Kathara		
Size (mm)	−0.5	−0.5+0.1	−0.1	−0.5	−0.5+0.1	−0.1	−0.5	−0.5+0.1	−0.1
n-Dodecane	8	9	−	1	15	0.5	1	5	2
Diesel oil	1	9	5	2	7	1	7	4	1

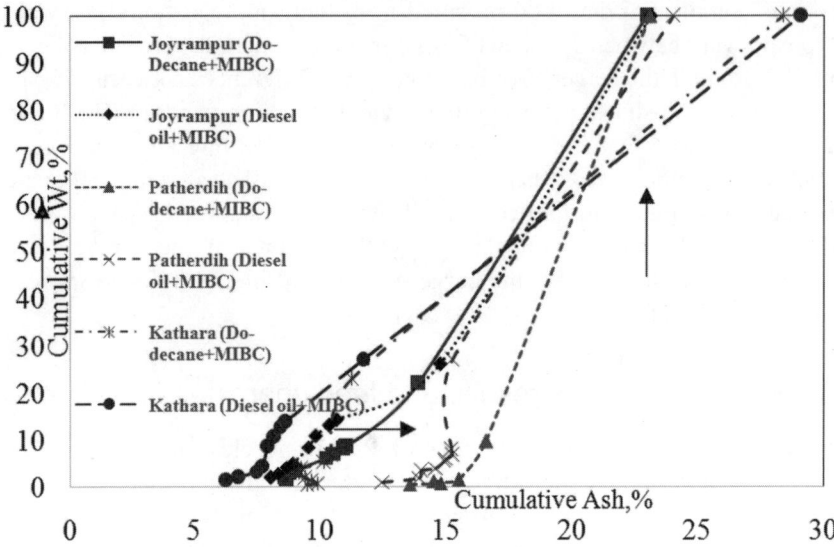

Figure 3: Release curves for the size −0.5+0.1 mm

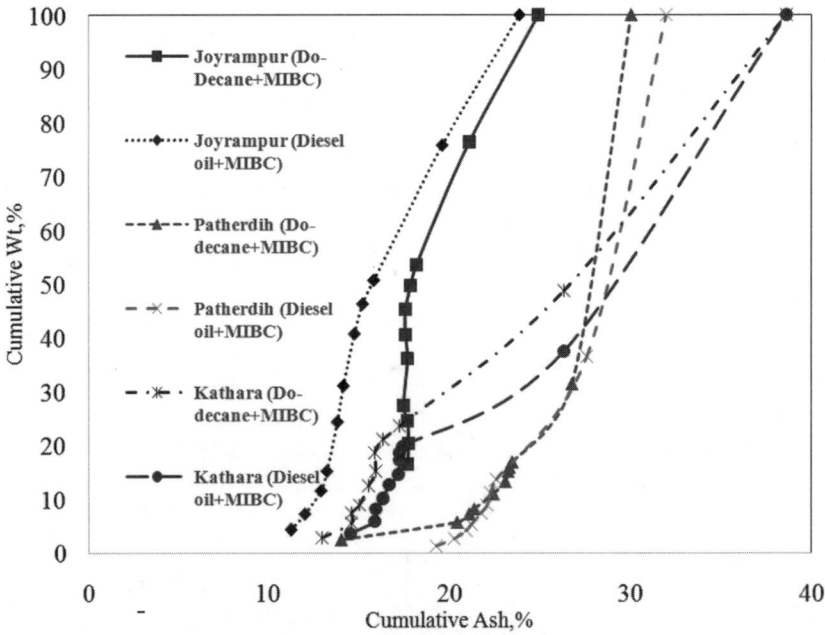

Figure 4: Release curves for the size −0.1 mm

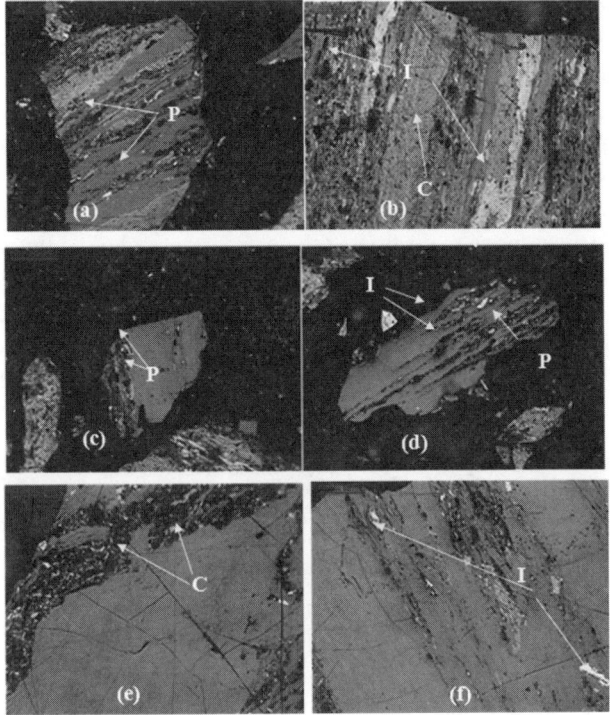

Figure 5: (a) Pyrite grains of <5 µm size embedded on vitrinite surface of −13+2 mm of Patherdih clean coal, (b) inertinite and clay interlocked with vitrinite in −13+2 mm Kathara clean coal, (c) pyrite grains of <5 µm size scattered on vitrinite surface of −2+0.5 mm Patherdih clean coal, (d) inertinite embedded in vitrinite in −13+2 mm Kathara clean coal, (e) clays filled pores in vitrinite of −2+0.5 mm Joyrampur clean coal and (f) very fine interlocking of inertinite and pyrite with liberated vitrinite in Patherdih −2+0.5 mm clean coal

In case of Patherdih clean coal pyrite grains of <5 µm size are embedded or scattered on vitrinite surface along with very fine interlocking of inertinite and pyrite with "liberated" vitrinite (Fig. 5a, c, f). Inertinite and clay interlocked with vitrinite and inertinite embedded in vitrinite are observed for Kathara clean coal (Fig. 5b, d). Although, some fairly liberated vitrinite and inertinite macerals are observed in the petrographic study of the ROM coal, macerals and mineral matters were so closely interlocked that it appears to be almost impossible to be liberated even at finest sizes. The best results in release analysis have been obtained for Joyrampur coal (Table 2), yet photomicrographs of the clean coal indicate inertinite embedded on vitrinite surface along with clay filled pores (Fig. 5d, e). Given the size consist of Kathara coal (Table 1), size fractions −0.5+0.1 mm and −0.1 mm show better clean coal yields than −0.5 mm (Table 2), that is difficult to explain. Virtually identical results were obtained after repeat tests, possibly because particles <0.1 mm particularly ultra-fines had acted as a kind of inhibitors in the flotation of coarser particles. FESEM images show that the ultra-fine particles inherently present in coal adhered on the surface of the coarser particles (Fig. 6). Adherence appears to be least for Joyrampur coal and most for Patherdih coal. Particles adhered could be of clay, other mineral matter and pyrite, the latter having a tendency to co-float with coal, further reducing the scope of ash reduction through flotation. Fractures and pores do not appear to be significant virtually excluding thereby the scope of size degradation and reagent loss in course of flotation. Minimum adherence of ultra-fines on coarser particles could be the reason for better results of Joyrampur coal in release analysis (Fig. 6).

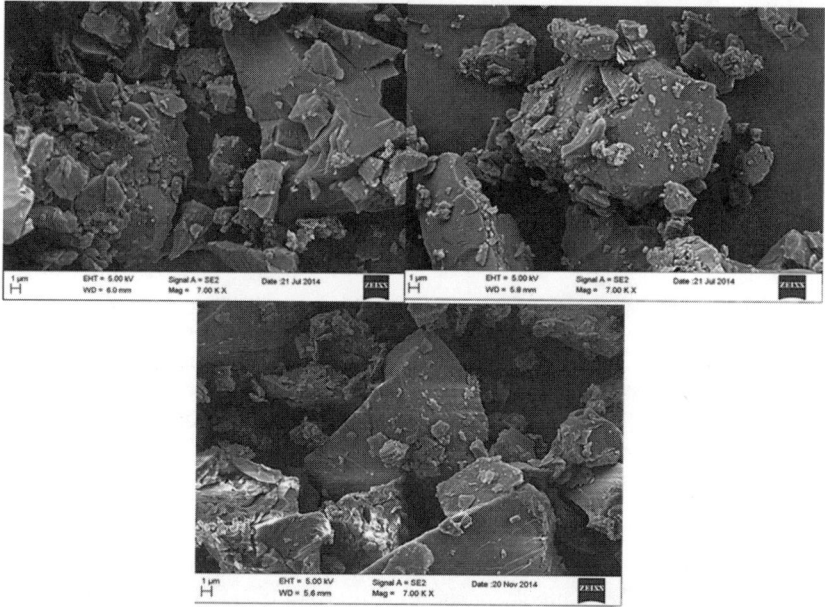

Figure 6: (a) FESEM Image of Joyrampur ROM coal, (b) FESEM image of Patherdih ROM coal and (c) FESEM image of Kathara ROM coal

4. Conclusion

Size by size release analysis carried out for three LVC coals by and large belonging to the same group of seams does not appear to show encouraging results. The curves obtained for +0.1 mm size fractions are quite close to 45° angles and are very similar to the standard washability curves of LVC coal (Jyoti et al., 2015). Release curves of −0.1 mm size fraction appear to be distinctly different for all the three coals. Only Joyrampur coal indicates the liberation as is usually expected for −0.1 mm particles, highest and a reasonable yield of 65% at 18% ash being obtained with diesel oil as collector. Compared to drift origin coal however, the cleaning potential remains difficult. The present work appears to confirm the observation from an earlier work that n-dodecane might not be the ideal collector for release analysis of un-liberated or partially liberated coal fines, though could be applicable for the significantly more liberated ultra-fines of the same coal. The present investigation demonstrates the collector dependency of release analysis results and that diesel oil can be used as a collector in release analysis of LVC coal depending on coal type as reflected by its petrography and feed size consist. Photomicrographs of macerals indicate that inertinite, clay, pyrite and other mineral matters of approximately 5 μm size and even smaller in large proportion are adhered to the vitrinite surface or embedded in it. Some of these particles could be larger than 5 μm. Minimum adherence of ultra-fines to the coarser particles was for Joyrampur coal, which delivered consistent and reasonable yields. Yield variation between the two collectors also appears to be coal, reagent combination and particle size and particle surface specific.

5. References

[1] F. F. Aplan: Mining Engineering, 1993, Vol. 45, pp. 83–96.

[2] A. Das, B. Sarkar, A. Vidyadhar, A. K. Singh, K. K. Bhattachraya: International Journal of Coal Preparation and Utilization, 2008, Vol. 28, pp. 189–200.

[3] C. C. Dell: Recent Developments in Mineral Dressing, Institute of Mining and Metallurgy, London, 1953, pp. 75–84.

[4] W. R. Forrest, G. T. Adel, R. H. Yoon: Coal Preparation, 1994, Vol. 14, pp. 13–27.

[5] G. Huang, H. Xu, L. Ma, L. Wu: International Journal of Coal Preparation and Utilization, 7, 2018, Vol. 38, pp. 361–374.

[6] M. S. Jena, S. K. Biswal, S. P. Das, P. S. R. Reddy: Fuel Processing Technology, 2008, Vol. 89, pp. 1409–1415.

[7] D. Jyoti, S. Bhattacharya, A Anupam, V. K. Saxena: Transactions Indian Institute of Metals, 2015, 4, Vol. 68, pp. 649–660.

[8] M. K. Mohanty, R. Q. Honaker, K. Ho: Coal Preparation, 1998, Vol. 19, pp. 51–67.

[9] H. Singh, S. Dey, S. Bhattacharya: International Seminar on Mineral Processing Udaipur, India, 2011.

12

Structural optimization and experimental study of jet-mixing apparatus based on CFD

Wei Zhou, Jin-bo Zhu, An-an Feng, Fan-fei Min, Chuan-chuan Cai, Yong Zhang

Anhui University of Science and Technology, Huainan Shi, China

Abstract: To optimize the structure of the jet-mixing apparatus, the numerical simulation scheme was adopted. The computational fluid dynamics (CFD) software was used to carry out that numerical calculation of the mixing performance and internal flow field distribution of jet-mixing apparatus. It was concluded that the optimization of the area ratio and nozzle exit position (NXP). Then four groups of jet-mixing apparatus were produced. Through the droplet diameter test experiment, the influence of area ratio on dispersive properties of agent was investigated. The results showed that the optimal range of area ratio a is 1.96–3.24, and the optimal range of the NXP Le is 0.2–0.6Dh in the jet device. The position of the suction tube has a great influence on the mixing of the agent and the working fluid. The optimal position of suction tube is positive against the nozzle exit (0−Le). According to the distribution characteristics of the flow field and the droplet diameter test experiment, when the area ratio a = 3.24, the distribution characteristics of the flow field are ideal, which provides the optimal conditions for the dispersion of the agent in the working fluid, and the dispersing effect is the most pronounced.

Keywords: Jet mixing, numerical simulation, structural optimization, agent absorption performance; agent dispersed

1. Introduction

Floatation is the most widely used dressing method in fine grained material separation. In recent years, the efflux technology has been widely used in various aspects of flotation process (Güney et al., 2002; Deng et al., 2013), such as the micro-bubble generator of the swirl micro-bubble flotation column (Gong et al., 2015; Wang et al., 2015a; Shen et al., 2009; Chen et al., 2008), the pneumatic mixing device of jet flotation machine (Hacifazlioglu, 2014; Salemsaid et al., 2013; Tasdemir et al., 2011), and the mixing device of the jet mixer (Ma et al., 2014; Dorpmund et al., 2012). At present, the research focuses on the optimization of structural parameters, flow field analysis and mixing process of agents and working fluids. Ni et al. (2016) studied the effect of jet bubble generator on flotation of coarse particle coal. Li et al. (2010) conducted numerical simulation of vapor and liquid in a self-priming bubble generator using FLUENT software, and obtained the optimum area ratio of the device was 3. Wang et al. (2015b) conducted a numerical simulation of the three-dimensional flow field of micro-bubble generator with different nozzle exit position (NXP), and studied the influence of the NXP on air flow, bubble diameter and distribution. Yang et al. (2013) designed a jet flotation column, and simulated the flow field distribution in the jet aeration device based on the standard k–ε turbulent mixing model and the multiphase flow model. Taşdemir et al. (2007) studied the effect of jet length and jet velocity on the recovery of different graded products in Jameson flotation machine. Vondricka (2007) used computational fluid dynamics (CFD) numerical simulation to study the distribution characteristics of mixed agent concentration in the flow field under different mixing ratios. Xu et al. (2012) used the Fluent software to simulate the distribution characteristics of axial and radial agent concentration in the mixed tube and studied the influence of area ratio on the uniformity of the mixture. In this paper, CFD software was used to calculate the mixing performance and internal flow field distribution of jet-mixing apparatus, and the area ratio and NXP were optimized. The droplet diameter test experiment was carried out, and the influence of area ratio on dispersion performance was investigated, and the experimental results were corroborated to provide theoretical guidance for structural optimization and improvement.

2. Theoretical analysis of jet mixing

The structure and working process of jet-mixing apparatus are shown in Fig. 1. The working fluid is ejected at high speed through the nozzle 2 of the jet chamber 1. Then the working fluid Q_g generates negative pressure at the mixing chamber 3. The volume produces by the flocculating diffusion effect which is generated by the fluid attracts the driving fluid Q_y. Then, the exchange of momentum and mass between the two fluids happen in the throat tube 4, and they eventually flow out of the diffusion tube 5. Usually, the jet flow can be divided into the initial and the jet sections along the flow direction of the fluid. In the jet initial section, there is a jet core area. The jet velocity is basically the same as the flow rate of the nozzle outlet in the jet initial section. Besides, with the involvement of the driving fluid, the jet boundary gradually widens and the flow rate decreases. After the cross section, the fluid enters the jet main section. As a result, the velocity of the jet on the axis begins to wane.

The working parameters and structural parameters of the jet-mixing apparatus are more. In order to describe the mixing performance under different working parameters and structural parameters, the performance of jet mixing is evaluated by using the relation curve about the non-dimensional parameter pressure ratio p, flow ratio q, mixed agents efficiency η, and structural parameters in engineering.

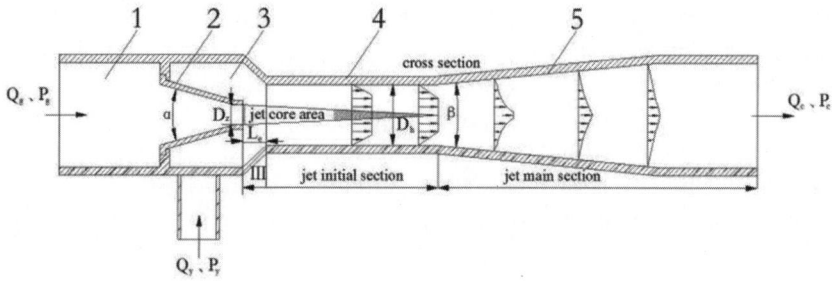

Figure 1: Structure and schematic diagram of jet device

$$p = \frac{P'_c - P'_y}{P'_g - P'_y}; q = \frac{Q_y}{Q_g}; \eta = pq.$$

P'_g, P'_y, and P'_c are the total pressure of the working fluid, the total pressure of the driving fluid, and the total pressure of the mixed fluid, Mpa, respectively.

3. Structure and parameters

Structural parameters are divided into fixed and variable parameters. Fixed parameters value in the related articles within the range of the optimal value: angle of nozzle convergence is 15°, angle of throat tube diffusion is 10°, length of throat tube is 120 mm, length of diffusion tube is 100 mm, diameter of feed tube is 30 mm, diameter of drug suction tube is 5 mm, and symmetrical way is adopted. Variable parameters are test parameters to be optimized. The research objects of this article are area ratio a (ratio of throat tube diameter D_h to nozzle diameter D_z), NXP L_e (distance of nozzle and throat tube inlet), and suction nozzle position. The change of area ratio is achieved by fixing the nozzle diameter D_z and changing the throat tube diameter D_h. The suction tube position is dynamically laid out along the mixing chamber. After optimizing area ratio a and NXP L_e, the position of the suction tube is optimized.

Working fluid enters the jet chamber at a speed of 2 m/s, sprays at a speed of 18 m/s from the nozzle, forms negative pressure in the inner chamber of mixing chamber to inhale medicament. Drugs are dispersed under the impact and cutting of jet action, and are ejected from the diffusion tube after mixing with working fluid.

4. Numerical simulation scheme

4.1 Grid type and subdivision mode

Use Gambit2.4.6 to grid the model. In other words, the hexahedral Hex grid was used to divide injection chamber, nozzle, throat tube, and diffusion tube. The tetrahedral T_{grid} was used to divide the absorbing tube and the mixing chamber.

4.2 Solver parameters and boundary condition setting

By choosing volume of fluid, the dispersion of each parameter, like convection, adopted the second order upwind with high precision. The coupling of speed and pressure adopted SIMPLEC mode and the convergence precision was set to 10^{-4}. The Realizable k–ε model was selected as turbulent model, $C_2 = 1.9$, $\delta_k = 1.0$, $\delta_\varepsilon = 1.2$, and standard wall functions was selected as near-wall treatment.

The inlet of feed tube adopted speed boundary conditions. The pressure boundary conditions were applied to the inlet of the suction tube, and the relative pressure was set to 0 Mpa. The pressure boundary condition was applied to the outlet of the diffusion tube and the wall surface was non-slip boundary condition. The diffusion tube outlet adopted pressure boundary conditions, and the wall surface adopted non-slip boundary condition, and the near wall area adopted standard wall function.

4.3 Theoretical verification of numerical solution

Structure of jet-mixing apparatus belongs to the type of venturi tube. Therefore, the venturi tube with no ejection tube can be calculated theoretically. Then the Bernoulli equation was applied to the injection chamber inlet and outlet of diffusion tube. So the results were calculated: the inlet pressure of the injection chamber was $P_{gI} = 119.609$ kPa, and the outlet pressure of the diffusion tube was $P_{gII} = -78.945$ kPa. In order to verify the correctness of the numerical solution, we removed the suction tube, output the grid model, set the same inlet and outlet boundary conditions, and used fluent to calculate $P_{gI} = 114.459$ kPa; $P_{gII} = -75.908$ kPa Compared with the theoretical calculation and fluent calculation, the deviation of P_{gI} was 4.50%. The deviation of P_{gII} was 4.00%. For numerical calculation, deviation less than 5% was within acceptable limits. Therefore, the calculation method was correct, and the numerical solution was reasonable.

5. Calculation results and discussion

5.1 Influence of drug absorption performance on structural parameters

Area ratio a is in the range of 1.0–10.0, and NXP L_e is in the range of $0.2D_h$ to $2D_h$. The suction tube is dynamically laid out along the mixing chamber, opposite the nozzle is at o-point. Moving to the left is in the $-X$ axis direction, and moving to the right is in the $+X$ axis direction. The calculation results are shown in Fig. 2.

As can be seen from Fig. 2, when area ratio a is in the range of 1.96–3.24, and throat spacing L_e is in the range of 0.2–0.6 D_h, agent absorption performance of the jet mixing apparatus is ideal. The position of the suction tube has little influence on the agent absorption performance of jet-mixing apparatus. In terms of the agent absorption performance, the ideal position is the region of the nozzle exit $1/4–1/2L_z$ (L_z is the nozzle length).

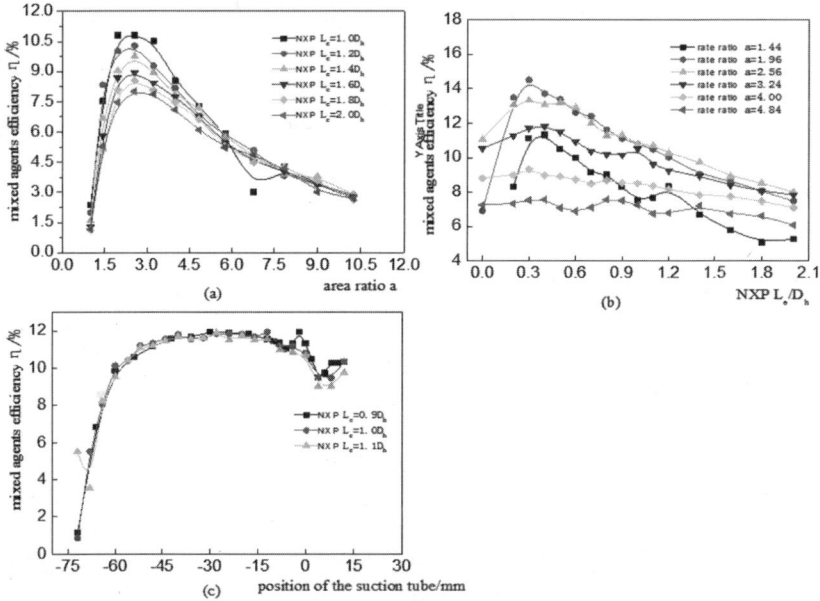

Figure 2: Influence of structural parameters on the agent absorption performance

5.2 Influence of structural parameters on flow field

The axial length coordinates of the numerical simulation calculation are $x \in$ (0 m, L_e + 0.22 m), and the axial coordinates of the nozzle exit $x = 0$ m. In the precondition of the structural parameter with the best agent absorption performance, test distribution characteristics of flow field in jet apparatus in the condition of the area ratio $a = 1.96$, 2.56, and 3.24, the NXP $L_e = 0.2$, 0.3, 0.4, 0.5, and $0.6D_h$. The area ratio $a = 2.56$, and the NXP $L_e = 1.0D_h$. The tube spacing is set in different positions in the mixing chamber, and the distance of the nozzle is 12 mm, 0 mm, −20 mm, −40 mm, −60 mm, −72 mm. Under the above parameters, the influence of the convection field is examined.

5.2.1 Analysis of pressure field

As shown in Fig. 3, in the contraction section of the nozzle, because the pressure energy of the working fluid can be converted into kinetic energy, the flow rate of the fluid reaches maximum, and static pressure reaches the maximum gradient pressure drop. There is the pressure nadir at the inlet of the tube. The smaller the area ratio is, the larger the flow velocity is in the nozzle, the less the static pressure is. From the nozzle outlet to the inlet of the throat tube, the flow rate of the mixed fluid decreases and the static pressure gradually increases. In the throat tube, the working fluid and the ejection fluid are initially mixed violently, and then tend to be uniform. The static pressure rises first and then stabilizes. The smaller the area ratio is, the smaller the static pressure is in the throat tube. In diffusion tube, the static pressure increases along the axis, flattening at the outlet. The static pressure of the outlet of different area ratio of jet-mixing apparatus tends to be consistent. This indicates that the length of the diffusion tube is reasonable. When the area ratio is the same, the NXP L_e is changed in the range of 0.2–$0.6D_h$, and the static pressure axial distribution tends to consistent and overlaps. This indicates that the NXP L_e has little influence on static pressure in the jet-mixing apparatus.

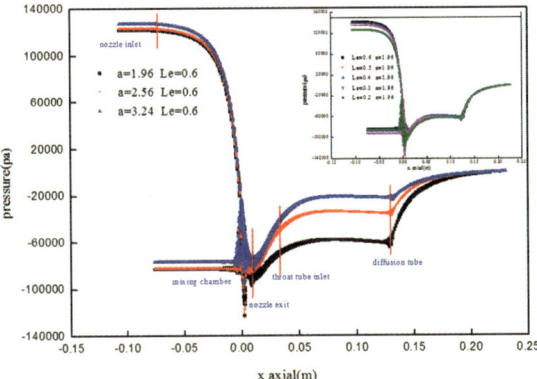

Figure 3: Axial distribution diagram of the pressure under different area ratio and NXP

5.2.2 Analysis of velocity field

As can be seen from Fig. 4, the effect of area ratio on center speed of flow is more obvious. The greater the area ratio is, the faster the decay of the center speeds is. The change of throat spacing is not very important to the decay of the center speeds, but the NXP is too small $L_e = 0.2D_h$, and the flow core remains at a very long distance. It is obviously unfavorable to the mixing of materials. From the symmetric speed distribution of driving fluid along the center line, the effect is obvious which the absorbent tube is arranged symmetrically.

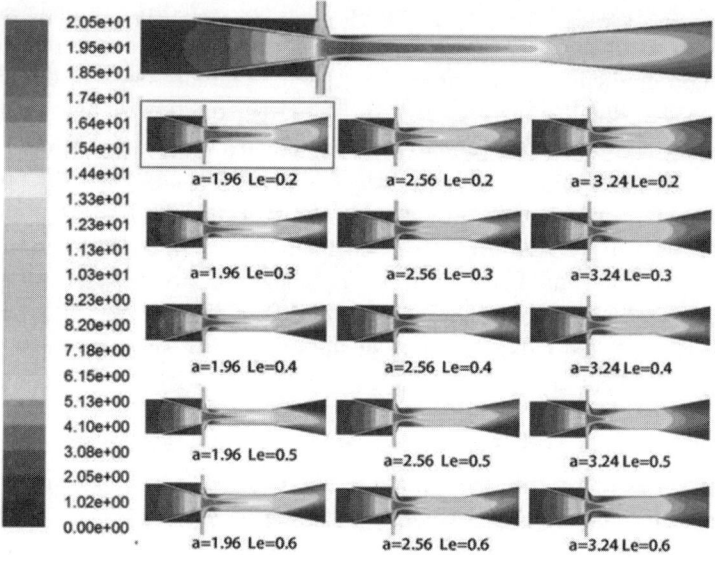

Figure 4: Axial speed cloud graph under different area ratio and NXP

5.2.3 Distribution of turbulence intensity

According to Fig. 5(1) and (2), when the suction tube is opposite to the jet flow, turbulence of throat inlet and 1/5 section increases along the radial firstly, and then decreases. The turbulence of the radial points of the other sections tends to be consistent. The closer the outlet of the throat tube is, the greater the turbulence is. This indicates that the momentum exchange is completed in the throat. In Fig. 5(3)–(6), the turbulence of each section presents a parabolic shape. The origin of the parabola of the throat tube inlet to 3/5 section is all at zero. This indicates that the flow core of the fluid is constant to the 3/5 section. Considering the turbulence distribution, the absorption tube should be opposite to the flow core zone.

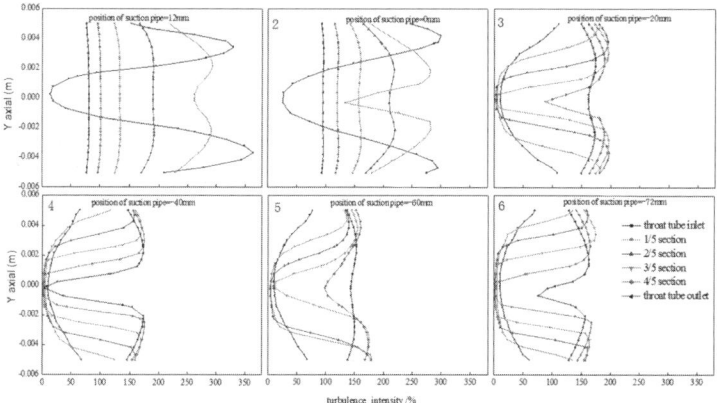

Figure 5: Turbulence intensity distribution of the radial section of the throat in the position parameters of different suction tubes

6. Test results and discussion

According to the model structure, four sets of jet devices, where area ratio a = 1.44, 3.24, 5.76 and 9.00, respectively, were produced. In each group, the throat tube diameter and the NXP were (D_h = 12 mm and L_e = 12 mm), (D_h = 18 mm and L_e = 18 mm), (D_h = 24 mm and L_e = 24 mm), and (D_h = 30 mm and L_e = 30 mm), and the experimental system was set up as shown in Fig. 6. Compressed air with pressure P = 0.08 MPa was used as working fluid. The driving fluid was n-dodecane, and flow rate Q = 24 L/h. The agent was dispersed by the jet device 7 and then formed a droplet. Then it was introduced into the container 9 through the communicating pipe 8. Used the Winner312 laser particle size analyzer, the droplet diameter test was carried out on the droplets in container 9. The test results are shown in Fig. 7.

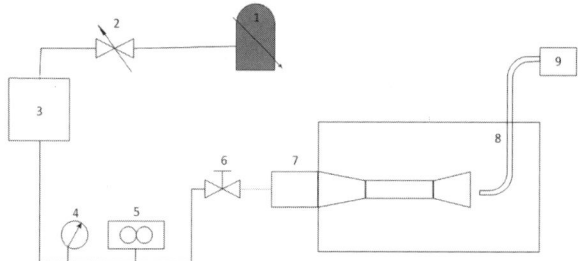

Figure 6: Droplet diameter test system. 1, Air compressor; 2, control valve; 3, air dome; 4, pressure gage; 5, flowmeter; 6, high-pressure pipe; 7, jet device; 8, communicating pipe; and 9, container

As can be seen from Fig. 7, the dispersion ability of the agents is consistent in the four sets of jet devices with different area ratio. But the dispersion effect is the best when area ratio $a = 3.24$, and the average diameter of the droplets decreases. Among them, the cumulative production rate of the particle size of 7.3 μm, 8.9 μm, 13.1 μm, and 20.1 μm is 50%, 70%, 80%, and 90%, respectively. The diameter of all droplets is less than 38 μm.

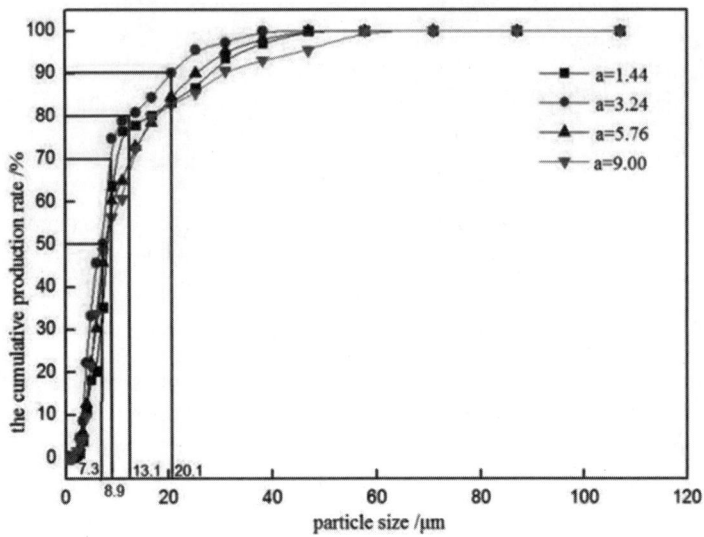

Figure 7: Result of droplet diameter test experiment

7. Conclusion

(1) Considering the agent absorption performance, the optimum range of the area ratio a of the jet device is 1.96–3.24, and the optimal value of NXP L_e is 0.2–$0.6D_h$. The position of the suction tube has little effect on the agent absorption performance of jet mixing apparatus. The ideal location is the area of nozzle exit on the left side of the $1/4$–$1/2L_z$

(2) According to means of flow field in the characteristics, the area ratio has a more significant impact. Along the direction of the jet flow, the larger the area ratio is, the greater the turbulence intensity is around the flow core, the faster the pressure drops in the throat tube, the more rapidly flow core disappears, and the better the mixing of the working fluid and the driving fluid are. The NXP has smaller influence on the distribution of the flow field. When the NXP $\leq 0.2D_h$

and the flow cores of a stream is longer, it is unfavorable to the fluid mixing. Therefore, the optimal value of NXP range is 0.3–0.6D_h. The layout that the suction tube is opposite to the nozzle exit is ideal. It is advantageous to complete kinetic energy exchange between working fluid and driving fluid in the throat tube.

(3) It is known from the test experiment of droplet diameter test that the dispersion effect is the best when area ratio a = 3.24, that is, the optimal range of area ratio is 1.96–3.24. When a = 3.24, the distribution characteristics of the flow field are ideal. This situation provides the optimal conditions for the dispersion of the driving fluid in the working fluid.

8. Acknowledgement

We sincerely thank the Natural Science Foundation of China (51374015) and Natural science Foundation of Anhui province (1708085QE128 and 1608085QE12), especially Prof. Zhu Jinbo from Anhui University of Science and Technology, thanks should also be given to all undergraduate students in the laboratory.

9. References

[1] Chen W Y, Wang J B, Jiang N, et al. Numerical simulation of gas–liquid two-phase jet flow in air-bubble generator. Journal of Central South University of Technology, 2008, 15(1):140–144.

[2] Deng X W, Liu J T, Wang Y T, et al. Velocity distribution of the flow field in the cyclonic zone of cyclone-static micro-bubble flotation column. International Journal of Mining Science and Technology, 2013, 23(1):89–94.

[3] Dorpmund M, Cai X, Walgenbach M, et al. The challenge of cleaning direct-injection systems for pesticide application. Transactions of the ASABE, 2012, 55(5):1643–1650.

[4] Gong M M, Li C S, Li Z X. Numerical analysis of flow in a highly efficient flotation column. Asia-Pacific Journal of Chemical Engineering, 2015, 10(1):84–95.

[5] Güney A, Önal G, Ömer Ergut. Beneficiation of fine coal by using the free jet flotation system. Fuel Processing Technology, 2002, 75(2):141–150.

[6] Hacifazlioglu H. Enrichment of Silica Sand Ore by Cyclojet Flotation Cell. Separation Science and Technology, 2014, 49(10):1623–1632.

[7] Li L, Liu J T, Wang L J, et al. Numerical simulation of a self-absorbing microbubble generator for a cyclonic-static microbubble flotation column. International Journal of Mining Science and Technology, 2010, 20(1):88–92.

[8] Ma J, Qiu B J, Yan R, et al. Study on the working state of jet-mixing apparatus. Applied Mechanics and Materials, 2014, 563:219–223.

[9] Ni C, Jin M, Chen Y, et al. Improving the recovery of coarse-coal particles by adding premineralization prior to column flotation. International Journal of Coal Preparation and Utilization, 2016, 37(2):87–99.

[10] Shen X J, Chen Q R, Li S Y. Velocity and pressure distributions characteristics of coal slurry in floatation cyclone. 2009, 1(1):814–818.

[11] Salemsaid A H, Fayed H, Ragab S. Numerical simulations of two-phase flow in a Dorr–Oliver flotation cell model. Minerals, 2013, 3(3):284–303.

[12] Taşdemir A, Taşdemir T, Öteyaka B. The effect of particle size and some operating parameters in the separation tank and the downcomer on the Jameson cell recovery. Minerals Engineering, 2007, 20(15):1331–1336.

[13] Tasdemir T, Tasdemir A, Oteyaka B. Gas entrainment rate and flow characterization in downcomer of a Jameson cell. Fizykochemiczne Problemy Mineralurgii – Physicochemical Problems of Mineral Processing, 2011, 47:61–78.

[14] Vondricka J. Study on the process of direct nozzle injection for real-time site-specific pesticide application. Bonn: University of Bonn, 2007.

[15] Wang A, Yan X, Wang L, et al. Effect of cone angles on single-phase flow of a laboratory cyclonic-static micro-bubble flotation column: PIV measurement and CFD simulations. Separation and Purification Technology, 2015a, 149:308–314.

[16] Wang L J, Jia Y, Yan X K, et al. Gas–liquid numerical simulation on micro-bubble generator and optimization on the nozzle-to-throat spacing. Asia-Pacific Journal of Chemical Engineering, 2015b, 10(6):893–903.

[17] Xu X C, Qiu B J, Deng B, et al. Analysis on online mixing performance of jet-mixing apparatus. Advanced Materials Research, 2012, 347–353:417–421.

[18] Yang R Q, Wang H F, Liu J C. Jet flotation column system structure design and numerical simulation of two-phase flow. Advanced Materials Research, 2013, 616–618:655–661.

13

Creation of flotation machines based on research patterns of coal pulp aeration

Antipenko Lina Alexandrivna

LLC "Sibniiugleobogashchenie", Prokopevsk, Russia

Abstract: The increasing degree of complex mechanization of mining causes an increase in the ash content of mined coal and in the content of crushed rock mass in it. In this regard, the value of the coal fines cleaning by flotation increases.

- The principle of operation of modern flotation machines is based on the use of pulp aeration process, carried out in various ways:
- mechanical dispersion of air by aerators of various designs operating in air exhausting or air injection mode;
- air supply through porous stoppings (perforated rubber, porous ceramics, etc.);
- airlift, which represents the collapse of air jets when lifting in liquid media;
- hydraulic, occurring when the supply of free jets to the surface of liquid media;
- pneumohydraulic, consisting in the joint supply of air and fluid in hydraulic devices.

Of particular interest is the aeration of liquid media with free jets as one of the aeration processes, allowing to intensify the flotation process. Studies show that the use of aeration with free jets in the flotation process is energetically promising. With aeration of liquid media by mechanical aerators, the achieved degree of aeration is 1 m3 of air per 1 m3 of pulp pumped by the aerator, and with aeration with free jets – 1 m3 of liquid can disperse up to 10 m3 of air, i.e. aerating capacity of free jets is much higher.

Keywords: Flotation machine, air dispersion, aeration of liquid media, free air jets

Improving the quality of coal is one of the main areas of development of the coal mining industry in the Russian Federation.

An important task in coal washing at the present time is to improve the quality of commercial products and reduce the loss of raw coal with coal washing waste.

At the coal preparation plants built after 2000 there is no flotation process for the sludge wash in which the solid particles are less than 0.5 mm.

Efficient operation of the processing plants is possible with a radical revision of the technical policy in the field of coal flotation – the introduction of low-flow, low-operational technological circuits, and the use of flotation equipment of high unit productivity.

Flotation machines are complex technological devices operating in several different functions simultaneously: providing a certain dispersion of the gas environment, creating an optimal contact surface for the interacting phases and forming stable air–mineral complexes that are elements of aeration processes.

Of particular interest is the aeration of liquid media with surface jets, which will increase the aeration capacity of flotation machines 10 times higher compared to machines of a mechanical type.

A study of the formation of air bubbles when a droplet strikes a liquid surface by several researchers has shown that a hemispheric cavern and a cylindrical wave along the perimeter of the cavern occur at the impact site [6,9,13]. An air film separating the droplet from the liquid remains between the surface of the drop and the bottom of the cavity for tens of milliseconds. At a certain critical speed ($u \sim 5$ m/s), a thin film of liquid under the action of surface tension forces closes above the cavity and forms an air bubble. The depth of the cavity, which determines the amount of trapped air, depends on the kinetic energy of the drop.

When the jets hit the surface of the liquid, as well as when the drops hit, a cavity is formed, which is a source of aeration of the near-surface layer of the liquid. Studies of fluid aeration with laminar free jets (Re < 1500) have established that a thin film of air forms at the point of jet entry into the fluid, which collapses in the cavity zone, forming bubbles [11]. The value of the critical rate V_{cr}, at which aeration begins, is determined by the diameter of the jet. When the jet diameter changes from 2.2 to 9.5 mm, V_{cr} decreases from 1.25 to 0.75 m/s.

With a jet diameter of 6 mm and $V_{cr} = 0.95$ m/s, the diameter of the bubbles formed when the laminar jet falls is 2.4 mm; at a rate of 2 V_{cr}, the diameter of the bubbles is 0.9 mm. However, it should be noted that the aerating capacity of laminar jets is insufficient for intensive process management.

Turbulent jets (Re > 1500), whose air trapping mechanism is different from that of the laminar ones, have a much greater aerating ability. This is because the cross section of turbulent jets is not constant. At the entrance of the deformed portion of it into the liquid, a funnel is formed. The thickening of the jet is absorbed, and the upper edge of the funnel is closed with the jet, forming a closed goidal space filled with air, which breaks down into small bubbles. According to [11], the flow rate of ejected air can be determined for turbulent vertical jets by the equation:

$$\frac{Qr \cdot X}{Q\text{ж}} = \left[\frac{dc \cdot X}{d\text{н}} \right] \tag{1}$$

where Q_r is the ejected air jet, m³/s; $Q_ж$ is the fluid flow rate, m³/s; d_c is the maximum jet diameter, m; $d_н$ is the nozzle diameter, m; and X is the jet length, m.

One of the most important parameters characterizing the aeration capacity of free jets is the depth of a two-phase "torch". Theoretical and experimental studies of the depth of a two-phase "torch" formed by turbulent round vertical jets for aeration were described in [10,13] and an equation was proposed that allows determining the depth of the "torch" depending on the operating parameters:

$$Z = \frac{1}{2tg(L/2)} \cdot \frac{Vc \cdot dH}{Vn} + Zo \tag{2}$$

where Z is the depth of the two-phase "torch", m; t_g is the "torch" taper angle, degree; V_c is the jet speed at the exit out of the nozzle, m/s; and $d_н$ is the distance from the surface to the top of a two-phase "torch", m.

The angle at which the jet falls on the surface of the liquid has a significant effect on the capture of air by a free jet. By the method of high-speed filming, it was found that when a jet falls at an angle L, a horizontal flow occurs in the surface layer of the liquid [12]. On the one hand, the edge of the cavern funnel runs onto the jet, on the other hand, there is an outflow of fluid. The free edge periodically interlocks with a jet that entrains air, which is subjected to fragmentation into small bubbles.

The amount of air Q_b ejected by the inclined jet, according to the model proposed in [12], is determined by the following relation:

$$Q_b = f(L, ?, P, G) \cdot V_y \cdot D^z \cdot (\sin L)^{1/3} \tag{3}$$

where L is the jet length, m; $?$ is the fluid viscosity, cp; P is the fluid density, kg/m³; D is the jet diameter, m; L is the angle of incidence of the jet, degrees; V is the jet flow rate, m/s; $y \approx 2.5$; and $z \approx 1.5 \div 2.5$.

The most important characteristic of liquids aeration for the flotation process is the dispersion of bubbles. In water, the average diameter of the bubbles varies from 1.3 to 3.6 mm. For an approximate calculation of the average particle size of the bubbles, the equation given in reference [7] can be used:

$$d_в = 4.34 \sqrt[3]{Qb/Qж} \tag{4}$$

where Q_b and $Q_ж$ are the volumetric flow rate of air and liquid, m³/s.

Reducing the surface tension of water by 20% leads to a decrease in the average particle size of the bubbles by 10%.

The shape of the free jet, determined by the geometry of the nozzle, significantly affects aerating and dispersing properties. Aerated volume of liquid with a flat jet is 3–4 times greater than with aeration with ainclined jet. Experimental studies of aerating and dispersing properties of free jets were carried out according to the previously developed technique [4].

Research on aeration with free jets was envisaged at the stand installation, developed based on the elements of the existing continuous flotation stand. The scheme of the stand installation is shown in Fig. 1.

The stand installation consists of the following elements:

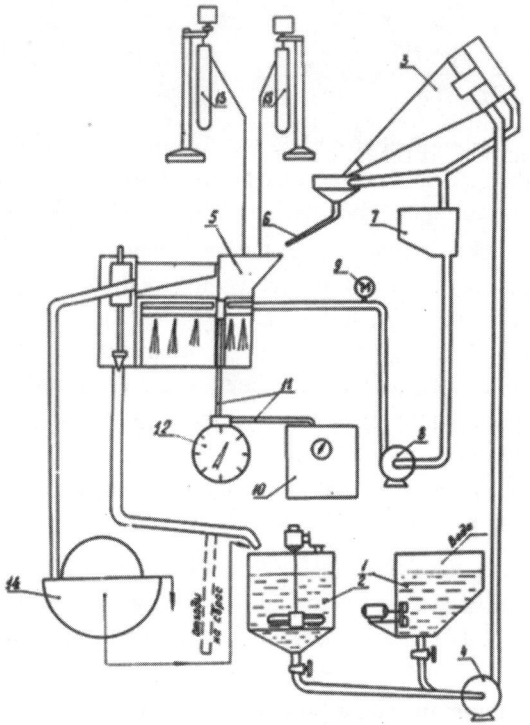

Figure 1: Scheme of the stand installation. 1 and 2, mixing tanks with a capacity of 20 and 40 dm³, respectively; 3, laboratory model of a hydrocyclone; 4, centrifugal pump; 5, laboratory model of flotation jet aeration; 6, pulp dispenser; 7, intermediate tank; 8, centrifugal pump; 9, pressure gauge; 10, compressor; 11, air ducts; 12, gas meter; 13, reagent feeders; and 14, vacuum filter.

During the development of the stand installation, it was possible to conduct experiments to study the aeration of free jets in both clean water and coal suspension.

The outflow of fluid through a round hole in a thin wall and cylindrical nozzles is accompanied by the jet compression.

Volumetric flow of fluid through the hole or nozzle in the vessel wall is determined by the formula in reference [3].

$$Q = U_{cnc} \cdot E \cdot F_0 = Y \cdot E \cdot F_0 \frac{\sqrt{2qH\text{uст}}}{\sqrt{1 - N_1(E \cdot Y \cdot F_0/F_1)^2}}$$

$$= M_0 \cdot F_0 \cdot \frac{\sqrt{2qH\text{uст}}}{\sqrt{1 - N_1(M_0 \cdot F_0/F_1)^2}}$$

$$= M_0 F_0 \sqrt{2q\Delta P/Y} \qquad (5)$$

where U_{cnc} is the fluid flow rate in a compressed jet section, m/s; E is the jet compression ratio; F_0 is the hole area, m²; $H\text{uст}$ is the total outflow pressure, kg/m²; N_1 is the kinetic energy coefficient; $M = EY$ is the outflow coefficient of the hole (nozzle); and ΔP is the pressure differential kg/m².

The flow coefficient is influenced by the same factors that affect the speed and compression ratios. In general, these coefficients depend on the Reynolds, Froude, and Weber numbers. In practice at the outflow of fluid through the holes or nozzles the Froude and Weber numbers can be neglected and considered as E and Y, and therefore M as a function of the number Re. For values of Reynolds number $(2.5 \div 4.0) \cdot 10^3$, the outflow coefficient takes the values of $0.65 \div 0.66$.

Water flow through the nozzles was determined depending on the pressure created by the pump. During the experiments, the pressure was changed within $1.0 \div 1.6$ kg/cm². To measure the pressure, a pressure gauge of OBMI-160 type, accuracy class 1.5, GOST 2405-72 was used.

Three types of jet sources were tested: a hole ø 3 mm in a thick-walled pipe, a nozzle cylindrical with a channel of ø 2.5 mm and a nozzle with an outlet section of rectangular section 3.5 × 1.5 mm².

Table 1 shows the results of measuring the flow rate at the outflow of water for the studied sources of free jets.

Table 1: Measuring results

Type of jet source	Pressure drop ΔP, kg/s m²						
	1.0	1.0	1.2	1.3	1.4	1.5	1.6
	Liquidflow, dm³/m						
Hole ø 2.5 mm	3.2	3.4	3.5	3.6	3.7	3.8	4.0
Nozzle ø 3.0 mm	4.2	4.5	4.7	4.9	5.0	5.1	5.2
Nozzle 3.5 × 1.5mm²	4.3	4.4	4.6	4.8	5.0	5.1	5.2

The aeration capacity of free jets was estimated by a value equal to the ratio of the air flow captured by the jet to the flow rate of water supplied as a free jet in the aeration chamber.

Air flow during aeration by jets was determined using a laboratory-type RDS rheometer, GOST 9932-75.

Since the diameter and depth of the cavity were determined by the pressure at the source of the jet, the air flow rate was selected so that the possibility of large air bubbles breaking into the chamber was excluded.

The flow rate of air ejected by the free jet was determined by the formula:

$$Q_\text{в} = Q_1 \cdot \sqrt{\frac{P_\text{ж}}{P_\text{в}}} \cdot \sqrt{1 + \frac{\Delta P}{B_1}} \tag{6}$$

where Q_1 is the air flow rate according to the reading of the rheometer, l/min; $P_\text{ж}$ is the density of the liquid used for filling the rheometer, kg/m³; $P_\text{в}$ is the density of water, kg/m³; B_1 is the barometric pressure at which performance graduation of the rheometer, mmHg; and ΔP is the overpressure of the air in the aeration chamber, mmHg.

The results of the study of the dependence of the air flow rate on the liquid flow when using a hole of ø 2.5 mm as a jet source are shown in Table 2.

Table 2: Results of dependence studies

Pressure drop ΔP, kg/s m²	1.0	1.1	1.3	1.5	1.6
Liquid flow $Q_\text{ж}$, l/min	3.2	3.4	3.6	3.8	4.0
Air flow $Q_\text{в}$, l/min	21.5	23.5	28.0	34.0	35.5

When changing the pressure drop in the source of a free circular jet from 1.0 kg/cm² to 1.6 kg/cm², the ratio of air and liquid flow rates $\dfrac{Q_\text{в}}{Q_\text{ж}}$ Qv/Qzh was 6.7 ÷ 8.9.

Table 3 presents data characterizing the aeration capacity of a free flat jet obtained by a nozzle with an outlet of rectangular cross section.

Table 3: Free jet aeration capacity

Pressure drop ΔP, kg/s m²	1.0	1.2	1.4	1.6
Liquid flow $Q_\text{ж}$, l/min	4.3	4.6	5.0	5.2
Air flow $Q_\text{в}$, l/min	19.5	28.0	34.0	35.5

The coefficient of ejection is equal to the ratio of air flow and liquid flow $Qv/Qzh = 4.5 ÷ 6.8$ for a flat jet. This is because the angle of expansion of a circular jet is greater than the angle of expansion of a flat jet. At the same time, it follows that the penetration range of a two-phase torch of a flat jet is much greater than the "long range" of a circular jet.

The dispersing ability of free jets was estimated by the size of air bubbles, the diameter of which was determined when processing photographs of aerated liquid in a transparent chamber. Figures 2 and 3 show air bubbles obtained in pure water without surfactants and COBS foaming agent.

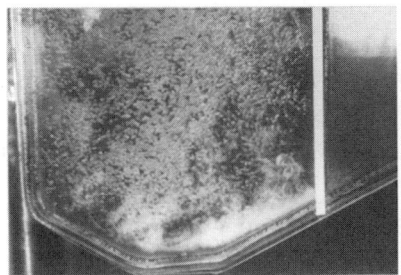 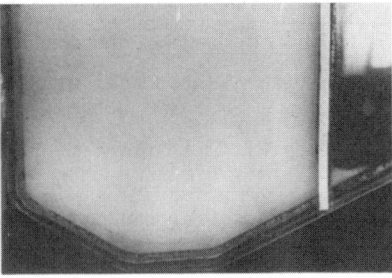

Figure 2   **Figure 3**

The use of bottoms from the production of butyl alcohols (COBS) as a foaming agent at optimum flotation concentrations made it possible to obtain air bubbles, the average size of which did not exceed 1 mm.

Studies [5,8] found out that the dispersion of air bubbles without surfactants during mechanical dispersion is 3–4 mm, and in the presence of surfactants – $0.9 \div 1.0$. The analysis shows that the dispersing ability of free jets and mechanical aerators are of the same order.

For the studied sources of free jets, the ejection coefficient, characterizing their aeration capacity, varies in the range of $4.5 \div 8.9$, which is higher than the degree of aeration by mechanical impellers.

The use of foaming agent (COBS) allows you to get air bubbles of sufficiently high flotation activity, the size of which does not exceed 1 mm.

To intensify the flotation process, it is advisable to use free flat jets with high aerating and dispersing abilities and considerable "long range".

Based on studies of the regularities of coal pulp aeration, a pneumatic flotation machine was designed with elements of jet aeration, which has been tested at the CPP "Sibir".

At present, based on the results of the tests and the operating experience of the pneumatic-type flotation machines, an updated design has been proposed and a patent has been obtained for the invention [1,2].

References

[1] Antipenko L.A. Pneumatic flotation machine with elements of jet aeration: Patent No. 2634313 Russian Federation, IPC B03D 1/14 (2006.01), IPC B03D 1/02 (2006.01)/L.A. Antipenko, A.M. Bulaeva, A.E. Kravchenko, N.G. Sarin; Applicant and patent holder LLC "Sibniiugleobogashchenie". № 2016139065; Appl. 10/4/2016; publ. 03/06/2017, Bull.№ 7.

[2] Antipenko L.A. Flotation machine with elements of jet aeration MFU – CA.1. Mining Information Analytical Bulletin (scientific and technical journal) (GIAB). M.: GornayaKniga, 2012. Separate issue No. 5. P. 167–178.

[3] Idelchik I.E. Handbook of hydraulic resistance. M.-L.: Gosenergoizdat, 1960.

[4] Investigation of the regularities of pulp aeration with surface jets to intensify the flotation process: Methodology. Kuzniiuglej, jgazhenije – Prokopyevsk, 1985.

[5] Klassen V.I., Mokrousov V.A. Introduction to the theory of flotation. M.: GITI, 1959.

[6] Meshcheryakov A.F. Flotation machines and devices. M.: Nedra, 1982.

[7] Some regularities of aeration of a liquid with jets/A.F. Meshcheryakov, Yu.V. Ryabov, M.A. Podvigin and others. In the book. Improvement of coal washing processes and facilities at the enterprises of the mining and chemical industry. Edited by Dr. of Technical Sciences A.F. Meshcheryakov. M., 1981.

[8] Perepelkin, K.E., Matveev, V.S. Gas emulsions. L.: Chemistry, 1979.

[9] Engel O.G. Aeronautical Systems Division Report No WaDD-TR-GO-475, Part II, 1962.

[10] Henderson J.B., Mc Garthy M.S., Molloy N.A. Entrainment plunging jets. Chemica 70. A conference convened by the Austral Nat Comm of Institution of Chemical End and the Austral Academy of Sci., 1970.

[11] Tong Yoe Lin, Harold G. Donnely. Gas bubble entrainment by plunging laminar liquid jets. AIChE, vol 12, № 3, 1966.

[12] Van de Sand E., Smith J.W. Eintragen von Luft in line Flussigkertdureheinen Wasserstrall ChemieJnd Tech, 44 Jnhrd 1972, № 20, 1177–1183.

[13] Van de Sande E., Smith J.W., van Oerd J.J.J. Energy transfer and cavity formation. J. Appl. Phys. vol 45, № 2, Feb. 1974.

14

Characteristics and mechanism research on hydrophobic aggregation of kaolinite particles

Chen Jun, Min Fan-fei, Liu Ling-yun

Anhui University of Science and Technology, Huainan Shi, China

Abstract: The adsorption of four different amine/ammonium salts of DDA (dodecylamine), MDA (N-methyldodecylamine), DMDA (N,N-dimethyldodecylamine) and DTAC (dodecyl trimethyl ammonium chloride) on kaolinite was investigated in the study through the measurement of contact angles, zeta potentials, aggregation observation, adsorption, and sedimentation. The results show that different amine/ammonium salts can adsorb on the particle surface of kaolinite to enhance the hydrophobicity and reduce the electronegativity of particle surface, and thus induce a strong hydrophobic aggregation of particles which promotes the settlement of kaolinite. To explore the adsorption mechanism of these four amine/ammonium salts on clay mineral particle surfaces in coal slurry water, the adsorptions of DDA+, MDA+, DMDA+ and DTAC+ on the surface of kaolinite are calculated with DFT (density functional theory). The DFT calculation results indicate that different amine/ammonium cations can strongly adsorbed on t kaolinite surface by forming N–H···O strong hydrogen bonds or C–H···O weak hydrogen bonds, and there are strongly electrostatic attractions between different amine/ ammonium cations and the kaolinite surfaces. The main adsorption mechanism of amine/ammonium cations on kaolinite surface is hydrogen-bond interaction and electrostatic attraction, and the electrostatic attraction plays the main role in different amine/ammonium cations adsorption on kaolinite.

Keywords: Coal slurry water, hydrophobic aggregation, kaolinite, amine/ ammonium salts, density functional theory, adsorption mechanism, sedimentation

1. Introduction

Kaolinite is the main clay in coal slurry water, it can easily turn into ultra-fine particles during coal preparation process and stably disperse in coal slurry water [1], which are serious and increase the difficulty of coal slurry water sedimentation. Hydrophobic aggregation is an effective process for kaolinite sedimentation [2,3]. Cui et al. [4] indicated that one of the important premises to achieve the effective aggregation of kaolinite is to adjust its surface properties. For the hydrophilic kaolinite particles, hydrophobic aggregation could be realised through rendering surface hydrophobicity by the adsorption of surfactants on the surfaces so as to reduce hydration repulsive force and enhance hydrophobic attraction [5,6], which can promote the effective settlement of fine kaolinite particles in aqueous solution. Cationic surfactants such as alkyl amine [7] and quaternary ammonium salts [8] were investigated and proved to be effective surfactants for the hydrophobic aggregation of kaolinite.

However, although there are lots of researches on hydrophobic aggregation behaviour of kaolinite particles, few of them are about its micro mechanism of kaolinite surfactants [7,8]. Therefore, it is great significance to study the adsorption mechanism of kaolinite surfactants. Generally, computer simulations have been widely used for the study of the surfactant adsorption on a solid surface recently [9,10]. The quantum chemical approach is more used to understand the atomic structure of adsorbed surfactant and its electronic structure when the molecular adsorption on clay surfaces is studied, and DFT (density functional theory) method find its extensive application in this fields recently [11,12].

The combined method of DFT calculation and experimental verification was used to lucubrate the characteristics and mechanism on hydrophobic aggregation of kaolinite particles in this paper, which can provide a theoretical basis to develop new technologies and design new drugs of sedimentation and clarification for coal slurry water.

2. Materials and methods

2.1 Materials

Kaolinite samples used in this study were originally collected from the Huaibei Jinyan Kaolinite Company (China). Figure 1 shows the particle size distribution and X-ray powder diffraction (XRD) pattern of the kaolinite. The d_{50} (average particle size) of the kaolinite sample was 2.004 μm. And there are three types of minerals in this sample, kaolinite-1 MD, kaolinite-1A and dickite-2 M1 belonging to kaolinite family, indicating that the kaolinite sample is a high-purity one. The specific surface area of the sample was 8.04 m^2/g, which was obtained by using a V-Sorb 2800P gas adsorption analyzer (China) based on the Brunauer–Emmett–Teller (BET) method.

DDA (dodecylamine, $CH_3(CH_2)_{11}NH_2$), MDA (N-methyldodecylamine, $CH_3(CH_2)_{11}NH–CH_3$), DMDA(N,N-dimethyldodecylamine, $CH_3(CH_2)_{11}NH–(CH_3)_2$) and DTAC(dodecyl trimethyl ammonium chloride, $CH_3(CH_2)_{11}NH–(CH_3)_3Cl$) used in this study were from Shanghai Sinopharm Chemical Reagent Co., Ltd, China. pH value of the aqueous solutions was adjusted by adding hydrochloric or sodium hydroxide solutions. All of electrolytes are analytical reagents. The water used in this study was first distilled and then passed through resin beds and a filter (0.2 μm). The residual conductivity of the water was less than 1 μs/cm.

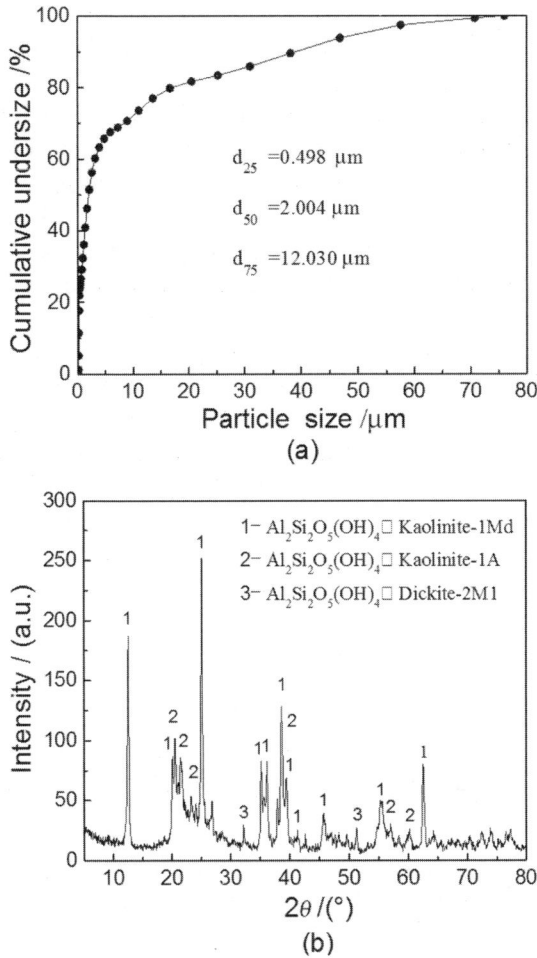

Figure 1: Particle size distribution (a) and XRD pattern of the kaolinite sample

2.2 Experimental methods

The specific measurement methods of zeta potential, contact angle, adsorption, sedimentation and aggregation observation according to the process detailed by Chen et al. [13]. After sedimentation test, the microstructure of the aggregations was observed on a glass slide by using HSA10 Monocular Zoom Microscope when the hydrophobic aggregation became stable.

2.3 DFT calculation

DFT (density functional theory) calculations of different amine/ammonium cations on kaolinite (001) surface and (00$\bar{1}$) surface were implemented in the CASTEP program Materials Studio version 8.0 software, BIOVIA Corporation [14]. And the other specific parameter settings, computational method and the calculation of adsorption energies according to the process detailed by Chen et al. [15]. The optimised cell parameters of kaolinite with the unit-cell formula of $Al_2Si_2O_5(OH)_4$ were a = 5.196 Å, b = 9.007 Å, c = 7.372 Å, and α = 93.029°, β = 105.983°, γ = 89.866, which is consistent with the simulation optimisation values [16] and experimental values [17], indicating that the calculation results are reasonable. The periodic supercells (2×1×1) of kaolinite (001) surface and (00$\bar{1}$) surface of the platelet were used with a vacuum thickness of 40 Å to prevent the interaction between the adjacent levels. The optimisations of different amine/ammonium cations were calculated in a 20×20×30 Å^3 cubic box with Brillouin zone sampling restricted to Gamma point, and the optimisation parameters were the same for primitive unit cell optimisation. Figure 2 shows the optimised geometry for configurations of different amine/ammonium cations. According the surface symmetry, three hollow adsorption sites of the single amine/ammonium cation on kaolinite (001) surface and (00$\bar{1}$) surface were examined, as shown in Fig. 3.

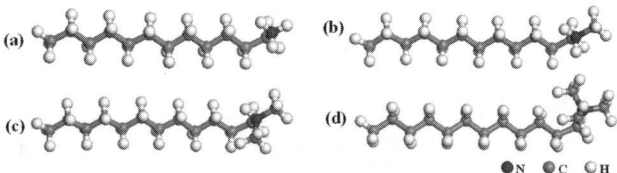

Figure 2: The optimised geometries of different amine/ammonium cations.
(a) DDA$^+$; (b) MDA$^+$; (c) DMDA$^+$; (d) DTAC$^+$

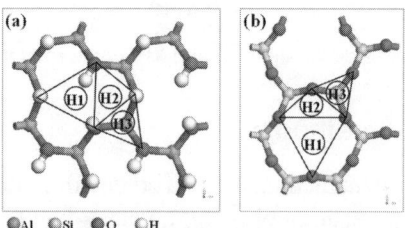

Figure 3: The initial adsorption sites of different amine/ammonium cations on kaolinite (001) surface (a) and (00$\bar{1}$) surface (b), where H1, H2, and H3 in the circle denoted the sites of hollow 1, hollow 2, and hollow 3, respectively

3. Results and discussion

3.1 Contact angles, zeta potentials and aggregation observation

Figure 4 shows contact angles and zeta potential of kaolinite particles as the function of reagent concentration. The contact angles of kaolinite samples were increased with the increase of reagent dosage of different amine/ ammonium salts, and the hydrophobic modification abilities of this four salts for kaolinite fell in the order of DTAC > DDA > DMDA > MDA. And the zeta potential increased with the increase of reagent dosage of different amine/ ammonium salts. Figure 5 shows the microscopic images of the aggregated kaolinite particles with reagent concentration of 6×10^{-4} mol/L at pH 7.0. The sizes of the aggregations fell in the order of DTAC > DDA > DMDA > MDA, which indicated that the aggregating abilities of this quaternary ammonium salts for coal slurry fell in the order of DTAC > DDA > DMDA > MDA. The sizes of the aggregations were about 0.5–1 mm, which were 250–500 times of the average particle size (see Fig. 1) of kaolinite particles.

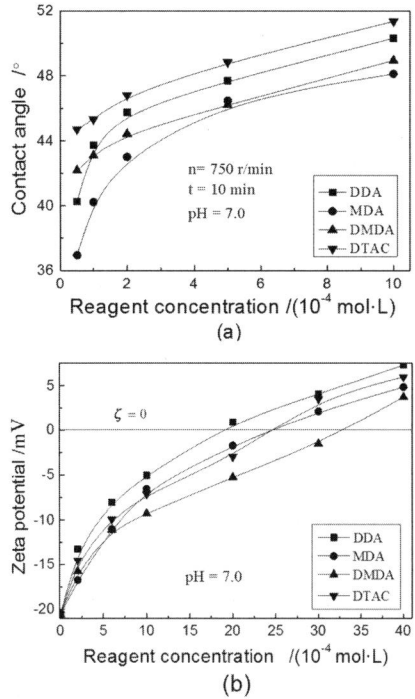

Figure 4: Contact angles and zeta potential of kaolinite particles as the function of reagent concentration

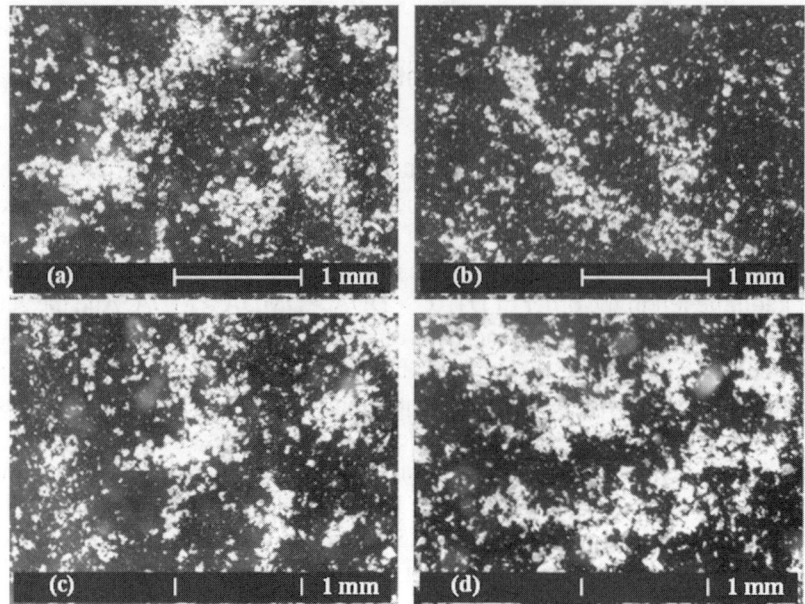

Figure 5: Microscopic images of aggregated kaolinite particles with reagent concentration of 6×10^{-4} mol/L at pH 7.0. a, DDA; b, MDA; c, DMDA; d, DTAC

These results show that these four kinds of different amine/ammonium salts can enhance the hydrophobicity and reduce the electronegativity of kaolinite particle surfaces, and promote the fine particles to form aggregation. This aggregation closely correlated with the adsorption of different amine/ammonium salts on kaolinite surface and thus the hydrophobicity of particle surface. Accordingly, it belonged to hydrophobic aggregation.

3.2 Adsorption and sedimentation

Figure 6 shows the adsorption quantity of different amine/ammonium salts on kaolinite particles as a function of reagent concentration and pH value. With the increase of reagent concentration the adsorption quantity presented an upward trend, and it tended to be stable when the reagent concentration reaches 8×10^{-3} mol/L. However, there was a slight increase of the adsorption along with pH value increasing. Figure 7 shows the influence of reagent concentration and pH value on hydrophobic aggregation settlement of kaolinite. The sedimentation efficiency increased with the increase of reagent concentration then gradually tended towards balance, but decreased steadily with the pH increasing from 3 to 11.

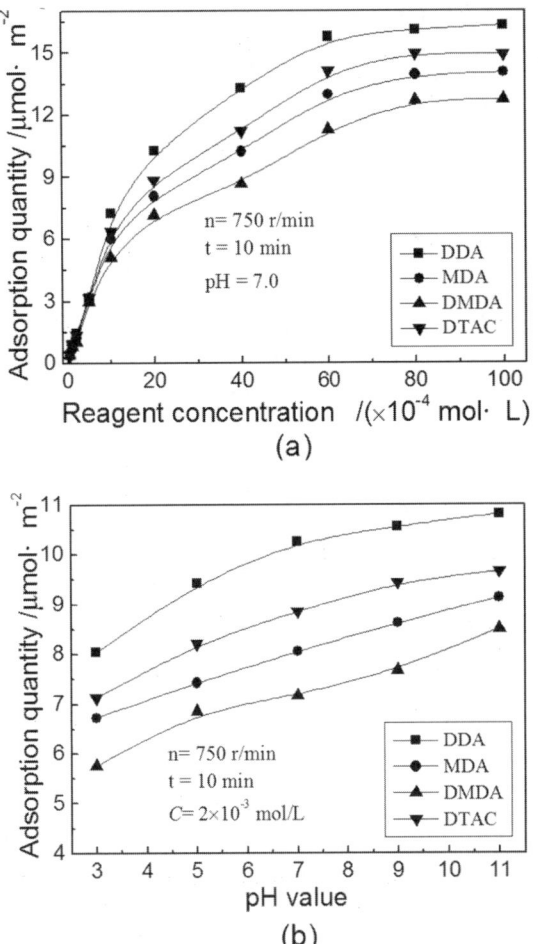

Figure 6: Adsorption quantity of different amine/ammonium salts on kaolinite particles as a function of reagent concentration (a) and pH value (b)

The results of adsorption and sedimentation show that these four different amine/ammonium salts can adsorb on kaolinite particles to promote the settlement of fine kaolinite particles, and that pH has a significance influence on the adsorption and sedimentation of kaolinite with different amine/ammonium salts. Namely that amine/ammonium salts can realise the effective hydrophobic aggregation settlement of fine kaolinite particles by adjusting its surface properties, which can provide a theoretical basis to develop new technologies of sedimentation and clarification for high muddied coal slurry water.

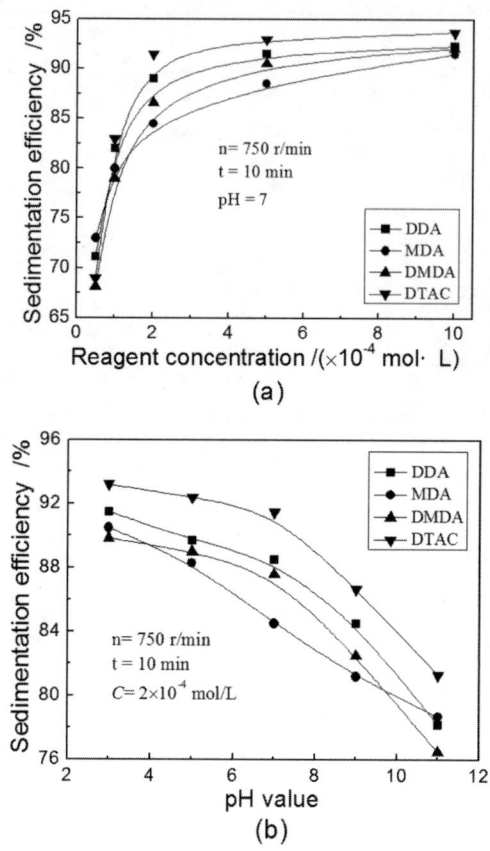

Figure 7: The influence of reagent concentration (a) and pH value (b) on hydrophobic aggregation settlement of kaolinite

3.3 DFT calculation

The configurations of the four different amine/ammonium cations adsorbed on different hollow positions of kaolinite (001) and (00$\bar{1}$) surfaces were optimised. Position changes and adsorption energies optimised are given in Table 1 which shows that the best adsorption sites for DDA$^+$, MDA$^+$, DMDA$^+$ and DTAC$^+$ adsorbing on kaolinite (001) surface and (00$\bar{1}$) surface are H3 site and H1 site, respectively. Adsorption energies corresponding to H3 site on kaolinite (001) surface are -104.808, -79.808, -86.538 and -103.846 kJ/mol, respectively. Similarly, adsorption energies corresponding to H1 site on kaolinite (00$\bar{1}$) surface are -122.692, -119.904, -118.654 and -104.808 kJ/mol, respectively.

Table 1: Summary of the different trial structure and resulting adsorption energies for different amine/ammonium cations on kaolinite (001) surface and (00$\bar{1}$) surface

Adsorption configuration	On (001) surface			On (00$\bar{1}$) surface		
	Initial position	Final position	E_{ads} (kJ/mol)	Initial position	Final position	E_{ads} (kJ/mol)
DDA+	H1	H1	−98.077	H1	H1	−122.692
	H2	H2	−95.192	H2	H2	−94.038
	H3	H3	−104.808	H3	H3	−88.461
MDA+	H1	H1	−67.308	H1	H1	−119.904
	H2	H2	−61.538	H2	H2	−99.712
	H3	H3	−79.808	H3	H3	−52.885
DMDA+	H1	H1	−72.501	H1	H1	−118.654
	H2	H2	−68.632	H2	H2	−101.346
	H3	H3	−86.538	H3	H3	−67.308
DTAC+	H1	H1	−96.273	H1	H1	−104.808
	H2	H2	−91.346	H2	H2	−101.923
	H3	H3	−103.846	H3	H3	−89.808

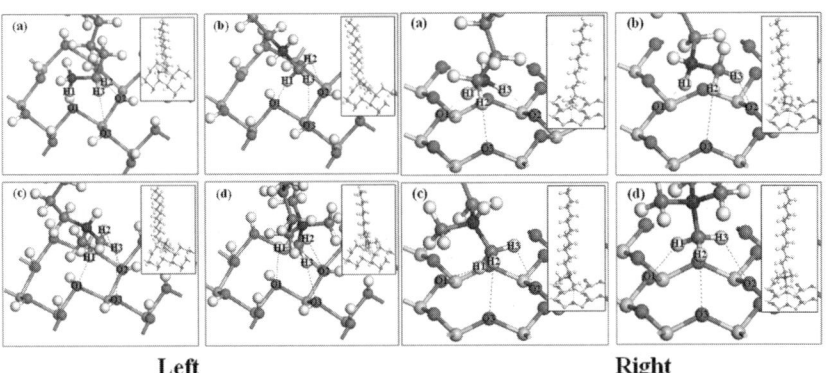

Left　　　　　　　　　　　　　　　Right

Figure 8: Optimised configurations of amine/ammonium cations adsorption on kaolinite (001) surface (left) and (00$\bar{1}$) surface (right). a, DDA⁺; b, MDA⁺; c, DMDA⁺; d, DTAC⁺

According to the results in Table 1, taking the optimised adsorption configurations of DDA⁺, MDA⁺, DMDA⁺ and DTAC⁺ on kaolinite (001) surface and (00$\bar{1}$) surface as examples to analyse bonding property. The optimised adsorption configurations of kaolinite (001) surface and (00$\bar{1}$) surface are given in Figure 8. (The strong hydrogen bonds of N–H⋯O are displayed by the blue dash line, and the weak hydrogen bonds of C–H⋯O are displayed by the black dash line.) We can see, the orientation of the amine/

ammonium cations with respect to the kaolinite surface were perpendicular with a slight tilt. The Mulliken bond populations of the optimized adsorption configurations of DDA$^+$, MDA$^+$, DMDA$^+$ and DTAC$^+$ on kaolinite (001) surface and (00$\bar{1}$) surface are given in Table 2, which shows that only DDA$^+$ of the four adsorption configurations of H3/(001) surface can form N–H···O strong hydrogen bond with the kaolinite surface, and the bond is 1.615 Å in length, and the Mulliken bond population is 0.16. While the other three cations only form C–H···O weak hydrogen bond with the kaolinite surface, their bond lengths are from 2.00 to 3.00 Å, and their Mulliken bond populations are between 0.0 and 0.03. On the one hand, only DDA$^+$ and MDA$^+$ of all the four adsorption configurations of H1/(00$\bar{1}$) surface can form N–H···O strong hydrogen bond with the kaolinite surface, their bond lengths are from 1.8 to 2.1 Å, and their Mulliken bond populations are from 0.06 to 0.09. On the other hand, the other three cations only form C–H···O weak hydrogen bonds with the kaolinite surface, their bond lengths are from 2.00 to 3.00 Å, and their Mulliken bond populations are 0.0–0.03. Contrasted with the average adsorption energies of different amine/ammonium cations adsorbed on kaolinite (001) surface and (00$\bar{1}$) surface (see Table 2), the adsorption stability of the four type of cations on kaolinite (001) and (00$\bar{1}$) surface are in the order of DDA$^+$ > DTAC$^+$ > DMDA$^+$ > MDA$^+$, this result is basically consistent with the results of experimental measurement.

Table 2: The Mulliken bond populations of different amine/ammonium cations adsorption on kaolinite (001) surface and (00$\bar{1}$) surface

Adsorption configuration	(001) surface/H3			(00$\bar{1}$) surface/H1		
	Bond	Length (Å)	Bond population	Bond	Length (Å)	Bond population
DDA+	N–H1···O1	1.615	0.16	N–H1···O1	1.727	0.09
	C–H2···O2	2.136	0.02	N–H2···O2	2.086	0.03
	C–H3···O3	2.907	0.00	C–H3···O3	1.983	0.01
MDA+	C–H1···O1	2.037	0.03	N–H1···O1	1.858	0.06
	C–H2···O2	2.581	0.00	C–H2···O2	2.444	0.00
	C–H3···O3	2.718	0.00	C–H3···O3	2.482	0.00
DMDA+	C–H1···O1	2.092	0.03	C–H1···O1	2.195	0.00
	C–H2···O2	2.785	0.00	C–H2···O2	2.599	0.00
	C–H3···O3	2.878	0.00	C–H3···O3	2.432	0.00
DTAC+	C–H1···O1	2.529	0.00	C–H1···O1	2.190	0.01
	C–H2···O2	2.452	0.01	C–H2···O2	2.640	0.00
	C–H3···O3	2.206	0.02	C–H3···O3	2.389	0.00

The mechanism of different amine/ammonium cations adsorption on kaolinite surface could be clarified by the electron density difference of the cation/kaolinite adsorption systems [18]. Figure 9 show the electron density difference of different amine/ammonium cations adsorption on kaolinite (001) surface and kaolinite (00$\bar{1}$) surface, respectively. The blue area in Fig. 9 indicates the electron accumulation, while the yellow area indicates the electron depletion. As seen in Fig. 9, a significant charge transfer has happened between atoms of the four adsorption system and the adjacent atoms of the kaolinite (001) surface and kaolinite (00$\bar{1}$) surface. What's more, electrons were transferred from the hydrogen atom of cations to the oxygen atom of kaolinite surface. In the adsorption systems, apart from the charge transfer between the hydrogen atoms and the surface oxygen atoms of hydrogen bonds, there is a certain amount of electron accumulation around the oxygen atoms approaching the hydrogen atoms of the cations. Due to the charge transfer from amine/ammonium cations to the surface of kaolinite, the kaolinite (001) surface and (00$\bar{1}$) surface displayed the negative charge. As the result of electrostatic attraction, the positively charged amine/ammonium cations adsorbed on the negatively charged kaolinite (001) surface and (00$\bar{1}$) surface.

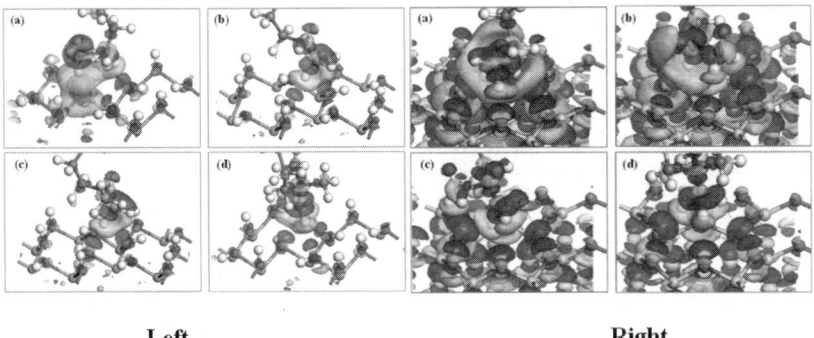

Left   Right

Figure 9: The electron density difference of amine/ammonium cations adsorbed on kaolinite (001) surface (left) and (0 $\bar{1}$) surface (right), the isosurface value is 0.006 electrons/Å³. a, DDA⁺; b, MDA⁺; c, DMDA⁺; d, DTAC⁺

The Mulliken atomic charges [19] before and after different amine/ammonium cations adsorption on kaolinite (001) surfaces and (00$\bar{1}$) surface, as displayed in Table 3. The amount of charge transfer from four cations to the kaolinite (001) surface is 0.39e, 0.30e, 0.37e, 0.38e, respectively. And the amount of charge transfer from four cations to the kaolinite (00$\bar{1}$) surface is 0.70e, 0.70e, 0.59e, 0.57e, respectively. Combined with the adsorption energy calculation results, the more electrons transferred from cations to kaolinite surface, the more negative the adsorption energy and the more stable the adsorption system. Due to a large number of electrons transferred to the kaolinite surface, the surface displays a large negative charges (−0.30e to −0.70e) and easily adsorbs the positively charged cations by the means of electrostatic attraction.

Table 3: Mulliken atomic charges before and after different amine/ammonium cations adsorption on kaolinite (001) surface and (00$\bar{1}$) surface

Atomic number Adsorption states		Mulliken charge (e)							
		H3/(001) surface				H1/(00$\bar{1}$) surface			
		DDA+	MDA+	DMDA+	DTAC+	DDA+	MDA+	DMDA+	DTAC+
H1	Before	0.30	0.27	0.27	0.25	0.30	0.27	0.27	0.25
	After	0.39	0.27	0.28	0.27	0.43	0.45	0.29	0.29
H2	Before	0.30	0.28	0.25	0.27	0.30	0.28	0.25	0.27
	After	0.31	0.28	0.26	0.28	0.44	0.30	0.28	0.29
H3	Before	0.26	0.28	0.27	0.25	0.26	0.28	0.27	0.25
	After	0.29	0.28	0.27	0.27	0.30	0.31	0.30	0.29
N	Before	−0.93	−0.68	−0.42	−0.16	−0.93	−0.68	−0.42	−0.16
	After	−0.85	−0.66	−0.40	−0.14	−0.80	−0.60	−0.39	−0.16
O1	Before	−1.06	−1.06	−1.06	−1.06	−1.06	−1.06	−1.06	−1.06
	After	−1.07	−1.06	−1.06	−1.08	−1.13	−1.15	−1.15	−1.15
O2	Before	−1.06	−1.06	−1.06	−1.06	−1.06	−1.06	−1.06	−1.06
	After	−1.06	−1.06	−1.07	−1.06	−1.16	−1.18	−1.18	−1.17
O3	Before	−1.05	−1.05	−1.05	−1.05	−1.05	−1.05	−1.05	−1.05
	After	−1.05	−1.05	−1.05	−1.06	−1.16	−1.16	−1.17	−1.16
Cations	Before	0	0	0	0	0	0	0	0
	After	0.39	0.30	0.37	0.38	0.70	0.70	0.59	0.57
(001) Surface	Before	0	0	0	0	0	0	0	0
	After	−0.39	−0.30	−0.37	−0.38	−0.70	−0.70	−0.59	−0.57

The DFT calculation results show that different amine/ammonium cations can strongly adsorbed on kaolinite (001) surface and (00$\bar{1}$) surface by forming N–H⋯O strong hydrogen bonds and C–H⋯O weak hydrogen bonds, and that there are strongly electrostatic attraction between different amine/ammonium cations and kaolinite surfaces. This indicates that the main adsorption mechanism of amine/ammonium cations on kaolinite is hydrogen-bond interaction and electrostatic attraction. Additionally, the adsorption is stronger on the siloxane surface (kaolinite (00$\bar{1}$) surface) than the hydroxyl surface (kaolinite (001) surface), which agrees well with those found in the study of Geatches et al. [20].

4. Conclusion

(1) Fine kaolinite particles strongly aggregates with amine/ammonium salts, which can enhance the hydrophobicity and reduce the electronegativity of particle surfaces. This situation belonged to hydrophobic aggregation. The aggregating abilities of the four amine/ammonium salts for kaolinite fall in the order DTAC > DDA > DMDA > MDA.

(2) Different amine/ammonium salts can adsorb on kaolinite particles, and realise the effective hydrophobic aggregation settlement of fine kaolinite particles. The different amine/ammonium cations can strongly adsorbed on kaolinite (001) surface and $(00\bar{1})$ surface by forming $N-H\cdots O$ strong hydrogen bonds or $C-H\cdots O$ weak hydrogen bonds, and there is a strongly electrostatic attraction between different amine/ammonium cations and kaolinite surfaces. Additionally, the adsorption is stronger on the siloxane surface than the hydroxyl surface.

(3) The main adsorption mechanism of amine/ammonium cations on kaolinite is a hydrogen-bond interaction and electrostatic attraction.

5. Acknowledgement

The financial supports for this work from the National Natural Science Foundation of China under the grant No. 51874011 and the National Natural Science Foundation of China under the grant No. 51804009 are gratefully acknowledged.

6. References

[1] J. Chen, F. Min, L. Liu, C. Peng, Y. Sun, J. Du, Study on hydrophobic aggregation settlement of high muddied coal slurry water, J. China Coal Soc. 39 (2014) 2507–2512.

[2] F. Min, C. Peng, S. Song, Hydration layers on clay mineral surfaces in aqueous solutions: a review, Arch. Min. Sci. 59 (2014) 373–384.

[3] M. Żbik, R.G. Horn, Hydrophobic attraction may contribute to aqueous flocculation of clays, Colloids Surf. A. 222 (2003) 323–328.

[4] J. Cui, Q. Fang, G. Huang, Crystal structures and surface properties of diaspore and kaolinite, Nonferr. Metal. 51 (1999) 25–30.

[5] S. Song, A. Lopez-Valdivieso, J.L. Reyes-Bahena, H.I. Bermejo-Perez, Hydrophobic flocculation of galena fines in aqueous suspensions, J. Colloid Interf. Sci. 227 (2000) 272–281.

[6] F. Merzel, J.C. Smith, Is the first hydration shell of lysozyme of higher density than bulk water? Proc. Natl. Acad. Sci. U.S.A. 99 (2002) 5378–5383.

[7] Y. Hu, L. Liu, F. Min, M. Zhang, S. Song, Hydrophobic aggregation of colloidal kaolinite in aqueous suspensions with dodecylamine. Colloids Surf. A. 434 (2013) 281–286.

[8] H. Jiang, Z. Sun, H. Long, Y. Hu, K. Huang, S. Zhu, A comparison study of the flotation and adsorption behaviors of diaspore and kaolinite with quaternary ammonium collectors, Miner. Eng. 65 (2014) 124–129.

[9] L. Wang, Y. Hu, W. Sun, Y. Sun, Molecular dynamics simulation study of the interaction of mixed cationic/anionic surfactants with muscovite, Appl. Surf. Sci. 327 (2015) 364–370.

[10] S.S. Rath, H. Sahoo, B. Das, B.K. Mishra, Density functional calculations of amines on the (101) face of quartz, Miner. Eng. 69 (2014) 57–64.

[11] X. Wang, Y. Huang, Z. Pan, Y. Wang, C. Liu, Theoretical investigation of lead vapor adsorption on kaolinite surfaces with DFT calculations, J. Hazard. Mater. 295 (2015) 43–54.

[12] X. Wang, P. Qian, K. Song, C. Zhang, J. Dong, The DFT study of adsorption of 2,4-dinitrotoluene on kaolinite surfaces, Comput. Theor. Chem. 1025 (2013) 16–23.

[13] J. Chen, F. Min, L. Liu, C. Peng, F. Lu, Hydrophobic aggregation of fine particles in high muddied coal slurry water, Water Sci. Technol. 73 (2016) 501–510.

[14] S.J. Clark, M.D. Segall, C.J. Pickard, P.J. Hasnip, M.I.J. Probert, K. Refson, M.C. Payne, First principles methods using CASTEP, Z. Kristallogr. 220 (2005) 567–570.

[15] J. Chen, F. Min, L. Liu, The interactions between fine particles of coal and kaolinite in aqueous, insights from experiments and molecular simulations, Appl. Surf. Sci. 467–468 (2019) 12–21.

[16] X. Hu, A. Michaelides, Water on the hydroxylated (0 0 1) surface of kaolinite: from monomer adsorption to a flat 2D wetting layer, Surf. Sci. 602 (2008) 960–974.

[17] D.L. Bish, Rietveld refinement of the kaolinite structure at 1.5 K, Clay. Clay. Miner. 41 (1993) 738–744.

[18] C. Peng, F. Min, L. Liu, J. Chen, A periodic DFT study of adsorption of water on sodium-montmorillonite (001) basal and (010) edge surface, Appl. Surf. Sci. 387 (2016) 308–316.

[19] R.S. Mulliken, Electronic population analysis on LCAO-MO molecular wave functions, I. J. Chem. Phys. 23 (1955) 1833–1840.

[20] D.L. Geatches, A. Jacquet, S.J. Clark, H. C. Greenwell, Monomer adsorption on kaolinite: Modeling the essential ingredients, J. Phys. Chem. C. 116 (2012) 22365–22374.

15

Mechanical flotation machines for the oxidized coal slurry treatment

I.V. Jeremejev, S.O. Fedoseeva, E.V. Rudavina, A.N. Voronov

OP «Ukrniiugleobogashchenie», Dnepropetrovsk, Ukraine

Abstract: This paper presents the flotation complex formation research results with representation of interaction between the coal particle having various surface oxidation degree OKn and air bubble by means of numerical simulation based on the discrete element method (DEM).

It is shown that the increase in degree of surface oxidation above the critical value at the equal reagent schemes causes decrease in every parameter with the most considerable decrease in separation selectivity. The surface oxidation results in the decreased contrast of properties of its individual areas and thus deterioration of separation results due to the increased mutual contamination of products.

It is found that surface oxidation OKn equal to 15% is critical with regard to providing the necessary duration of coal particle contact with air bubble in order to enhance the flotation complex formation probability. At the OKn = 15–24% and above, a particle loses contact with a bubble. The low degree of surface oxidation facilitates formation of steady flotation complex – a coal particle is reliably engaged to the air bubble surface.

The research results indicate that the increase in the rate of removal of medium size and most grainy oxidized particles with less ash content to the froth product requires the use of the most efficient collecting and frothing agents and special feeding mode including slug feed and emulsification of reagents. At the same time, increased degree of slurry aeration must be provided.

In order to obtain the required quality froth product from oxidized aged slurries, the improved flotation machine is developed with the enhanced bubble mineralization zone formed due to the unique impeller design.

The special design of impeller blades facilitates the flotation process intensification, maximum air inflow and efficient dispersing of aerosol of flotation agents due to ejection thereof through an air duct to the slurry aeration zone.

Keywords: Coal, flotation, modes, slurry, degree of oxidation, flotation machine, impeller

1. Introduction

The process of foam separation of fine carbonaceous materials is quite complex since it depends on the great number of not only physical and chemical parameters but also mechanical ones [1–4]. For determination of coal particle surface oxidation degree impact on the flotation complex existence conditions, the methodology of this process simulation is developed, the flotation ability of oxidized particles is investigated, and the rational flotation reagent schemes are determined [5].

It is found that probability of flotation success is provided by the following flotation complex formation factors: the coal particle collision with bubble, particle engagement to bubble at a contact, keeping of engaged on bubbles and retention of particles by the froth.

The research found that the increase in duration of coal air storage results in coal surface oxidation increase from 0.4–0.8% (operational sample) to 9–14% (fresh-mined coal passing through the transport routes) and to 24–26% (aged slurry from slurry settler or external slurry pond), and for some grades, coal oxidation degree reaches 32–36% at storage in the external slurry pond [5].

The impact of phase surface properties on the flotation complex formation is investigated by means of numerical simulation with the use of discrete elements. The best detailed description of simulation methodology and summary of method are presented in the works [6,7]. The time of particle residence on bubble surface τ was determined based on the numerical simulation data from the variation of distance S between elements.

The variation of time of elements interaction and flotation complex existence at the simulation of different coal particle surface oxidation degrees is shown in Figs. 1 and 2.

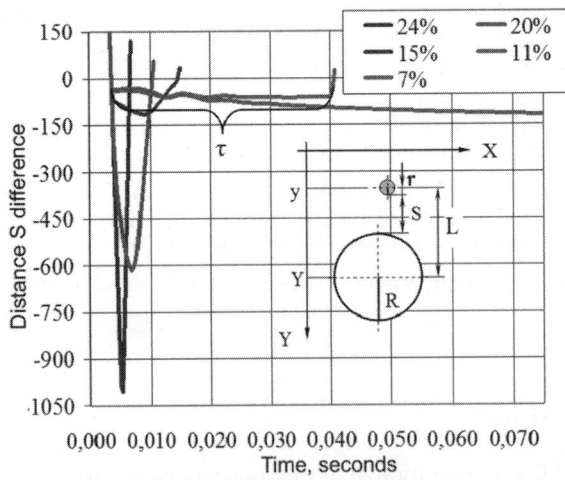

Figure 1: The effect of oxidation of the surface of coal particles on the lifetime of the flotation complex

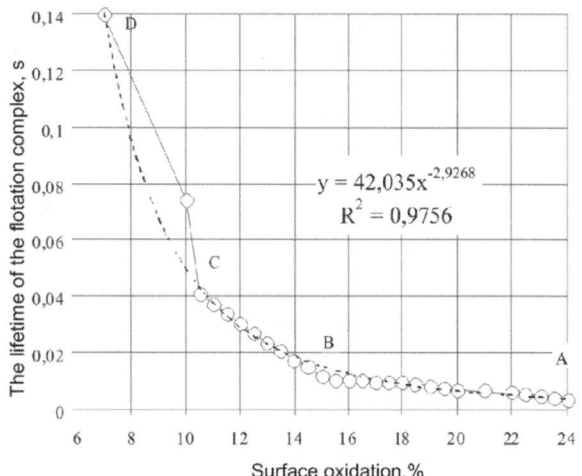

Figure 2: Change in the lifetime of the flotation complex at different oxidation of the surface of the coal particle

The conducted research found that dependence is of power-law type and is expressed as the following function:

$$\tau = 42.04 OK\text{п}^{-2.93} \tag{1}$$

where τ is the flotation complex existence time, s and OKп is the coal particle surface oxidation degree, %.

Here, three zones are distinguished in the diagram: (1) at high surface oxidation degree from 15% to 24% (AB zone), flotation complex existence time is insignificant; (2) oxidation degree decrease from 15% to 11% (BC zone) facilitates the elements contact time increase and particle engagement probability enhancement; and (3) at oxidation degree below 11% (CD zone), reliable particle to bubble engagement is most probable.

Thus, surface oxidation degree equal to 15% can be considered as critical value for providing the flotation complex formation probability. At the high surface oxidation values (over 24%), a coal particle slides over an air bubble without engagement. Particle velocity in relation to the bubble surface varies continually due to the variation of resultant of forces arising from phase interaction. At the moment of collision with a bubble, particle velocity increases sharply, whereupon its decrease occurs. This decrease is more considerable and occurs in lesser time, the lesser is degree of surface oxidation.

Results of research for flotation ability of coal slurries with different degrees of surface oxidation conducted with the use of flotation chemicals currently applied at the coal preparation plants in Ukraine and most efficient reagent schemes are summarized to Table 1.

As can be seen, the increase in degree of surface oxidation above the critical value at the equal reagent schemes causes decrease in every parameter with the most considerable decrease in separation selectivity. The surface oxidation results in the decreased contrast of properties of its individual areas and thus deterioration of separation results due to the increased mutual contamination of products.

Therefore, the flotation concentration of oxidized aged carboniferous slurries from technogenic deposits requires development of technology with the use of the most efficient flotation agents, schemes and feed methods, flotation process charts, and the use of improved flotation equipment.

The analysis of obtained results allows for the conclusion that at the same flotation modes, process performances for coals with the degree of oxidation above critical value are much worse than those for the same grade coals with the low degree of oxidation. Generally, regularities of change of flotation separation estimation parameters continue to apply to the oxidized coals, as well as to the low-oxidized coals. However, absence is noted of sharp process parameter fluctuations at the change of the reagent scheme. Typical of the performances of aged slurries with surface oxidation degree above the critical value is lower fluctuation of values at the change of the reagent scheme.

The use of the most flotation-active agents allows obtaining of the waste ash content above 70% at the ash content of concentrate suitable for utilization for power production even from aged oxidized slurries.

It is found that the selection of the reagent scheme for successful flotation of the oxidized coals requires increased consumption of frothing agent and the use of collecting agent with high content of heteropolar compounds.

Low rate of oxidized aged slurry flotation is related to extended slurry surface, since it requires increased amount of small air bubbles.

For this purpose, the institute developed an aerator with a radial–axial impeller. Combination of two opposing axial and radial wheels in single impeller provides conditions for maximum air inflow with subsequent dispersing and distribution within a liquid phase, and allows increased intensity of slurry mixing with reagents and degree of slurry aeration.

During the top axial impeller wheel rotation on a shaft, wheel blades provide air suction from an atmosphere through the air duct made of casing pipe and suction tube with the air flowrate controller. Lead edges of top axial impeller blades execute initial dispersing of air bubbles entering the impeller. The lower axial impeller wheel blades suck the slurry from bottom part of flotation machine cell and mix it with air entering the blade passages.

Table 1: Development of flotation regimes of coal sludge of various degrees of oxidation of their surface

Coal grade	Sludge sample	Ash content (%)	Reagent modea reagent consumption (g/t)	Coal oxidation, OK p (%)	Selectivity	Removing the combustible mass (%)	Effectiveness
Г	Ordinary coal 1	48.4	DT 480, KETGOL 60	13	26.4	84.8	70.0
			RSL 480, KR (50:50) 80		27.4	86.0	70.2
			KR (92:8) 770		30.9	87.9	71.8
			CR (92:8) 730		30.8	89.5	70.1
	Ordinary coal 2	43.5	TC 1500, KR (50:50) 150	14	21.9	84.0	67.8
	Ordinary coal 3	34.5	TS 1500, KR (50:50) 150	20	14.4	69.6	60.0
	Ordinary coal 4	20.3	TS 3500, T-66 240	12	31.4	92.6	66.5
			DT 2060, POD 580		43.4	95.6	63.6
	Sludge tank 5	61.4	KR (92:8) 600	26	11.9	54.5	62.0
	Sludge tank 6	52.1	KR (92:8) 600	24	9.9	72.6	57.3
Ж	Ordinary coal 7	26.2	TS-1: 1100, COBS 100	14	36.3	95.7	56.4
			KR (92:8) 850		29.5	94.0	61.3
	Sludge tank 8	55.4	DT 750, KR (50:50) 150	25	7.6	42.0	38.3
OC	Common coal 9	28.6	TC-1 1500, KETGOL 100	15	24.1	93.4	58.7
	Open warehouse 10	26.7	TC-1 1500, KETGOL 100	20	20.1	92.4	57.7
A	Ordinary coal 11	29.5	KR (92:8) 470	15	19.0	91.4	58.9
	Clarifier 12	42.7	KR (92:8) 700	26	5.5	80.6	45.8
			DT 1955, COBS 155		5.8	57.9	46.9
	Sludge tank 13	41.4	KR (70:30) 480	20	65.4	92.6	68.3
			DT 720, COBS 100		29.8	92.9	64.5

aCollectors: diesel fuel, TS-1 fuel, catalytic cracking gas oil – RSL; foaming bodies: brand oxal T-66, KETGOL, COBS, composite reagent KR with a different ratio in it blowing agent and collector.

Combination of two opposing axial wheels and radial blades in single impeller allows decreasing of the vigorous mixing zone height and increasing of the quiet mineralized air bubbles floating-up zone in the flotation machine cell. This provides conditions for pulsating turbulence combined with wave slurry motion, conditions of bubbles mineralization, conditions of the mineralized bubbles floating-up and demineralization, and froth layer maintenance.

Figure 3 shows the diagram of distribution of air and liquid streams in the radial–axial type aerator impeller, which is the basic element of the developed mechanical radial flotation machine.

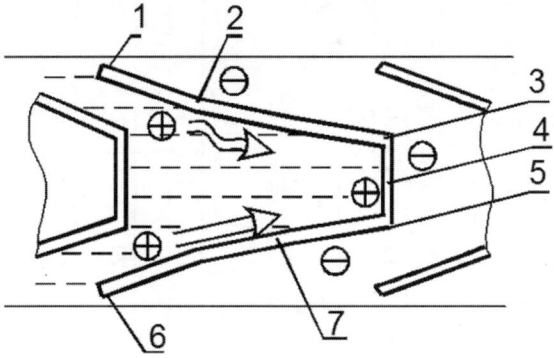

Figure 3: The pattern of distribution of air and fluid in the impeller. 1 and 6, input edges of the blades; 2 and 7, axial blades; 3 and 5, output edges of the blades; and 4, radial blade

Liquid pressure on the internal pressure faces of blades 1 and 6 is designated with (+) sign, and that on the external faces of these blades is designated with (−) sign. The radial–axial type aeration unit operates as follows.

At the start, liquid levels in the cell and air duct are the same. When the impeller rotation starts, top and bottom axial blades start to suck liquid from the air duct and cell, respectively. Figure 3 shows that slurry is driven over the internal pressure faces of blades 2 and 7 from the lead edges 1 and 6 to the tail edges 3 and 5 of each blade and is fed to the radial blade 4 action zone. Here, liquid motion direction changes from axial to radial one, and its pressure increases. The inter-phase boundary in the bowl gains oscillatory motion, which causes wave motion of liquid at the pressure face (+) of top axial blade 2 directed from the lead edge 1 to the tail edge 3 of each blade. Atmospheric air is captured from bowl into the trough between wave crests, transferred to the level of centrifugal blade 4 and discharged from the impeller flow passage to the stator blades by means of the liquid stream driven by bottom axial blades 7. The special design of impeller blades facilitates the intensification of process of coal particles interaction with air and reagents. Air with reagents and slurry are separately fed to impeller and are only mixed up at its outlet, and this provides stable (pulsing-free) aerator capacity both by air and by slurry, and high dispersion of reagents.

For concentration of oxidized slurries stored in external slurry pond, the institute developed technology and parametric series of new-generation radial flotation machines with single aeration unit [8,9] of 2 m^3, 16 m^3, 25 m^3, and 36 m^3 in volume. With regard to the specific indicators (solids intake) and technological capabilities (aeration degree, stable, and sufficient air supply to process), the radial flotation machine with single radial–axial type aerator exceeds the mass produced models of flotation machines with six aerators. The flotation machine design provides for the aeration degree variation by varying the electric drive rotation speed by means of thyristor converter and by the method of air throttling through the air duct tube. The variation of air flowrate through the air duct tube allows regulation of air content in the cell volume without the change of slurry stirring intensity.

2. Conclusion

Oxidized aged slurry processing by the flotation method is a complex challenge requiring the use of the improved methods, optimal selection of the reagent schemes, and special design of the flotation machines developed by the institute in Ukraine on a level with the highest world standards.

3. References

[1] Filippenko Yu.N., Morozova L.A., Mavrenko G.A., Fedoseeva S.O., Processing of the waste coal stored in external slurry ponds. Collection of Papers. International Ecological Forum «Environment for Ukraine»: June, 23–24, Donetsk, 2011, P. 113–115.

[2] Garkovenko E.E., Nazimko E.I., Samoilov A.I., Papushin Yu.L., Particularities of flotation and dewatering of the finely dispersed carboniferous materials. Donetsk: Nord-Press, 2002, 266 p.

[3] Filippenko Yu.N., Morozova L.A., Fedoseeva S.O., Particle-size analysis of mined coals and the coal slurry treatment technology status analysis. Coal of Ukraine, 2013, No. 3, P. 12–14.

[4] Fedoseeva S.O., Research of flotation ability of the coal slurries from technogenic deposits in Ukraine. Mineral processing: Collection of Science and Technology Articles. Iss. 41(82)–42(83). Dnipropetrovsk: NMU, 2010, P. 186–192.

[5] Fedoseeva S.O., Justification for parameters of flotation of the finely dispersed carboniferous materials from technogenic deposits. Author's abstract of thesis, Krivoy Rog: KNU, 2014, 25 p.

[6] Golikov A.S., Naumenko, Fedoseeva S.O., Nazimko E.I., Simulation of the water–slime system operation in dewatering and flotation processes at coal slurry treatment. Mining Informational and Analytical Bulletin, 2013, № 2, P. 75–81.

[7] Nazimko L.I. Kinetics of Phases Interaction during Mineral Processing Simulation. L.I. Nazimko, E.E. Garkovenko, A.N. Corchevsky et al., Proceedings XV international Coal Preparation Congress. China, 2006, Vol 2, P. 785–798.

[8] Ukrainian utility model patent № 58790. Flotation Machine. Mavrenko G.A., Kocheshkov B.O., Spineev V.A., Tiutiareva V.V., Fedoseeva S.O., 4 p.; Publ. 26.04.2011, Bull. No. 8.

[9] Development of integrated process of the secondary fuel recovery from waste coal slurry: Research report. Luhansk: Ukrndivuglezbagachennia. 2007, 236 p.

16

Research, development and application of TDS in coal preparation industry in China

Shaolei Zhou and Taiyou Li

DADI Engineering Development Group Corporation, Beijing Shi, China

Abstract: It is of great significance to study the water-free dry coal separation technology with simple system for clean utilization of coal. Research institutions in China have always been exploring and studying the dry coal separation technologies. As the big data analysis technologies develop in recent years, Tianjin Meiteng Technology Co., Ltd., an affiliate of DADI Engineering Development Group, has successfully developed a high-precision Telligent dry separator (TDS). TDS adopts X-ray recognition technology and assistant image recognition technology to digitally recognize coal and reject, and then separate them with compressed air. In comparison with the traditional washing process, TDS features high intelligence degree, high precision in separation, free of water or medium, no coal slime, no adding product moisture, simple system, small equipment, low production cost, and investment. In addition to normal application, TDS is especially suitable for using in arid area, underground to reject ejection and back-filling and suitable for lignite separation. TDS can effectively separate 300–25 mm sized lump coal. Since the first TDS was applied in Zhaozhuang Coal Preparation Plant, Shanxi Province in 2016, more than 40 units of TDSs have been sold in China, and applied in underground reject ejection effectively, with promising prospect in the future.

Keywords: X-ray, Telligent dry separation, underground reject extraction

It is stated in the China's Energy Outlook 2030 issued by China Energy Research Society that the proportion of coal in China's primary energy structure in 2020 and 2030 is 68.8% and 58.7%, respectively. In the Research Report on Sustainable Energy Development Strategy of China, more than 20 academicians of Chinese Academy of Sciences and Chinese Academy of Engineering reached the consensus that the proportion of coal in energy will not be less than 50% until 2050, and the status of coal as main energy will keep unchanged in short term, which vividly represents the important position of the clean utilization technology of coal.

About 78% of coal resources in China are reserved in the dry and water-starved northwestern region. The shortage of water resource restricts the application of coal washing technology. In addition, the annual output of lignite is about 500 million tons with serious sliming, making it improper to adopt the coal washing process. As a result, the water-free dry coal separation technology with simple system becomes a key solution. Research institutions in China have always been exploring and studying the dry coal separation technologies. As the big data analysis technologies develop, the intelligent dry coal separation technology based on X-ray recognition has been rapidly developed and extensively promoted and used.

1. Development of telligent dry separation in China

The study on X-ray-based dry separation technology began in England in the 1960s. The X-ray Coal Separator released by the 10th International Mineral Processing Congress in 1973 introduced that the X-ray coal separator was co-developed by the UK National Coal Board and Gangsheng Corporation, with 12.2–22.6% of coal content in reject and 23.2–31.8% of reject content in coal.

China began to study the X-ray based dry separation technology for coal separation in 1988, manufactured the dry separator based on γ-ray recognition and applied it in the industrial field in 2005, and installed the imported X-ray dry separator in 2014. By 2014, the overall separation accuracy was 5–10% of the coal content in reject and 10–20% of reject content in coal. Due to low separation accuracy and processing capacity, the application of X-ray dry separator in coal industry was very limited. DADI Engineering Development (Group) Corporation began to study X-ray dry separation technology in 2008. With the development of big data technology, substantial breakthrough was made in 2016. The X-ray high-precision Telligent dry separator (TDS) produced by Tianjin Meiteng Technology Co., Ltd. was successfully applied in Zhaozhuang Coal Preparation Plant, Shanxi Province, accelerating the promotion of the intelligent dry separation technology.

2. Theory and development of TDS

2.1 Theory of TDS

TDS adopts intelligent recognition methods such as X-ray and image recognitions, establishes corresponding analysis model for different coal features, digitally recognizes coal and reject through big data analysis, and then divert the determined objects by compressed air from high-frequency solenoid valve, thus realizing the accurate separation of coal and reject. The intelligent dry separation system is composed of three major systems of feeding, recognition, and execution. TDS auxiliary systems include air supply, dust remove, power distribution, and control. See Fig. 1 for the theory of TDS.

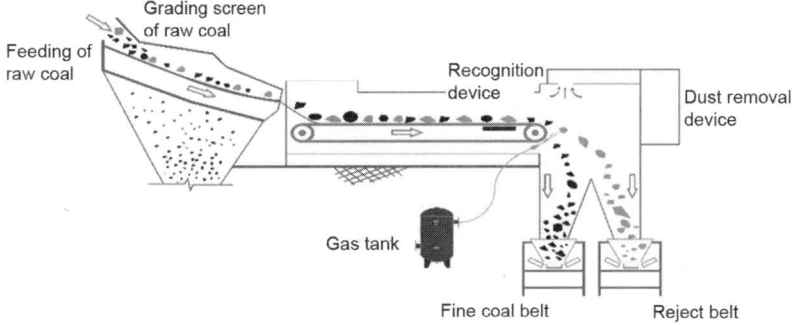

Figure 1: Schematic diagram of TDS

2.1.1 X-ray and image double pattern recognition system

(1) X-ray recognition

TDS dry separation technology adopts dual-energy X-ray transmission and recognition technology; the Lambert–Beer's law is used as the main recognition theory, namely detecting the high-energy X-ray intensity I_{0high} and low-energy X-ray intensity I_{0low} before X-ray transmitting the object, as well as the high-energy X-ray intensity I High and low-energy X-ray intensity I Low after X-ray transmitting the object respectively with a X-ray detector, then putting them into the formula of Lambert–Beer's law:

$$\frac{\mu_{low}}{\mu_{high}} = \frac{\ln(I_{0low} / I_{0low})}{\ln(I_{0high} / I_{0high})} = K$$

where I and I_0 is the ray intensities before and after transmitting the detected object and μ is the mass absorption coefficient.

A value X can be set; when $K > X$, the detected object can be judged as reject, obtaining the reject product; when $K \leq X$, the detected object can be judged as coal, obtaining the coal product; X-ray recognition technology is suitable for the separation of common type coal.

(2) Image recognition

It is found from the raw coal sample testing in China that the TDS system has low accuracy in recognition of special middling; therefore, the image recognition technology is introduced. The X-ray transmission and CCD image recognition are combined to give play to respective characteristics of both technologies; the X-ray transmission is based on the internal information detection of object; the CCD image is based on surface information detection of object, such as size, shape, luster, texture, pattern, color, etc.; and the co-recognition is realized by giving different confidence levels to the X-ray and the image detection values in line with characteristics of the object, improving the detection accuracy.

(3) Technology for recognizing of +200 mm large-sized materials

The poor penetration of X-ray to +200 mm sized, especially to +300 mm sized objects restricts the improvement of upper separation limit of X-ray separator, and results in the upper separation limit below 200 mm, even below 150 mm for a long time, which limits the development and application of X-ray separation technology.

The "large-sized material recognition technology" of TDS realizes the recognition of +300 mm sized or even 1000 mm-sized objects, breaks through the technical bottleneck of X-ray separation, and provides a condition for the improvement of upper separation limit.

2.1.2 Array actuator

In TDS, the arrangement form of single channel and air nozzle is designed as multi-row and multi-specification array arrangement. After the accurate position of material is given by the recognition device, the target object may be aimed accurately by optimizing the combination order of jetting and firing through dynamic optimization and matching algorithm. The array actuator solves the adaptability problems of non-uniform size and different shapes of coal, and improves the separation accuracy and processing capacity.

2.2 Technical level of TDS

With constant efforts made on research and development, now TDS can reach the following technical levels:

(1) Separation accuracy

The separation accuracy of this technology is up to 1–3% of coal content in reject and 3–5% of reject content in coal, better than that of jigging washing method.

(2) Processing size

With wide range of processing sizes, this technology can process 300–25 mm sized lump coals; if the execution mode of "coal ejecting" is adopted, the maximum top size may be up to 1000 mm. The small-size intelligent dry separator is under study, and the lower size is expected to be reduced to 10 mm.

(3) Processing capacity

The series model spectrum has been designed. The processing capacity of the largest model TDS40-300 series is up to 380 t/h. If the yield of 300–50 mm sized lump coal is calculated as 25%, one intelligent dry separator will be enough for a coal separation plant with processing capacity of 8.5 Mt/a to process 300–50 mm sized lump coals.

(4) Reliability

The first TDS has been put into operation for more than 2 years in Zhaozhuang Coal Preparation Plant of Shanxi Jincheng Anthracite Coal Mining Group. Since the operation, it has experienced the test of heavy production tasks with stable production and reliable system.

2.3 Secondary development of TDS

(1) Three-product TDS

The three-product TDS is developed on the basis of coal quality in China to solve the problem of difficult coal separation. The three-product TDS adopts the intelligent recognition method to recognize the raw materials into fine coal, middling and reject, and then control the actuator to jet the fine coal and reject with different energies. The rejects jetted with maximum energy enter in the farthest chute; the middlings jetted with small energy enter in the middle chute; and the fine coals not jetted enter in the nearest chute along the original movement path; the three-product process can classify the lump coal into three kinds of products of fine coal, middling, and reject.

(2) Underground TDS

90% of coal mines in China are underground coal mines; the underground reject extraction and filling technology has huge market in terms of avoiding safety hazards in gob, effective protection of natural environment, and improvement of resource recovery. For the traditional washing process is complex in system and difficult to be applied underground, the water- and medium-free intelligent dry separator with simple system is suitable for the underground application. The redeveloped TDS meets requirements of the *Interim Measures for Mining Product Safety Signs* issued by the State Economic and Trade Commission and National Coal Mine Safety Administration in 2007, and has been awarded the coal mine safety certificate for underground application equipment. The first underground TDS has been used in Wanglou Coal Mine of Linyi Mining Group in 2018.

3. Application advantages of TDS

The separation accuracy of TDS is higher than that of washing jigging process. Compared with the traditional washing method, TDS has the following advantages.

3.1 Low investment in construction

The TDS intelligent coal separation method separates the coals without water and medium, and does not need to remove water and medium, or recover medium, or concentrate and recover the slime water, significantly simplifying the process flow and the system.

The TDS intelligent dry separation process is smaller than the washing process in terms of system configuration, equipment quantity, building volume and floor area, etc., and can effectively reduce the input and shorten the erection period.

3.2 Low production cost

Compared with the traditional washing system, the TDS intelligent dry separation system can save the production cost from the following three aspects:

(1) Without water and medium consumption
The TDS intelligent dry separation system does not need or consume water or medium, with the installed power and electricity consumption per ton of raw coal lower than those needed in washing process.

(2) Saving labor cost
The TDS intelligent dry separation process is featured by high intelligence degree, simple staffing, fewer post workers, production workers, and management personnel required, compared with washing system, saving labor cost, and improved efficiency of workers and staff.

(3) Saving maintenance cost
The TDS intelligent dry separation process requires simple system, few equipment, small amount of maintenance, and small demand of spare parts; compared with traditional washing process, the TDS intelligent dry separation process can effectively reduce maintenance costs.

3.3 No coal slime produced

For traditional steam coal cleaning plants adopting washing process, their coal products are sold separately or in blended at an unsatisfying price, or even cannot be sold, but only discharged to the reject pile yard, due to large amount of coal slimes produced by the washing process with water as medium; this will pollute the environment and reduce the economic benefits of enterprises. In addition, the solid-liquid separation of slime water requires lots of investment and production costs. TDS does not require water for the process without coal slime produced, and can effectively reduce the environmental pollution, improve the yield of lump coals, and increase the economic benefits of enterprises.

For traditional coking coal preparation plants adopting washing process, TDS can realize the pre-extraction of rejects in the lump coals, so as to extract the large rejects in advance, reducing the content of coal slime, lowering the load of slime water system, and saving the floatation cost.

3.4 No moisture in product increased

For the coal preparation plant pricing the products with calorific value, the TDS intelligent dry separation process will not use water during the separation process, increasing the calorific value of products, compared with traditional washing process.

3.5 High intelligent level

With high intelligent level, TDS can realize the intelligent learning of system, self-improvement of separation accuracy, self-detection of fault, and unattended operation to certain extent.

4 Application prospect of TDS

TDS has broad prospect in application.

4.1 Pre-extraction of reject

TDS can replace manual separation and moving screen to realize reject pre-extraction of lump coal. Coal preparation plants in China often design the manual separation or moving screen to pre-remove reject and impurity of lump coal; however, the manual separation has low extraction rate of reject, poor labor environment, and high labor intensity. The moving screen has high failure rate and high coal content in rejects. The above problems can be solved by using TDS instead of manual separation and moving screen.

4.2 Lump coal separation

TDS has simple system and can improve the quality of 300–25 mm sized lump coals, and save investment.

4.3 Combined process of full size fraction separation

For the coal preparation plants requiring full size fraction separation of raw coal, it is possible to adopt the combined process of lump coal separation by TDS and smalls separation by other processes, thus washing all the raw coals.

4.4 Technical reform

For existing coal preparation plants using lump coal jigging process or shallow chute process, the TDS may be adopted for technical reform and replace the jig or shallow chute, solving the slime water problem, lowering the product moisture, improving the calorific value of products, and increasing the yield of lump coals.

4.5 Mobile separation

Due to its small volume and simple system, TDS intelligent dry separation system is suitable for mobile arrangement. The mobile TDS can be conveniently used for reject extraction of lump coal in small strip mines and coal storage centers.

TDS dry separator and distribution room are integrally arranged on one sweeping board; the air supply system is arranged on the other sweeping board; and the materials are delivered with a movable belt for convenience of movement. See Fig. 2 for the mobile TDS separation device.

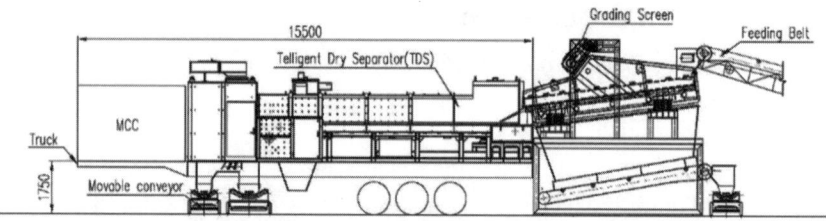

Figure 2: Mobile TDS system –TDS sweeping board

4.6 Underground reject extraction

Compared with traditional washing technologies, such as shallow chute and moving screen, the TDS has advantages of small volume, simple system, and free of water, medium, slime water treatment, and heavy medium recovery. Therefore, the TDS has obvious advantages in underground application.

4.5 Other applications

(1) Suitable for the separation of lignite and coal lignite in water-deficient area.

(2) Separation of other non-coal minerals, such as separation of metal ore, kaolinite rock, nickel-soil ore, and garbage.

5. Application process of TDS

5.1 Lump coal separation process

TDS is used for separation of 300–25 mm sized coals. Fine coals separated by TDS can be sold as a separate product or in mixed with 25–0 mm smalls after breaking. See Fig. 3 for the specific flow.

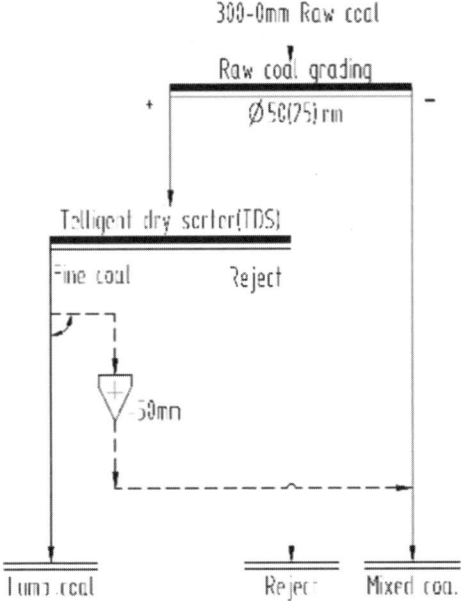

Figure 3: Diagram of 300–25 mm sized lump coal separation process

5.2 Fines removal process

The process comprises of intelligent dry separation for 300–50 (25) mm sized lump coals, 6 mm sized fines removal from 50 (25)–0 mm smalls, washing for 50 (25)–6 mm sized coals, and by-pass process of 6–0 mm powdered coals, in which 50 (25)–6 mm smalls can be separated by shallow chute separator, dense-medium cyclone, jig, and other washing devices or air separation devices. See Fig. 4 for the specific flow.

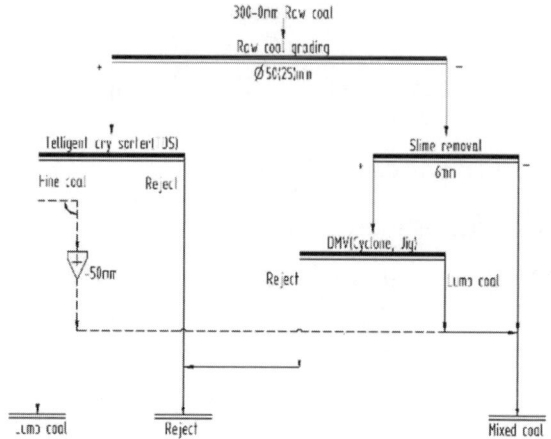

Figure 4: Diagram of 300–6 mm sized separation process

5.3 Full size fraction separation process of raw coal

300–50 (25) mm sized lump coals may be separated by adopting intelligent dry separation process, and 50 (25)–0 mm sized coals may be separated by dense-medium cyclone. See Fig. 5 for the specific flow.

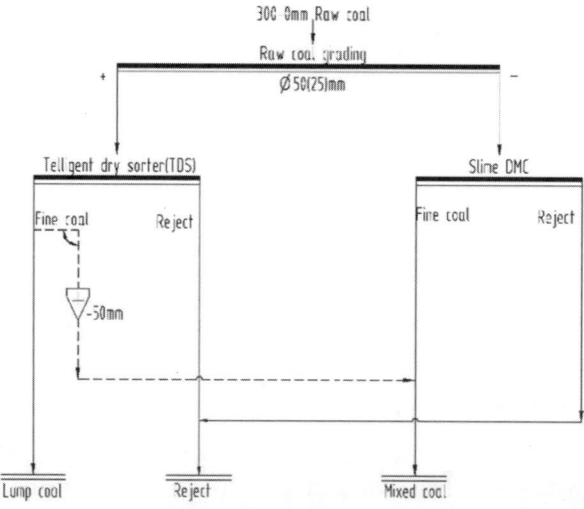

Figure 5: Diagram of full washing process of 300–0 mm sized raw coals

5.4 Reject pre-extraction process

For coals with high content of reject, the TDS can be used for pre-extracting the rejects from 300–50 (25) mm sized lump coals; the fine coals shall be broken and separated by dense-medium cyclone with 50 (25)–0 mm smalls. See Fig. 6 for the specific flow.

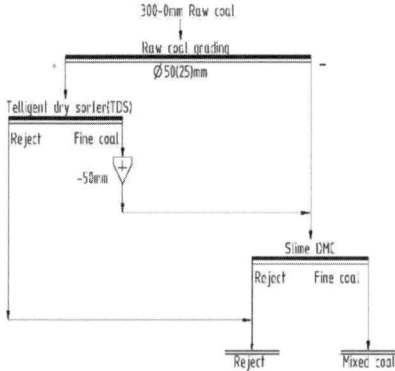

Figure 6: Diagram of TDS reject pre-extraction process

6. Industrial application of TDS

Since 2016, TDS has been applied to projects of Zhaozhuang Coal Industry, Pingshang Coal Industry, Huashengrong Coal Mine, Jinkeer Coal Mine, Huozhou Coal Industry, Yanzhou Coal Industry, Zhunneng Group, ENN Group, Linyi Mining Group, Xinwen Group, Huaibei Mining Group, etc.

Table 1: Application of TDS

Name of owner	Capacity (t/h/unit)	Separation size (mm)	Separation accuracy
Zhaozhuang Coal Mine of Jincheng Anthracite Coal Mining Group	150	300–100	Coal content in reject: 0.85% Reject extraction rate: ≥95%
Pingshang Coal Mine of Jincheng Anthracite Coal Mining Group	100	100–25	Coal content in reject: 1–3% Raw coal ash content: 20–40% of fluctuation, Fine coal ash content: Ad≤14%
Huashengrong Coal Mine	160	300–50	Coal content in reject: 0.54% Reject extraction rate: 97.80%
Jinkeer Coal Mine	50	100–30	Coal content in reject: 0.39% Reject content in coal: 1.4%

Contd...

Contd...

Name of owner	Capacity (t/h/unit)	Separation size (mm)	Separation accuracy
Huozhou Ganhe Coal Preparation Plant	180	300–50	Coal content in reject: 0.30%
Shenhua Ningdong Lingxin Coal Preparation Plant	240	200–40	Coal content in reject: 1.16% Reject content in coal: 2.25%
Sanhekou Coal Preparation Plant of Zaozhuang Mining Group	100	300–50	Coal content in reject: 1% Reject extraction rate: ≥95%

It can be seen from actual operation results of the above projects that the TDS has characteristics of high separation accuracy, wide range of separation sizes, large processing capacity, high intelligence level, stable operation, etc. TDS is widely promoted and used in China for separating 300–25 mm sized lump coals.

7. Conclusion

(1) TDS adopts the X-ray and image dual-source recognition system and array actuator, resulting in high separation accuracy, wide range of processing sizes and large processing capacity.

(2) Compared with traditional washing process, TDS does not need water or medium, and has advantages of simple system, investment saving, low production cost, high intelligence level, and little environmental pollution without coal slime generated or product moisture increased.

(3) TDS can be used for reject pre-extraction and separation of 300–25 mm sized lump coals, and realize the full size fraction separation of raw coals, underground reject extraction, and non-coal mineral separation; therefore, TDS is suitable for the separation of coals and lignite in water-deficient areas, and has broad prospect in application.

(4) To date, more than 40 sets of TDSs have been successfully used with mature technologies, and are now proactively promoted on the market.

8. References

[1] Kuang Yali. Process Design and Management of Coal Preparation Plants. Xuzhou: China University of Ming and Technology Press, 2006.

[2] Dai Shaokang. Design Idea and Method of Coal Preparation Process. Beijing: China Coal Industry Publishing House, 2003.

[3] Zhou Shaolei, Zeng Zhiyuan. Discussion on Utilization of New Technology and Process in Clean Processing of Coals in Western China. Coal Processing & Comprehensive Utilization, 2016(7):5–10.

[4] Qiao Zhizhong, Liu Libo. Study on Application of Telligent Dry Separator (TDS) in Zhunenng Haerwusu Coal Preparation Plant. Coal Preparation Technology, 2017(5):35–37.

[5] Duan Fushan, Sun Youbin. Application of Telligent Dry Separator (TDS) in Ganhe Coal Mine. Coal Preparation Technology, 2018(02):68–72.

[6] Yao Guanghua, Yang Hongzhi, Hao Jun. Analysis on Application Necessity of Intelligent Dry Separation in Low-rank Lump Coal. Coal Processing & Comprehensive Utilization, 2016(9):13–16.

17

Simulation of a horizontal flow coal winnower

Quentin Campbell, Marco le Roux, Loo Roy Morgan

North-West University, Potchefstroom, South Africa

Abstract: Dry coal processing is an attractive beneficiation technique, especially in arid areas. Where water is abundant, dry processing still has environmental benefits regarding water pollution. There are a number of dry processing options available, but these are more suited to coarser and easy-to-beneficiate feeds. Attempts to process dry fine coal industrially are still mostly in the developmental stages with limited practical application.

Air winnowing has the potential to be used for dry fine coal processing, and it has been used for millennia in agriculture. A proof of concept experimental setup, consisting of a closed box through which a horizontal air stream flows, has been developed. Coal or tracer particles are dropped into the air stream and are displaced horizontally by a distance depending on the particle size and density, thus establishing separation based on these parameters.

Modelling and simulation of the system using fundamental equations of motion was done to assist in designing and optimizing the dimensions and operation of a laboratory scale test apparatus. Using tracers, the simulation results as well as the size dependency of the separation was confirmed, and the partition data for narrow size ranges were promising. After the next stage of operating on real coal samples with realistic wash ability characteristics, the information will be used in the scale-up towards a pilot- and full-scale prototype.

Keywords: Dry processing, fine coal, winnowing, simulation

1. Introduction and background

Dry fine coal beneficiation is becoming increasingly important in South Africa due to the decline of high-quality coal resources coupled with the depletion of suitable process water sources. Fourie et al. (1980) states that it has always been custom to either discard the coal smaller than 5 mm or blend it unwashed into a steam coal product, but it is becoming imperative, both from an environmental and financial point of view, to treat the fine material. South African coal operations are mostly beneficiating fine coal using wet separation methods such as the dense medium cyclone (van Houwelingen and de Jong, 2004), but, the handling and processing of fine coal is much more water-intensive. This has a significant environmental impact (de Korte, 2015).

Water scarcity and pollution is a global risk and given that South Africa is a water scarce county, the impact on coal beneficiation in future will be significant. The country is currently faced with challenges where the demand for water exceeds the availability (Kanjere et al., 2014). Several arid and semi-arid regions in South Africa have very rich coalfields, however due to unpredictable climate changes, unforeseen droughts and seasonal rainfall these coalfields cannot be completely utilized because of the lack of water (Jeffrey, 2005).

Thus, there is a need for research into dry processing methods, especially for fine coal. One alternative to consider is air driven separators that can be viable substitutes for wet processes, if correctly applied. Several air driven dry beneficiation methods are being investigated and even implemented, such as the FGX and X-ray sorter (de Korte, 2015), and although some have been commercialized these methods cannot in general compete with the efficiencies achieved by dense medium separation.

The agricultural sector has been using wind for hundreds of years to separate wheat from chaff by using a technique called winnowing. The concept of winnowing is based on the difference in density between particles. The method is traditionally carried out by throwing the mixture of particles up in the air and allowing the wind to blow through the particles. The lighter particles are carried away by the force of the air along the horizontal axis, while the heavy particles will stay behind or in this case fall back down.

If the principle of winnowing is applied to fine coal particles a similar approach can be taken. By allowing air to pass through a falling or suspended stream coal particles, the lighter coal will be transported away leaving the heavier particles behind. The question remains as to whether or not efficient separation can be achieved for fine (say, -6 mm) particles. Winnowing is being investigated by a number of researchers to find practical and economic ways to apply this principle to fine coal (Sakhre et al., 2018).

2. Experimental equipment and procedure

2.1 Design of the winnowing box

A laboratory size horizontal winnowing box was designed using first principles to determine the dimensions, as well as the required airflow velocity to achieve a certain horizontal displacement for a range of particle sizes and densities. The design for the winnowing box started with the concept of a closed box (Fig. 1) with a single particle inlet, an air inlet, and an air outlet sufficiently large not to have a pressure build up.

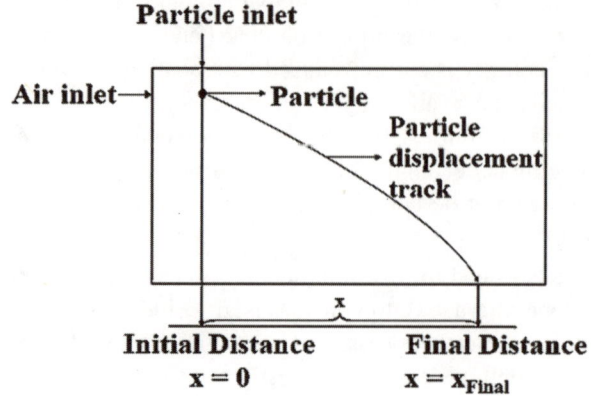

Figure 1: Initial design concept of the winnowing box

Fundamental laws of motion incorporating an air drag factor were used to predict the behaviour of ideal particles under ideal conditions.

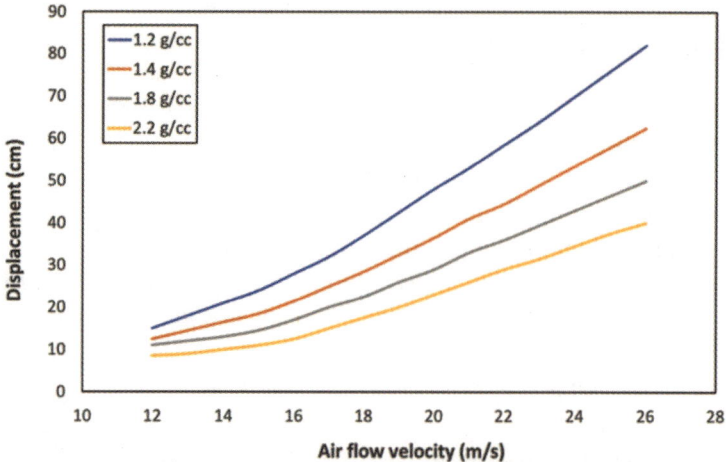

Figure 2: Predicted horizontal displacement of 4 mm tracer particles dropping from a height of 0.63 m above the bottom of the box

Figure 2 shows the displacement of 4 mm spherical particles, for example, if introduced into a horizontally flowing air stream at a height of 0.63 m above the point of impact. Table 1 lists a selection of these values for three particle sizes. The flow velocities to be considered in the project were chosen within the range possible with a 1.6 kW household vacuum cleaner in reverse (blower) mode.

From this information, a flow velocity of 21 m s^{-1} was selected to be used for all tests to ensure that a practically sized box length of about 1 m can accommodate the horizontal movement of any of the particles in question. The motion calculations did not take into account the flow regimes, so the particle Reynolds numbers (Rep) were used to determine whether flow was laminar or turbulent (Table 2). It was found that turbulent flow existed for all cases. While it would not affect the operation of the winnower significantly, it was expected to introduce some variability in the subsequent results.

The final design of the experimental winnowing box, constructed from plywood and transparent acrylic sheet, is shown in Fig. 3. Moveable separating wedges were placed to create variable sized collection bins on the floor of the box. The vacuum cleaner was connected to the air feed port of the box, and the air flow velocity and patterns were measured inside the box using a FLUS Professional ET-961 hot film anemometer placed at different points in the box. The velocity just inside the feed port was correlated with the power setting on the vacuum cleaner, and it was set at 21 m s^{-1} for all tests in this experiment – a setting considered to be suitable to achieve separation for most sizes and densities within the dimensional confines of the box.

Table 1: Horizontal displacement distances of particles for selected air flow velocities

Particle size (mm)	Air flow velocity (m s^{-1})	Particle density (g cm^{-3})					
		1.3	1.4	1.5	1.6	1.8	2.0
6	18	21	19	18	16	14	12
	20	24	22	21	19	16	14
	22	28	26	24	23	19	16
	24	34	32	30	28	24	20
4	18	33	29	27	26	23	20
	20	42	37	35	33	29	26
	22	52	45	42	40	36	33
	24	62	54	51	48	43	39
2	18	67	55	54	53	50	45
	20	84	70	68	66	62	56
	22	101	84	82	79	75	68
	24	121	101	98	95	89	81

Table 2: Particle Reynolds numbers for a range of velocities and particle sizes

Particle size (mm)	Air flow velocity	
	20 m s⁻¹	**22 m s⁻¹**
6	7656	8421
4	5103	5614
2	2552	2807

2.2 Tracer tests

To simulate the separation of coal particles by winnowing according to size and density, a series of density tracers were used. The plastic cubic tracers were color-coded according to density to facilitate the determination of partition curves after separation. The tracers were introduced one-by-one by hand via a small port in the top of the box and were collected in the collection bins at the bottom.

During phase 1 scoping tests were done to confirm the expected to the actual impact points at the bottom for single sized tracers. The predictions were found to be adequate, taking into account the variability introduced by operating in the turbulent flow regime. Hence the position of the separating wedge from the datum point directly below the feed port (defined by the point $x = 0$ on Fig. 3.1) could be established to produce target partition cut points using the data in Table 1 for each tracer size.

Figure 3: Experimental horizontal flow of winnowing box

The main investigation (phase 2) was performed using 36 tracers of each size (2 mm, 4 mm, and 6 mm) and each density (1.3–2.0 g cm⁻³, in intervals of 0.1 g cm⁻³) were introduced together, giving a total of 1620 tracers. The airflow rate was set at 21 m s^{-1}. From the results obtained in phase 1, two wedges were placed at positions $x = 28$ cm and 50 cm, respectively, to create three convenient product bins. The tracers reporting to each bin were counted, classified according to size and density, and partitions curves were constructed.

Finally, phase 3 was done to determine the predictability of cut points using calculated wedge positions for 6 mm tracers only. A run was done for each of two cut points (1.7 g cm−1 and 1.5 g cm^{-1}) with corresponding wedge positions at $x = 20$ cm and then $x = 23$ cm, respectively at a flow velocity of 21 m s^{-1}. Again, tracers reporting to either side of the wedge, mimicking the light "product" and the heavy "discard" were collected, counted, classified, and the partition data determined.

3. Results and dicussion

3.1 Feed densimetric data

Since the same number of tracers (36) of each density was introduced in all cases, a linear feed densimetric curve was defined, resulting in a separation difficulty classification of "formidable" according to the Bird's scale. Whilst it is acknowledged that this is unrealistic compared to real coal wash ability characteristics, it was decided that the effect of variations in feed wash ability would be considered at a later stage in the investigation. This can easily be achieved by manipulating the numbers tracers per density, or by using coal as a feedstock. This is, by definition, not expected to influence the mechanistic operation of the winnower, but only the yield, e.p.m., and cut point of separation.

3.2 Size and density responses

The results from phase 2 confirmed, as expected, a preference to size separation rather than density. Figure 4 shows the number of tracers of a particular size and all densities reporting to each of the three bins, and it is evident that size separation is the primary effect. Bin 1, closest to the feed point, contained mostly 6 mm tracers, while bin 3, the furthest away, contained mostly 2 mm tracers. Bin 2 in the middle contained a mixture of sizes, but with 4 mm tracers prevalent. This indicates that only very closely sized feed material can be practically separated according to density – a result completely in line with wet and dry gravity separation methods. The distributions of densities within each of the three bins for selected sizes are shown in Figs. 5–7. For 2 mm particles (Fig. 5), most of the light material was concentrated in bin 3, while the central bin 2 received the heavier particles.

Virtually no particles reported to bin 1, since the small 2 mm tracers were easily carried to the furthest part of the box. For 4 mm particles, most of the separation was accomplished between bins 1 and 2 (Fig. 6). For both 2 mm and 4 mm tracers there was a significant amount of misplaced material that would translate into a low separation sharpness or e.p.m. in the practical case, but the poor densimetric characteristic of the feed population was the cause of this. The 6 mm particles (Fig. 7) did not carry very far and reported mainly to bin 1, with some of the lighter fraction landinginbin 2. Again, there was a significant inefficiency for the same reason. The size sensitivity of air winnowing is confirmed by these observations, and density separation using a wide feed size distribution is indeed problematic.

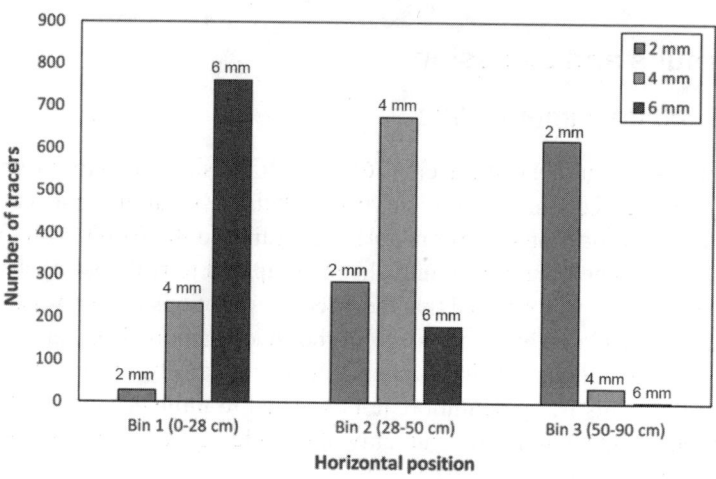

Figure 4: Size responses of tracers of all densities at 21 m s−1 air velocity

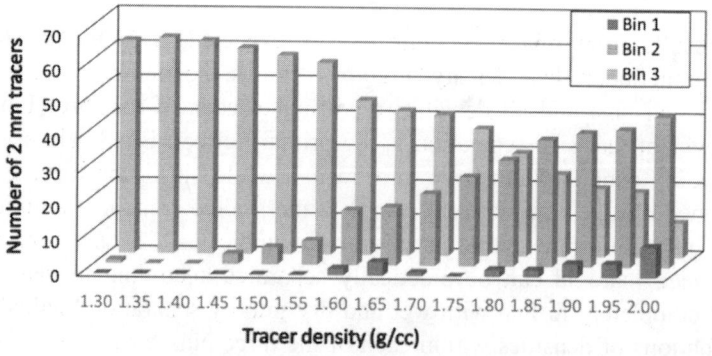

Figure 5: Density distribution of 2 mm tracers in the three product bins

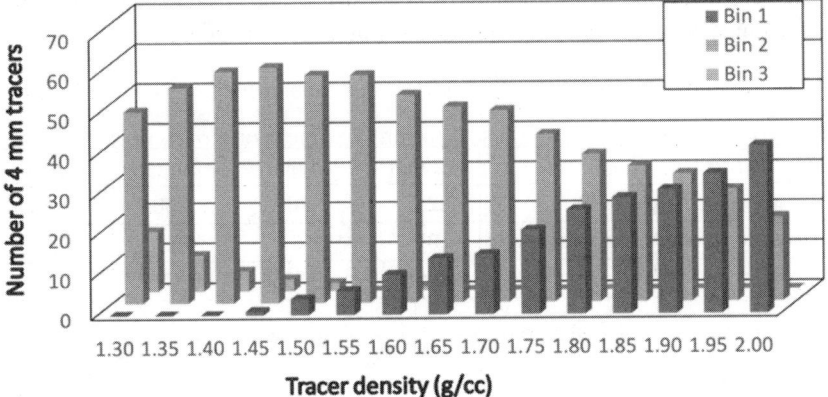

Figure 6: Density distribution of 4 mm tracers in the three product bins

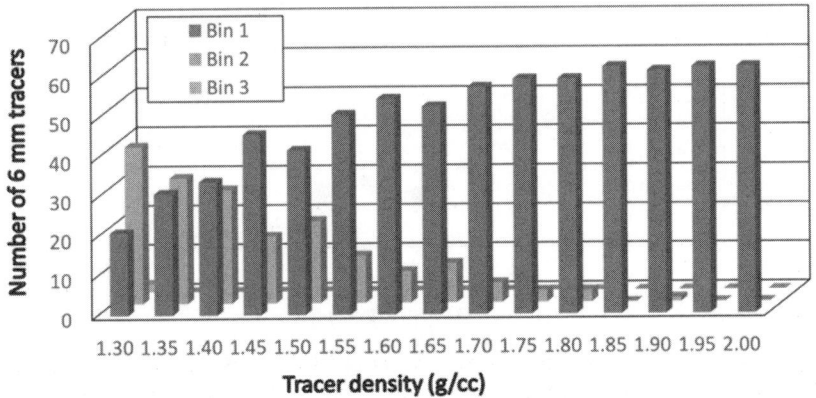

Figure 7: Density distribution of 6 mm tracers in the three product bins

To obtain an indication of the expected yields, cut points and separation sharpness for the mixed sized feeds, partition curves were calculated for the arbitrary case where bin 1 was considered the product, while the combined bin 2 and bin 3 contents constituted the discard. The Scott and Napier-Munn (1992) theoretical partition function was fitted to the experimental points, and the e.p.m. and cut points were determined by curve fitting. Figure 8 shows the individual partition curves for each size, as well as the combined overall partition curve. It can be seen that there are distinct partition curves, each with a unique cut point and e.p.m. for each size. It is also evident that the overall efficiency of separation is much worse than for anysingle size, which again confirms the need for narrow feed sizes. The operational parameters determined from the partition data are summarized in Table 3.

Table 3.3: Performance data for the simultaneous separation of all sized particles at 21 m s⁻¹

Particle size	2 mm	4 mm	6 mm	Overall
Actual cut point (g cm−3)	2.08	1.89	1.38	1.85
Yield (by number of tracers) (%)	97.2	75.2	19.4	63.9
E.p.m.	0.054	0.177	0.136	0.384

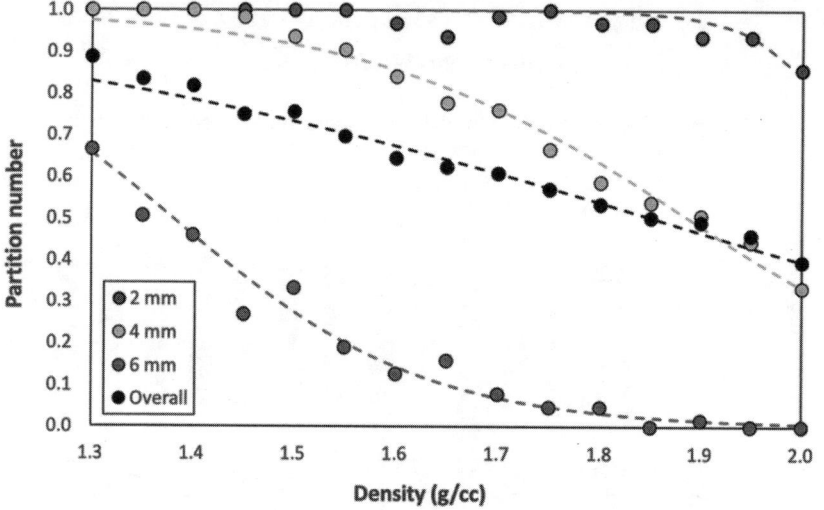

Figure 8: Partition curves for all sized tracers at 21 m s⁻¹. Separation at x = 23 cm. Partition curves per size for the wedge at x = 28 cm, and an air flow velocity 21 m s⁻¹. The dotted lines signify the fitted Scott and Napier-Munn partition function

3.3 Partition data for 6 mm tracers

The purpose of the phase 3 tracer tests was to demonstrate that it was possible for the operator to have some predictive control over the cut point by modifying the position of the separating wedge. Yield and e.p.m. information for specifically the 6 mm tracers were generated, which gave an indication of the order of magnitude of the e.p.m. that could be expected in future applications. Obviously, the tracers had regular shapes, and a very narrow size, hence it is expected that irregularly shaped real coal particles would not perform as well. This would be the aim of the investigation in the next stage of the project.

Table 4 shows the results for the two cases where the target cut points were set at 1.5 g cm^{-3} and 1.7 g cm^{-3} and Fig. 3.9 shows the partition curve for these target cut points.

Table 4: Performance data for the separation of 6 mm particles at 21 m s^{-1}

Target cut point (g cm^{-3})	1.70	1.50
Actual cut point (g cm^{-3})	1.72	1.58
Wedge position (cm)	20	23
Yield (by number of tracers) (%)	58.7	42.2
E.p.m.	0.17	0.16

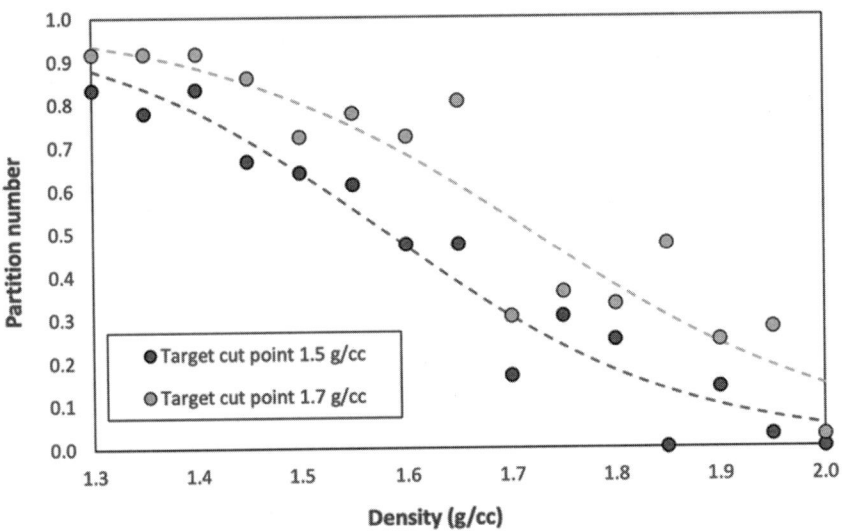

Figure 9: Partition curves for 6 mm tracers with the separating wedges set at 23 cm (for 1.7 g cm^{-3}) and 20 cm (for 1.5 g cm^{-3})

Again, the scatter is the partition data was due to turbulence in the winnowing box, and the fact that some tracer breakage was starting to appear after repeated use. Despite this, it was possible to show that targeted cut points were possible by an adjustment of the separating wedge, and an adjustment in the air flow velocity, even though the latter was not tested during this stage on the project.

4. Conclusion

Dry fine coal separation by winnowing has been shown as a feasible option that warrants further investigation. It was shown, initially by studying tracers, that the size sensitivity of this gravity process makes it imperative to use narrow size distributions for adequately sharp separations. It was shown that with narrow distributions, good separation can be obtained (with e.p.m. values in the order of 0.16 or 0.17) but this would require very good control over the air flow velocity and flow regime inside the box. In all cases linear feed densimetric distributions were used, but efficiencies will improve significantly if more realistic wash ability material is used.

It was possible to accurately control the cut point by changing the position of the separating wedge to create the two product bins.

The next stage of development of the winnowing process would be to introduce real coal samples, that will have realistic densimetric characteristics, irregular shapes, and non-uniform size.

The major aspect to address would be the size-density interdependency. To alleviate this effect, it is planned to have a two-stage process in the next generation of winnowing boxes, where the first stage will be used for size separation only into any number of fractions (like in Fig. 3), followed by a second stage density separation on each of the narrow size products from the first stage. It is clear that there is a potential for density separation of the winnowing process.

5. Acknowledgement

The authors wish to acknowledge the financial and technical contribution made by The Coaltech Research Association NPC Ltd., and the Centre of Excellence in Carbon Based Fuels, North-West University, South Africa.

This work is based on the research supported in part by the National Research Foundation of South Africa (Coal Research Chair Grant Numbers: 86880; and UID85643, UID85632). Any opinions, findings and conclusions or recommendations expressed in any publication generated by the NRF supported research is that of the author(s) alone, and that the NRF accepts no liability whatsoever in this regard.

6. References

[1] de Korte, G., 2015. Processing low-grade coal to produce. The Journal of The Southern African Institute of Mining and Metallurgy, Volume 115, pp. 569–572.

[2] Fourie, P. J. F., Van der Walt, P. J. & Falcon, L. M., 1980. The beneficiation of fine coal by dense-medium cyclone. The Journal of The South African Institute of Mining and Metallurgy, pp. 357–361.

[3] van Houwelingen, J. & de Jong, T., 2004. Dry cleaning of coal: review, fundamentals. Geologica Belgica, 7(3–4), pp. 335–343.

[4] Jeffrey, L., 2005. Characterization of the coal resources of South Africa. The Journal of The South African Institute of Mining and Metallurgy, pp. 95–102.

[5] Kanjere, M., Thaba, K. & Lekoana, M., 2014. Water shortage management at Letaba water catchment area in Limpop Province, of South Africa. Mediterranean Journal of social Sciences, 5(27), pp. 1356–1360.

[6] Sakhre, D. et al., 2018. Coal winnowing: an innovative technique for dry coal beneficiation. CPSI, 10(28), pp. 3–13.

[7] Scott, I.A. & Napier-Munn, T.J. 1992. A dense medium cyclone model based on the pivot phenomenon. Transactions of the Institutions of Mining and Metallurgy (Section C: Mineral Processing and Extractive Metallurgy) Volume 101, pages C61–C76.

18

Recovery of clean coal from 300-0 mm high ash dirty coal by using both gamma-ray sorter and ZM dry separator at dafeng coal mine

Yunkai Xia and Gongmin Li

Tangshan Shenzhou Manufacturing Co., Ltd

Abstract: A large amount of dirty coal is produced annually during open-pit mining operation at Dafeng coal mine. The high ash dirty coal not only occupies the mine area but also causes environmental issues, such as easily spontaneous coal combustion and landslide disaster of coal pile. Owing to the high raw coal ash, high rock hardness, there are problems with conventional dense medium separation process in handling dirty coal, such as difficulty to meet high deshaling requirement, low clean coal yield, severe wear and tear on equipment and pipes, wasting reject transportation cost, etc. The newly designed process, made of gamma-ray coal sorter processing 300-70mm coal, ZM type dry separator handling 70-0mmcoal and fine removal from clean coal by screening, can recover good quality clean coal with an average yield of 35% and overall ash below 28%. After successful implementation of the project, the coal mine can improve coal quality after pithead deshaling, it utilizes the coal resources reasonably and creates economic & social benefits for the enterprise.

Keywords: Dry coal cleaning, ZM dry separator, Gamma-ray sorting, High ash coal

1. Introduction

The anthracite coal produced by Shenhua Ningxia Coal Group Company is known as "the king of coal" in China, it has the characteristics of low ash, low moisture, low sulfur, low phosphate, high calorific value, high resistance ratio, lump coal rate, high chemical activity, high plant recovery, and high mechanical strength, it enjoys a high reputation both at home and abroad. To fully recover scarce resource and prolong the service life of coal mine, the coal mine changed the shaft mining into open-pit mining. With the increase of coal mining mechanization degree, a lot of dirty coal diluted with rocks has been made during operation of surface coal seam mining and in the process of recycling coal pillars in the underground mined-out area. Millions of tons of dirty coal have been piled up in coal yard and more will be produced annually with the mining operation.

The mined raw coal is transported to Taixi Coal washery for processing. The joint separation process, consisting of heavy medium cyclone + coal slime processing + froth flotation, is designed to produce clean coal product with ash content below 8%. Due to the particularity of the joint washing process, they need to control raw coal ash. If the raw coal ash is too high and washed directly by the wet process in Taixi coal washery, it will cause problems such as high quantity of discarded rejects, low clean coal yield, severe wear and tear of equipment, unstable product quality and repeat deshaling, etc. The transportation of dirty coal from open pit to coal washery will increase coal resource management cost, invalid transportation cost of waste rock and cost of subsequent deshaling cost. The overall economic benefits of the coal mines will be seriously influenced.

The annual dirty coal output is about one million tons, it cannot be directly processed in Taixi coal washery due to its extreme high ash content. The dirty coal pile occupies much of the land at the mine site and it even affects the progress of the mine project construction. In view of the scarcity of and high added value of Taixi coal, if properly handled, the washing of low yield dirty coal can produce high quality low ash clean coal and create huge economic benefits. Therefore, the dirty coal shall be processed and recovered rather than being sold at a low price. Restricted by available coal washery process at the moment, the raw coal ash content in washery must be controlled under 28%. So the on-site deshaling and upgrading of dirty coal in pithead to meet required clean coal ash content requirement in coal washery and then transporting lower ash coal to downstream coal washery for separation is the best way to solve the current dirty coal problem.

Dry coal preparation technology has advantages of no water usage, electricity saving, low investment and operation cost, it has been widely used in China for thermal coal deshaling and pre-deshaling of coking coal. The well-known commercial dry coal separators widely used worldwide are Chinese FGX and ZM type separators (Li and Yang, 2006; Honaker et al. 2006) and Germany All air jig of Allmineral Company (Kelly and Snoby, 2002. Weinstein and Snoby, 2007; Kip Alderman et al, 2013). These equipment are used mainly for the processing of <80mm mixed coal and fine coal（Ghosh T., Patil D.et al., 2013; Patil, D.; Parekh, B., 2011）. The R&D of compound separator has made significant progress in past 30 years in the area of unit capacity, separation accuracy, automation level, environmental protection performance, modular design, individualized design, product seriation and equipment performance reliability, etc. It is the preferred choice for processing of high ash thermal coal and easy degradation low rank coal.

For ZM type dry separator, the commercial unit module capacity is increased from 30 tons/hour to 600 tons/hour, its separation accuracy in handling easy-to-wash thermal coal can match dense medium vessel coal separator and jig in term of clean coal heating value increases. The emitted dust content and noise level of the dry separator is also in line with the latest environmental protection law requirements. When separating easy degradation coal, the dry separation can solve the problem that no wet process can be used for easy degradation coal and low rank coal. The development of ZM separator has formed series equipment which can be used for processing of mixed coal, lump coal and fine coal separately.

Coarse coal intelligent sorting machine such as x-rays and gamma rays separator can be used for > 80 mm lump coal separation to replace rock hand-picking operation, the mechanized coarse coal sorting will also save rock crushing cost (Cristoper Robben et al, 2013,), Gamma ray based sorting is good for large particle separation and is very flexible in terms of achievable density cut-points. For easy-to-clean coal processing, it offers an easy and viable separation option. As the auxiliary system of production in the coal mine, the dry cleaning equipment is chosen for pithead coal deshaling, it has characteristics of the short construction period, low investment cost, simple process, reliable operation performance. The lower the operation cost, the greater economic benefit created for coal mines.

2. Raw coal washability analysis

Sampled at some points in the dirty coal yard, raw coal quality analysis shows an ash content in a range of 38-65%, low moisture, low calorific value, and high content of rocks. It is shown from raw coal size distribution analysis as shown in Tables 1 and 2 that raw coal is one of special high ash coal with ash content of 63.98%, raw coal sample also contains general lump coal rate with 50.82% of +13 mm material.

For +13mm lump raw coal, the ash content increases with decreasing of coal particle size; on the contrary; for -13mm fine raw coal, the raw coal ash content decreases with decreasing of coal size. This explains that fine coal is fragile and lump coal contains relatively more rocks than fine coal.

Raw coal sample contains 22.67% of <6mm fine coal, raw coal is of fine size and its powder coal content is high. There exists a large quantity of minus 3 mm fine coal content and its ash content is high, the fine high ash coal powder shall be screened out in separation process. The main purpose in coal preparation is deshalling of lump coal while the problems of a large quantity of high ash fine coal powder and easy degradation property of raw coal shall be reminded. If raw coal is handled by wet process, it is predicted that fine clean coal product will have high moisture content which will adversely affect clean coal heating value, therefore, this raw coal shall not be separated directly in the wet process without pretreatment.

Table 1: Raw coal size analysis

Size, mm	Individual yield,%	Plant Yield,%
+60(coal)	20.44	5.13
+60(rock)	79.56	19.96
-60		74.91
Total		100

Table 2: Size distribution of -60mm raw coal after crushing

Size mm	Wt %	Individual			Cumulative				
		Moisture Mt%	Ash Ad%	LHV Qnet.ar, MJ/kg	Size mm	Wt%	Moisture Mt%	Ash Ad%	LHV Qnet. ar, MJ/kg
+25	29.80	3.69	66.88	8.85	+25	29.80	3.69	66.88	8.85
25-13	21.02	3.78	69.33	7.99	+13	50.82	3.73	67.89	8.49
13-6	18.83	3.94	61.58	10.75	+6	69.65	3.78	66.19	9.10
6-3	7.68	3.92	59.93	11.32	+3	77.33	3.80	65.57	9.32
-3	22.67	6.70	58.58	11.32	Total	100.00	4.46	63.98	9.77
Total	100.00	4.46	63.98	9.77					

Table 3: Float-sink testing of +25mm raw coal

S.G g/cm³	Wt %	Individual			Cumulative				
		Moisture Mt%	Ash Ad%	LHV Qnet.ar, MJ/kg	Size mm	Wt %	Moisture Mt%	Ash Ad%	LHV Qnet.ar, MJ/kg
-1.4	11.49	5.1	4.66	32.15	-1.4	11.49	5.10	4.66	32.15
1.4-1.6	10.34	5.1	7.58	30.68	-1.6	21.83	5.10	6.04	31.45
1.6-1.8	4.42	4.8	15.54	26.84	-1.8	26.25	5.05	7.64	30.68
+1.8	73.75	3.2	87.96	1.08	Total	100	3.69	66.88	8.85
Total	100	3.69	66.88	8.85					

Table 4: Float-sink testing of 25-13mm raw coal

S.G g/cm³	Wt %	Individual			Cumulative				
		Moisture Mt%	Ash Ad%	LHV Qnet.ar, MJ/kg	Size mm	Wt %	Moisture Mt%	Ash Ad%	LHV Qnet.ar, MJ/ kg
-1.4	9.15	5.3	5.07	31.89	-1.4	9.15	5.30	5.07	31.89
1.4-1.6	11.62	4.9	7.94	30.62	-1.6	20.77	5.08	6.68	31.18
1.6-1.8	2.47	4.5	20.25	25.46	-1.8	23.24	5.01	8.12	30.57
+1.8	76.76	3.4	87.86	1.15	Total	100	3.78	69.33	7.99
Total	100	3.78	69.33	7.99					

Table 5: Float-sink testing of 13-6mm raw coal

S.G g/cm³	Wt %	Individual				Cumulative				
		Moisture Mt%	Ash Ad%	LHV Qnet.ar, MJ/kg	Size mm	Wt %	Moisture Mt%	Ash Ad%	LHV Qnet.ar, MJ/kg
-1.4	9.66	5.4	3.93	32.04	-1.4	9.66	5.40	3.93	32.04
1.4-1.6	17.24	5	7.37	30.36	-1.6	26.90	5.14	6.13	30.96
1.6-1.8	5.52	4.7	18.83	25.93	-1.8	32.42	5.07	8.30	30.11
+1.8	67.58	3.4	87.14	1.46	Total	100	3.94	61.58	10.75
Total	100	3.94	61.58	10.75					

Table 6: Float-sink testing of 6-3mm raw coal

S.G g/cm³	Wt %	Individual				Cumulative				
		Moisture Mt%	Ash Ad%	LHV Qnet.ar, MJ/kg	Size mm	Wt %	Moisture Mt%	Ash Ad%	LHV Qnet.ar, MJ/kg
-1.4	8.99	5.2	3.37	32.18	-1.4	8.99	5.20	3.37	32.18
1.4-1.6	16.81	5.0	6.74	30.14	-1.6	25.80	5.07	5.57	30.85
1.6-1.8	7.15	4.6	15.16	27.10	-1.8	32.95	4.97	7.65	30.04
+1.8	67.05	3.4	85.63	2.12	Total	100	3.92	59.93	11.32
Total	100	3.92	59.93	11.32					

It is shown from raw coal float-sink testing analysis in Tables 3~6:

(1) The dominant density component in each size fraction is >1.8 g/cm³, the low ash (<5%) clean coal for chemical industry can be made at low density cutting point of 1.4 g/cm³, however, the clean coal yield is very low.

(2) The ash content of +1.8 g/cm³ density fraction is above 85%, it explains that raw coal contains very high ash rocks with low heating value.

(3) The 1.6-1.80g/cm³ intermediate density fraction content is low and raw coal shows easy washability. This coal is suitable for dry deshaling at high density (>1.80g/cm³)

(4) The top quality clean coal with a heating value above 29.3 MJ/kg can be recovered from high ash lump coal.

3. Design of dry coal cleaning process flowsheet

In 2017, Tangshan Shenzhou Manufacturing Company designed and built a 1.0 Mtpa dry coal preparation plant. The try separation of raw coal with different ash contents were completed from January 13 to 15, 2017. The testing purpose is to inspect equipment performance and product quality. It also provides basic data for production line acceptance and system improvement in the future.

The coal preparation process flowsheet is shown in figure 1. The inferior coal from coal storage yard is classified at 70mm, the oversize and undersize products are sent to the intelligent separator and ZM200 type separator respectively. The +70mm oversize coal is separated into +70mm lump clean coal and reject. The ZM200 type separator will handle -70mm coal and produce clean coal, middlings and reject. The -70mm clean coal is further classified into 70-3mm and -3mm coal; the middlings is also classified at 8mm. The 70-8mm becomes final middlings product and -8mm fine middlings is blended and discarded with reject. The coal dust from dust collecting system is also discarded with reject.

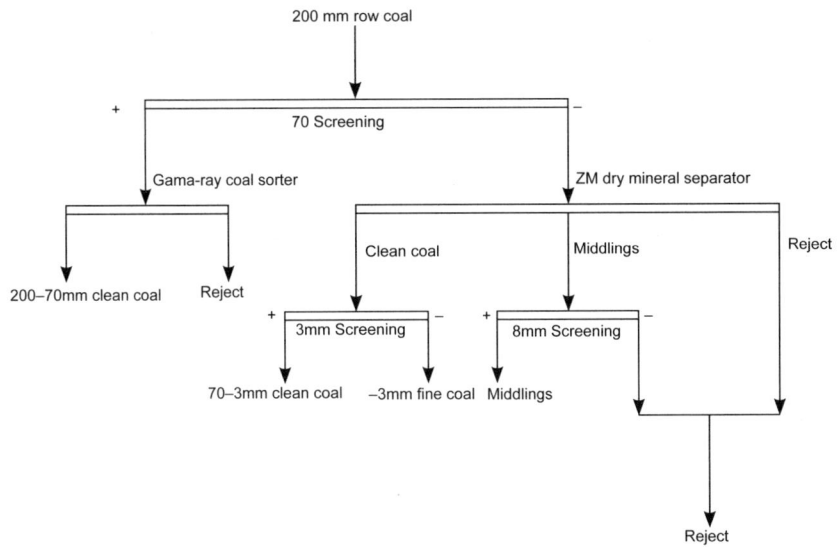

Figure 1: Process flowsheet of dry separation

4. Results and discussions

4.1 Simple one-stage dry separation

The product balance of dry separation of high ash raw coal after one-stage primary processing is listed in Table 7. Deducting the screen out -3 mm fine coal, the clean coal yield reaches 35.63% and clean coal ash content is 31.75% which is above the clean coal target ash content of 28%. The existing problem is mainly the low classification efficiency and screen plugging when screening at 3mm.

Table 7: Product balance of primary dry separation of high ash raw coal

Product	Size, mm	Wt, %	Ash, %
Clean coal	>70	8.89	17.36
	70-13	6.99	13.55
	13-3	10.95	41.77
	<3	8.80	48.29
	Total clean coal	35.63	31.75
Coal powder	<3	1.80	48.29
Middlings		10.54	50.04
Reject of zm separator	70-0	32.92	65.08
Reject of gamma-ray sorter	>70	19.11	78.99
Raw coal	Total	100.00	53.98

The main cause of cleaned coal ash content exceeding target ash is that there is high content (24.7%) of -3mm high ash fine coal in clean coal product. Theoretically, calculation shows that is 75% of -3mm fine coal can be screened out and the screen oversize product will meet clean coal ash target(<28%). As shown in Table 8, the clean coal yield is as low as 29.04% after clean coal screening. The need for screening out a large amount of fine coal puts the high requirement for screening system.

Table 8: Product balance of primary dry separation of high ash raw coal with improved clean coal screening

Product	Size, mm	Wt, %	Ash, %
Clean coal	>70	8.89	17.36
	70-13	6.99	13.55
	13-3	10.95	41.77
	<3	2.21	48.29
	Total clean coal	29.04	28.00
Coal powder	<3	8.39	48.29
Middlings		10.54	50.04
Reject of ZM separator	70-0	32.92	65.08
Reject of Gamma-ray sorter	>70	19.11	78.99
Raw coal	Total	100.00	53.98

4.2 Two stage dry separation (primary and secondary separation)

The other option to control overall clean coal ash content is to re-clean and lower ash of -70mm primary clean coal product. The testing shows that 6.97% of high ash reject (ash = 71.24%) can be removed from primary clean coal product in the secondary separation process, the clean coal ash is lowered by 2.72 percentage points.

Table 9: Product balance of re-cleaning of 70-0mm primary clean coal

Product	Size, mm	Wt, %	Ash, %
Clean coal	70-13	17.18	9.01
	13-3	35.04	35.77
	<3	39.31	43.90
	Total clean coal	91.53	34.24
Screen out coal powder	<3mm	1.50	43.90
Reject from ZM separator	70-0mm	6.97	71.24
Total		100.00	36.96

After two stage separation and deducting fine coal from primary clean coal, the product balance of processing high ash raw coal is shown in table 10. The clean coal yield is 33.36% and clean coal ash is 29.34% which still can't meet the < 28% ash content target. The main cause of cleaned coal ash content exceeding target ash is that there is high content (25.2%) of -3mm high ash fine coal in clean coal product after two stage separation.

As shown in Table 11, theoretically calculation shows that if 30% of -3mm fine coal can be screened out and the screen oversize product will meet clean coal ash target(<28%). Therefore, the requirement of screening out -3mm fine coal from clean coal product can be loosened if the two-stage dry separation process can be utilized.

Table 10: Product balance of primary and secondary dry cleaning of high ash raw coal

Product	Size, mm	Wt, %	Ash, %
Clean coal	>70	8.89	17.36
	70-3	16.07	26.06
	<3	8.40	48.29
	Total clean coal	33.36	29.34
Screen out coal powder	<3	2.20	48.29
middlings		10.54	50.04
Reject of Primary ZM separator		32.92	65.08
Reject of Secondary ZM separator		1.87	71.24
Gamma-ray sorter reject		19.11	78.99
Raw coal	Total	100.00	53.98

Table 11: Product balance of primary and secondary dry cleaning of high ash raw coal with improved clean coal screening

Product	Size,mm	Wt, %	Ash, %
Clean coal	>70	8.89	17.36
	70-3	16.07	26.06
	<3	6.20	48.29
	Total clean coal	31.16	28.00
Screen out coal powder	<3	4.40	48.29
middlings		10.54	50.04
Reject of Primary ZM separator		32.92	65.08
Reject of Secondary ZM separator		1.87	71.24
Gamma-ray sorter reject		19.11	78.99
Raw coal	Total	100.00	53.98

4.3 Classifying of clean coal at 5mm

The screening opening size is changed from 3mm into 5mm to increase clean coal classifying efficiency. After one-stage dry separation of high ash content raw coal, the clean coal yield is 30% and ash content is of 39.29%. The high ash -5mm fine coal content in clean coal product is 49.08%. In order to meet clean coal ash requirement, the cutting density of intelligent gamma ray sorter is adjusted to ensure that the high density rejects content in clean coal product is controlled below 10% and low density fraction misplaced in reject product is below 3%. Meanwhile, for meeting clean coal ash target of 28%, it is needed to improve the screening system and use 5mm opening screen to remove part of < 5 mm high ash coal powder.

Table 12: Product balance of high ash raw coal dry separation

Product	Size mm	Wt %	Product	Wt,% Individual	Plant yield	Ash,%
Raw coal	+70	22.47	coal	50.47	11.34	12.00
			rock	49.53	11.13	79.71
			total	100.00	22.47	45.53
	-70	77.53				55.42
Total		100.00				53.20

Contd...

Contd...

Separation of 70-0mm raw coal by ZM200 dry separator						
Product	Size mm	Wt %	Product	Wt,%		Ash,%
				Individual	Plant yield	
Clean coal	70-0	30.00	70-5	47.07	14.12	24.24
			5-0	52.93	15.88	49.08
			Total	100.00	30.00	37.39
	5-0	15.88	5-3	22.25	3.53	47.95
			3-0	77.75	12.35	49.29
Middlings		18.00				52.21
Reject		29.53				75.68

Table 13: Product balance of high ash raw coal dry separation with 5mm clean coal screening

Product	Size, mm	Wt%	Ash%
Clean coal	>70	12.47	18.14
	70-5	14.12	24.24
	5-0	8.35	49.08
	Subtotal	34.94	28.00
Removed coal powder	5-0	7.53	49.08
Middlings		18.00	52.21
Reject of zm separator		29.53	75.68
Reject of gamma ray sorter		10.00	79.71
Total		100.00	53.20

Based on the above testing data, the process flowsheet shown in figure 1 is improved and adjusted, the proposed dry separation flowsheet in handling high ash raw coal from Dafeng coal mine is shown in figure 2. The main change is the addition of one ZM70 type dry separator as re-cleaning to improve deshaling rate. One double-deck flip-flop screen is used to replace the exiting 3mm linear vibration screen. Through classification, the low ash +13mm lump clean coal can be recovered timely and high ash -5mm fine coal removed. Only 13-5mm primary clean coal is processed in ZM70 type dry separator.

5. Economic benefit analysis of dirty coal dry separation

It is of great significance for Dafeng coal mine to adopt pre-deshaling dry separation before delivering dry clean coal product to wet process washery for further processing. Using dry separation can save costs of coal resources management, coal transportation and washing. At the same time, pre-deshaling by dry separation will lower the rock content in the coal feeding to wet process washery. The benefits of washing lower ash raw coal include lower downstream reject handling cost, improved heavy medium separation performance, more stable process equipment performance and less wear and tear of equipment and pipes.

When the final clean coal ash is set up at 28%, the clean coal yield is 34.94%, on this basis, the corresponding economic benefit analysis is calculated:

(1) Product price

The prices of dirty coal, middlings, removed coal powder and clean coal are 100, 100, 100 and 600 RMB /ton respectively. After dry separation, the total salable coal price is 235.07 rmb per ton of raw coal processed。

(2) Processing cost

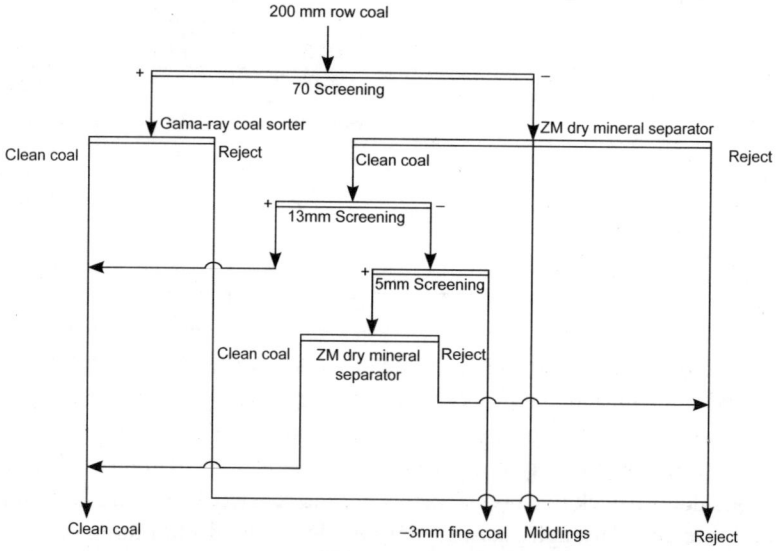

Figure 2: Recommended process flowsheet

The processing cost of wet process and dry process separation are 20 and 12 rmb/ton respectively. When the composite process, made of predeshaling by dry separation + heavy medium separation, is adopted, the clean coal yield is 34.94% and the comprehensive processing cost is 18.99 RMB/ton (12+20 * 0.3494 = 18.99).

(3) Transportation

The transportation cost before dry separation is 50 RMB/ton，after dry separation, only clean coal product is shipped to wet process washery. Since clean coal yield in dry separation is 34.94%，the transportation cost is lowered to 17.47 rmb/ton (0.3494*50=17.47)。

(4) Economic benefit

The dry separation plant has a capacity of 1 mtpa, the gamma-ray intelligent sorter can remove 0.1Mtpa of rock and ZM dry separation can discard 0.2953 Mtpa high ash reject, the total -5mm coal power screened out is 0.0753 Mtpa.

If no dry separation is used, all dirty coal is shipped to Taixi coal washery and total cost occurred in wet process washery is:

Total cost without dry separation = raw coal sale price + transportation cost+ wet process handling cost = 120+50+20 = 190 (RMB/ton);

Total cost after dry separation = clean coal sale price + transportation cost + composite processing cost = 235.07 + 17.44 + 18.99=271.5 (RMB/ton);

Total economic benefits: 1Mtpa*(271.5-190) RMB/ton =81.50 millions (RMB)

It can be concluded that, for every 1 million tons dirty of mixed coal processed by dry separation technology, the economic benefit will be about 81.9 million RMB, not including indirect benefits such as reduced equipment wear and tear and water consumption saving.

6. Conclusion and recommendations

The successful operation of dry separation system at Dafeng coal mine has shown that the dry pre-deshaling of high ash coal will fully recycle the dirty coal resources and improve the quality of raw coal feeding wet process washery.

The separation testing results showed that:

1. Through using the exiting processing system, we can obtain 35.63% of clean coal with ash content of 31.75%if we deshaling the high ash dirty coal (ash content:53.98%)by dry separation technology. The clean coal ash content exceeds clean coal ash target of 28% except for 75% of -3mm coal powder can be removed from primary clean coal product. This will put a high challenge for screening system since low classification efficiency at 3mm even flip-flop screens are used.

2. Although Classification of primary clean coal product at 5mm will produce qualified clean coal product, the clean coal yield is lowered and the quantity of discarded coal powder increased.

ZM mineral high efficiency separator is a new generation of dry separation equipment, it shows its remarkable superiority in handling different high ash dirty coal and in the comprehensive utilization of low quality coal resources. Due to most of the high density rock has been removed before feeding raw coal to regular heavy medium separation plants, the separation performance of wet process washery can be improved, capital cost, operation cost and coal transportation cost saved. At the same time, recycling of dirty mixed coal will reduce coal piling up and avoid possible gangue spontaneous combustion. Dry separation of dirty coal can not only recover good quality coal but also improve the mining area environment and bring remarkable economic benefits to the coal enterprises.

There is a need to continually renovate and improve dry coal preparation system in the areas of capacity scaling-up, automation level, multi-application design, etc to meet the different needs of customers at home and abroad.

7. References

[1] Li, G. and Yang, Y., 2006, "Development and Application of FGX Series Compound Dry Coal Cleaning System," China Coal, Technology Monograph of the Tangshan Shenzhou Machinery Co., Ltd., pp. 17-28.

[2] Honaker, R. Q., Luttrell, G. H. and Lineberry, G. T., 2006, "Improved Coal Mining Economics Using Near- Face Deshaling," Minerals and Metallurgical Processing Journal, Vol. 23. No. 2, pp. 73-79.

[3] Kelley, M. and Snoby, R., 2002, "Performance and Cost of Air Jigging in the 21st Century," Proceedings, 19th Annual International Coal Preparation Exhibition and Conference, Lexington, Kentucky, pp. 175-186.

[4] Weinstein, R. and Snobly, R., 2007, "Advances in Dry Jigging Improves Coal Quality," Mining Engineering, Vol. 59, No. 1, pp. 29-34.

[5] Kip Alderman, Richard J Snoby and Heribert Breuer, 2013, "Dry jigging of coal-10 years allair Technology-Operating results and test work data", Proceeding of the 17th international coal preparation congress, Istanbul, 315-319.

[6] Ghosh T., Patil D., Parek B.K., Honaker R. Q., 2013: Upgrading Low Rank Coal Using A Dry, Density-Based Separator Technology. Proceedings of the 17th International Coal Preparation Congress. Istanbul, p. 295-308。

[7] Patil, D.; Parekh, B., 2011, Beneficiation of fine coal using the air table, International Journal of Coal Preparation and Utilization, 31/3-4, 203-222.

[8] Cristoper Robben, Hohan de Korte, Hermann Wotruba and Mathilde Robben, 2013, "Experience in dry coal separation using X-ray –transmission-based sorting."Proceeding of the 17th international coal preparation congress, Istanbul, 321-325.

19

Combination of wet and dry jigging for high ash non coking coal beneficiation in India

K S Ashvani[1], Mustafi Gurdas[2], Banerjee Chiranjib[3]

Allmineral Asia Pvt. Ltd., Kolkata, India

Abstract: Degradation in the quality of non-coking coal is the current reality in India. Every year we see poorer and poorer coal being excavated. On the other hand, environment rules are getting more and more stringent. Coal reserves of Jharkhand, Chhattisgarh, and Orissa are approaching towards 50% ash after open cast mining whereas power coal requirement for coal fired boilers is 33.5–34.5% ash which can only be fulfilled after beneficiation or blending of coal. After a period of 3–4 years crude oil prices are again rising slowly which in future may escalate cost of imported coal due to increase in transportation cost. In this situation coal beneficiation again becomes vital and necessary. For such high ash Indian, coal cut point density lies above 1.8. Heavy media above 1.8 is not cost effective and also difficult to stabilize which ultimately results in to loss of yield. In such cases wet and dry jigging of coal is wise solution. In this paper we have discussed the combination of wet and dry jigging of coal for beneficiation of high ash non coking coal in 0–50 mm size range. For coarse fraction 13–50 mm, wet jigging is utilized whereas for finer fraction (0–13 mm), dry jigging is used. A comparative study is made between completely wet circuit with wet jig for both the size fractions and combination of wet and dry jigging for coarse and fine size fractions, respectively.

Keywords: Washability, Bag house, Optimisation, Coarse, Rejects, Concrete

1. Introduction

Jigging technology is the classical method of coal preparation. Until the commercialization of the dense medium process, it was the backbone of coal washing. If we consider the special case of high ash non coking coals of India, jigging emerges as one of the important tool for coal beneficiation. Wet jigging plants can be built at 10–12% less capital investment and operation cost also is around 35% less than heavy media plants of same capacity. If the density of separation is high, imperfection and organic efficiencies of wet jigging approaches towards heavy media. A small loss due to difference in yield can be compensated by long term gains in terms of investment and production cost.

Apart from wet jigging dry jigging is also possible for coal. A new air jig design was introduced to the coal industry by allmineral. The new air jig employed a different air fluidisation system, a different air distribution system, and an instrumentation and control system for controlling the separating gravity. Since 2001, new air jig plants have been installed in the USA, Colombia, Spain, Ukraine, and India for both thermal coal and coking coal. Dry processing eliminates the clean coal moisture penalty. For small capacities up to 2 mtpa, dry jigging plant can be installed at around 12–15% less capital investment with 35–40% less production cost because of the elimination of dewatering, filtering, and water circulation equipment. With 1.3 billion population water is an increasingly scarce and expensive commodity. Water conservation and optimisation is a matter of concern in every plant design.

In this paper we have discussed the combination of wet and dry jigging for beneficiation of 0–50 mm high ash non coking coal in a 3 mtpa coal washery. For coarse fraction (13–50 mm) wet jigging is utilized whereas for finer fraction (0–13 mm) dry jigging was used. A comparative study is made between completely wet circuit with wet jigs for both the size fractions and combination of wet and dry jigging for coarse and fine coals, respectively.

2. Wet jigging beneficiation technology

Separation of gangue in Alljig is based on the fact that particles will stratify in pulsating water. The upward and downward currents fluidize and compact the grains into relatively homogenous layers. Low density pieces stratify on the surface while specifically heavy grains settle to the lower level of the bed. The most precise stratification of particles requires that the frequency and amplitude of the water pulsation – which may be adjusted during operating – will be optimized according to feed characteristics. After stratification the discharge of the concentrate is done by a bottom gate discharge system into the jig hutch hopper. The concentrate/reject (depending on mineral or coal) is then extracted from the hopper by a frequency controlled dewatering screen or bucket elevators.

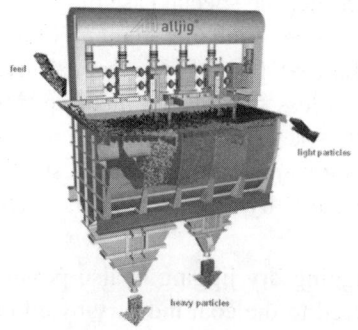

3. Dry beneficiation technology

The allmineral allair® utilises the basic principles of jigging; it stratifies the feed material by specific gravity and subsequently measures and discharges the high density (and high ash) strata. To stratify the material, the allair® uses pulsating and constant air flow through a perforated jig bed. Vibrating mechanisms assist the transport of material across the bed. The feed star gate provides an even feed distribution over the width of the bed, and the discharge star gate provides an even removal of the heavy particles (rock) from the jig, thereby maintaining a residual layer of refuse below the light particles (coal). Continuous sealing at both ends for allair® decreases the consumption of air.

To measure the coal–rock interface, a nuclear density measuring device is located towards the discharge end of the jigging bed. This device automatically controls the discharge mechanism (the star gate) of the jig. If more dilution enters the jig, the star gate will speed up and discharge more rock. If more coal enters the jig, the star gate will slow down thereby discharging less rock and minimizing the loss of clean coal in the refuse. To complete the allair® jig process, dust particles are removed via a baghouse filter (Fig. 1).

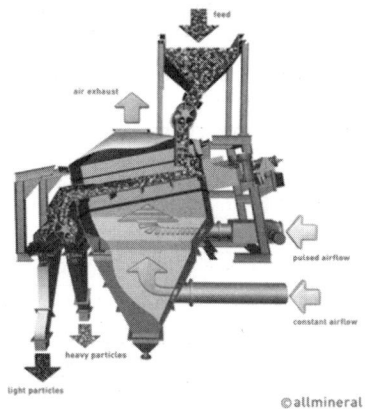

Figure 1: Schematic of allair® dry jig

The allair® is designed to handle material up to a maximum particle size of 50 mm. The maximum feed capacity of the allair® is 100 tph per machine.

4. Results and discussion

The coal sample was received from some open cast mine location of SECL. Received coal sample was at size <300 mm. Sample was crushed down to size 50 mm and was screened at sizes 50 mm, 25 mm, 13 mm, 6 mm, 3 mm, and 1 mm and determination of ash% and moisture% of each screened fraction was done. Ash content and moisture (air dry basis) are shown in Table 1.

Table 1: Size wise ash analysis (0–50 mm)

Size (mm)	Wt%	Ash%	Moist%
50–25	42	51.37	4.3
25–13	22.2	48.48	5.2
13–6	14.3	44.78	5.6
6–3	8.1	40.66	6.1
3–1	5.4	38.25	6
–1	8	43.28	5.1
Total	100	47.56	5.0

Objective of the test work was to upgrade the coal quality to <36% target ash with good yield and organic efficiency. Client wanted to setup a 3 mtpa coal washery with 6000 h/year operating hours. Since it was a mercantile washery where final product was to be sold either by blending or alone, yield and reject ash were most important factors.

Float and sink tests of RoM (run-of-mine) coal sample crushed down to 50 mm size was carried out at 1.4, 1.5, 1.6, 1.7, 1.8, and 1.9 specific gravities for different size fractions of 50–1 mm. The results are given in Table 2.

Table 2: Washability data for different size fractions

Size (mm)	50–25		25–13		13–6		6–3		3–1	
Wt % raw	42		22.2		14.3		8.1		5.4	
Wt % 100	45.65		24.13		15.54		8.80		5.87	
Density (g/cm³)	Wt %	Ash %	Wt %	Ash %	Wt %	Ash %	Wt %	Ash %	Wt %	Ash %
<1.4	10.8	18.8	14.1	17.1	18.3	11.4	21.7	5.9	23.2	5.3
1.4–1.5	16.8	31.6	17.2	29.5	14.7	25	13.5	19.9	15.3	18.3
1.5–1.6	15.5	42.5	18.5	39.3	18.5	36.8	13.4	32.6	11.3	31.8
1.6–1.7	8.9	51.5	4.6	49.2	4.1	46	6.1	40.2	4.9	38.7
1.7–1.8	10.1	53.9	8.6	52.2	8.6	51.5	5	45.3	5.1	42.5
1.8–1.9	11.3	62.8	5.9	58.2	6.8	57.5	8.7	51.4	7.6	47.5
>1.9	26.6	76.4	31.1	75.7	29	75.8	31.6	73.2	32.6	70.4
Total	100	51.373	100	48.48	100	44.78	100	40.66	100	38.25

Fractions 50–25 and 25–13 were merged to obtain the washability analysis of 50–13 mm size faction which is shown in the Table 3. 50–13 mm size fraction will be subjected to wet jigging.

Table 3: Washability analysis for 13–50 mm size fraction

[1]		[2]	[3]	[4]	[5]	[6]	[7]	[8]	[9]	[10]	[11]	[12]
Relative density fractions						Size fraction basis						
		Mass	Ash (air dry basis)	Propor-tion of ash	Relative density (ρ)	Cumulative floats			Cumulative sinks			Percent-age mass of NGM (ρ ± 0.1)
						Mass Σ[2]	Proportion of ash Σ[4]	Ash [7]/[6]	Mass Σ[2]	Propor-tion of ash Σ[4]	Ash [10]/[9]	
Sinks	Floats	%	% (m/m)	[2]×[3]		%		% (m/m)	%		% (m/m)	
									100	5037.41	50.37	
–	1.4	11.94	18.11	216.20	1.4	11.94	216.20	18.11	88.06	4821.21	54.75	
1.4	1.5	16.94	30.86	522.76	1.5	28.88	738.97	25.59	71.12	4298.45	60.44	
1.5	1.6	16.54	41.26	682.37	1.6	45.42	1421.33	31.30	54.58	3616.08	66.25	23.95
1.6	1.7	7.41	51.01	378.12	1.7	52.83	1799.45	34.06	47.17	3237.96	68.64	16.99
1.7	1.8	9.58	53.37	511.38	1.8	62.41	2310.83	37.03	37.59	2726.59	72.54	19.01
1.8	1.9	9.43	61.81	582.99	1.9	71.84	2893.81	40.28	28.16	2143.60	76.13	
1.9	–	28.16	76.13	2143.60	2.1	100.00	5037.41	50.37				
Total		100.00										

Note: NGM (near gravity material) = percentage mass for ρ + 0.1 in [6] minus percentage mass for (ρ − 0.1) in [6].

Washability curves were drawn (Fig. 2) to obtain ideal yields at various target ash% 35% and 36% (Table 4). Ideal yield are necessary to calculate the organic efficiency after knowing about practical yields in various jigging process. It is noticeable that values of NGM are below 20 for selected specific gravity of separation.

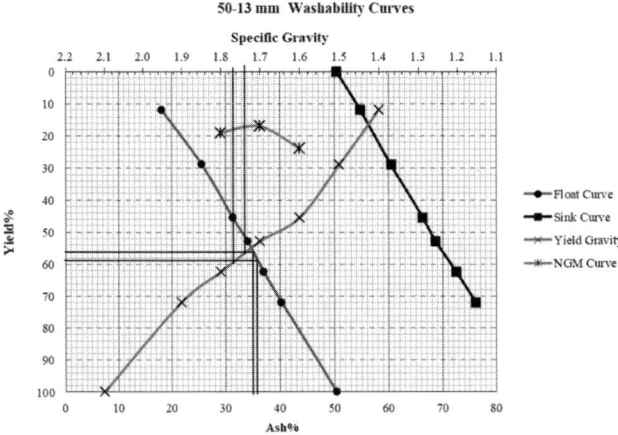

Figure 2: 50–13 mm Washability curves

Table 4

Target ash %	35	36
Cut point	1.74	1.765
Ideal yield	56	58.5

Fractions 13–6, 6–3, and 3–1 mm (Table 2) were merged to obtain the washability analysis of 1–13 mm size faction which is shown in the Table 5. This size fraction will be subjected wet jigging and also dry jigging after mixing with 0–1 mm fraction.

Table 5: Washability analysis for 1–13 mm size fraction

[1]		[2]	[3]	[4]	[5]	[6]	[7]	[8]	[9]	[10]	[11]	[12]
Relative density fractions		Size fraction basis										
		Mass	Ash (air dry basis)	Proportion of ash	Relative density (ρ)	Cumulative floats			Cumulative sinks			Percentage mass of NGM (ρ ± 0.1)
						Mass Σ[2]	Proportion of ash Σ[4]	Ash [7]/[6]	Mass Σ[2]	Proportion of ash Σ[4]	Ash [10]/[9]	
Sinks	Floats	%	% (m/m)	[2]×[3]		%		% (m/m)	%		% (m/m)	
									100	4230.73	42.31	
–	1.4	20.24	8.32	168.50	1.4	20.24	168.50	8.32	79.76	4062.23	50.93	
1.4	1.5	14.47	22.24	321.70	1.5	34.71	490.20	14.12	65.29	3740.53	57.29	
1.5	1.6	15.62	35.05	547.28	1.6	50.32	1037.48	20.62	49.68	3193.25	64.28	20.45
1.6	1.7	4.84	42.43	205.30	1.7	55.16	1242.77	22.53	44.84	2987.96	66.64	11.71
1.7	1.8	6.87	48.89	335.92	1.8	62.03	1578.69	25.45	37.97	2652.04	69.85	14.38
1.8	1.9	7.51	53.47	401.54	1.9	69.54	1980.23	28.47	30.46	2250.49	73.89	
1.9	–	30.46	73.89	2250.49	2.1	100.00	4230.73	42.31				
Total		100.00										

Note: NGM (near gravity material) = percentage mass for ρ + 0.1 in [6] minus percentage mass for (ρ – 0.1) in [6].

Washability curves (Fig. 3) were drawn to obtain practical yields at target ash% 34.5 (Table 6). It is noticeable that around 30% material is above 73.89% ash level. For target ash% of 34.5, cut point density will be around 2 with NGM level more than 20%. High NGM and finer size drops down the imperfection values for fines jigging.

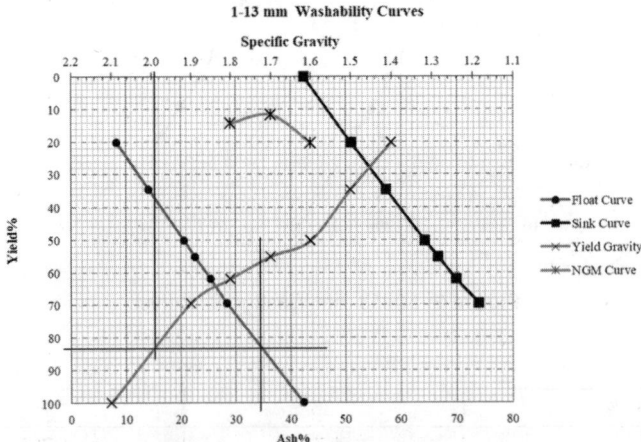

Figure 3: 1–13 mm Washability curve

Table 6

Target ash %	34.5
Cut point	1.99
Ideal yield	83

Washability analysis of 0–1 mm faction is also represented in the Table 7.

Table 7: Washability analysis for 0–1 mm size fraction

[1]		[2]	[3]	[4]	[5]	[6]	[7]	[8]	[9]	[10]	[11]	[12]
Relative density fractions		Mass (air dry basis)	Ash ash	Proportion of ash	Relative density (ρ)	Cumulative floats			Cumulative sinks			Percentage mass of NGM (ρ±0.1)
						Mass Σ[2]	Proportion of ash Σ[4]	Ash [7]/[6]	Mass Σ[2]	Proportion of ash Σ[4]	Ash [10]/[9]	
Sinks	Floats	%	% (m/m)	[2]×[3]		%		% (m/m)	%		% (m/m)	
									100	4326.71	43.27	
–	1.4	20.16	7.59	153.01	1.4	20.16	153.01	7.59	79.84	4173.70	52.27	
1.4	1.5	11.51	22.16	255.06	1.5	31.67	408.07	12.89	68.34	3918.64	57.34	
1.5	1.6	15.39	34.84	536.29	1.6	47.06	944.37	20.07	52.94	3382.35	63.89	23.29
1.6	1.7	7.90	43.18	341.12	1.7	54.96	1285.49	23.39	45.04	3041.23	67.52	15.68
1.7	1.8	7.78	49.03	381.49	1.8	62.74	1666.98	26.57	37.26	2659.74	71.38	16.62
1.8	1.9	8.84	56.72	501.41	1.9	71.58	2168.38	30.29	28.42	2158.33	75.94	
1.9	–	28.42	75.94	2158.33	2.1	100.00	4326.71	43.27				
Total		100.00										

Note: NGM (near gravity material) = percentage mass for ρ + 0.1 in [6] minus percentage mass for (ρ − 0.1) in [6].

5. Development of flowsheet and mass balance

The flowsheet for the project was developed in two schemes to compare and optimise the yield, Capex and Opex.

Scheme-1: Utilizing wet jigging for both 1–13 mm and 13–50 mm size fractions and blending of fines (0–1 mm) with jig product to increase yield.

Scheme-2: Utilizing wet jigging for 13–50 mm and dry jigging for 0–13 mm size fraction. Fines recovered from jigging operations both wet and dry were mixed with the product. Slimes < 150 mic partially discarded to tailing dam.

5.1 Scheme-1: Wet jigging for both the size fractions

In this scheme two size fractions were subjected to wet jigging for ash reduction. Overall circuit was developed utilizing wet jigs for coarse and fines coal. 0–1 mm material was passed through hydrocyclone for slimes removal at 150 micron, rest of the material was dewatered in centrifuge decanter. Slimes from thickener were filtered in belt press filter. 0–1 mm material recovered from belt press and centrifuge was mixed with product from jig to increase the overall yield. Target ash% for coarse jig was 35% whereas same for fines jig was 34.5%. Flow chart (Fig. 4) represents the whole scheme with mass balance.

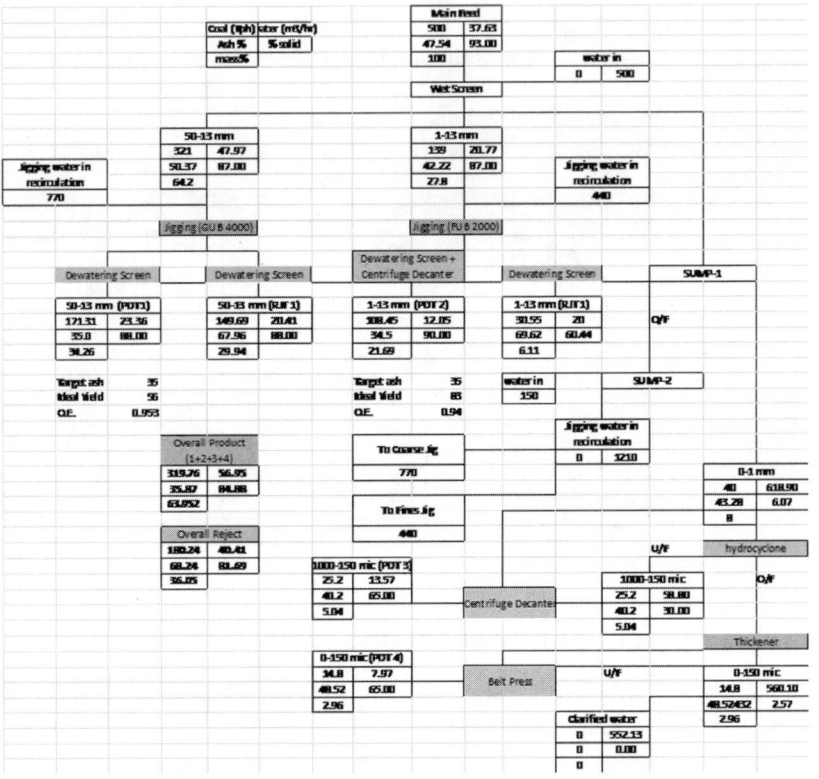

Figure 4: Scheme-1 water and mass balance

Organic efficiencies obtained in coarse and fines jig was 95.3% and 94%, respectively. Imperfection for the coarse jig lies in the range of 0.13–0.14 whereas same for the fines jig lies in the range of 0.16–0.17. High NGM and finer size drops down the imperfection values for fines jigging (Fig. 3). Overall product obtained in this method was 63.95% with 35.87% ash. Overall organic efficiency of the circuit was found to be more than 95%. Some better organic efficiency could have been achieved if we beneficiated fines material also through floatation or spiral circuit. Flow sheet for scheme-1 is shown in the Fig. 5.

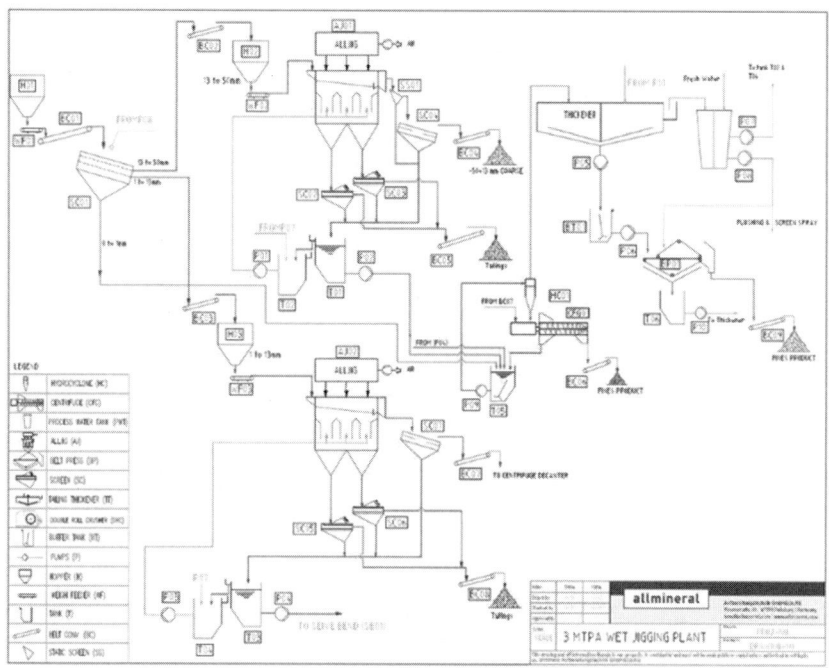

Figure 5: Flowsheet for scheme-1

5.2 Scheme-2: Combination of wet and dry jigging

Dry screening was carried out in a flip flop screen at 13 mm and 0–13 mm dry material was fed to a dry jig which is equipped with bag filter for fines recovery. Fines from the bag filter were mixed with the product of the dry jig. Coarse fraction 13–50 mm after screening is fed to wet jigging. Target ash in dry jig was considered 35% whereas for wet jig it was considered as 36% to achieve the overall target ash in product as 36%.

Flow chart (Fig. 6) represents the whole scheme with mass balance.

Organic efficiencies obtained in wet coarse and dry fines jig was 96% and 90%, respectively. Imperfection for the coarse jig lies in the range of 0.13–0.14 whereas same for the fines jig lies in the range of 0.17–0.18. Overall product obtained in this method was 62.99% with 35.62% ash. Overall organic efficiency of the circuit was found to be 94%. Flow sheet for scheme-2 is shown in the Fig. 7.

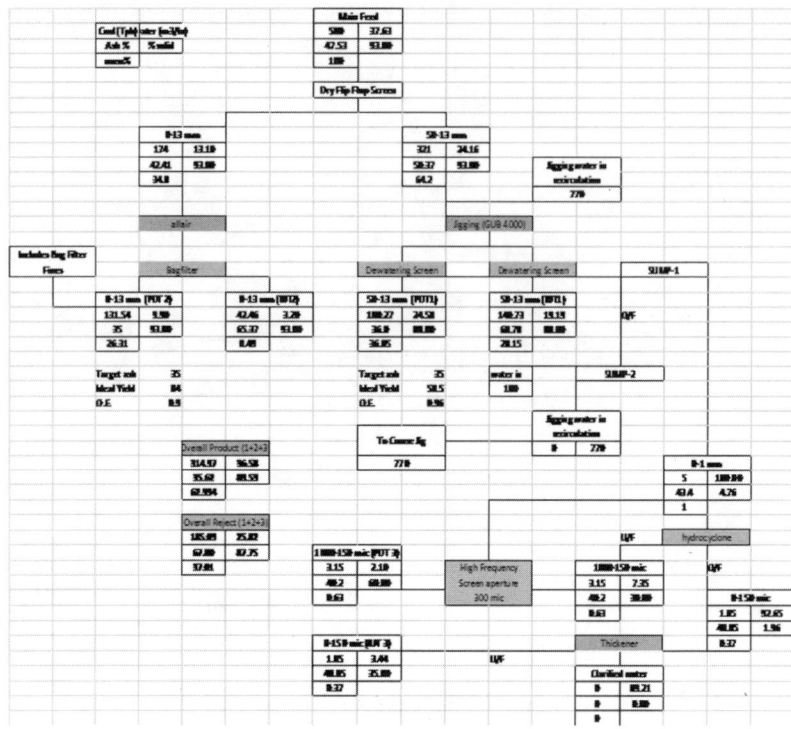

Figure 6: Scheme-2 water and mass balance

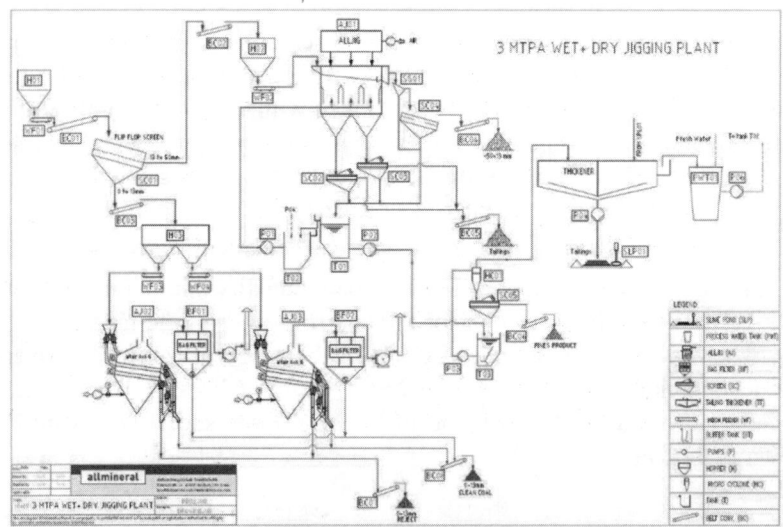

Figure 7: Flow sheet for scheme-2

5.3 Comparison between scheme-1 and scheme-2

If we compare scheme-1 and -2 on the basis of various parameters of yield, water, and efficiency we can represent it in the form of Table 8 as below.

Table 8

Parameters	Scheme-1 (two wet jigs)	Scheme-2 (combination of wet and dry jigs)
Yield%	63.952	62.994
Product ash%	35.87	35.62
Organic efficiency (O.E.)	95.31	94
Product (tph)	319.76	314.57
Makeup water (m3/h)	59.73	24.77
Difference in yield (%)	0.96	
Product loss t/h in scheme-2	4.79	
Water quantity saved m3/h in scheme-2	34.96	
Power saved per ton of coal fed (%) in scheme-2	18	

6. Conclusion

(1) Washability data for 13–50 mm size fraction coal suggests that specific gravity of separation is around 1.78 but the NGM is below 20. Since the NGM was within 20%, we could obtain 14–15% up gradation with organic efficiencies 95–96 in wet jigging. Dry jigging was not employed in this range because the particle size was big and range of ash reduction was also high which may reduce the organic efficiency of separation.

(2) For size 1–13 mm specific gravity of separation was as high as 2.0 but the desired target ash% reduction was in the range of 7–8 %. NGM is also high (>20%) at cut point density. High NGM and finer size drops down the imperfection values for fines jigging (Fig. 3). Since the % reduction in ash is not very high, we can utilize dry jigging for this size fraction. For small coal size it not very difficult for air to lift the particle and efficiency of separation is improved.

(3) In scheme-1 where wet jigs were used for two size fraction, many extra equipment were involved for dewatering, filtering, and water balance which increases the capital investment and power consumption. Extra equipment used were (a) Centrifuge – for fine coal dewatering,

(b) belt press – for filtering of slimes, (c) additional pumps – for water and slurry handling. In scheme-2 where dry jig was used, we can eliminate these equipment because fines handling and recovery can be done directly in bag filter which is the part of dry jigging system.

(4) Because of more volume of water and slurry handling size of thickener will be more than double in scheme-1. Bigger de-sliming cyclones will be used in scheme-1 as compared to scheme-2.

(5) In scheme-2 we can save around 35 m^3/h water consumption as compared to scheme-2 which is vital and scarce commodity for countries like India.

(6) Around 18% less power consumption may occur in scheme-2 as compared to scheme-1

(7) For certain cases with this type of coal characteristics we can utilize suggested mode of beneficiation combining wet and dry jigging.

7. References

[1] Mitchell, D.R., 1942, Progress in Air Cleaning of Coal, Transactions of the American Institution of Mining and Metallurgical Engineers.

[2] Heribert, Breuer, Andreas, Hees and Hakan, Oezdemir, South Africa Coal Preparation Society (SACPS), 2015, Dry jigging of coal.

[3] Singh, Harish, Biswal, Dinesh, Mishra, Sabyasachi, Wet Jigging – Economical & Efficient Beneficiation of Coal, 2013, National Seminar on Coal Preparation Technology-CPT-2013, CIMFR-India.

[4] Oren, E., Van Wyk, G., Ashvani, K.S., Advantages of Utilising Dry Jigging and X-ray Sorting Technology for Coal Beneficiation in Australia, 17th Conference on Coal Preparation 2018; Australian Coal Preparation Society.

20

TDS dry sorting benificiation applied for underground coal mine

Xingguo Liang and Yunfeng Li

Tianjin Meiteng Technology Co., Ltd., Tianjin, China

Abstract: As the TDS (Telligent Dry Separator) technology is successfully developed and applied, TDS provides a new idea for the coal mining and underground reject backfilling. TDS adopts X-ray recognition technology to separate coals, and is very suitable for underground application due to its small volume, simple system and free of water and medium. The coals mined from the underground working face will be firstly graded; the lump coals will be separated by TDS and then delivered to the ground; and the rejects will be delivered back to the working face and directly discarded in the gob by the support stowing machine.

Keywords: Underground backfilling, TDS underground reject extraction, energy conservation and emission reduction

1. Coal mining status in China

In recent years, the quality of raw coal in China has generally declined; the ash content of run-of-mine coal is more than 40%, even up to 50–60% in some mines; and the content of reject in raw coal is particularly high. It is explicitly stated in the Measures for the Management of Comprehensive Utilization of Coal Reject that the area of reject hill shall be matched to the coal production, washing and processing capacity. In principle, such area shall not exceed three years of refuse storage quantity; and the subsequent comprehensive utilization plan must be developed. In addition, with the advance of the policy of "Road-to-Railway Transport" in Hebei, Shandong, and other regions, the coal-carrying trucks will gradually fade out of the stage of history. As the TDS (Telligent Dry Separator) technology is developed and applied successfully, TDS provides a new idea for the coal mining and underground reject backfilling.

2. Significance of underground reject extraction and backfilling by adopting TDS

If TDS can drive the development of underground reject backfilling technology and realize the green mining mode of reject not be lifted to the ground, it could bring great economic and social benefits to the development and utilization of coal resources [5–7].

2.1 Increasing coal recovery rate

(1) With the popularization and application of caving coal mining technology in China, coal enterprises may set the seam floor at a higher elevation and abandon part of bottom coals, for the purpose of lowering the production cost of raw coal per ton and ensuring the quality of raw coal lifted to the ground.

(2) During the mining of thin seam, lots of reject may be mixed in the raw coal due to the inevitable top cutting and bottom digging.

It is clearly stipulated in the *Interim Provisions on Administration of Coal Mine Recovery Rate* that the recovery ratios shall not be less than 75% for thick seam panel, 80% for medium-thick seam panel, and 85% for thin seam panel, respectively. Saving energy consumption and reducing production cost The productive rate of +50 mm sized lump coal in raw coal is 31.61%; the reject content in lump coal is 51.19% (+1.9 g/cm^3 is reject); and the predicted annual output of reject is about 1,618,100 t. If the underground reject backfilling by TDS is adopted, +50 mm sized rejects will be discharged underground, thus reducing the content of rejects in annual output by 16% and creating huge economic benefits to the enterprises (Table 1).

(1) Saving underground transport cost

According to the field survey results, the costs of transporting raw coal from the underground to the ground are different due to the difference in mining depth and transport mode. If the average transport cost is RMB 6/ton, the transport cost of the 1,618,100 t rejects lifted to the ground is RMB 9,708,600.

(2) Lowering the washing and processing costs

After elevating to the ground, the rejects will be delivered to the systems of raw coal preparation, storage, washing, and water treatment of the coal preparation plant. Due to the complexity of ground system, the average reject processing cost per ton of the whole set of washing and processing system is RMB 40/t; however, the average reject processing cost per ton of underground TDS separation system is only RMB 20/t. Therefore the processing cost of 1,618,100 t rejects will be reduced by RMB 32,362,000 in each year.

Reject disposal cost

The sum of cost of transporting the washed rejects to the reject hill and pile and subsequent disposal cost is about RMB 10/t. If the TDS technology is adopted for reject extraction and backfilling, the cost of reverse transporting and backfilling of rejects in the underground will be calculated as RMB 6/t; the disposal cost of RMB 4/t will be saved, and about RMB 6,472,400 will be saved for the underground extraction of 1,618,100 t rejects.

Table 1: Composite table of economic benefits

Item	Project indicator	Economic benefits increased (RMB 10,000/a)
Saving underground transport cost	Saving RMB 6/t	9.7086
Lowering the washing and processing costs	Reducing by RMB 20/t	32.3620
Reject disposal cost	Saving RMB 4/t	6.4724
Total	RMB 48.5430 million	

In summary, for a 10 Mt/a mine, if the underground reject extraction and backfilling by TDS is adopted, the enterprise will gain the economic benefit of RMB 48,543,000 in terms of underground transport, ground washing, reject disposal, and tax reference, and realize the cost reduction and benefit increase.

2.2 Environmental protection

Currently, coal mining and reject piling have a great impact on the environment, such as surface subsidence, spontaneous combustion of reject, farmland occupation, etc. The application of TDS technology in underground reject extraction and backfilling can effectively prevent surface subsidence and reduce the environmental pollution caused by spontaneous combustion of reject. If the TDS technology is combined with comprehensive utilization of reject, the problem of environmental protection can be thoroughly solved.

3. Underground reject backfilling and separation technology

3.1 Introduction to existing technologies

The underground reject backfilling technologies widely used at present includes: paste and slurry backfilling, cemented fragmental rock backfilling, loose discharge of reject on working face, and discharge of reject in abandoned roadway (Table 2).

Table 2: Comparison of underground reject backfilling technologies

Filling type	Paste and slurry backfilling	Cemented fragmental rock backfilling	Loose discharge of reject on working face	Discharge of reject in abandoned roadway
Filling material	Crushed reject + cementing agent	Lump reject + cementing agent	Lump reject	Lump reject
Filling mode	Grinding tailings into fine particles, mixing with the cementing agent to form past slurry, and pumping to the backfilling area through pipeline	Mixing the lump tailings with cementing agent, compacting and filling the materials in the gob under the pressure effect of hydraulic support	Directly discharging the separated tailings to the working face	Transporting and discharging the separated tailings to the abandoned roadway by underground transporting system
Advantage	Uniform strength of backfilling body, high tight-filling ratio, effectively controlling the strata movement and preventing surface subsidence	Reducing the energy consumption due to grinding of tailings unnecessary	Low backfilling cost and simple backfilling mode due to direct discharge on working face upon separation	Little impact to production and low backfilling cost due to direct backfilling to abandoned roadway through transporting system
Disadvantage	High backfilling cost, small backfilling scale, and not suitable for large-scale mines Relatively high backfilling cost and certain requirements for hydraulic support		Low backfilling density and tight-filling ratio, and partial control to the surface subsidence	Low backfilling density and tight-filling ratio, and partial control to the surface subsidence
Application	Mainly applied in nonferrous mine and few coal mines	Applied in few coal mines	Mainly applied in coal mine	Mainly applied in coal mine

Due to the high backfilling cost, the paste and slurry backfilling technology is not widely used in coal industry; however, other technologies of cemented fragmental rock backfilling, loose discharge of reject on working face, and discharge of reject in abandoned roadway are relatively low in backfilling cost, and can be promoted and applied in combination with the underground reject and lump coal separation by TDS.

3.2 Underground reject extraction and backfilling by TDS

(1) Underground TDS

Currently, there are three technologies used for underground reject extraction in raw coal: moving sieve jig, dense-medium shallow chute, and TDS. The moving sieve jig and shallow chute technologies are limited in the underground application due to limitations of underground space and working conditions, and complex system configuration (Table 3).

TDS is a new type of lump coal separation equipment based on X-ray recognition and high-pressure air gun. Compared with jig and shallow chute, TDS does not need any water, medium or agent, as well as slime water treatment system and medium purification, recovery and recycle system, and realizes the separation by using high-pressure blast without coal slime or slime water generated.

Table 3: Comparison of underground separation processes

Item	Moving sieve jig	Underground shallow chute	TDS
Separation accuracy	Coal content in reject: 5–8%	Coal content in reject: 0.5–1%	Coal content in reject: 1–3%
Size	300~50 mm	150–13 mm	300~50 mm and 100~25 mm
System configuration	Equipped with systems of grading, dehydration, and slime water recycle	Equipped with systems of grading, dehydration, medium removal, medium recycle, slime concentration	Equipped with grading system
Installation space/L × W × H	80 m × 7.5 m × 8.5 m	Shallow chute and slime water system: 102 m × 12.46 m × 12.95 m	41 m × 5.5 m × 7.4 m
Power consumption	1.80 kW h/t	2.90 kW h/t	0.86 kW h/t
Medium consumption	None kg/t	0.8 kg/t	None kg/t
Agent consumption	<1 kg/t	<5 kg/t	None kg/t
High-pressure blast	Little	Little	6.8 m3/t
Slime water	0.15 m3/t	0.92 m3/t	None
Coal slime	2 m3/t	4 m3/t	None
Processing cost	RMB 18.2/t	RMB 33.8/t	RMB 11.8/t

Note: The processing cost of lump coal separated by jig, shallow chute, and TDS includes labor cost, electric charge, material cost, depreciation expense, maintenance cost, and others

(2) Underground reject extraction and backfilling by TDS

The underground lump coal and reject separation by TDS can be well combined with the technologies of cemented fragmental rock backfilling, loose discharge of reject on working face, and discharge of reject in abandoned roadway. This system can be installed near the working face to move with the working face, or in the fixed roadway to deliver the separated reject to the working face for backfilling or to the abandoned roadway for piling.

• *Mobile dry separation and backfilling*

For the application of the "mobile dry separation and backfilling process", TDS will move with the system as the mining on working face progresses, and the extracted rejects will be discarded on the coal mining face, integrating the system to a certain extent and reducing the step of reject transfer (Fig. 1).

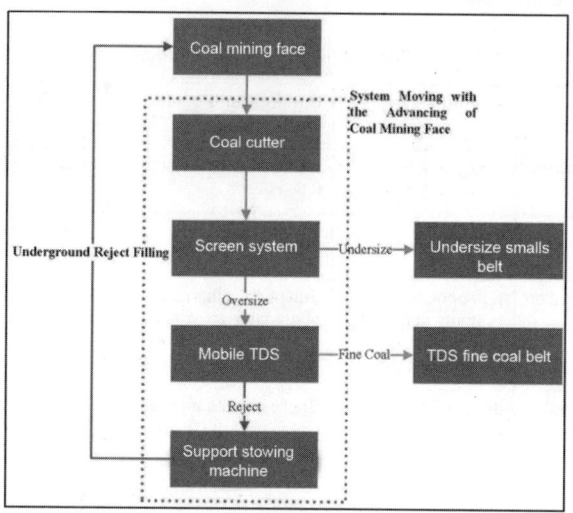

Figure 1: Diagram of mobile dry separation and backfilling process

• *Dry separation + working face backfilling*

For the application of the "dry separation + working face backfilling process", TDS will be fixed in a specific roadway, and the separated rejects will be delivered back to the mining face through reject transport system (Figs. 2 and 3).

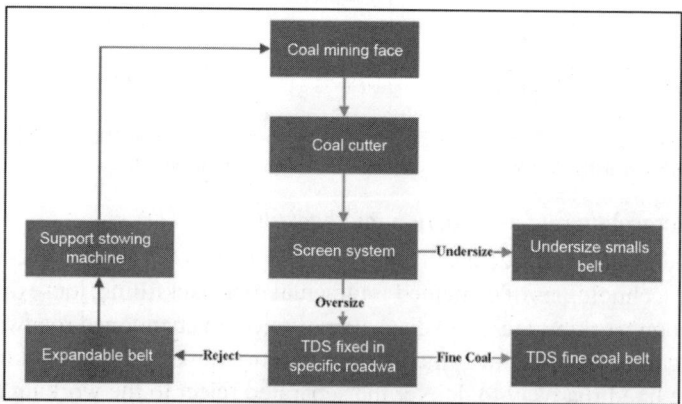

Figure 2: Diagram of dry separation + working face backfilling process

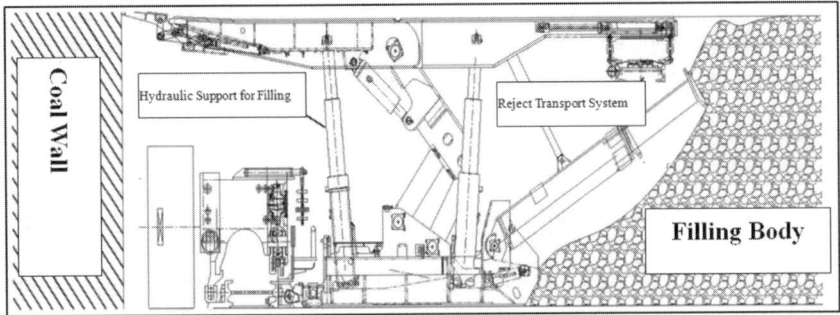

Figure 3: Diagram of reject discarding and backfilling on working face

• *Dry separation + reject discarding in old roadway*

For the application of the "dry separation + reject discarding and backfilling in old roadway process", TDS will be fixed in a specific roadway, and the separated rejects will be delivered back to the abandoned roadway for discarding through transport system. This process is similar to the "dry separation + working face backfilling process", and suitable for the mines with abandoned roadways, which may be specified according to the mining plan or construction of the coal mining enterprises (Fig. 4).

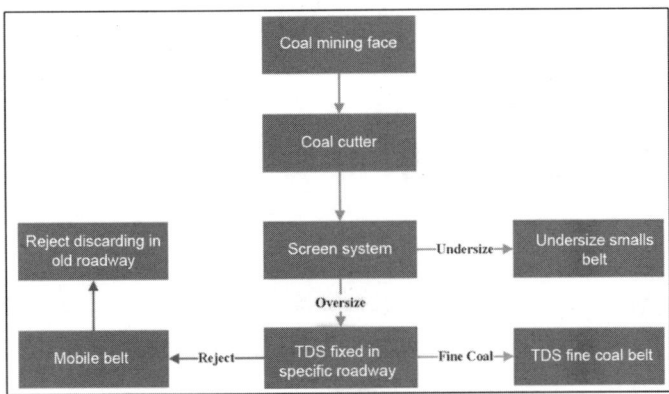

Figure 4: Diagram of dry separation + reject discarding and backfilling on special working face

4. Application of "mining, dry separation and filling system" in Wanglou coal mine in Shandong

With limited lifting capacity of main shaft, Wanglou Coal Mine has applied the TDS technology in underground reject extraction and backfilling to realize the objective of "More Coal and Little Reject", improve the quality of raw coal lifted to the ground, and achieve the strategic target of no reject lifted to the ground, replacing reject with coal, and green mining.

4.1 Overview

Wanglou Coal Mine adopts the "dry separation + working face backfilling process". The raw coal mined on the working face is transported to new TDS reject extraction system and graded by 50 mm grading screen; the +50 mm sized raw coal is fed to TDS for separating the reject and lump fine coal. The lump fine coal is broken and delivered back to the raw coal system together with the undersize of 50 mm grading screen. The large reject is delivered to the specific backfilling face through transport system and discarded by the support stowing machine (Figs. 5 and 6).

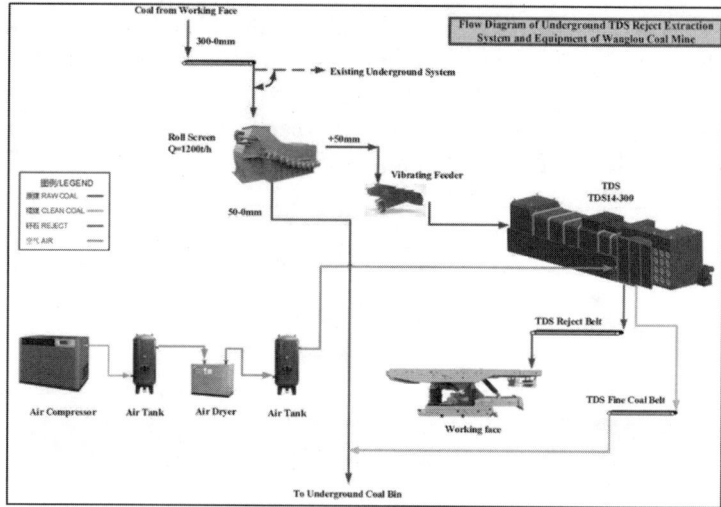

Figure 5: Underground process flow of Wanglou Coal Mine

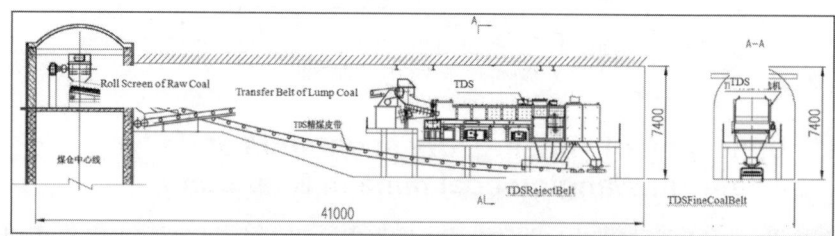

Figure 6: Layout of TDS lump coal separation system

4.2 Expected effects

The actual annual output of Wanglou Coal Mine is 2.30 Mt/a. By applying the "underground reject extraction and backfilling by TDS", the annual capacity of reject extraction is about 0.30 Mt/a. According to the previous calculation, the costs of transport, washing and reject disposal, and tax can be saved by RMB 44/t; and the income of company will be increased by RMB 13.2 million each year.

In addition, if the underground TDS reject extraction system is installed, the lifting capacity of main shaft is up to 2.30 Mt/a; and the run-of-mine coal output of working face can reach to 2.60 Mt/a, with actual capacity of the mine increased by 0.30 Mt/a. According to the research results, the cost of raw coal mining of Wanglou Coal Mine is RMB 370/t; if the price is RMB 450/t, the economic benefit will be increased by RMB 24 million each year.

In summary, after applying the "underground reject extraction and backfilling by TDS", Wanglou Coal Mine will increase its economic benefit by at least RMB 37.2 million each year. In addition, the new mining mode will prevent the surface subsidence and reduce the damage of reject pile on the ground to the ecological environment. It is an important measure in response to the national call green coal development.

5. Conclusion

The process of "underground reject extraction and backfilling by TDS" is highlighted by its simple system, and can be well combined with existing mining technologies, which helps to realize the reject extraction and separation by connecting to the high-pressure blast. The underground TDS reject extraction and backfilling system can be designed as either movable or fixed type. It can realize the dense backfilling by adding cementing agent in reject to prevent the surface from subsidence, or only discard the reject to realize the objective of reducing the quantity of reject lifted to the ground. This process increases the number of underground machines to four (including coal cutter, support stowing machine, dry separator, and conveyor).

The first set of underground coal mining, dry separation, and backfilling system has been put into construction in Wanglou Coal Mine in Shandong. If the system can be successfully operated, the design and operation mode of mine will be changed. The system will contribute to increasing the mine capacity, lowering energy consumption, reducing environmental pollution, and improving the economic benefits of enterprises.

By the end of 2017, there are 3907 coal mines in China with a capacity of 3.34 billion tons; while the underground mining accounts for 80%. If the process of "underground reject extraction and backfilling by TDS" is promoted throughout China, it will reduce the annual discharge of reject on the ground by 260 million tons, bringing great economic and social benefits.

6. References

[1] Li Guixuan, Li Xinguo. The Opportunity and Challenge of Revitalizing Coal Mine Machinery in China. China Coal, 2003(02):9–11.

[2] Yuan Jianguang. Research on Underground Mining Technology and Underground Mining Development. Energy and Conservation, 2017(05):120–121.

[3] Tan Weifeng. On the Application of Underground Mine Ventilation and Energy Conservation Technology. Mechanical & Electrical Engineering Technology, 2015, 44(12):147–149.

[4] Yan Lei. Analysis on Development Trend of Underground Mining Technologies. Technology & Market, 2015, 22(12):222.

[5] Chen Qiao. Development and Application of Series Technology and Equipment for Underground Reject Filling. Management & Technology of SME (Midmonth), 2015(03):109–110.

[6] Liu Yan. Development Status and Trend of Coal Mine Machinery in China. Value Engineering, 2013, 32(04):28–29.

[7] Li Youzhong. On Technical Innovation and Reliability Improvement of Coal Mine Machinery in China. Mechanical Engineering & Automation, 2008(03):198–200.

[8] Huang Bingxiang, Liu Changyou, Cheng Qingying. Relation between Top-coal Drawing Ratio and Refuse Content for Fully Mechanized Top Coal Carving. Journal of China Coal Society, 2007(08):789–793.

21

Fundamental errors in sampling, the impact on cost and decision making

J. Bekker[1] and D. Sharma[2]

[1]Cotecna, Durban North, South Africa
[2]Cotecna Inspection India Pvt Ltd., New Delhi, India

Abstract: The simple act of taking a sample implies that someone will use the information contained in the analytical results to make a decision about a course of action. In the coal mining and processing industry, the decision may involve huge capital commitments for the opening and closing mines, the design of coal washing operations and the producing of the right specification to meet contract specifications.

Sampling is among the most primary activities in a mining and processing operation, and this paper aims to examine some of the fundamental errors that industry is encountering that has a major impact on cost and decision making.

The sampling errors most encountered includes inadequate or in-proper design of sampling systems that create the lack of compliance of equipment installed or sampling plants to meet the comprehensive standards requirements. In majority cases, the samplers or sampling plants face failed validation or bias tests, as result of the incorrect sampling equipment integration in the plant including commissioning can be devastating and costly.

On the operational front the understanding of the product to be sampled goes unattended; in-sights into the scale of product variability require thorough investigation and can only be minimized by precision techniques.

The coal industry would profit from use of methodology for routinely monitoring the overall measurement precision and the precision of the individual components of measurement: sampling, sample preparation, and laboratory testing. This would not only give necessary credence to the measurement results, but would also be no doubt, lead to more efficient sampling systems design.

Keywords: Sampling, Quality, Variability, Limitations, Tailings, End user, Compliance, Theory

1. Introduction

The business of managing coal is complex and crucial to success. Adopting the right quality protocol and practices can help control value loss. Coal is the biggest single cost component for example: thermo power generation with players facing an array of issues across the sampling chain from sourcing to logistical management, bulk handling, stock management, and overall quality management. In emerging markets such as India, the addressable value loss can reach 7–12% of the total cost.

Quality management is one of the major components of the value stream where value-loss can occur. Sampling is among the most fundamental activities in coal mining and processing operations, and this paper aims to examine some of the technical issues that can assist in ensuring that samples are representative and correct. Insight into sampling systems design features and protocols are highlighted and first step in establishing an optimal protocol that complies with the latest standard requirements. This must be followed by the appropriate measures to ensure the protocol is implemented such that minimal bias is introduced in the sampling process. Even with these insights the possibility exists for large unseen and hidden costs can accumulate as result of sampling errors. These hidden costs arise due to misunderstanding of the principle factors that affects the size of the sampling errors, such as the mass of the sample, the effects of crushing, splitting to reduce the mass and the influence of nominal particle size.

In-sight into the scale of variability requires that we disaggregate sampling variance into its component parts and evaluate each individually. Implementation of appropriate sampling protocols in the coal value chain remains a challenge throughout the value process the growing understanding of sampling theory and methods means we stand on the threshold of a new era for implementing and understanding appropriate sampling procedure, compliance and protocols.

2. Design of sampling systems

Much has been written, published, and presented in recent years in terms of the theory of sampling (TOS) principles of sampling correctness, grade control, process quality control, independent sampling standards, etc., but there is a relative paucity in terms of best practice implementation guidelines for sampling equipment manufacturers (SEM) and necessary sampling project execution or *modus operandi*. This is especially true in respect of suitable and acceptable equipment designs in the mining and processing industries.

What are the key elements that will ensure a successful sampler installation in these sectors from a SEM's point of view and something that could be readily validated by a knowledgeable third party? Does a SEM have a contribution to make here, and if so, what credentials or experience need accompany any of the SEM input for it to be meaningful and add value to the customer's value stream? SEM's have a voice that is often muted, because they wage continuously struggle to secure business based on lower cost procurement, like many other equipment suppliers trying to make a living. What responsibility do SEM's need to take and how do they prepare themselves to ensure they are taken seriously when providing proposals for sampling solutions? How does lower cost procurement by end-users contribute to or aggravate the implementation of a lasting and reliable bias free sampling solution?

Many of us may have seen presentations on incorrect sampler designs (Holmes, 2009) at various conferences, providing a source of amusement and diversion from some of the "meatier" aspects of conference agendas, but what are the circumstances that have contributed to the woeful sampler designs or installations, causing great commercial cost and loss of reputation to all concerned?

This paper endeavors to shed light some light on "correct" sampler design and installation practice, not from an expert point of view, but rather based on many years of practical experience. A new paradigm is needed to overcome the major challenges faced by the end-user project companies and SEM's themselves. Surmounting these obstacles is critical if future sampling solutions for mining operations, processing plants and sampling systems where samples and their analysis are used for commercial purposes are optimized, providing potentially huge benefits for all stakeholders. We demonstrate that even the best sampling solution are difficult to implement if the importance and position of sampling in the overall scheme of plant operation and quality product delivery are not considered early in the design and construction of the plant.

2.1 Sampling knowledge

Sampling knowledge is the first important starting point from the SEM. A deficiency in knowledge about sampling practice and the principles of the TOS can be very costly when planning and implementing a sampling solution for a process plant. This applies to plant feed, tailings, and product applications and even for the monitoring of unit process efficiencies. Unfortunately, many critical and financially expensive decisions in industry are based on poor quality data that is generated by improper sampling techniques, improper equipment, or improper sampling practices.

Much has been written on the subject of sampling and the various related aspects by specialists and consultants over the years. This information is readily available as books, sampling course notes, and conference papers and cannot over emphasize the importance of this subject presented at coal conferences to play a very important role in making key people aware of the principles and aspects of correct sampling and making them aware of its critical important role. If we do not know what the potential sources of sampling errors are, how on earth can we begin to address them and prevent people from repeatedly making the same mistakes? SEM's must themselves keep abreast of developments and be aware of sampling theory.

Various ISO sampling standards must be used by SEM's as guidelines for the design, and sampling of coal, gives reference to comprehensive information regarding frequency of sampling for a desired precision level, minimum sample sizes for chemical and physical increments and composite final sample sizes and even basic guidelines on equipment design.

2.2 Sampling standards

Sampling standards come in many forms, SEM's, plant quality and technical staff must be familiar with the critical aspects of their contents, particularly relating to the minimum increment masses, quality variations, minimum number of increments per sub lot, lot or consignments, etc. SEM's must know what standards are available for the required application.

ISO sampling standard for coal 13909, 2016, hard coal, and coke – mechanical sampling part 1–8 are very comprehensive and detailed in Table 1

Table 1: ISO 13909 part 1–8

Part 1: Coal	General introduction
Part 2: Coal	Sampling from moving streams
Part 3: Coal	Sampling from stationary lots
Part 4: Coal	Preparation of test samples
Part 5: Coke	Sampling from moving streams
Part 6: Coke	Preparation of test samples
Part 7	Methods for determining the precision of sampling, sample preparation and testing
Part 8	Methods of testing for bias

2.3 Sampler design and improvements

This is an activity which is driven largely by combination of a new sampling applications, feedback from clients on existing equipment limitations (including failures), and the continual pressure to reduce project and hence equipment costs without compromising the integrity of what is supplied. Plants often need a custom-designed sampler to fit into an existing plant and into headroom that is invariably restricted. This is the 'Number One Enemy' to good sampling solutions and shows no sign of going away, despite detailed sampling consultants reports at inception and early project discussions with SEM's about optimum design layout for a satisfactory solution (Fig. 1). The fraction of sampler projects thus affected is as much as 75% and often they have little chance of providing a reliable or consistent result. They turn out to be poor investments and does nothing more than temporarily appease the conscience of management into believing that the interests of sampling have been adequately served at minimum cost.

Figure 1: Conveyor discharge points onto product stockpiles do not easily lend themselves to cross-stream sampling

3. Quality control on sampling systems

Quality control is an increasingly important component of daily process plant operations as managers endeavor to further optimize either final product yields to within agreed specifications and/or grades to ensure long-term operational sustainability and acceptable profitability for their shareholders. SEM's play their role in trying to provide mechanical sampling equipment that is both compliant and suitable for the various metallurgical applications as well as mechanically reliable.

The process QC department is normally headed up by a knowledgeable individual who ensures there is compliance with client-specified sampling standards in terms of equipment selection and sampling installation *modus operandi*. This person needs to be freed from direct laboratory management responsibilities for him to be most effective in the sampling arena, otherwise sampling practice does not receive the priority it so critically deserves.

These in-house sampling champions play a key role in keeping SEMs 'on their toes' in terms of their product and service offering, ensuring the samplers supplied for metallurgical accounting are 'correct' in design and application and hence the more affective and reliable in the longer term, and providing needed integrity to daily plant operations results.

Sampling audits are normally done on an annual basis for samplers situated at contractual payment stations or at coal export terminals, where there are formal agreements in place between suppliers and customers. Guidelines for inspections of coal sample stations prior to the audit are clearly provided in *ISO 21398: Guidelines for the inspection of mechanical sampling*. Visual inspections and identifying bias triggering mechanisms have now become relatively simple for coal systems.

3.1 Equipment installation

Why should installation be of interest? Users and project managers have a lot of influence on some of the outcomes, which in turn can greatly influence the final result and its value add. Increased awareness is never a bad thing, but whether you do something, or can do something, about it is another matter (Figs. 2 and 3).

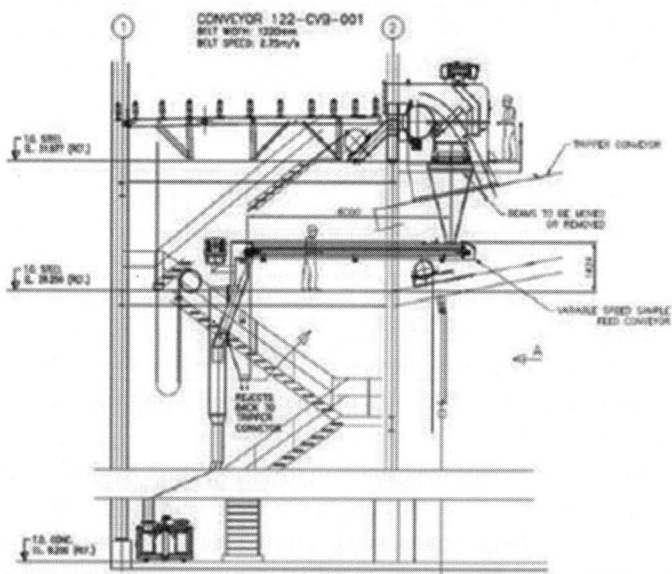

Figure 2: Cross stream sampler retrofit installation at conveyor discharge at a new plant, showing structural design interference requiring modification

Figure 3: Generous inspection hatch for primary slurry sampler cutter assembly

Validation of the sampler installation is necessary after sampler hot commissioning and includes a bias test, the methodology of which is clearly documented in *ISO 13909 Hard coal and coke – Mechanical sampling – Part 8: Methods of testing for bias.* Bias testing is required to be conducted annually on any sampling system that is used for commercial purposes. Third party laboratories or in-house quality control department conduct these bias tests to ensure that no consistent errors are indicated in composite sample results and that bias are within minimum tolerable levels specified in agreements between customers and suppliers.

Continued longer reliability of sampler installations and their worth to clients is largely dependent on service contracts provided by SEMs. If maintenance is done in-house, samplers are often neglected relative to, say, production units such as mills, crushers, filters, screens, etc. It is wrongly perceived that samplers do not directly contribute to the bottom line of a company and are simply additional mechanical units requiring repetitive maintenance. Producers have spoken about a 6-month ROI based on subsequent reduction delivered specifications and improving yields due to better control on supplier contract specifications. They had previously over-compensated during production by providing above-specification consignments. Valuable applications ad sampler performance feedback from site allowed the SEM to continuously improve their production offering (in terms of grade and yield obtained) and allow producers to be confident of the measurement accuracies or at least optimum operability of a sampler installation and their own process plant.

4. Coal variability and errors related to this

Coal is a variable product, as result of the coal geology, where factors such as *in situ* coal qualities, coal testing, coal reactivity, and coal weathering will have a major impact on mining activities and coal processing.

It is up to the sampling systems to extract representative samples through this value chain to assist decision makers to direct the process based on its chemical composition. Any non-representative sample could lead to incorrect decisions with enormous financial implication.

It is imperative that any sampling system, subjected to these requirements must comply with the required standard measures by taking a representative sample, in respect of its variability at that stage.

4.1 Implementing the required ISO 13909 mechanical sampling standard

Once a sampling system has been designed and installed, the precision which is being achieved on a routine basis should be checked. An estimate of the precision can be obtained from the primary increment variance, the numbers of increments, and sub-lots, and the preparation and testing variance. The preparation component is made up of on-line sample processing and off-line sample preparation.

Sampling variance is a function of product variability, so the same number of increments, sub-lots, and preparation and testing errors will yield different precision with fuels that exhibit different product variability.

An estimate of the precision actually achieved can be obtained by taking the sample in a number of parts and comparing the results obtained from these parts. There are several methods of doing this, depending on

(a)   the purpose of the test, and

(b)   the practical limitations imposed by the available sampling procedures and equipment.

Where a sampling system is in existence, the purpose of the test is to check that the scheme is in fact achieving the desired precision. If it is not, it may need to be modified and rechecked until it meets the precision required. In order to do this, a special check scheme will have to be devised which may be different from the regular scheme, but which measures the precision of the regular scheme.

4.2 Equations relating to factors affecting precision and precision

Precision is a measure of the closeness of agreement between the results obtained by repeating a measurement procedure several times under specified conditions and is a characteristic of the method used.

If a large number of replicate samples, j, are taken from a sub-lot of fuel and are prepared and analyzed separately, the estimated precision, P, of a single observation is given by equation as follows:

$$P = 2s = 2\sqrt{V_{SPT}} \qquad (1)$$

where s is the sample estimate of the population standard deviation and V_{SPT} is the total variance.

The total variance, V_{SPT}, in Eq. (1) is a function of the primary increment variance, the number of increments, and the errors associated with sample preparation and testing.

4.3 Precision and total variances

In all methods of sampling, sample preparation and analysis, errors are incurred, and the experimental results obtained from such methods for any given parameter will deviate from the true value of that parameter. While the absolute deviation of a single result from the "true" value cannot be determined, it is possible to make an estimate of the precision of the experimental results. This is the closeness with which the results of a series of measurements made on the same coal agree among themselves, and the deviation of the mean of the results from an accepted reference value, i.e. the bias of the results.

The theory of the estimation of precision is discussed in ISO 13909-7. The following equation is derived:

$$P_L = 2\sqrt{\frac{(V_1/n) + V_{PT}}{m}} \qquad (2)$$

where P_L is the estimated index of overall precision of sampling, sample preparation and testing for the lot, expressed as a percentage absolute, V_1 is the primary increment variance, n is the number of increments per sub-lot, m is the number of sub-lots in the lot, and V_{PT} is the preparation and testing variance.

4.4 Primary increment variances

The primary increment variance, V_1, depends upon the type and nominal top size of coal, the degree of pre-treatment and mixing, the absolute value of the parameter to be determined and the mass of increment taken.

The number of increments required for the general-analysis sample and the moisture sample shall be calculated separately using the relevant values of increment variance and the desired precision. If a common sample is required, the number of increments required for that sample shall be the greater of the numbers calculated for the general-analysis sample and the moisture sample respectively.

4.5 Preparation and testing variance

The value of the preparation and testing variance, V_{PT}, required for the calculation of the precision using Eq. (1) can be obtained by either:

(a) Direct determination on the coal to be sampled using one of the methods described in ISO 13909-7; or

(b) Assuming a value determined for a similar coal from a similar sample preparation scheme.

If neither of these values is available, a value of 0.5 for ash content can be assumed initially and checked, after the preparation and testing has been carried out, using one of the methods described in ISO 13909-7.

4.6 Number of sub-lots and number of increments per sub-lot

The number of increments taken from a lot in order to achieve a particular precision is a function of the variability of the quality of the coal in the lot, irrespective of the mass of the lot. The lot may be sampled as a whole, resulting in one sample, or divided into a number of sub-lots resulting in a sample from each. Such division may be necessary in order to achieve the required precision, and the necessary number of sub-lots shall be calculated using the procedure given.

4.7 Calculating of number of sub-lots and increments

The number of sub-lots and number of increments required per sub-lot are established using the following equations.

$$n = \frac{4V_1}{mP_L^2 - 4V_{PT}} \qquad (3)$$

A value of infinity or a negative number indicates that the errors of preparation and testing are such that the required precision cannot be achieved with this number of sub-lots. In such cases, or if n is impracticably large, increase the number of sub-lots by one of the following means.

(a) Choose a number corresponding to a convenient mass, recalculate n from Eq. (2) and repeat this process until n is a practicable number.

(b) Decide on the maximum practicable number of increments per sub-lot, n_1, and calculate m from the following equation:

$$m = \frac{4V_1 + 4n_1 V_{PT}}{n_1 P_L^2} \tag{4}$$

Required when measuring the increment variance, the preparation and testing errors are included more than once.

5. Conclusion

Quality management in one of the most primary activities in mining and processing being sampling cannot serve this industry if not adopting the right quality protocol. Appropriate design of sampling systems, correct sampling equipment and installation, quality control on sampling systems and coal variability with the implementation of required standards cannot be under estimated.

Industry need to understand the principles and aspects of correct sampling and the critical role it plays in the coal quality and value stream, by implementing this quality protocol will not only add value in exploiting valuable mineral assets but, limit fundamental errors to achieve by adding value to this precious commodity.

6 References

[1] J. Bekker "Importance of Standards". SA Coal Preparation Conference, 2009.

[2] J. Bekker "The importance of sampling in coal processing". CPSI, 2017.

[3] P.E. Hand "Sampling the coal chain". Journal of the Southern African Institute of Mining and Metallurgy. January 2014, P 6–8.

[4] R.J. Holmes "Sampling mineral commodities – the good, the bad and the ugly". SAIMM, 2009.

[5] R.C.A. Minnitt "Sampling: the impact on cost and decision making". SAIMM, 2007.

[6] R.C. Steinhaus, R.C.A. Minnitt "Mechanical sampling – a manufacturer's perspective". SAIMM, 2014.

[7] ISO 13909 (2016) Hard Coal and Coke – Mechanical Sampling Standard, Part 1 to 8.

22

Coal quality scenario: The Singareni Collieries Company Limited (SCCL)

K. Nagabhushana Reddy and A. Ravikumar

SCCL, Bhadradri Kothagudem Dist. Telangana, India

Abstract: Coal reserves in the command area of SCCL are predominantly of high ash and low heat value thermal coals. In this situation necessary steps are required to be during mining so that contamination with overburden debris is avoided and consistency in the quality of coal dispatches is maintained. In order to ensure compliance of government mandate the ash content must be at ash content 34% or below, standard operating guidelines are issued to the open cats mining personnel. These inter alia include, carefully designed drilling and blasting of coal and OB, cleaning of coal benches, maintaining proper sequence of mining of different coal seams etc. Authors have attempted to elaborate the steps taken to ensure uniform quality of coal dispatches.

Keywords: Jigs, OB, Picking, MGR, CAGR, Deshaling, Barrrels, Crushing

SCCL is a Joint venture of Govt. of Telangana and the Govt. of India with an equity participation in the ratio of 51:49, respectively with current proven coal reserves of 10,846 Mt. SCCL is operating 19 opencast and 29 underground mines in 6 districts of Telangana and contributing about 9% of domestic coal production.

SCCL being the only coal producing company in southern India is vested with the responsibility of meeting the coal needs of Thermal Power Plants, Cement industries, Sponge Iron industries, Fertilizer industries, etc. of the region. The coal production in SCCL has increased from 50.47 Mt in 2013–14 to 62.01 Mt in 2017–18 with a Compound Annual Growth Rate (CAGR) of 5.28% in the last 4 years. Out of 64.6 Mt of coal dispatched during the year 2017–18, about 82% was supplied to Power sector and this would continue in the present financial year as well.

Coal reserves in the command area of SCCL are predominantly of high ash and low heat value thermal coals. Around 65% of the coal produced is having more than 34% ash content. Current environmental directives of Government of India restrict the movement of high ash coal (ash > 34%) beyond 500 km from the coal dispatch points of mines. To guard against the contamination and maintain consistency in quality of coal supplies, following steps are being taken in addition to face level management, coal preparation through proper crushing, sizing, blending and washing wherever it is required.

- Drilling and blasting of coal and OB layers separately.
- Ensuring the coal bench clean before blasting.
- Keeping OB bench at least 5 Mts ahead of coal bench.
- Separate loading of stone to avoid contamination.
- Use of magnetic separators for separating the metallic objects.
- Maintain grade wise separate stocks so as to blend proportionately before dispatch.
- While working OCPs on developed galleries the contaminated/fiery coal is being separately dumped and sent to washeries.
- Shale/stone picking rigorously at all the mines/OCPs/CHPs.
- Introduction of surface miners at suitable locations for selective mining of coal preventing contamination with dirt bands. Currently, one surface miner is producing 3.0 Mty at KOC-II mine and another surface miner is envisaged be introduced in KOC-III.
- Working of the seams as per grade declaration plan to produce the planned ratios of coal from different seams so as to maintain the overall declared grade.
- To ensure consumer satisfaction, third party sampling has also been resorted to for coal supplies to both power sector and non-regulated sector.
- **Power sector:** Third party sampling and analysis started from 14.07.2016 with NTPC which has been extended to all the Customers of Power sector M/s CIMFR is the Third Party Sampling Agency.
- **Non Power Sector:** Third party sampling and analysis started from 01.09.2018 for the coal supplied to non-regulated sectors in the same lines of power sector. M/s IICT, Hyderabad is the Third Party Sampling Agency.

A separate organizational setup is established with three general managers one in each region assisted by the area level quality managers for continuous monitoring of coal quality at all mines/OCPs/CHPs. With increasing environmental restrictions over the transport of high ash coals, SCCL is taking various measures to ensure and enhance the quality of coal to comply with these environmental concerns. SCCL is supplying coal to different power plants of NTPC, TSGENCO APGENCO, Karnataka Power Corporation – KPCL, MAHAGENCO, TANGEDCO, NTPC-SAIL Power Company Ltd and MSEB of different states in this region. Coal produced from the mines is dispatched to different customers through different modes as given below.

- **Pit head dispatch through MGR:** Low grade coal with high ash content is supplied to the nearby customers through MGR system. Around 15% of the coal is being supplied to Ramagundam Super Thermal Power Plant of NTPC from the mines of Ramagundam region through MGR rail system.
- **From the coal handling plants**: Blended coal of different grades produced from different mines is being dispatched to different consumers from the coal handling plants through rail for transport over long distance.
- **Pit head road dispatch**: High grade coal with low ash content is being supplied to the consumers directly from the mines.
- **From the coal washeries**: The low grade coal with high ash content is cleaned at the coal washeries established near the mines. Around 3 Mt of craw coal is being washed and supplied to customers either by rail of by road.

Demand for less than 34% ash coal for power sector away from the pit heads is envisaged to increase from current level of 24 Mt to 40 Mt by the year 2027–28. This requirement is being met through blending of coal largely and washing to a limited extent. However, it is now proposed to enhance the washing capacity from current level of 3 Mty to 10 Mty gradually by 2026–27 through establishing three more washeries.

The envisaged yield from the washeries using the presently operated jigging technology or barrel technology would be in the range of 45–75% for the likely enrichment of grade from G16/G13 grade to washed G9 grade. The envisaged improvement of calorific value in the washed coal and the corresponding notified price of enriched coal would not offset the cost involved in washing these coals. As a result, the washed coal (after considering carbon loss in terms of rejects, etc.) would cost higher than the notified price for corresponding grade on heat value basis. Whereas, the consumers are reluctant to bear such additional cost of washing although the economic/environmental benefits of using the washed coals for power generation is in favor of them. Once the pricing issue is addressed, supply of washed coal would increase many fold. If the additional cost of washing is allowed as a pass through in the tariff system by regulators, it would help the consumers using washed coal for power generation.

Disposal of washery rejects is another issue related to washing of coal. Since 25–40% of the raw coal gets converted into rejects in the present system coal washing, handling of rejects becomes a major issue from environment point of view. While some FBC based boilers are able to consume rejects with high ash content of more than 65% and GCV of around 1200 Kcal/kg, however, it is not possible to make use of all the rejects produced during the course of washing. Therefore, it is important to allow the rejects to be used for back filling of mine voids.

The technology in vogue for washing thermal coals is mainly jigs and barrels with water. Dry washing technology is also proven commercially in other countries. However, its application in Indian conditions is awaited. Such closed circuit deshaling plants should also be able to cater to the requirements of power sector. The industry players need to establish the most economical technologies suitable for thermal coal washing in India.

It would be desirable to do away with the present distance restrictions for use/supply of washed coal for thermal power generation on ash content basis. This would impact the power off take on merit order basis as pit head power plants need not pay any additional cost compared to the power plants located beyond 500 km from the coal dispatch points. Rather it is important to restrict the movement of coal over distances on the basis of heat content of coal. This would bring commercial and environmental discipline in the market.

23

Uncoaler/activater feeder for coal extraction

¹K.S.Nalwaya, ²Jogesh Narula

¹ Managing Director, KSN Tech Ventures Pvt. Ltd, New Delhi, India
² Vice President Projects, KSN Tech Ventures Pvt Ltd, New Delhi, India

Abstract: This paper is focused on the Un-coaler /Activator feeders. It explains the working principals of the Uncoaler, Its Application for Extracting Coal and Limestone, etc from under Stockpiles, Track Hoppers/Wagon Tipplers, Silos and The Advantages of this equipment, vis-à-vis, Apron feeders, Belt Feeders, and Vibrating Feeders in terms of Ease of operation and maintenance and more compact plant layout.

Keywords: Uncoaler, 2-Mass Theory, Compact Equipment, Compact Installation and, Compact Layout, Ease of Operation, Minimal Maintenance, Low Power Requirements and Consumption

1. Introduction

India is Primarily a Coal Driven economy and coal-based power plants would remain the dominant part of India's Energy Mix for a Long Time

- In an effort to make Coal Burning more efficient and more environmentally acceptable India is. increasingly adopting Coal Washing for minimizing ash, at source,
- Storing Coal in Sheds and Silos controlling moisture and pollution
- **Adopting better coal handling technologies** to be able to feed ever increasing Plant requirements.

For improved coal handling we introduce a very important equipment, the **Uncoaler activator feeder** which enables large capacity Coal extraction from Coal Stockpiles, Wagon Tippler/ Track Hoppers and Silos and makes possible compact layouts of the Coal Stockyards.

Uncoalers are based on 2-Mass vibration system, where the exciter mass is connecting to the Conveying mass through a spring system, as opposed to a Single Mass Vibrating System where the Exciter is Rigidly connected to the Conveying Mass.

- The 2-Mass design ensures High amplitude vibration under the loaded condition with very low motor ratings and protects the equipment body and supporting structure from stress fractures.

- The Motor ratings for a 2-Mass system is hardly 20-25% of the Motor in a rigid connection system.

- The 2-Mass system also ensures that it is easily possible to design higher capacity equipment safely, which is beyond the capacity of rigid connection systems.

- Shown below is a 5000 tph Uncoaler based on the 2-Mass Design.

In the 2-Mass Vibration system, the equipment always operates below its frequency of resonance (The left side of the curve in the slide below as shown below) This ensures safety of the equipment and the mounted structure from stress fractures

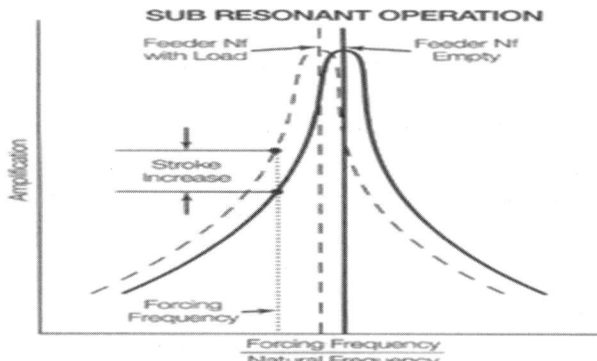

2. Working principal of uncoaler

The Uncoaler is used for extraction of Coal and other Bulk Material, and is installed under Stockpiles, Silos, Track Hoppers, Wagon Tipplers and Reclaim Hoppers. The Material rests directly on the Uncoaler and the Uncoaler with the Projected cone vibrates and activates the pile and induces discharge from opposing sides into the belt conveyor installed below. Uncoalers are easy to install and perform effectively even while supporting a full load of material in the Hopper or Silo above. There is no need whatsoever to install rod gates, feed control gates or isolation gates.

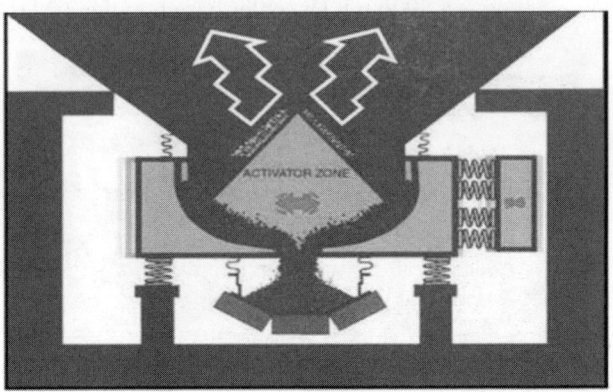

2.1 Advantages of uncoalers

The Uncoaler Activator/Feeder is a Compact, Sturdy and Low Maintenance equipment, installed under Coal Stockpiles, Track Hoppers, Wagon Tippler Hoppers, Silos and Reclaim Hoppers. More than 3000 Uncoalers are in operation worldwide

Uncoalers offers the following advantages over equipment like Apron Feeders, Belt Feeders and Vibrating feeders

- Uncoalers are Compact Height- for example, the height of 1000 tph unit is only 0.9 m
- Uncoalers have High Extraction Capacity and range from 150 tph to-8000 Tph for single unit
- Uncoalers have High Catchment/Inlet Area-For a 1000 tph Uncoaler this is3.0 m x 3.0 m and a compact centralized discharge, this enables an Uncoaler based installation to have a lower height/shallow depth and leads to a compact layout. This result is substantial depth reduction for Wagon Tipplers and Track Hopper System as well reduction of the conical height of silos.
- Uncoaler/Activator Feeders are a versatile equipment, which works effectively for a variety of material ranging for a lump size of -450mm to fine powders like Potash and Calcine at the other extreme.

Shown below is a sketch and Photographs of Uncoalers installed under Coal Stockpiles, where Coal is extracted from the Piles and discharged on to the Belt Conveyor below. The Discharge Capacities Possible are from 500 to over 20,000 tph.

- Uncoalers under Stockpiles eliminates the Bucket Wheel Reclaimer Yet High Volume coal extraction is possible. Grizzly on the top of Hopper will ensure the correct size of the material is extracted from Coal stock Piles at Mines and oversize material can be eliminated
- In many cases different Grades of Coal may be stocked at different locations and maybe extracted by Uncoalers and be discharged on a common belt conveyor system installed below, this makes coal blending possible.
- Uncoalers have a large catchment area and can draw in coal over a wide area and minimizes mechanical handling.
- As the Uncoalers are not placed inside stock Piles unlike other equipment and does not have any moving parts except the motor as such there is no wear and tear of moving parts.
- Due to the compact design and large opening, depth of Tunnel for Stock Piles is reduced substantially.
- Uncoalers are especially useful in coal Mines, Coal terminals and Power Plants where high-volume coal extraction is required Uncoaler

2.2 Uncoalers under track hoppers

- Track-hopper installations with Uncoalers are generally 65 m long and 9 m deep as against, 320 m long and 15 m deep with paddle feeders.
- Uncoaler based track hoppers have high extraction rates up to8000 tph as against 2000 tph extraction rate for Paddle Feeder based track hoppers.

- Uncoaler based Track hoppers are only 20% of the cost of a Conventional Track Hopper and have accordingly less operation and maintenance cost.

In India, Uncoalers have been installed under the Track hopper system at the Dhamra Port in Odisha. The Picture of the Track Hoper is shown. The Adjacent Picture is of Uncoalers installed under Track Hopper of 3000 MW Erraring Power Station in Australia. This is s 25-year-old Installation and evacuates a 56-wagon rake /4500-ton coal within 45 minutes and with negligible maintenance cost

2.3 Uncoalers under wagon tipplers

Shown below is the sketch of an Uncoaler Installation under Wagon Tippler

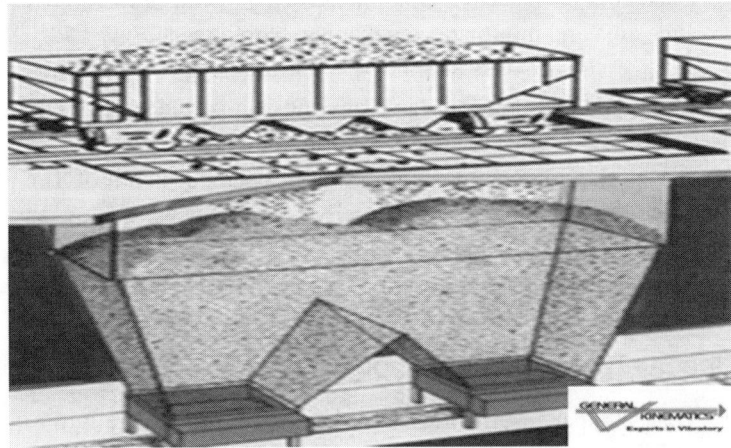

Shown below is the Most important Installation of Uncoalers under wagon Tippler is at the Shinhua Coal Terminal in China which handles over 300 Million Tons of Coal annually. Here 32 Uncoalers are installed under 4 x 4 wagon Tippler line. Each Line is 8000 tph Capacity

At the same Shenhua Terminal we have world largest Uncoaler installation of 576 Uncoalers under 96 Coal Silos. Each Silo is 30,000 T Capacity

2.4 Uncoalers under Silo

- Due to its large inlet cross-section Uncoalers under Silos, reduces the height of the Silo Conical Portion and thus the Overall Silo Height.
- Since the Coal rests directly on the Uncoalers, the projected vibrating cone eliminates, Rat holes, Bridging and Arches and enables better Coal Extraction Performance

 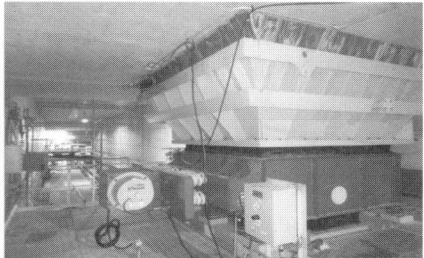

2.5 Uncoalers under reclaim hoppers

Uncoalers are very useful under the Mobile hoppers, where they make Low height Hoppers possible. Doing so makes the loading of the hopper possible with wheel loaders. No ramp is required for accessing the hopper top opening as described below

Reclaim Hoppers, with VibroFeeder /Belt Feeders are normally so high that these are above the operating height of front-end loaders and a ramp has to built up to the hopper so that the wheel loader bucket can reach the hopper opening This leads to the reduction of the storage yard and the reduction of the Coal storage and also is more strenuous on the wheel loader.

Reclaim Hoppers with wheel loaders are low height and can be accessed with front end loader easily and the ramp is not required.

3. Conclusion

Where ever these may be installed, the Uncoaler installation is very simple and the following are eliminated:

- Inlet Chute
- Isolation Gate
- Feed Control Gate
- Outlet Chute
- Sector Gate
- The discharge from the Uncoaler is opposed balanced and centralized onto the central section of the Belt Conveyor thus no tracking of belt conveyor and also eliminates any spillage.

To summarize, we may conclude that, in a coal handling system, Uncoaler is always a winner, because of the following:

- The Uncoaler is compact
- The Uncoaler enables low height and shallow installations, thus reduces civil cost substantially for all these applications
- Uncoaler based Layouts are compact and lead to savings in Land Cost
- Uncoalers are Largely Maintenance free
- Uncoalers consume Minimal Power.
- Uncoalers are available in capacities ranging from 150 tph to 8000 tph for a single unit.

24

Economic benefits and risks of optimized homogenization in coal stockpiling systems

Michael P. Cipold[1,2], Pradyumn K. Shukla[1], Claus C. Bachmann[2]

[1]*Karlsruhe Institute of Technology, Institute AIFB, Karlsruhe, Germany*
[2]*J&C Bachmann GmbH, Pforzheim, Germany*

Abstract: Homogenization of coal in blending beds is standard for coal processing worldwide to reduce naturally occurring fluctuation of raw material parameters like calorific value or ash content. Standard stacking methods like Chevron or Windrow stacking are used to deposit the material in stockpiles. This however leads to an uncontrolled process resulting in non-optimal unreliable homogenization. Material parameters fluctuations can lead to inefficient downstream processing, waste of material or use of expensive compensation material. In extreme cases violation of process or regulatory constraints can cause huge outages or fines.

A software solution developed at KIT can be integrated into existing blending bed machinery using material parameter information already available in the plant. The software builds a quality prediction model and simulates the stacking process to evaluate the expected homogenization efficiency. It then applies evolutionary optimization to maximize this efficiency.

The benefits from an increase in homogenization efficiency are numerous: enabling tighter constraints, reduction of wasted material, reduction of emissions and overall negative environmental impact, and reduction of process outages due to parameter fluctuations. This leads to immediate financial benefits to the user of the system.

Integration of the software into existing blending systems is cost-effective and allows the use of hidden potential of already installed machinery. The software can also be installed in a passive monitoring mode which allows economic evaluation by the plant operator before application of changes minimizing potential risks. This paper evaluates the economic benefits and risks of integrating this novel system into an existing blending bed system.

Keywords: Optimized homogenization, rolling horizon optimization, evolutionary optimization under uncertainty, stockpile simulation

1. Introduction

Quality parameters of run of mine coal are highly variable. Coal processing plants like coal fired power plants are designed for specific mean quality parameters like calorific value, ash or sulfur content. It is standard procedure to use stockpiles for homogenization of mined material. Material is regrouped by stacking the input material in layers and reclaiming it in cross sections. This way the variance of material parameters is reduced.

Ideally material is regrouped in such a way that the variance of all relevant parameters is as close to zero as possible resulting in optimal homogenization. Practical limitations of the involved equipment define the stockpiling system parameters like stacker travel speed, maximum tonnage or limits for stockpile dimensions. Material parameters like particle size, density or moisture also influence the stacking and reclaiming process e.g. by affecting the angle of repose. Given these parameters, a specific material input stream and a standard stacking technique the homogenization efficiency can be estimated (De Wet, 1983) or measured using an online analyzer and a belt scale.

Additional to homogenization input material can be blended with (ideally homogeneous) material from a different source (other mine, bench, stockpile, etc.) to compensate for the deficiencies in the mean quality of the input material. Blending has to be used in cases where the mean quality does not meet the plant's requirements. Additionally, blending also has to be applied in cases where the variance of mined or poorly homogenized material leads to periods in which material with critical parameters is supplied to further processing. For example, this can be sulfur not meeting the emission requirements for an extended period of time.

However, blending because of material variance issues with material specifically purchased or refined for compensation is expensive. The option of discarding poor quality material is expensive as well. Material has been processed that far and needs to be further handled. If no action is taken emissions might exceed limits, forced outages might occur, maintenance could be required and the plant could not be able to meet full load. This leads to the conclusion that poorly homogenized material leads to process risks and high cost.

Homogenization systems mostly apply standard stacking techniques like Cone-Shell, Chevron or Windrow stacking following a fixed material deposition schedule. The schedule might have been optimized using a dynamic number of layers but is still limited to the stacking technique. As such and due to natural fluctuations in material quality homogenization using these techniques is still suboptimal, unreliable, and might lead to poorly homogenized material.

The institute AIFB of the Karlsruhe Institute of Technology in cooperation with the J&C Bachmann GmbH developed an optimization system specifically targeting the improvement of homogenization in existing stockpiling systems. It makes use of real-time information already available in the plants like exploration data, sampling information, material data sheets, and online analyzers. This unlocks the hidden potential of existing equipment and optimally uses the hardware and information present in current plants without the need for expensive hardware modifications.

This work is related to research dating back to the fundamental works of Gerstel (1977) and Gy (1981) extensively describing homogenization efficiency, its mathematical modeling and limitations. Research by Lu and Xu (2010) and Loubser and Korte (2015) described voxel-based simulation which is similar to the simulation methods used in this paper. This simulation software is however not available to other researchers and thus the simulation used in this paper was independently developed and presented by Cipold et al. (2012, 2013). Stevanovic et al. (2014) describe optimization of material grouping at the stockpile level. In contrast the methods described in this paper do not influence the material flow within the plant besides optimizing homogenization by changing the regrouping of the material within the stockpile. The methods described in this paper could be combined with research on optimized reclaiming described by Lu and Myo (2010).

As far as we know our approach is unique and leads to unmatched improvements in homogenization efficiency, reliability and thus plant performance especially for existing plants dealing with homogenization issues.

2. Methodology

In order to optimize homogenization, the system is composed of multiple interlinked components which are controlled in real time. This chapter briefly describes the components involved. A detailed description of the methodology can be found in the proceedings of the 9th World Conference on Sampling and Blending.

2.1 System components

One of the core components is a material prediction system which uses all plant provided momentary information to create a virtual material quality curve used for optimization. It can be fed automatically using sensors, belt scales, online analyzers, transport schedules, mining schedules, geostatistical information, probing of trains or trucks as well as with manual input of lab information or material data sheets. Using all available information combined a material quality curve is generated on every information update and provided to the optimization system. This curve can be incomplete at the beginning of optimization and updated continuously during the stacking process even after material covered by the prediction has already been stacked (e.g. using latest lab information or instrument recalibration). This way knowledge about material in the future can be used and the model of already stacked material can be updated in real time.

A rolling horizon optimization based on the current material knowledge horizon and information from the stacking system (belt scales, stacker motion recordings) is performed. The optimization system is supplied the desired maximum stockpile dimensions by the plant operation system or by a manual controller and is preconfigured with the knowledge about the homogenization system like maximum stacker travel speed or estimated angle of repose. An optimization is triggered in intervals defined by the plant control system. Evolutionary optimization is used to compose a material deposition which is feasible for the homogenization system and optimal regarding all optimization objectives. The objectives can be various with the most important one being the best homogenization regarding all relevant material quality parameters for the material knowledge given under the constraints of the system. Additional optimization objectives can be specified like minimal variance in reclaiming tonnage, minimal stockpile length or minimal stacker movement.

In order to perform an optimization, step each deposition calculated by the optimization system needs to be evaluated. For this evaluation a novel stockpile simulation system (Cipold et al., 2012, 2013) was integrated. The simulation is particle based and can be applied for various types of homogenization systems like longitudinal or circular blending systems supporting many types of machinery. Simulation of a particular deposition using momentary material information results in a material reclaims quality curve prediction. Afterwards automatic evaluation of all optimization objectives is performed. The optimization system combines and evaluates hundreds of depositions per second in high detail and high-speed simulation until the optimization converges or predefined stop criteria are met (e.g. a time limit).

As this is multi-objective optimization there is never a single best solution but a Pareto front is approached giving the plant operator the choice between various objective trade-offs to make an informed decision. The system can also be supplied with parameters for automated decision making (e.g. by weighting objectives or specifying constraints).

During the stacking process numerous standard stacking techniques are also taken into account and evaluated automatically in comparison with the optimized solutions found by the system. This gives the operator additional information about system performance and enables its self-evaluation.

2.2 System requirements

Integration of this novel system has standard industry requirements. The stacker needs to be computer-controlled. Online measurement of tonnage is required.

Furthermore, relevant material parameters require online measurement, e.g. using an Online XRF Analyzer, PGNAA or another appropriate measurement method which is often already in place.

Data for predicting the future material curve and be acquired from the above-mentioned sources. The system itself operates from an industry computer interfacing with the plant management system to acquire material information and current stockpile requirements as well as control the deposition schedule.

3. Evaluation

3.1 Experimental setup

We composed multiple realistic scenarios to evaluate the performance of the described real-time optimization system. They are designed with specific characteristics to provide challenging realistic environments for evaluation. The equipment used in the scenarios is a single stacker using only its travel capabilities for stacking in Chevron like motion with a maximum of 6 m/min. Reclaiming is done with a bridge reclaimer as it is the least flexible option and thus the most difficult to optimize for.

Scenario A shows the operation of stacking and reclaiming supplied by a single mine. This is a difficult optimization scenario as prediction of long-term averages does not reflect the high frequency variance of the material. Based on exploration 2 h averages are predicted for 24 h of operation. In order to give a minimum of information for the optimization system this plant uses probing of trucks which the excavators load onto 45 min before the material is actually fed to the stacker. An online measurement is installed at the belt 10 min before the stacker which gives average parameter values every 60 s. The relevant material parameters in this scenario are ash and sulfur. Exemplary parameter curves for this scenario are shown in Fig. 1.

Scenario B shows operation supplied by two mines. Each homogenization system is optimized independently of other homogenization systems in the plant using only the schedule and mine averages for 24-h prediction. Again, these averages are based on exploration. This time no probing is done but the analyzers are installed 30 min before stacking giving the same 60 s material parameter averages. Here the sharp changes in material parameters when switching between the currently supplying mine pose a challenge for the system. This way material parameters have a low frequency fluctuation due to the schedule only changing every 20 min to 4 h. Figure 2 shows an exemplary parameter curve for this scenario where the only relevant parameter for homogenization is calorific value.

In scenario C the unloading of a train is shown. This plant is partially supplied by train and unloads material to stockpiles before further processing. Besides the short-term online measurement 15 min before stacking only average material parameters per wagon (e.g. acquired by an XRF push probe) and their weights are known. The train in this scenario has 86 wagons with a capacity of 92 tons each. Figure 3 shows the typical fluctuation of material parameters unloading the wagons where only ash is regarded as the relevant homogenization parameter.

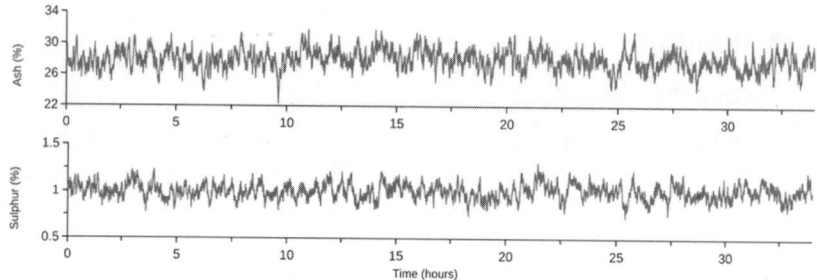

Figure 1: Exemplary material quality parameter curves for the scenario A

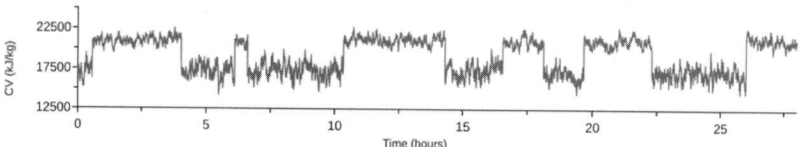

Figure 2: Exemplary material quality parameter curve for the scenario B

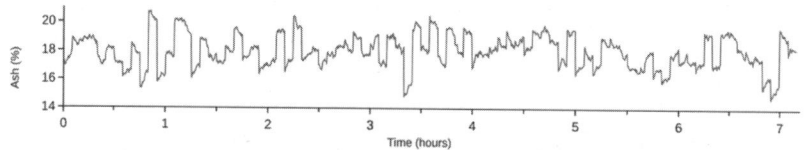

Figure 3: Exemplary material quality parameter curve for the scenario C

In Table 1 more details for the described scenarios are shown to give insights in the nature of the problems in each case.

These scenarios show the flexibility of the system and the wide range of applications. The system is however not limited to these types of situations and can be used with a wide range of customer specific hardware, material parameters, volumes, and blending bed types.

All scenarios are combined with various material curves derived from real material parameter recordings at a coal processing plant. Material curves include combined low and high frequency and periodic material parameter fluctuation, abrupt or steady change in material parameters and tonnage.

The optimization system was provided with the information described above and optimization was based on prediction data in real time. The best-known material deposition with a balanced trade-off between low parameter variance and low tonnage variance in reclaimer output material was automatically chosen. In the end the actual material curves were combined with the deposition which was calculated in real time based on the prediction and compared with a reference deposition. This reference deposition was Chevron stacking at maximum stacker travel speed for all scenarios.

For each of the scenarios, five different material instances were evaluated. To show consistency and reliability in results one of the materials was real-time optimized five times independently with the same settings.

Table 1: Additional parameters describing the three scenarios

	Scenario A Single mine	Scenario B Two mines	Scenario C Train
Tonnage	2100–2600 tph	3150–3850 tph 2530–3170 tph	1050–1150 tph
Total material	80,000 t	89,000 t	7900 t
Max. stockpile dim.	270 m × 40 m × 20 m	300 m × 40 m × 20 m	110 m × 20 m × 10 m
Stacking time	34 h	28 h	7 h
Max. stacker speed	6 m/min	6 m/min	6 m/min
Material parameters	Ash (20–36%) Sulfur (0.5–1.5%)	CV (13–21 MJ/kg) CV (18.8–23.2 MJ/kg)	Ash (13–23%)
Optimization every	30 min	30 min	15 min

3.2 Performance

After real-time optimization is finished an evaluation of actual performance compared with standard stacking techniques is performed. While Chevron stacking creates full stockpile width layers, the optimized depositions look more complex as they are composed by many linear pieces as Fig. 4 shows for all three scenarios.

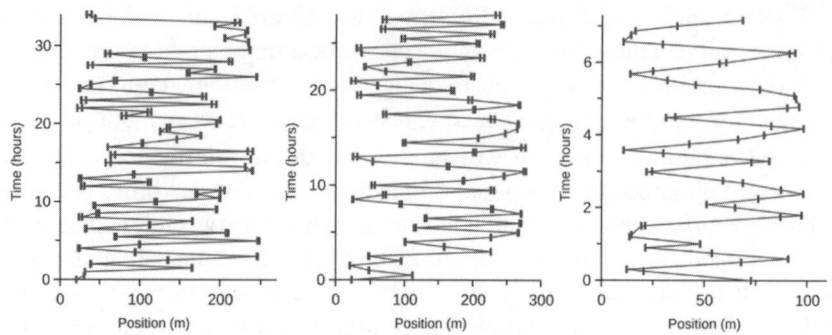

Figure 4: Exemplary depositions graphs computed for scenarios
A (left), B (middle), and C (right)

While the deposition might look chaotic to a human, result in homogenization efficiency and reclaim tonnage clearly show the desired effect. Figure 5 shows the reclaimer material output fluctuation using optimized deposition and the output using maximum stacker speed reference deposition. In ash clearly a trend over time can be observed for the reference deposition which is undesirable in homogenized material. But also, in tonnage stacking with optimized deposition achieves a more reliable output material tonnage for reclaiming with constant travel speed.

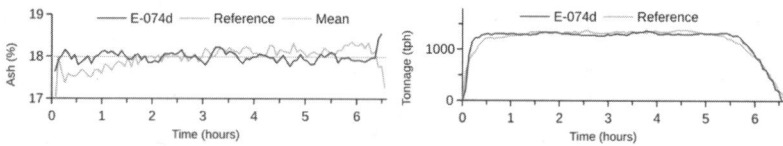

Figure 5: Exemplary material output graphs for optimize and reference deposition

The results shown this far were exemplary to illustrate the effects of real-time optimized homogenization. Figures 6–8 now show the results of all optimization system executions for the three scenarios. Each figure consists of two types of graphs showing the material parameter standard deviations achieved with optimized deposition (lower bars, blue) and reference depositions (max. speed Chevron stacking, upper bars, orange). The first graph type shows comparison for different materials (indicated by the "E" identifiers). The second graph type shows the reliability of optimized stacking with only one reference run and five independent real-time optimization runs for the same material.

Figure 6 shows results for scenario A, the single mining scenario. As expected, the results achieved using the optimization system are similar to the results achieved by standard stacking techniques because the fluctuations are in no way covered by the prediction system and there is near to no low frequency fluctuation. Still the optimization system manages to improve the homogenization noticeably.

In Fig. 7 results for scenario B with two mines show a different picture. We can see multiple material instances where the low frequency fluctuation but abrupt changes of the material source changes can be dealt with in a much better way using the optimized deposition. In the five runs high consistency and improvement in homogenization efficiency by more than a factor of two can be seen.

In the train scenario C with the results displayed in Fig. 8 we can see very similar performance. All optimized depositions clearly outperform the reference deposition and the low variation in multiple optimization executions and thus high reliability can be observed here as well.

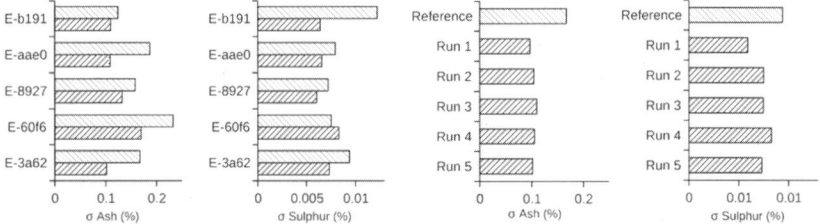

Figure 6: All results of optimization runs for scenario A including reference runs

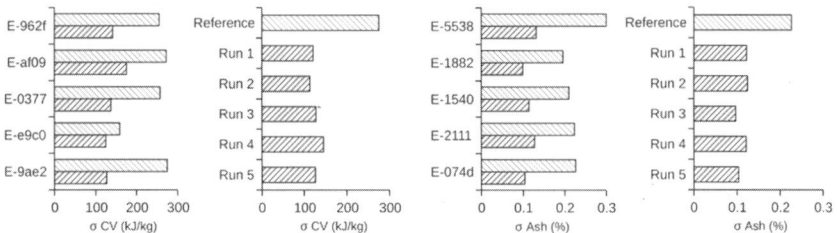

Figure 7: All results for scenario B   **Figure 8:** All results for scenario C

4. Conclusion

The optimization system presented in this paper is designed to improve homogenization efficiency in existing stockpiling systems. The results show a huge improvement in reliability and homogenization efficiency using existing equipment. Furthermore, the system makes use of the information already available in plants and does not require expensive hardware investments. The requirements of the system contain are industry equipment and thus present a low hurdle for test integration monitoring the current homogenization efficiency and evaluating the optimization potential in a plant. This approach ensures low risk integration before using this novel system to actively control the stacking process.

As shown, increase in homogenization efficiency and reliability leads to increase in process stability and thus to less outages, lower maintenance cost and increase of full load operation. Furthermore, blending or discarding costs are lowered significantly especially if the mean material quality from mining already meets the process requirements and variance is the largest issue.

This paper focused on longitudinal blending beds in specific configurations. However, the optimization system is not limited to the scenarios and blending systems presented in this paper. It can be used to optimize homogenization systems with circular blending beds, other stacking and reclaiming machinery and multiple blending systems in parallel as well.

5. References

[1] M.P. Cipold, P.K. Shukla, C.C. Bachmann, K. Bao, H. Schmeck (2012) An Evolutionary Optimization Approach for Bulk Material Blending Systems; Lecture Notes in Computer Science vol. 7491, pp. 478–488.

[2] M.P. Cipold, C.C. Bachmann, P.K. Shukla (2013) Stacker Traverse Path Evolution Towards Optimized Blending Efficiency; Proceedings of the 17th International Coal Preparation Conference, pp. 723–728.

[3] A.W. Gerstel (1977) Bed-Blending theory; Stacking Blending Reclaiming of Bulk Materials, pp. 321–341.

[4] P.M. Gy (1981) A New Theory of Bed-Blending Derived from the Theory of Sampling; International Journal of Mineral Processing 8, pp. 201–238.

[5] Z. Loubser, J.d. Korte (2015) Investigation of factors influencing blending efficiency on circular stockpiles through modelling and simulation; The Journal of The Southern African Institute of Mining and Metallurgy 115, pp. 773–780.

[6] T.-F. Lu, S. Xu (2010) SPSim: A Stockpile Simulator for Analyzing Material Quality Distribution in Mining; Proceedings of the 2010 IEEE International Conference on Mechatronics and Automation, pp. 299–304.

[7] T.-F. Lu, M.T.R. Myo (2010) Optimal stockpile voxel identification based on reclaimer minimum movement for target grade; International Journal of Mineral Processing vol. 98, pp. 74–81.

[8] D.R. Stevanovic, B. Kolonja, R. Stankovic, D. Knezevic (2014) Application of stochastic models for mine planning and coal quality control; Thermal Science vol. 18, no. 4, pp. 1361–1372.

[9] N. De Wet (1983) Homogenizing/blending plant applications in South Africa with special reference to Gencor's Hlobane and Optimum plants; Bulk Solids Handling vol. 3, no. 1, pp. 49–59

25

Assessment of coal washability using dual energy: X-ray transmission system

Yi Ran Zhang[1], Dr. Maria Holuszko[2], Nawoong Yoon[3]

[1]*University of British Columbia – NBK Mining Engineering, 517-6350 Stores Road, Vancouver BC, Canada V6T 1Z4*

[2]*University of British Columbia – NBK Mining Engineering, 517-6350 Stores Road, Vancouver BC, Canada V6T 1Z4*

[3]*Scare Davey Engineering Inc., 315 Mountain Highway, North Vancouver, BC V7J 2K7.*

Abstract: Coal washability characteristics have traditionally been assessed using the float-sink (washability) tests. These analyses determine how much of the coal can be separated from the associated rock (mineral matter) in liquids of different densities. However, the test work requires a wet processing where organic liquids, referred to as heavy liquids, are used. The test work is very time consuming, as it requires a significant amount of time to complete full washability for a wide spectrum of coal particles. To avoid using toxic chemicals and to save time, the dual-energy X-ray transmission (DE-XRT) can easily become an alternative method to assess the washability of coarse coal.

In this project, in order to assess the effectiveness of DE-XRT to derive the washability data for the coarse coal, the test work has been designed to include samples representing different coal lithotypes (pieces of coal of certain petrographic composition), and density measurements using both DE-XRT and pycnometer to determine the specific gravity of each coal specimen and ash analysis. The results are compared to the simulated float-sink tests for coarse coal fractions. Similar to float-sink tests or pycnometer measurements, the DE-XRT can estimate a relative density by measuring the attenuation of X-rays from two different energy levels. The paper explains the basics of the DE-XRT analysis and its data processing methods to determine the relative densities and how to derive washability curves from DE-XRT analysis. In addition, testing of different coal lithotypes will allow to build a correlation between the seam lithology and washability characteristics for the prediction of ease of washing of any particular seam at the active mine site.

Keywords: Dual-energy X-ray transmission (DE-XRT), washability, coal characterisation, coal lithotypes

1. Introduction

In coal preparation plants, washability of coal is assessed from a series of sink-and-float tests. The underlying principle behind the sink-and-float test is the density difference between particles. In the traditional sink-and-float test, material is grouped into a range of specific gravity from each round of sink-

and-float test (Leonard, 1979; Laskowski and Walters, 1987). This process can be time consuming for operations where quick turnaround of multiple washability results in a given day is required. Since the dual-energy X-ray transmission (DE-XRT) is able to measure particle by particle and if we are able to predict the ash content and specific gravity of a given particle, it should be possible to estimate the washability for any given assembly of coal particles.

In order to derive the washability data with DE-XRT, it is important to investigate the correlation between DE-XRT response and specific gravity as well as ash content. Another critical parameter in obtaining the DE-XRT result is the development of the H–L (high and low energy) curves which determines the relative density of a given pixel based on its X-ray attenuation. Further details on DE-XRT output will be described in the next section.

2. Data collection method

When conducting this experiment, there are four main parameters that are recorded for an individual coal piece. These coal pieces are macroscopically examined and categorised using classification derived by Diessel (1965) and are referred to as lithotypes, their specific gravity is determined by pycnometer and followed by DE-XRT and ash analysis. It is critical to ensure that the same sample is measured for each parameter to obtain correct data for correlation analysis. In this experiment, 100 coal specimens in the particle size range of +50–100 mm have been studied. Figure 1 illustrates the mass distribution of each lithotype and each lithotype's average ash content.

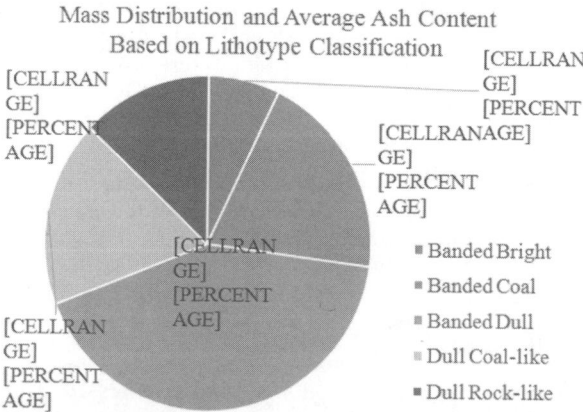

Figure 1: Mass distribution and average ash content based on lithotype classification (data label – top: ash content, %; bottom: mass distribution, %)

2.1 Coal classification – Lithotypes

The individual coal specimen has been classified based on lithotypes. The five lithotypes found in the coal sample set studied here include: banded bright coal, banded coal, banded dull coal, dull (coal-like) and dull (rock-like) (Diessel, 1965; Holuszko, 1994; Rusnak, 2018) (Fig. 2).

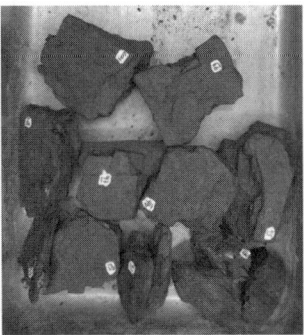

Figure 2: Banded bright coal (left) and dull (rock-like) (right)

2.2 Specific gravity by pycnometry

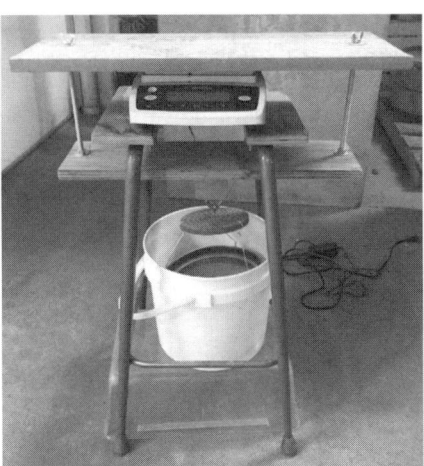

Figure 3: Pycnometer-for density measurement device developed at UBC

The specific gravity of each coal sample is derived from the dry weight of each sample measured by scale, and the submerged weight measured from the submerged weight-measuring device in Fig. 3, using the following two equations from the buoyancy principle shown below:

$$W_d - W_s = \rho_w V_o g \qquad (1)$$

$$\rho_o = \frac{md}{V_0} \qquad (2)$$

where W_d is the dry weight of the object, W_s is the submerged weight of the object, V_o is the volume of the object, g is the gravity, ρ_w is the density of water, ρ_o is the density of object and m_d is the mass of the object in air.

2.3 Dual-energy X-ray transmission (DE-XRT)

The dual-energy X-ray transmission (DE-XRT) technology measures the X-ray attenuation according to Beer's law for monochromatic narrow X-ray beams (Zhang et al., 2006) which is expressed in the equation below:

$$H = e^{-\mu} H^t \qquad (3a)$$

$$L = e^{-\mu} L^t \qquad (3b)$$

where μ_H is the mass attenuation coefficient for high energy (cm^2/g), μ_L is the mass attenuation coefficient for low energy (cm^2/g) and t is the mass thickness (g/cm^2).

The relationship between the high and low energy X-ray attenuation can be expressed by rearranging Eqs. (3a) and (3b), which is the basis of H–L curve generation:

$$H = L^{\frac{\mu H}{\mu L}} \qquad (4)$$

In this experiment, COMEX's MSX-400-VL-XR system, installed at UBC, is illustrated in the figure below (Fig. 4).

Figure 4: COMEX's MSX-400-VL-XR system in UBC's coal and mineral processing laboratory

2.4 Ash content

Measurement of ash content is one of the most crucial parameters in order to generate the washability curves. The individual coal specimen's ash content was determined following the ASTM (D7582) standard method.

3. Results and discussion

In order to compare the coal washability from the sink-and-float test (simulated from S.G. and ash content of each coal piece) with DE-XRT analysis, DE-XRT image processing must be completed. In this project, the 100 coal samples have been processed four times to obtain four sets of processed images based on several density targets. Then various correlations are compared before generating the washability curve to evaluate the predicted ash content as well as predicted specific gravity. One set of the processed image demonstrated a relatively good prediction for both ash content and specific gravity, where the r-squared value is 0.7470 and 0.8435, respectively.

Furthermore, when the washability curve is generated, the general trend of the best DE-XRT wahsability curve follows the simulated sink-and-float washability curve. When the S.G. versus mass yield to clean coal curve is compared, the low-density regions of the DE-XRT S.G. curve is modelled quite well with the simulated sink-and-float test. However, above 1.45 S.G,the model is unreliable for predicting the true specific gravity.

3.1 DE-XRT image processing

Using the MSX-400-VL-XR system, four different sets of images have been produced by changing the high-density and the low-density bounds within the COMEX's CXR software to create various colours, representing different relative density. Figure 3 shows an example of two different lithotypes, banded bright and dull (rock-like), based on the four different output images from the DE-XRT system. The dotted red line represents the density range that had the highest correlation with ash content.

As illustrated in Fig. 5, when the D3 density range is used, the banded bright coal and the dull rock-like coal samples can easily be distinguished via DE-XRT system. However, when comparing the banded bright, banded and dull coal, the DE-XRT system faced difficulty in identifying the various lithotypes. This could be due to the fact that dullness of coal can be a result of either high ash content or certain petrographic composition (Holuszko and Grieve, 1991; Holuszko, 1992, 1994).

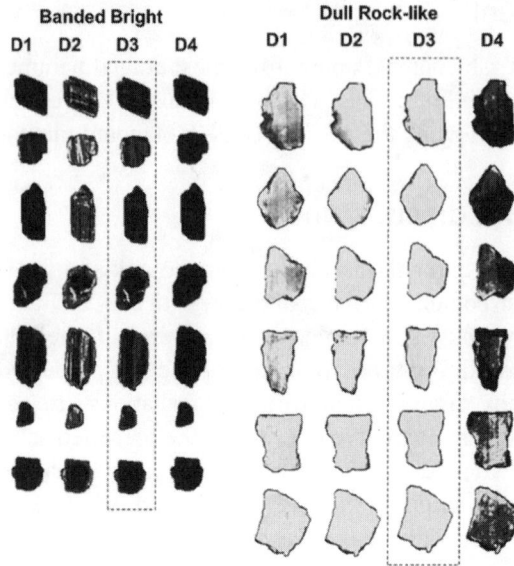

Figure 5: DE-XRT processed image with four different sets of images with varying density range comparing two different lithotypes (high ash and low ash)

Although the recognition of lithotypes is not possible from obtained DE-XRT image outputs, the processed images are able to predict ash content and specific gravity. In the next section, the relationship between DE-XRT responses, and both ash content and specific gravity is explained in detail.

3.2 Correlation analysis

There are three main correlations that are explored in this project which include the following:

1. S.G. versus ash content,
2. S.G. versus DE-XRT density signal,
3. Ash content versus DE-XRT density signal.

The first correlation is able to ensure that the data collected is not erroneous and demonstrates a reasonable correlation between S.G. and ash content. The second correlation is investigated since DE-XRT measures the atomic density, and it may be correlated with S.G. Then finally, the third correlation is compared since S.G. and ash content correlate well. The three correlation analyses have been conducted on the four different density ranges used in DE-XRT analysis. It is important to keep in mind that these comparisons still utilise the same 100 coal specimens.

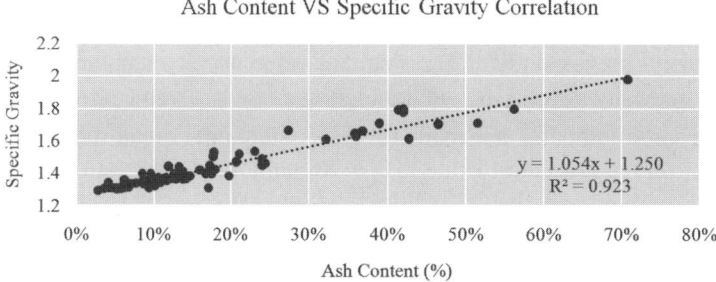

Figure 6: Plot of ash content and specific gravity with linearly fitted line

As expected, a strong correlation can be observed between the ash content and specific gravity. This provides confidence in the ash content and specific gravity measurements of the 100 coal samples (Fig. 6).

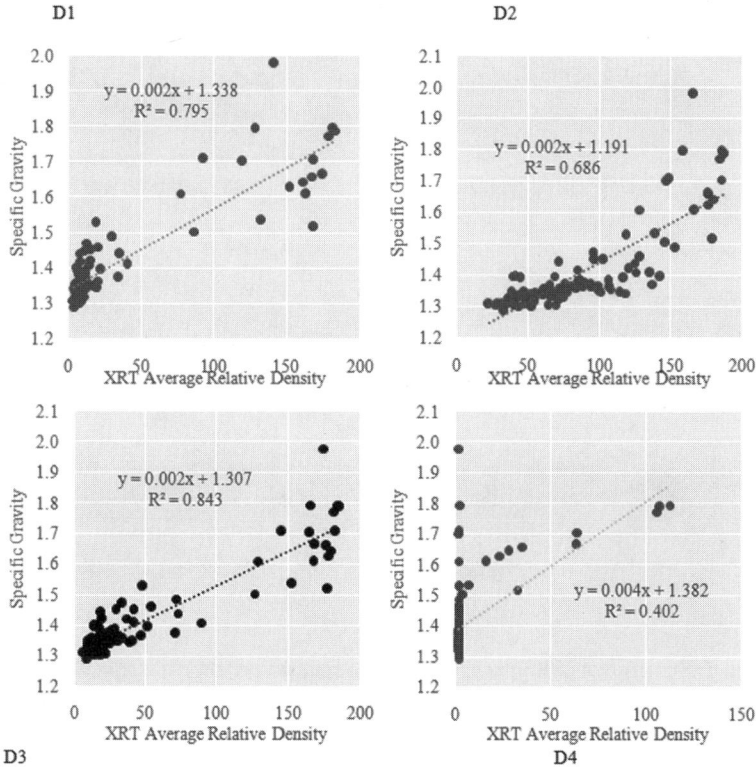

Figure 7: Plot of XRT output (average relative density) and specific gravity for four different XRT density ranges studied with linearly fitted line

Figure 7 graphically compares the average relative density values extracted from DE-XRT with the measured specific gravity for the four different DE-XRT density ranges studied. The r-square value is calculated and compared. The third density range, D3, had the highest r-squared value of 0.8435.

The same comparison is made for ash content rather than specific gravity. In this case, the r-squared value is 0.7470, which is lower than the previous comparison, however, the D3 still had the highest r-squared value out of the four density ranges studied. This is illustrated in Fig. 8.

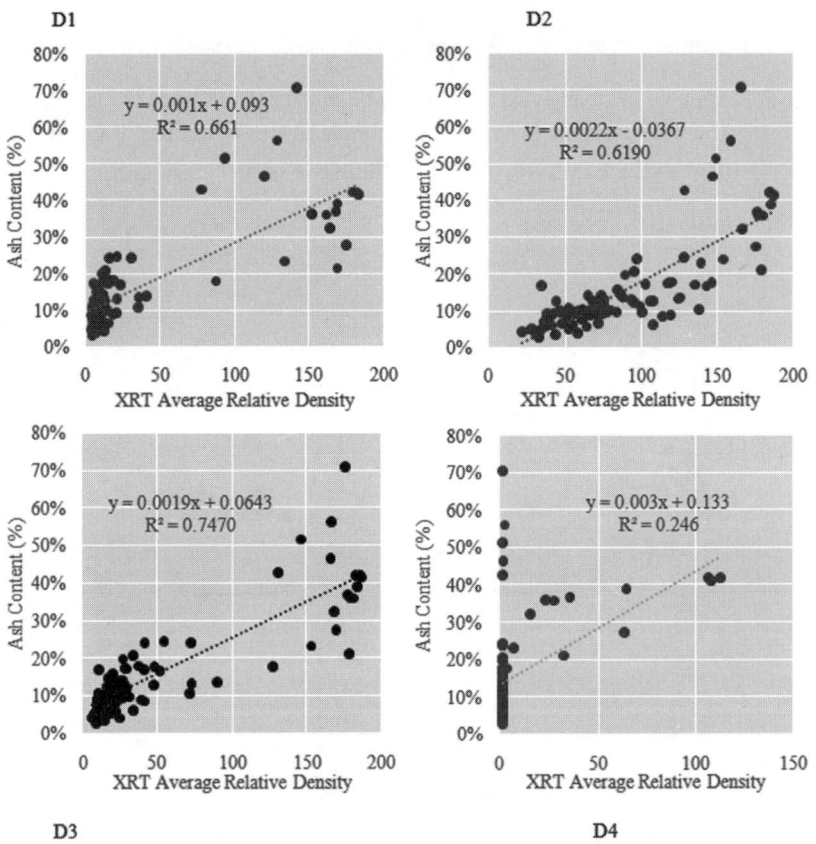

Figure 8: Scatter plot of XRT output (average relative density) and specific gravity for four different XRT density ranges studied with linearly fitted line

All the r-squared value comparisons are summarised in Table 1. S.G. and ash content had the highest correlation and, as mentioned previously, D3 had the highest correlation for both ash content and specific gravity.

Table 1: Summary of *r*-squared value comparison between four different XRT density ranges

Category	r-Squared	
	XRT with ash content	**XRT with specific gravity**
D1·	0.662	0.795
D2·	0.619	0.687
D3·	0.747	0.844
D4·	0.246	0.402
SG versus ash content	0.9234	

3.2 Washability comparison

To develop the washability curve and replicate the sink-and-float test using an individual particle's S.G., the coal lithotypes specimens are put in ascending order by S.G. The solid line represented in Fig. 9 is the washability curve based on each particle's S.G., ranked from lowest to highest.

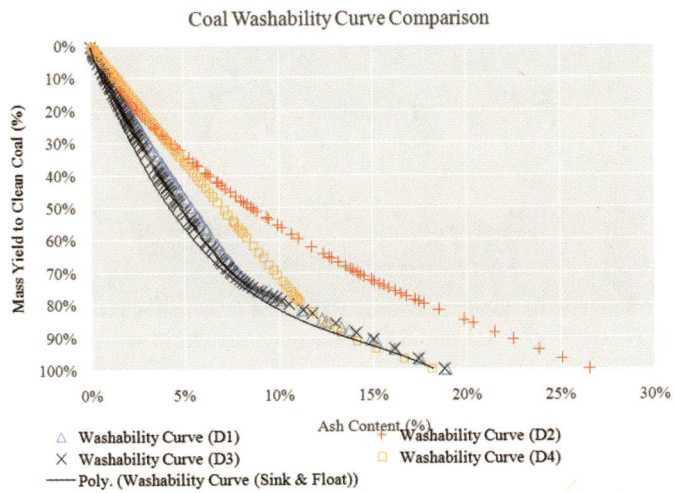

Figure 9: Washability curves generated from XRT output (average relative density) for four different XRT density ranges and washability curve generated from sink-and-float analysis

To then build a washability curve derived from DE-XRT, the linear equation found from the correlation analysis is used to calculate the predicted ash content. When using the DE-XRT data to model coal washability, instead of separating coal particles by S.G. values, the data is now ranked by lowest to highest XRT average relative density. From the four comparisons in Fig. 9, it is clear that D3 is the most suitable to model the washability curve, as derived from the sink-and-float analysis. This also emphasises the importance to selecting the optimum density range to model ash content.

Like the washability curve, the S.G. curve also uses the linear equation from the correlation analysis to predict the S.G. The solid line in Fig. 10 represents the simulated sink-and-float results. Once again, D3 models closest to the sink-and-float results, where it is able to model the lower specific gravity ranging from 1.3 to 1.65, but the S.G. predictions become unreliable beyond 1.65 S.G. Other XRT density ranges seem to have difficulty modelling the S.G. curve.

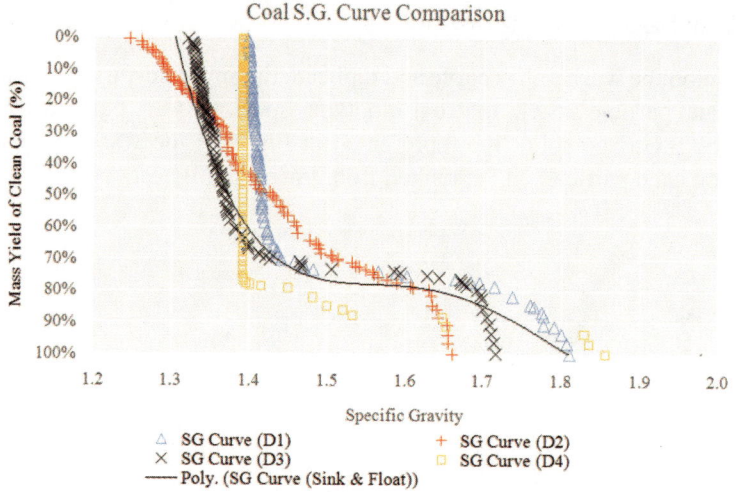

Figure 10: S.G. curves generated from XRT output (average relative density) for four different XRT density ranges and S.G. curve generated from sink-and-float analysis

4. Conclusion

The coal washability curve derived from DE-XRT results can model washability similarly to the one that is derived from the sink-and-float test. Since the r-squared value between DE-XRT and S.G. and XRT and ash content is 0.844 and 0.747, respectively, it seems logical for the model to use the data with the highest correlation to fit the washability curves best.

Coal's ash content commonly has a direct correlation with specific gravity. Since the DE-XRT density data correlate relatively well with S.G., it is also possible to observe DE-XRT output correlating with ash content. Although the DE-XRT density results can predict ash content and S.G. to a certain extent, the DE-XRT system was not able to distinguish various lithotypes in a similar range. However, it was possible to distinguish the bright-banded coal from dull rock-like coal by the colour-assigned density ranges, which are at the opposite end of the lithotype spectrum.

In future work, testing more coal lithotype samples from the same coal deposit or the same coal seam using the same equations to predict the ash content and S.G. should be conducted to validate the robustness of the model predictions. Moreover, if the correlation between other lithotypes and washability data could be validated, the prediction of the washability from seam lithology could be possible and could be used as a predictive tool during exploration and mine planning stages and this will facilitate more efficient coal resource extraction and downstream processing.

5. References

[1] Diessel, C.F.K., (1965). Correlation of macro amd micropetrography of some New South Wales coals: in Proceedings, Volume 8th, Commonwealth Mineralogical and Metallurgical Congress, Melbourne, Woodcock, J.T., Madigan, R.T. and Thomas, R.G., Editors, pages 669–677.

[2] Holuszko, M.E. and Grieve, D.A., (1991). Washability characteristics of British Columbia coals; in Geological Fieldwork 1992, Grant. B. and Newell, J.M., Editors, B.C. Ministry of Energy, Mines and Petroleum Resources, Paper 1991–1, pages 371–379.

[3] Holuszko, M.E., (1992). Washability of lithotypes from selected seams in the East Kootenay Coalfield, Southeast British Columbia; in Geological Fieldwork 1992, Grant. B. and Newell, J.M., Editors, B.C. Ministry of Energy, Mines and Petroleum Resources, Paper 1993–1, pages 517–525.

[4] Holuszko, Maria, E., (1994). Washability characteristics of British Columbia coals. B.C. Ministry of Energy, Mines and Petroleum Resources, Paper 1994–2, pages 1–36.

[5] Kolacz, J. (2012). Advanced sorting technologies and its potential in mineral processing. AGH Journal of Mining and Geoengineering, 36(4).

[6] Laskowski, J.S. and Walters, A.D., (1987). Coal preparation; in Encyclopedia of Physical Science and Technology, Volume 3, Academic Press, pages 37–61.

[7] Leonard, J.W., (1979). Coal Preparation, 4th Edition; American Institute of Mechanical Engineers, New York.

[8] Rusnak, John. (2018). Coal strength variation by lithotype for high-volatile a bituminous coal in the central Appalachian basin. International Conference on Ground Control in Mining, Page 10.

[9] Zhang, G., Chen, Z., Zhang, L., & Cheng, J. (2006). Exact reconstruction for dual energy computed tomography using an H–L curve method. IEEE Nuclear Science Symposium Conference Record.

26

Application of the direct optical method for determination of granulometric composition for quality control processes in coal dewatering

O. Onyshchenko, O. Kohanuk, T. Oleynik

ANA-TEMS LLC, Dnipropetrovsk, Ukraine
Kryvyi Rih National University, Dnipropetrovs'ka oblast, Ukraine

Abstract: Evaluation of the effectiveness of photo-optical measurement methods was carried out for optimization of the dewatering of fine classes of enrichment products from the coal factory. According to existing scheme dewatering step (centrifuge) feed consists of hydrosizers concentrate, flotation concentrate, and hydrocyclones concentrate.

The photo-optical SOPAT measuring technique for particle sizing was applied to process and measure the sizes by means of automated image analysis.

Coal sludge dewatered in filtration centrifuges is a fine material, the average particle diameter of which is 306–21 μm (according to testing), which predetermines its high water holding capacity.

Using the data of the technical specifications, a variant of the technology for the dewatering of enrichment products (hydrosizer concentrate, flotation concentrate, and product of the hydrocyclone) was considered. Based on the studies and performed calculations, the requirements of the process for ash content (12%) and humidity (12%) of the sediment in existing process are not met.

Keywords: Coal, dewatering, granulometric composition, coal particle size, photo-optical method

1. Introduction

The exact knowledge of particle size distributions plays a major role in various fields of applications to control and optimize processes as well as reduce waste knowledge of particle size distributions is thus essential for process optimization [14]. This study demonstrates the capability of a photo-optical measurement method for enrichment and centrifuging of enriched coal [1]. The moisture content of the centrifugation products is determined (in addition to the mode parameters) by the dispersion of the dehydrated material: the fine the starting material, the higher the humidity of the centrifuging output, while other conditions are equal [10,11,15].

Besides of the size distribution parameter, shape of the particles also plays significant role, certain share of coarse parties should be present in the initial slurry to provide final wet content for centrifuged dewatered slurry [13].

Evaluation of the effectiveness of optical measurement methods was carried out as part of the project for replacement of equipment for dewatering of fine classes of enrichment products from the coal factory at the coal preparation plant No. 1 of Private enterprise "Avdiivka Coke Plant". For correct equipment selection, the study of the particle size distribution in the dewatered products was required.

Aim of the study was to evaluate the performance centrifuge for the enrichment products of the coal directed to dehydration based on the laboratory studies of the samples.

The work included the following steps:

- the study of coal slurries entering centrifuging step;
- calculation of dewatering parameters;
- conclusions on the type of centrifuges and required capacity based on obtained measurement data.

2. Materials and methods

Photo-optical methods according to ISO 13322-1:2014 and ISO 13322-2:2006 were applied in this study to provide measurements of coal fines sizes using photo-optical imaging technology to perform real-time in-situ image acquisition and analysis of particle properties, including particle size, color, shape, concentration, distribution, and mobility [4,5].

The photo-optical SOPAT measuring technique for particle sizing was applied to process and measure the sizes by means of automated image analysis. Selected system consists of HD lenses surrounded by a protection tube (material: stainless steel 1.4571 or Hastelloy C-22) and they represent the front part of the probe. The camera is controlling a xenon flash and ensures high illumination intensities in a very short time. The flash time varies, depending on the intensity, from 2 ls for the lowest intensity up to 8 ls for the highest intensity. This ensures sharp image capture even from fast moving particles. The flash is inside the central box, the light from the stroboscope is transported by a fibre-optic. The image analysis is carried out on a work station. Intel CoreTM i7 processors. Software developed for quantification using photo-optical methods with image (Fig. 1).

Analysis is fully automated method based on MATLAB. The software employs a Normalised Cross-Correlation procedure algorithm, which is explained in detail along with the prefiltering which was employed by [1]. Additionally, it avoids human generated bias by different observers, also shown in [1,6].

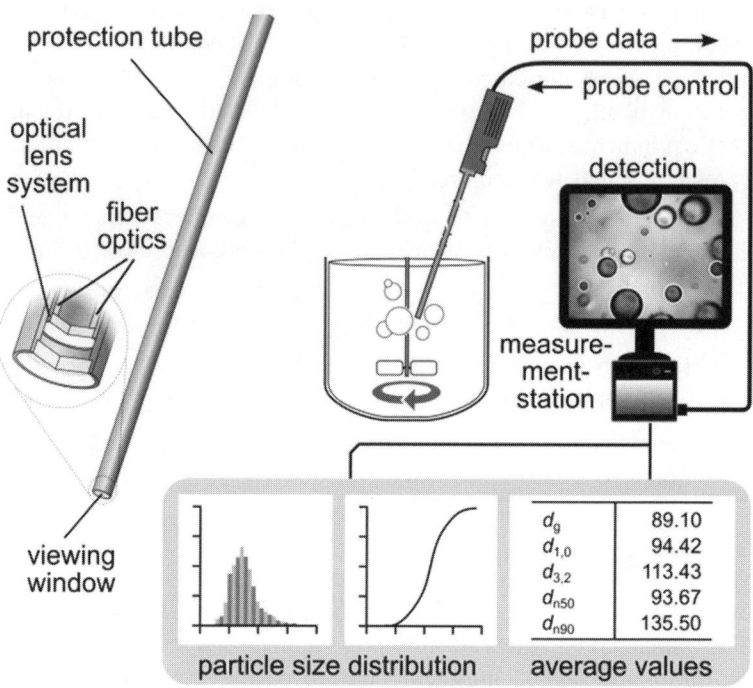

Figure 1: Automated particle size analyzer schematic view

In order to ensure robust and accurate particle detection, a series of images is first pre-filtered to remove irrelevant and misleading image information. This is done with image subtraction using the integrated sequence as difference image. Then the particle recognition consists of three steps: pattern recognition by correlation of pre-filtered gradients with search samples, the pre-selection of plausible circle coordinates, and the classification of each of those circles by an exact edge examination. The software employs a NCC algorithm to evaluate possible particle matches [1,6].

2.1 Calculation of dewatering parameters

Granulometrical composition of the slurry was determined employing optical method and other pulp parameters were calculated based on factory monitoring data.

Volumetric feed of the centrifuge (V) was calculated as:

$$V = P - Q \div \delta \qquad (1)$$

where P is the pulp load (m³/h) and δ is the coal actual density (t/m³).

Concentration of solids in pulp (C) in kg/m³ was calculated as:

$$C = 1000 \times Q \div P \qquad (2)$$

where Q is the mass load (t/h) and P is pulp load (m³/h).

According to existing scheme centrifuge feed consists of hydrosizers concentrate, flotation concentrate and circulation filtrate of the centrifuges. Ash content and density for all classes are taken as average obtained from factory laboratory monitoring data. δ_0 is the actual density of the organic part of coal, kg/m³ for cokes type and A^d is the ash content of coal (%) [8–10].

3. Results and discussion

Centrifuge is fed by following products: concentrate of hydrosizers; flotation concentrate; condensed product of hydrocyclones. Figures 2–4 show the results of particles Feret diameter distributions for three feed types.

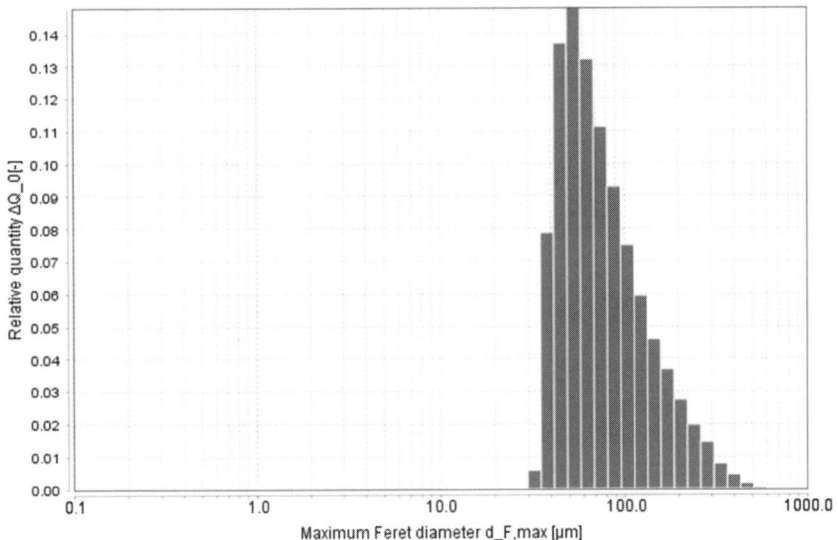

Figure 2: Maximum Feret diameter distribution for hydrosizer output

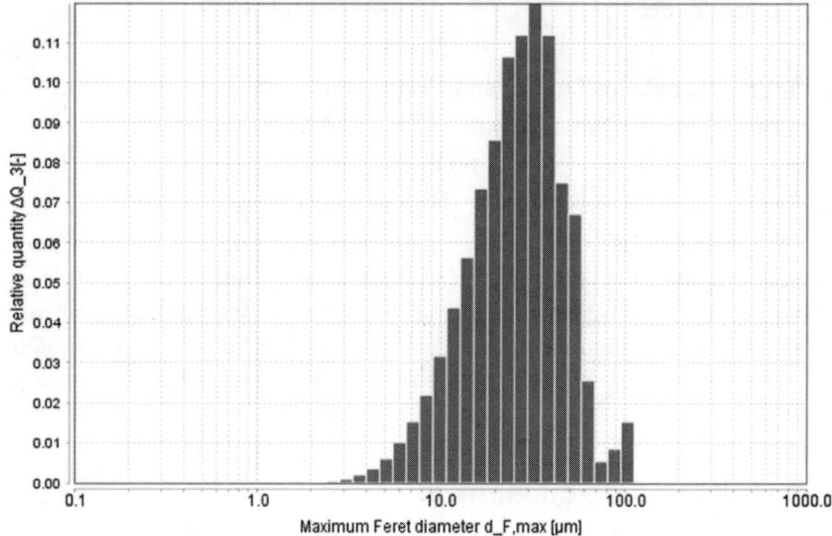

Figure 3: Maximum Feret diameter distribution for flotation concentrate

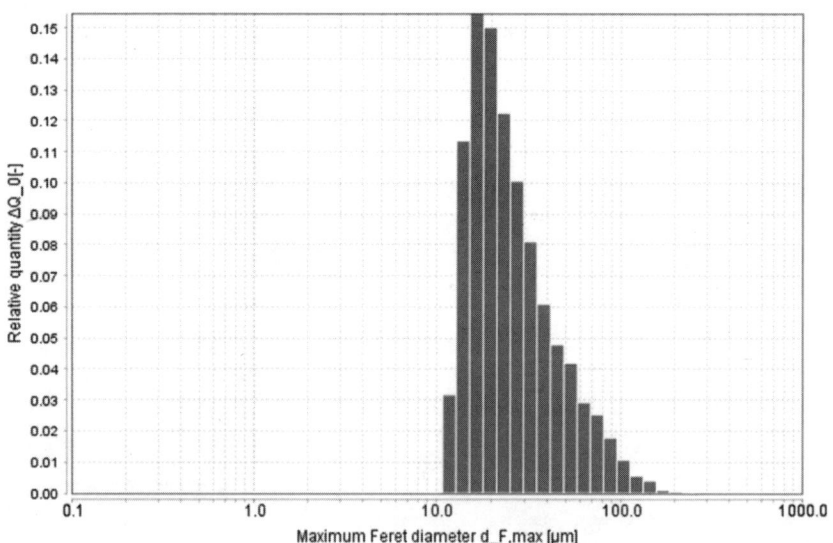

Figure 4: Maximum Feret diameter distribution for condensed product of hyderocyclones

Sensitivity analysis plots are presented in Figures 5–7.

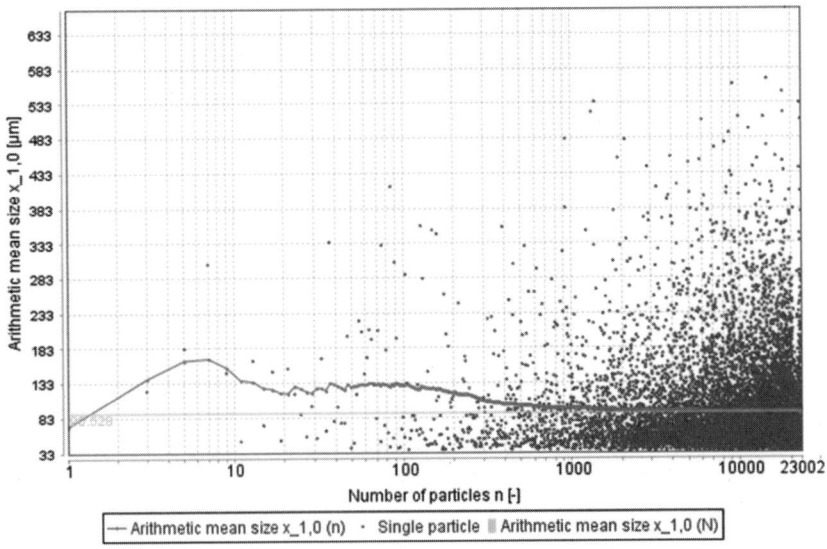

Figure 5: Sensitivity analysis for hydrosizer output sample

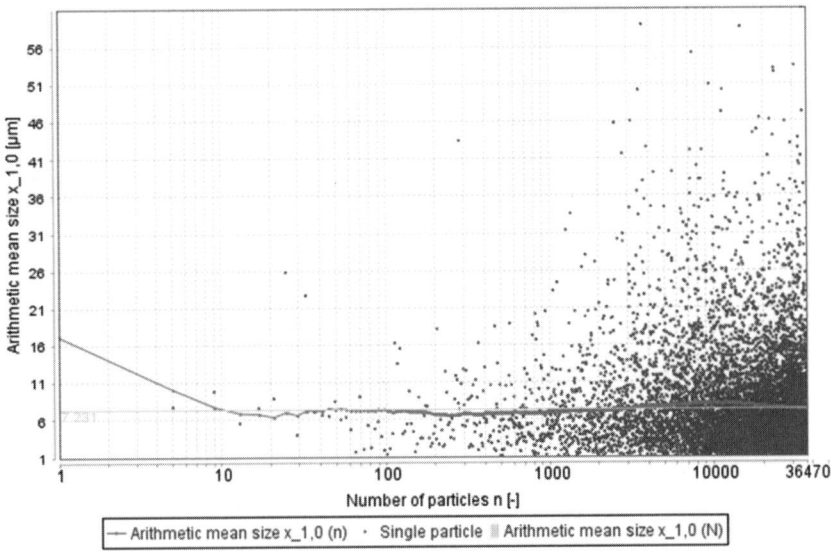

Figure 6: Sensitivity analysis for flotation concentrate sample

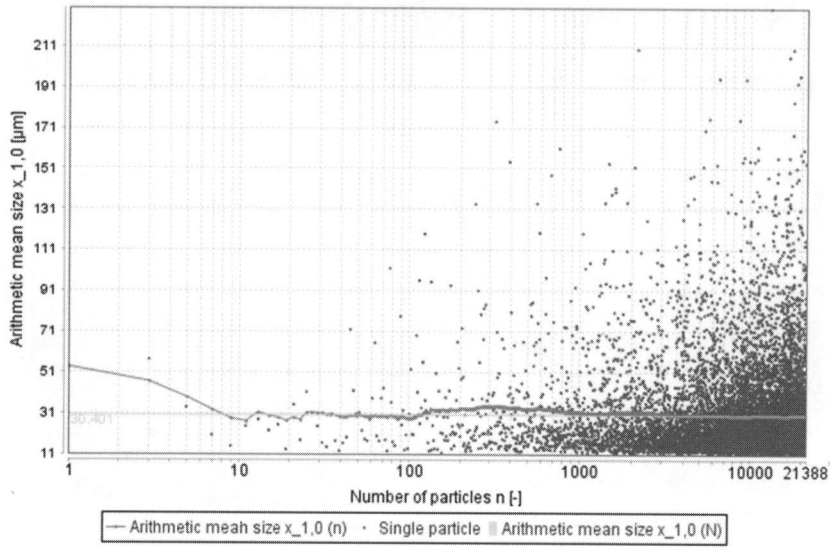

Figure 7: Sensitivity analysis for condensed product of hyderocyclones sample

Table 1 shows the calculation of the original product directed to the centrifuging step. The calculation of the indicators of the granulometric composition of the filtrate was carried out according to Eqs. (1) and (2).

Table 1: Coal particles size distribution data and centrifuging process main parameters calculated

Particle maximum diameter (mm)	Products							
	Hydrosizer output		Flotation concentrate		Hyderocyclones output		Total	
	γ_n (%)	Ad (%)	γ_n (%)	Ad (%)	γ_n (%)	Ad (%)	γ_n (%)	Ad (%)
0.5–1	13.37	2.1	0	1.5	0	2.0	4.46	2.0
0.25–0.5	48.19	3.4	0	2.0	5.96	2.7	18.05	2.7
0.16–0.25	17.01	16.5	0	3.3	10.12	5.0	9.04	5.0
0.074–0.16	21.37	27.8	2.98	4.4	49.21	5.9	24.52	5.9
0.04–0.074	4.73	31.1	27.91	5.7	21.16	6.2	17.93	6.2
–0.04	0.01	34.8	69.87	11.7	8.11	13.4	26.00	13.4
Total	100	12.8	100	7.4	100	11.9	100.00	8.6
Q (t/h)	25.0		143.1		20.4		188.5	
C (kg/m³)	269		175		450		197	
P (m³/h)	92.9		817.7		45.3		955.9	
V (m³/h)	74.2		707.8		30.0		812.0	

Coal sludge dewatered in filtration centrifuges is a fine material, the average particle diameter of which is 306–21 μm (according to testing), which predetermines its high water holding capacity.

Using the data of the technical specifications, a variant of the technology for the dewatering of enrichment products (hydrosizer concentrate, flotation concentrate and product of the hydrocyclone) was considered. Based on the studies and calculations performed, the requirements of the process for ash content (12%) and humidity (12%) of the sediment in this variant are not met.

4. Conclusion

1. Direct automated optical method of particle measurement allows obtaining accurate data of size distributions in fine classes of coals and provides quality prediction of dewatering step performance in the coal enrichment process.

2. Performed sample analysis indicated domination of very fine classes of coals which explains high water holding capacity of pulp and high final wet content of the obtained centrifuged products.

3. Additional dewatering step of application of flocculent allowing to aggregate very fine particles is highly recommended.

4. Further sturdies aimed to determine the impact of fine particles presence on the dewatering process are required to create automated control system.

5. References

[1] Automated drop detection using image analysis for online particle size monitoring in multiphase systems, S. Maaß, J. Rojahn, R. Hänsch, M. Kraume. Computers & Chemical Engineering, 2012, No. 45, P. 27–37.

[2] Dewatering of fine coal by screen-bowl centrifuges, E. Gallagher, J. Post, A. Swanson, L. Armstrong. Proceedings of the 1st Australian Coal Preparation Conference, Vienna, 1981.

[3] Dewatering ultrafine clean coal, A. Patwardhan, Y. Chugh, B. Arnold, A. Terblanche. Proceedings of the 21st International Pittsburgh Coal Conference, Osaka, 2004, P. 127–138.

[4] ISO 13322-1:2014 Particle size analysis – Image analysis methods – Part 1: Static image analysis methods.

[5] ISO 13322-2:2006 Particle size analysis – Image analysis methods – Part 2: Dynamic image analysis methods.

[6] Optical inline analysis and monitoring of particle size and shape distributions for multiple applications: Scientific and industrial relevance. J. Emmerich, Q. Tang, P. Naubauer, S. Junne. Chinese Journal of Chemical Engineering, 2018, No. 1, P. 138–146.

[7] Development of the screen bowl centrifuge for dewatering coal fines, N. Policow, J. Orphanos. Mining Engineering, 1983, No. 4, P. 333–336.

[8] Workshop for Computation of Mined Coal Quality Parameters: Training Aid, Polulyakh A.D, Polulyakh D.A. D.: National TU, 2016, 144 p.

[9] Distribution of Size Grades in the Process of Coal Preliminary Dry Screening, Polulyakh A.D., Berlin A.M., Polulyakh O.V. Збагачення корисних копалин: Research Collection Book, 2017, No. 66(107), P. 64–73.

[10] Distribution of size grades in the process of coal preliminary wet screening with vibrating screens, Polulyakh A.D., Polulyakh D.A. Machine and Technology Vibrations, 2017, No. 3(86), P. 102–109.

[11] Distribution of size grades in the process of run-of-mine coal aqua screening, Polulyakh A.D., Polulyakh D.A. Збагачення корисних копалин: Research Collection Book, 2017, No. 66(107), P. 48–56.

[12] Distribution of size grades in the process of coal preliminary wet screening with aqua screen machine with vibrating screens, Polulyakh A.D., Polulyakh D.A. Geomechanics: IGTM Research Collection Book, 2018, No. 138, P. 212–217.

[13] The Design and Development of the Derrick CIP/CIL Interstage Screen intermountain, Reinhofer. Mining and Processing operators Symposium, Reinhofer. Nevada: Elko, 1988, 18 p.

[14] The screen bowl centrifuge for dewatering froth floated fines, Shaw S. Mine and Quarry Magazine. 1980, P. 60–64.

[15] Centrifugal dewatering and reconstitution of fine coal: the GranuFlow process, W. W. Wen, R. Killmeyer. Coal Preparation, 1996, No. 17, P. 89–102.

27

Development of an online coal washability analyser for treating Indian coals

K M P Singh[1], T Gouricharan[1], G V Ramana[2], Anandaya Sinha[3], Pradeep K Singh[1]

[1]CSIR-Central Institute of Mining & Fuel Research, Dhanbad, India
[2]Ardee Hi-Tech Private Limited, Visakhapatnam, India
[3]Ministry of Coal, New Delhi, India

Abstract: The coal washability analysis is usually achieved by conducting traditional float and sink test. This test is time consuming and uses chemicals like bromoform, tetrachloroethylene, zinc chloride and benzol which are not environment friendly and hazardous in nature. With the improvement of modern 2D and 3D imaging technologies, the scanned digital images from duel energy X-ray based system are being processed to analyse different required parameter for coal washability and the developed methods, may be used for online evaluation of coal washability. Indian coals are of drift origin and vary widely in their washability characteristics. The coal washeries are being fed with coals from multiple sources/seams, which have varied washability characteristics, and to evaluate the performance of any washer, the standard procedure of float and sink test is the only alternative, and by the time the results are obtained there is a chance of losing good coal in the rejects. To overcome this, an automated coal washability analyser is being under developmental stage which will give near real-time float and sink data that would maximise the yield at desired ash level and impact positively on the profitability of the plant operation. The results are very promising with the actual standard float and sink data to the coal analysed with the X-ray analyser. The experimental set-up with process methodology and the data thus obtained is presented in the paper.

Keywords: Coal washability, float and sink analysis, hazardous chemical, X-ray based analyser, online coal washability analyser

1. Introduction

The Indian are known for its high ash due to its drifted origin and the ash forming minerals are highly inter-mixed. To meet the MoEF (Ministry of Environment and Forests) mandate/requirements thermal coal needs to be washed. In Indian contest the coarse coal +13/+10 mm are being washed while −13/−10 mm is directly mixed with washed fraction. For assessing the washing potentials prior to actual coal washing in coal washery, laboratory float and sink method is applied for determination of the washability characteristics of coal to evaluate the washing performance. Till date the conventional float and sink method is followed which is time taking, using hazardous chemicals and moreover the running plants cannot be assessed instantly, ultimately the desired quality is still a challenge for coal preparation engineers. The challenges become more intense with the drifted origin Gondwana coal.

There are no on-line techniques available for assessing this most fundamental aspect of coal washing characteristics. The peoples are working by using available advance technology and it appears possible to determine the coal washability on-line using dual energy X-ray techniques. The successful development of such a device is critical to the establishment of process control and automated coal washing systems.

Since discovery of X-rays in 1895, among the numerous important uses it has also been applied to examine the coal. The first investigation reported was by Couriot (1898), he submitted anthracite, bituminous coal and other fuels to X-rays to obtain their radiographs and nearly every detail of the intimate structure of the mineral matter. Latter on Mahadevan (1929), followed by Dhar and Niyogi (1942), have X-ray studies on Indian coal macerals and minerals.

The literature on on-line washability is very widely dispersed. The basic issue is that comparable results of on-line system vis-à-vis the standard float and sink method. In 1970s, 3D X-ray has been introduced and since then, 3D structure and mineral study were focussed to get the acceptable result. The literature indicting that the coal washability was attempted in 1990s by applying X-ray. The researchers like Lin et al. (2001) (USA) Schena et al. (2007), Bachmann et al. (2012) are working on washability. Later on Germany, South Africa, USA, etc. are working for online washability analyser. Shamaila et al. (2012) has reported for a prototype X-ray transmission washability monitor. Atkinson and Swanson (2016) are also working on further development in washability prediction using coal grain analysis to arrive washability of the coal using X-ray. Albert Klein, Sven Reuter and Audy Zein are also working on rapid coal analysis with the online X-ray elemental analyser. Bachmann et al. (2016) in his article argues that dual parameters lead to high inaccuracies with changing ash content in coal as has been the experience with dual energy ash gauges and proposes a triple parameter using an optical measurement technique to compensate for the changing variation of the third dimension of the particle.

The development of coal washability analyser needs the following characteristics of the coal to be measured instantaneously, i.e.: particle size distribution, ash content and density, for the development of washability curves. The system must be reliable, work in an on-line fashion, minimal intervention of personnel rather to say automated, which provide nearly real-time analysis.

2. Laboratory model

To stablish the concept a prototype laboratory model of the coal washability analyser is designed and was developed at CSIR-CIMFR in association with M/S Ardee Hi-Tech Pvt. Ltd. as depicted in Fig. 1.

Figure 1: Laboratory model coal washability analyser at CSIR-CIMFR

The laboratory model setup is very simple wherein coal of selective size is poured in the hopper and feed through vibratory feeder which ensures monolayer of coal particles spread on small moving conveyor belt and passes between X-ray generator and the detector. Multi-energy X-ray generator source is used which consists of an integrated X-ray tube, dual output high voltage power supply and a filament supply with control circuitry. Once initialised and enabled, the generator emits X-rays in a continuous spectrum of energies. The X-rays are collimated into a straight beam. All reflected and scattered rays are filtered out to reduce measurement errors.

The electrically generated X-ray applied to the broad-band X-ray radiation which falls on the moving conveyor caring feed material at a rate of 3 m/s to be scanned which works like a line-scan camera, record transmitted X-rays passing through the material. The formats of recorded data are in digital image data. The image captured of the material are in two different X-ray energy levels (low and high). The low/soft energy X-ray region falls from few hundred eV to 20 KeV, direct detectors image sensors are utilized, which provide high detection efficiency and high energy resolution. The high/hard X-ray region with energy higher than soft X-ray, having higher penetration efficiency for the object is utilised for industrial application. The attenuation depends on both the thickness and atomic density of the material. Images of different atomic densities are transformed into images of different spectral ranges, which make it possible to classify different pseudo colour pixels according to specific atomic densities which are regardless of material thickness. The connected computer is used as data acquisition system; the data are recorded in the required format which is finally converted to particle's area, mass and ash of the particles.

2.1 Dual energy X-ray transmission sensor (weight and density determination)

X-rays have the ability to penetrate into matter and interact with atomic species. The material under investigation is irradiated with X-rays of known incident energy and the attenuation is accounted for coherent, incoherent scattering and absorption. This has been well described by Zou et al. (2008) and applied for the determination of densities of materials which can be expressed as degree of X-ray transmission with material density and thickness.

$$I = I_0 \times e^{-\mu\rho d}$$

Where, I_0 is the incident radiation, I is the transmitted radiation, d is the absorption path length, μ is the absorption coefficient and ρ is the product density.

Knowing the incident radiation and the constant absorption coefficient of the material and measuring the transmitted radiation using an X-ray line detect or the weight per area (ρd) may be derived.

3. Online assembly of the system

After realisation of the concept was applied o 40 TPH coal washing pilot plant at CSIR-CIMFR Digwadih campus as shown in Fig. 2a and b. The online system is setup in between coal crushing unit and coal washing circuit in such a way that when before feed to actual washer one can have the washability data.

Figure 2: (a) Online system detector and (b) sampling and screening section

4. Results and discussion

It is a well-known fact that attenuation depends on both the thickness and atomic density of the material. Recorded data of different atomic densities are transformed into images of different spectral ranges, which make it possible to classify different colour (pseudo) pixels according to specific atomic densities regardless of the material thickness. The acquisition of data than processed for required format as needed for output of the washability curve.

The system requires several levels of calibrations, some automatic in nature which is a part of the equipment start-up routine and the others are manually activated. It start with switching on the system, each of the cards is exposed to collimated multi-energy X-rays, depending on sensitivity of each pixel, there is scintillation generated which is translated into a digital count. These are raw counts and are likely to vary quite drastically from pixel to pixel and are equalised through an internal gain mechanism. The second level of calibration is used to measure and analyses a combination of Etalons (calibration blocks) with different heights and densities. Based on the calibrated data system is given a logic for inclusion of the calibrated value, this enables higher level of accuracy for the size range. Third level of re-calibration is used to compensate X-ray hardening. X-ray hardening happens when mass per surface area exceeds a critical level. After this level, the attenuation curve becomes flatter due to which there can be an erroneous reading of data at higher densities and higher heights of the particles and it will have to be compensated.

4.1 Data generation for washability curve

The feed to coal washing circuit coal is diverted to screening system and passed through online analyser. The online analyser system continuously record the data passing through analyser and after completion of the state recorded data processed for data in required format i.e. particles size with densities, weight and the ash, on rearrangements in tabular format and the entire washability curves may be plotted as desired.

4.2 Comparative analysis of laboratory float and sink test with system generated data

Raw coal crushed to 75 mm feed through screen deck where – 13 mm screed out and +13 mm size fed to coal washability analyser (Figs. 3–5).

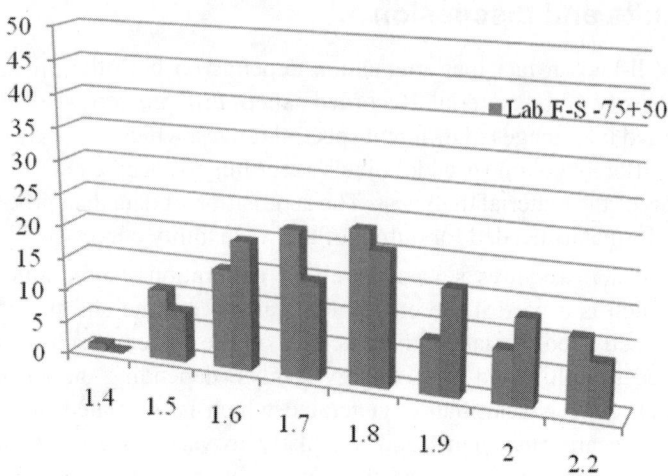

Figure 3: Comparative results of online and lab F–S bar graphs of coal size −75 + 50 mm

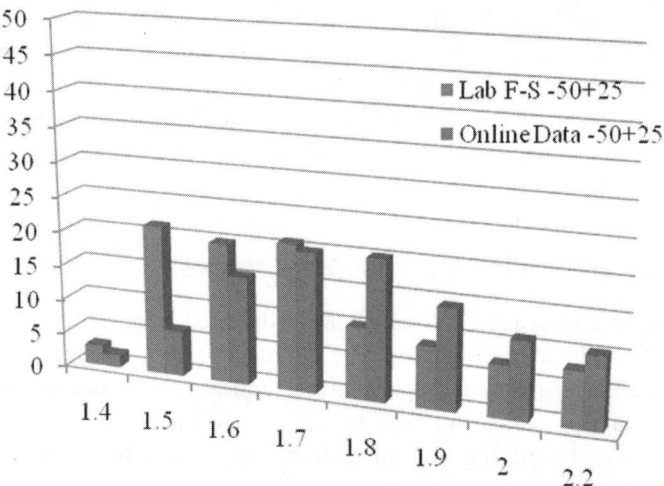

Figure 4: Comparative results of online and lab F–S bar graphs of coal size −50 + 25 mm

The software classifies each of the particle sizes into its respective categories and does the averaging for the category in terms of density, weight% and ash%. The data is then processed washability curve. The outputs of the analyser result and standard float and sink results of coal analysed through are presented.

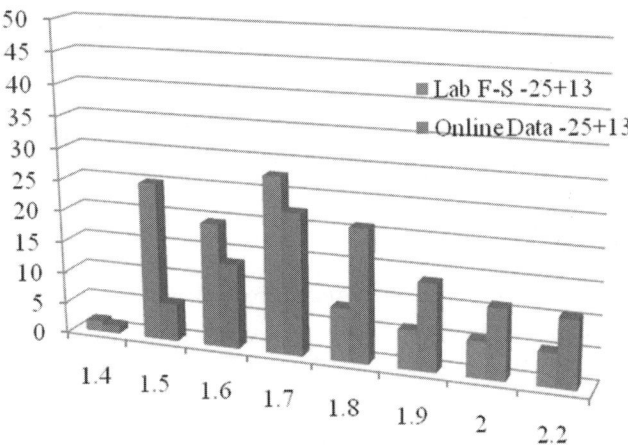

Figure 5: Comparative results of online and lab F–S bar graphs of coal size −25 + 13 mm

The standard washability method, float and sink (F–S) test is being conducted using specific gravity ranging from 1.40 to 2.00 with an interval of 0.1 to get yield and ash%. Comparing the washability data obtained from both, the laboratory float and sink and X-ray based analyser for the coal size faction 50–75, 25–50 and 13–25 mm, respectively, with respect to their size it can be observed that over all ash for the larger size (50–75 mm) data and float and sink results is in good agreement with the laboratory standards float and sink results, while going to relative smaller size fraction this deviated towards higher side from laboratory F–S.

5. Conclusion

The X-ray based coal washability analyser is not using any hazardous chemicals, at the same time very fast in processing to get the results almost in real time data analysis. This results in improvement of coal preparation plant strongly and maximum possible recovery can be obtained during the process of coal washing. Working on a large number of data may compensate the laboratory results. Further data is having each and every particles detail like size, density and ash which on processing one can set the class-limits as per the desired requirements. The fact that the results are stored in a data base offers the possibility to re-evaluate entire datasets under different aspects for in-depth analysis and associated physical and chemical properties, which cannot be done with the lab results.

6. Acknowledgement

This work is funded by Ministry of Coal, Government of India. Authors are thankful to Director, CSIR-CIMFR, Dhanbad for his kind permission to publish the paper and all the contributors of coal preparation and carbonization research group for their cooperation.

7. References

[1] Science note, OTAGO WITNESS, August 25, 1898, page 54.

[2] B. Atkinson and A. Swanson, Further development in washability prediction using coal grain analysis XVIII international coal preparation Congress 28 June–01 July, 2016, 73–78.

[3] Jan F. Bachmann, Claus C. Bachmann, Michael P. Cipold, Helge B. Wurst, Hauke Springer (Germany) and Mel J. Laurila (USA) Washability Monitor for Coal Utilizing Optical and X-Ray Analysis Techniques, 2012.

[4] Jan Bachmann, et al., Online Washability: Comparison of Dual Parameter and Triple Parameter Analysis, XVIII International Coal Preparation Congress, 2016, pp. 255–260.

[5] Douglasw Brown and Brian P. Atkin, A method for the rapid on-site assessment of handle ability, Coal Prep. 21 (2000) 299–313.

[6] J. Dhar and B.B. Niyogi, X-ray studies in Indian coals Part I, Volume VIII, No 1, Published on march 18th 1942, pp. 127–138.

[7] C.L. Lin, J.D. Miller, Cone beam X-ray microtomography for three-dimensional liberation analysis in the 21st century, Int. J. Miner. Process. 47 (1996) 61–73.

[8] C.L. Lin, J.D. Miller, Development of an on-line coal washability analysis system using X-ray Computed Tomography, Coal Prep. 21 (2000) 383–409.

[9] C.L. Lin, J. D. Miller, Advances in X-ray computed tomography (CT) for improved coal washability analysis, 16th International Coal Prep Congress (ICPC 2010), April 25–29, 2010 Lexington, Kentucky, USA.

[10] C.L. Lin, J.D. Miller, G.H. Luttrell, G.T. Adel, Development of an on-line coal washability analyzer, Final Report, University of Utah, Department of Metallurgical Engineering, June, 2001.

[11] C.L. Lin, J.D. Miller, G.H. Luttrell, Evaluation of a CT-based coal washability analysis system under simulated on line conditions, Miner. Metall. Process. 19(1), 2002.

[12] C.L. Lin, J.D. Miller, T. Nguyen and A. Nguyen, 2017, Characterization of breakage and washability of ROM coal using X-ray computed tomography, Int. J. Coal Prep. Utiliz., DOI: 10.1080/19392699.2017.1305364.

[13] Hongping Liu, Sandra Rodrigues, Fengnian Shi, Joan Esterle and Emmanuel Manlapig, Coal washability analysis using X-ray tomographic images for different lithotypes, Fuel 209 (2017) 162–171.

[14] Gianni Schena, Luca Santoro, Stefano Favretto, Conceiving a high resolution and fast X-ray CT system for imaging fine multi-phase mineral particles and retrieving mineral liberation spectra, Int. J. Miner. Process. 84 (2007) 327–336.

[15] S. Shamaila, B. Ntsoelengoe, J. Bachmann, H. Wurst and M. Cipold, Development of a prototype X-ray transmission washability monitor, J. South. Afr. Inst. Min. Metall. 112 (March 2012) 179–184.

[16] Hayley Strydom, The application of dual energy X-ray transmission sorting to the separation of coal from torbanite; M S Thesis, Johannesburg, South Africa 2010.

[17] A.R. Videla, C.L. Lin, J.D. Miller, 3D characterization of individual multiphase particles in packed particle beds by X-ray microtomography (XMT), Int. J. Miner. Process. 84 (2007) 321–326.

[18] L. Von Ketelhodt and C. Bergmann, Dual energy X-ray transmission sorting of coal, J. South. Afr. Inst. Min. Metall., 110 (2010) 371–378.

[19] Helge B. Wurst, Jan F. Bachmann, Claus C. Bachmann, Michael P. Cipold, Washability Monitor for Coal Utilizing Optical and X-ray Analysis Techniques, 17th International Coal Preparation Conference 1–6 October 2013, Istanbul, Turkey.

[20] Qian Zhu, Coal sampling and analysis standards, Clean Coal Center, IEA publisher April 2014.

28

Predictive maintenance system in coal preparation plant: Based on wireless IIoT and big data technology

Deyong Wu[1] and Xingyu Chen[2]

[1]Chengdu ALPHA Industrial Intelligence Co., Ltd., Chengdu, China
[2]Chengdu ALPHA Industrial Intelligence Co., Ltd., Room 1102, Building A, AVIC International Plaza, No. 777 Yizhou Avenue, High-Tech Zone, Chengdu, Sichuan Province, PR Chinaau

Abstract: Based on the analysis of the necessity and difficulties of predictive maintenance intelligent system in coal preparation plant, authors have proposed a specific integrated solution with wireless IIoT (industrial internet of things) and AI (artificial intelligence) big data platform. Through the comprehensive application of wireless communication, industrial IoT (internet of things), low-power circuit design, big data analysis, AI, Mobile APP push technology, etc., the complex factors in coal preparation plant can be well overcome, and the predictive maintenance intelligent system in coal preparation plant can be established economically and effectively. Further, authors have proposed key steps of predictive maintenance intelligent system construction combined with practical cases.

Keywords: Wireless IIoT (industrial internet of things), big data, predictive maintenance, coal preparation intelligent solution, AI (artificial intelligence)

1. Why predictive maintenance is important to coal preparation plant?

The performance and continuous operation of equipment have a direct impact on coal preparation plant, which is heavily equipment concentrated and dependent. Generally, equipment operation efficiency is that of the coal preparation plant, especially of the plants with old or even outdated equipment. Frequent unplanned downtime is inevitable, which means longer downtime, energy waste and higher maintenance personnel cost, etc. According to the traditional maintenance procedures, the equipment is regularly maintained, which means that operators are likely to maintain some equipment that does not need to be maintained, which will cause a waste of time and resources, and even replace the parts that are still valuable. Meanwhile, unplanned downtime probably occurs for reasons of reducing maintenance cost, extending maintenance period or negligence of maintenance personnel, etc. Traditional maintenance management cannot guarantee complying all the time, and may miss up the sign of big failures, which would bring tremendous lost to mining companies.

Therefore, for maintenance on demand, and to better extend the service life of equipment and machinery, effective prevention of enterprise unplanned downtime, effective reduction of maintenance costs in safe ways, improve the reliability of equipment operation and economic benefits, predictive maintenance came into being, and quickly caused attention.

2. Challenges of predictive maintenance in coal preparation plant

1. The foundation of predictive maintenance is big data, and the precondition and source of big data is industrial Internet of things. The complex environment of the coal preparation plant is characterized by heavy dust, high humidity, large surface temperature difference, heavy vibration, large noise, strong signal interference and shielding, etc., and with requirements of explosion-proof, waterproof, short downtime, etc. Therefore, predictive maintenance of coal preparation plants has very high requirements for the performance and transmission of the IIoT (industrial internet of things) devices.

2. Big data is the foundation of predictive maintenance, but predictive maintenance is more than just big data. On the basis of big data analysis and filtering, accurate and effective predictive results are the most critical. However, there's difference of equipment, technological process, coal property, working conditions, and other parameters between every coal preparation plant. As a result, the same equipment has different data property in different coal preparation plants, and the data analysis results of one coal preparation plant cannot be copied to another coal preparation plant. However, even in the same coal preparation plant, the data of the same equipment in different periods are different, so the standard parameter conditions cannot be used as the judgment basis of predictive maintenance. Therefore, each coal preparation plant needs specialized data analysts proficient in big data and equipment performance to analyze corresponding data in real time according to the actual situation on the site. However, such data analysts are too scarce, and even with such talent, analyzing the data of only one coal preparation plant is a waste of the ability.

3. PdM solution for coal preparation plant based on wireless IIoT and big data

In order to overcome the difficulty of predictive maintenance in coal preparation plant, an overall solution based on wireless industrial internet of things and AI (artificial intelligence) big data platform has been proposed by ALPHA team after years of experience accumulation, analysis, and test.

3.1 Wireless IIoT (industrial internet of things)

(Structure Ref. to Fig. 1)

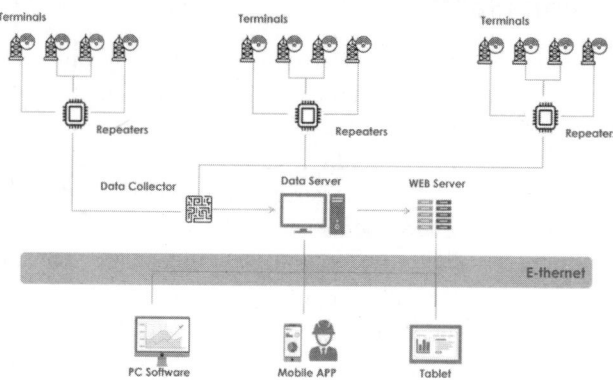

Figure 1: Structure of wireless IIoT

Terminal hardware is the AD001 wireless integrated sensor manufactured by ALPHA. With battery power, low power circuit, ZigBee wireless transmission, measurement of three-way vibration, acceleration, and other technologies, AD001 has achieved stable data transmission and easily installation and maintenance, overcame the abnormal influence brought by complex environment (Fig. 2).

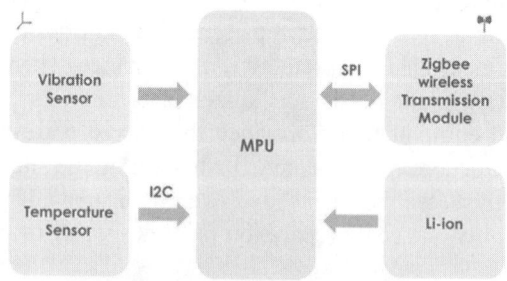

Figure 2: Structure of wireless terminal

3.2 Big data platform

On the basis of the wireless industrial internet of things, we can easily collect the relevant data of equipment status and constitute the big data of equipment operation. Combined with AI intelligence and machine learning, we can process and analyze the data that need to be analyzed manually through the automatic learning ability of the machine, and then directly push the analysis results to relevant users. In this process, data is the foundation, and AI and machine learning are the core.

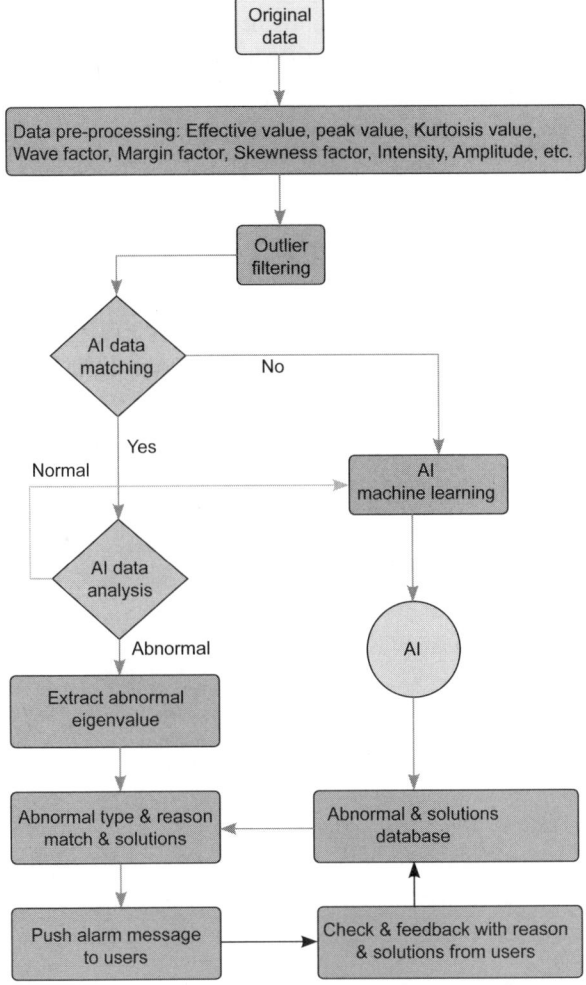

Figure 3: Process of PdM in coal preparation plant based on AI and big data

The first step is to collect temperature and vibration original signal data through the terminal, and then obtain the effective value, peak value, kurtosis value, waveform factor, margin factor, skewness factor, intensity, amplitude, and so on after preliminary analysis and processing. Due to the complexity of the coal preparation plant site and various interfering factors, the equipment condition cannot be judged directly by these parameters, so it is necessary to compare these parameters with the normal historical state of the equipment. Then judge the performance of equipment and predict through the fluctuation and development trend. Furthermore, clarify the type, reason, and solutions of early warning failures after matching these warning metrics to larger cloud data centers. The initial data of equipment operation status in data center is regarded as normal temporarily and then will be automatically corrected for more accurate diagnosis (Fig. 3).

The data preprocessing algorithm is as follows:

$x = [x_1, x_2, \ldots x_n]$: Original signal

ma = max(x): Maximum value

mi = min(x): Minimum value

me = mean(x): Average value

pk = (ma−mi)/2: Peak value

av = mean(abs(x)): MAD/mean absolute difference (RMEAN)

va = var(x): Variance

st = std(x): Standard deviation

ku = kurtosis(x): Kurtosis

rm = rms(x): Root-mean-square

S = rm/av: Waveform factor

C = pk/rm: Peak factor

Kr = sum((x).^4)/sqrt(sum((x).^2)): Kurtosis factor

I = pk/av: Pulse factor

xr = mean(sqrt(abs((x))))^2

L = pk/xr: Margin factor

The artificial intelligence algorithm adopts the learning vector quantization (LVQ) pattern matching algorithm. First of all, the data of starting operation is regarded as normal data for initial training, and then data will be judged according to the training results, normal or not, and followed with AI self-correction and determine the weight vector of neurons. Therefore, it can adapt to the monitoring of different equipment in coal preparation plant, learn independently according to different equipment, and adjust its neuron weight vector gradually, so as to reach the optimal decision effect of each equipment.

This can solve the problem of effective analysis of big data, without human intervention and adjustment of parameters, just need to ensure that the equipment is maintained before the start of monitoring equipment, so that the equipment has started to run normally. The follow-up system automatically learns and adjusts itself accordingly, finding and determining anomalies with increasing precision.

4. Case studies of PdM in coal preparation plant

After two months of on-site installation and commissioning, ShiYaoDian coal preparation plant (5 MTPA) has adopted the PdM based on wireless IIoT and AI technology. Sixty-four monitoring points have been placed on critical equipment in coal preparation plant, and continuously upload data to cloud center. AI data analysis platform (Fig. 4) has been established as well. Based on the submitted data, the analysis platform automatically learns the normal working state of the device, and then, according to the learning experience, pushes the information of the working device which is judged as abnormal to the mobile communication device of the relevant person in charge of the device (Fig. 5).

This PdM has helped with ShiYaoDian coal preparation plant avoiding a huge lose (Figs. 6–8).

Figure 4: AI big data platform of ShiYaoDian coal preparation plant

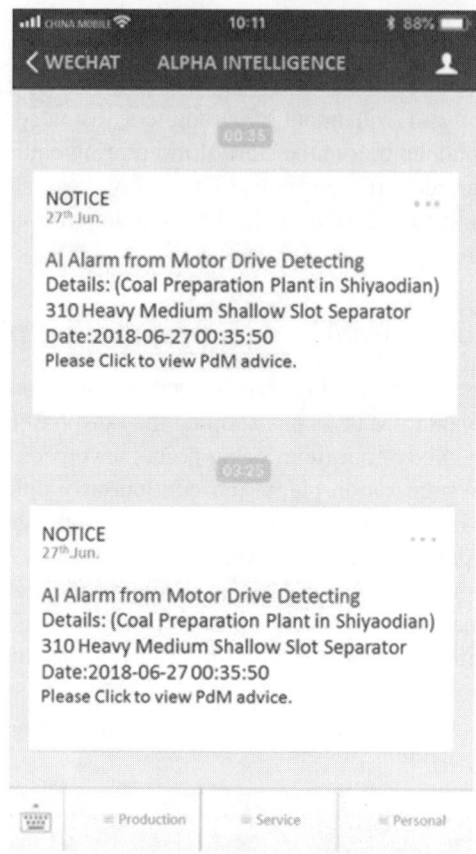

Figure 5: Msg push on mobile phone

Figure 6: ALPHA PdM hardware on-site

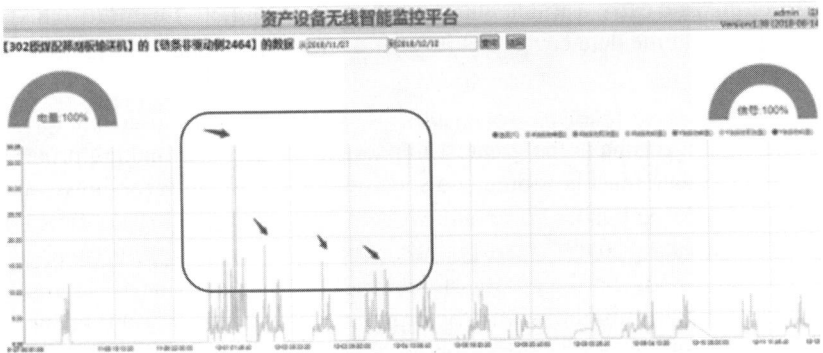

Figure 7: ALPHA PdM data – predicted failure curve 1

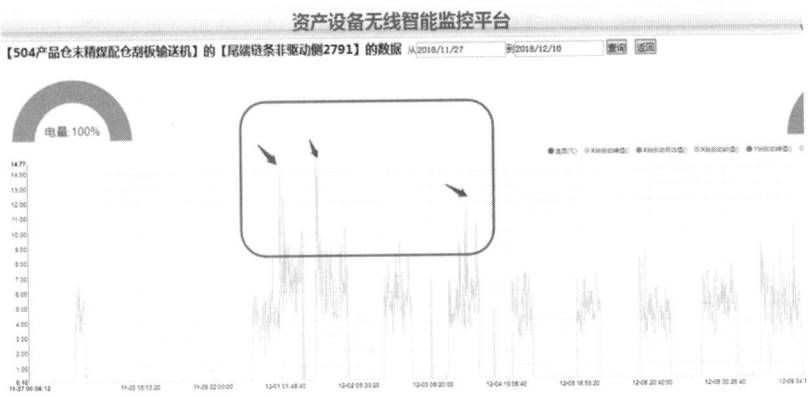

Figure 8: ALPHA PdM data – predicted failure curve 2

Therefore, the key steps of predictive maintenance in coal preparation plants are summarized as follows:

1. Building the Internet of Things Infrastructure: In order to have the predictive maintenance, we must have the basis of data collection, that is, select the monitoring terminal and networking equipment suitable for the coal preparation plant.

2. Building an AI Data Center: Once the device is connected to the network, the operation data will be stored in a special virtual knowledge base in the data center. Then, through AI intelligence, the data can be quickly checked, analyzed and sorted out, and customized diagnosis can be made for the current state of the device. The combination of data center and AI intelligence can effectively avoid the disadvantages of manual parameters adjustment and real-time data analysis. It can

provide early warning and prompt more accurately and more quickly, so that the data center has its own professional brain.

3. Push the right data to the right people: A key aspect of achieving predictive maintenance is pushing data across the organization so that it can have the greatest impact on the decision-making process. The data must be stored at a specific organizational level; however it should be pushed to on-site workers. Just like pushing notifications and data through smart phones, production and operation managers must consider the clarity of data transmission when trying to ensure data transmission within the organization and from various channels to operators in the workshop.

5. Summary

Combination of wireless IIoT and AI big data can help coal preparation plant with achieving intelligent management in fast and economic way while overcoming complex factors and labor cost challenges.

6. Reference

[1] Xia Yimin, Mei Shunliang, Jiang Yi. ZigBee-based wireless sensor networks, Microcomputer Information, 2007, 23(4).

[2] Zhang Yu, Fault diagnosis of rolling bearing based on vibration amplitude parameter index, Mechanical manufacturing and automation, 2011(3):47–51.

[3] Yang Mingbo, Liu Hua, Guo Xianchang, Digital factory + station maintenance service system, China Machine Press, 2017.

[4] Liu Zhen, Lin Hui, Si Liyun. Intelligent BIT fault diagnosis method based on learning vector quantization network. Measure and Control Technique, 2005(11):60–63.

29

Research on the fault diagnosis of coal preparation equipment based on artificial neural networks

Pan Yongtai[1], Li Zekui[1], Zhu Changyong[2], Liu Wenchang[3], Lang Jun[4,5], Wei Yinghua[2], Yao Fuqiang6, Liu Zhen[6]

[1]School of Chemical and Environmental Engineering, Engineering Research Center for Mine and Municipal Solid Waste Recycling, China University of Mining and Technology, Xuzhou, China
[2]Shenhua Ningxia Coal Industry Group Co., Ltd, Taixi CPP, Shizuishan, China
[3]Yangquan Coal Industry (Group) Co., Ltd, Yangquan, China
[4]School of Chemical Engineering and Technology, China University of Mining and Technology, Xuzhou, China
[5]National Engineering Research Center of Coal Preparation and Purification, Xuzhou, Jiangshu, China
[6]Top Crusher Co., Ltd, Xuzhou, China

Abstract: In the process of coal preparation, the abnormal sound signal of equipment often indicates the occurrence of equipment failure, and crusher is one of the equipment with the highest failure rate. This paper takes crusher for an example; the monitoring and diagnosis of the feed fault are studied. In the process of coal crushing, if the impurities like iron or wood are mixed in the feed, it will cause instantaneous and severe impact on the equipment, further the bending deformation of the roller, and even serious accidents. In this paper, the process of crushing single iron, wood and coal in the crusher cavity is regarded as the study subject. Collect the audio signals respectively and calculate the power spectrum by using MATLAB. Through the comparative analysis of the power spectrum curve, differences among signals are determined and feature quantities are extracted. Finally, the artificial neural network model is established according to feature quantities. Through the model verification, the correct recognition rate of iron, wood and coal reaches 80%, 65% and 60%, respectively. It preliminarily proves the feasibility of the method and the rationality of feature quantities selection, and provides a novel idea for intelligent monitoring and fault diagnosis of coal preparation equipment.

Keywords: Coal preparation equipment, crusher, sound recognition, artificial neural networks, fault diagnosis

1. Introduction

In the process of coal preparation, the equipment tends to make an abnormal sound when the failure occurs, and the crusher is one of the equipment with the highest failure rate in this process. Generally the main feeding material of crusher is coal. However, due to the complexity and unpredictability of the composition of the raw ore, the feeding material tends to be mixed with impurities like wood or iron occasionally. Especially in the situation that the larger size iron enters the crushing cavity, the iron will jump between the crushing roller since it cannot be occluded effectively by the crushing teeth, which will cause instantaneous and severe impact on the equipment, and further the bending deformation of the roller, and even trigger a huge safety accident. Therefore, it is of great significance to monitor and diagnose the feeding fault of the crusher.

This paper, mainly focus on the materials of coal, wood and iron as the study subject, firstly collects the audio signals in the process of material crushing, then analysis the power spectrum of each signal and compare the three differences and extract the characteristic. Finally, establish the artificial neural network model, by which it becomes possible to recognize the material types, and provides theoretical guidance for subsequent feeding fault diagnosis research.

2. Apparatus

Signal is the basis of equipment fault diagnosis, in order to obtain the characteristic signal in the crushing process, the corresponding test system should be established. Figure 1 is the layout diagram of the test system, which is composed of the crushing part, the sound test part and the data acquisition part.

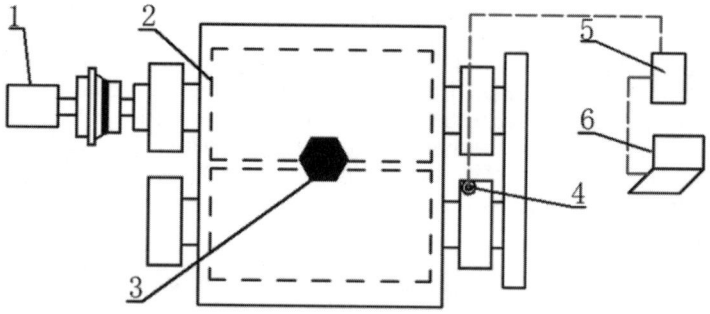

Figure 1: Layout diagram of the test system. 1, Motor; 2, teeth roller; 3, raw material; 4, sound sensor; 5, data acquisition instrument and 6, computer

In the test system, the type of ZKB-II shear crusher is used, with motor current frequency of 30 Hz and motor speed of 864 r/min. The sound signal acquisition device contains a YSY5000 IEPE sound pressure sensor with frequency response range of 20 Hz–20 kHz and a YSV8008 data acquisition instrument with a maximum sampling frequency of 51.2 kHz.

3. Signal acquisition

Signal acquisition is an extraordinarily important step, which is related to the accuracy of data used in the subsequent processing of the system [1], Fig. 2 shows the flow chart of signal collection in the test process. As a fundamental test, the testing material is arranged with the single particle of coal, wood and iron, respectively, and a total of 60 groups of data are collected.

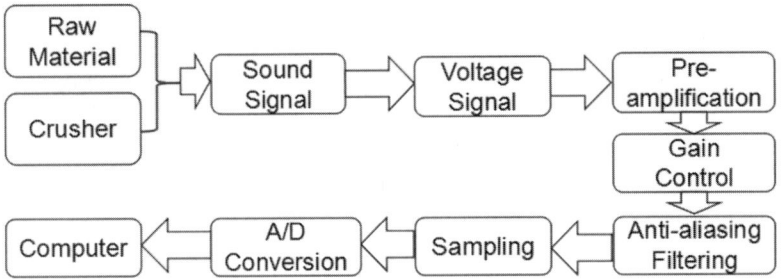

Figure 2: Flow chart of audio signal acquisition

In order to ensure that the sampling value can reflect the change rule of the waveform, it is required that the sampling frequency of waveform should be high enough. The higher the sampling frequency is, the more samples can be obtained per unit time and the higher the fidelity of the digital audio signal will be. However, the higher the sampling frequency is, the larger the data volume will be and the more storage space will be required [2]. In this acquisition process, the sampling frequency is fixed at 10,240 Hz, which is appropriately lower than the highest sampling rate of the signal. In this way, on the one hand, the length of data to be processed will be as large as possible, and useful information in the data will not be lost. On the other hand, it will also improve the processing speed of data, making the system process faster and more accurate [3].

4. Signal processing and analysis

MATLAB is one kind of computer language with powerful function, high efficiency, good interactive numerical calculation and visualization [4]. In this paper, MATLAB will be used for the analysis in time domain and frequency domain.

In time domain, spectrum subtraction is used in the noise reduction of the three audio signals. Spectrum subtraction is a method of speech noise reduction. Suppose the time series of the speech signal is $x(n)$, and the speech signal at the ith frame which is obtained after the processing of windowing and framing is $xi(m)$ and the frame length is N. After FFT (fast Fourier transform), the average spectrum value of the leading noise segment and the spectrum value of each frame signal in the frequency domain are calculated, and subtract the average spectrum value of noise from the spectrum value of each frame, then get the spectrum value of noise reduction [5]. Figure 3 shows three audio signals and their noise reduction signals.

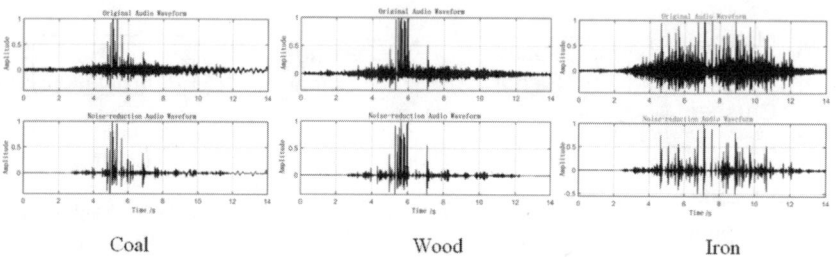

Coal   Wood   Iron

Figure 3: Three audio signals and noise reduction signals

As shown in Fig 3, there are significant differences between the three audio signals after noise reduction. The audio amplitude and distribution of coal and wood are similar over time, with short-term energy concentration and locally prominent amplitude. The main reason is that coal and wood are brittle materials compared with iron, and they are more likely to be damaged and fracture, in which process the instantaneous crushing energy is more concentrated. In contrast, the distribution of the iron audio is more uniform with a certain periodicity, caused by the periodic impact from the continuous jump of the iron in the crushing chamber since the iron cannot be completely occluded. It can be seen that there are certain differences among the three in terms of time-domain characteristics. However, due to the limitations of time-domain analysis, in-depth analysis will not be conducted here.

In speech signal processing, signal analysis and processing in frequency domain play an important role. The study of speech signals in the frequency domain makes some features that cannot be shown in the time domain greatly apparent [5]. Power spectrum analysis is one of the key technologies in digital signal processing, and it is an important statistic for describing random signals [2]. In this paper, the power spectrum analysis method in speech signal processing is used to analyze the audio signals of the material crushing in the frequency domain. As shown in Fig. 4, the power spectrum of audio signals of coal, wood and iron are obtained by using Pwelch function of mean period diagram method.

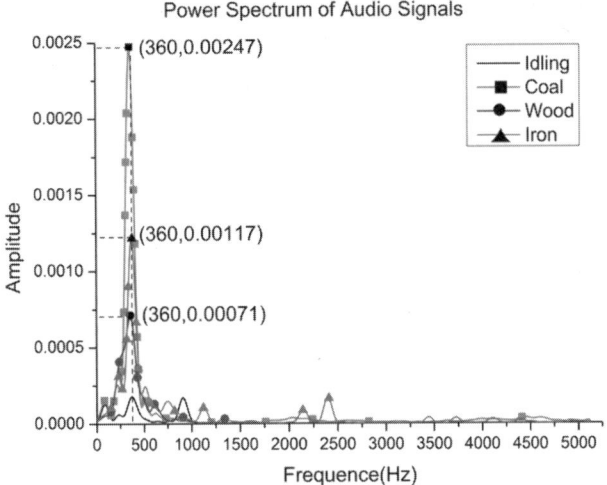

Figure 4: Power spectrum of three audio signals

It can be seen from the comparison in Fig. 4 that the power spectrum curves of the three materials are all prominent in the low frequency band (0 Hz–1000 Hz), indicating that the spectrum component in this frequency band of the three signals is quite rich, and the frequency spectrum energy of the audio signal of the coal is significantly higher than that of iron and wood. In addition, the main frequency peaks of the three are mainly concentrated around 70 Hz (1× frequency), 210 Hz (3× frequency) and 360 Hz (5× frequency), corresponding to the main peak of the no-load spectrum. It can be seen that in the low frequency stage, the frequency of material crushing audio shows up as the odd times of the frequency of the no-load equipment. Though the main frequency of every material is around 360 Hz, the amplitude corresponding to this frequency varies greatly.

In the middle frequency band (1000—3000 Hz), the power spectrum amplitudes of coal and wood are evenly distributed, while the power spectrum amplitude of iron fluctuates more significantly than those of coal and wood, indicating that the audio frequency of crushing coal and wood is less distributed in this frequency band and lower contribution.

In the high frequency band (above 3000 Hz), the spectrum amplitudes of the three are not obvious. In order to find out the difference among the three in this frequency band more accurately, the logarithmic amplitude of the power spectrum is adopted to eliminate the influence of the order of magnitude, and zoom in on the differences between the curves, as shown in Fig. 5. As can be seen from the figure, the power spectrum curves of the three in the high frequency band fluctuate distinctively. Among them, the power spectrum curve of wood is similar to the no-load condition, with the smallest fluctuation which is mostly made up of instantaneous small amplitude fluctuation. The power spectrum curve of iron fluctuates the most violently than the other two, with the continuous frequency peaks. And for the coal, the fluctuation of the power spectrum curve is relatively gentle, with a longer period.

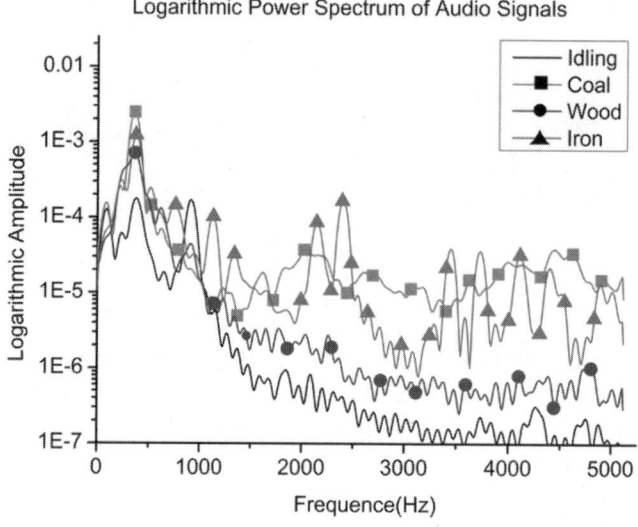

Figure 5: Logarithmic power spectrum of audio signals

According to the power spectrum analysis above, the power spectrum curves of the three materials are different in the low, middle and high frequency band. It is defined that the audio signal of coal crushing is normal signal, while the other two audio signals are fault signals, and their differences are taken as fault characteristics. In this paper, the amplitude corresponding to 360 Hz in the low frequency band (0—1000 Hz) is taken as the characteristic quantity A, the frequency value corresponding to the wave peak of the amplitude in the middle frequency band (1000—3000 Hz) is taken as the characteristic quantity B, and the fluctuation value (standard deviation) of the logarithmic amplitude of the power spectrum of each material in the high frequency band (above 3000 Hz) is taken as the characteristic quantity C.

5. Establish recognition model

Artificial neural network (ANN) is a nonlinear dynamic model established by simulating human nervous system based on the understanding of human brain structure and operation mechanism [6]. Because of its self-organization, self-learning, associative memory and other functions, it has been widely used in pattern recognition (including speech recognition and image recognition), optimization control and other aspects [7].

According to the fault characteristics above, 60 groups of audio signal data are trained, verified and identified, and an artificial neural network model is established. The distribution of characteristic values of 60 groups of audio signals can be seen in Fig. 6. Compared with the distribution of iron sample data, the distribution of coal and wood sample data is more concentrated and partly intersected, which predicts that the similarity of coal and wood characteristics is relatively high, and the recognition difficulty will be also higher than that of iron samples.

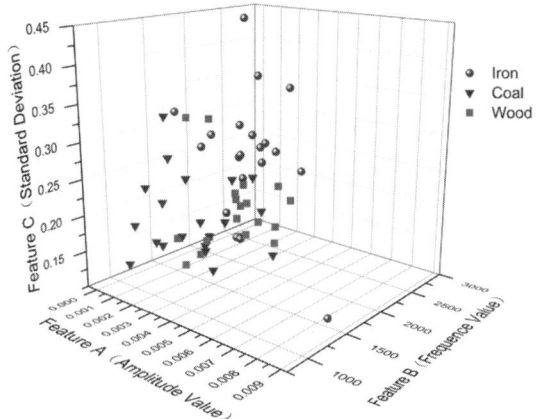

Figure 6: Three kinds of audio signals characteristic value distribution diagram

Generally, the forward network forms of artificial neural network include BP network, perceptron, linear neural network, etc. BP network greatly improves its computing capacity by using error back-propagation method, and the network can realize the approximation of any continuous function by adjusting the relationship between neurons in the hidden layer [8]. Based on these advantages, in this paper, BP artificial neural network model is selected to recognize the feeding fault.

As shown in Fig. 7 is the model structure diagram that is established, the neural network adopts a 3-layer structure, with 3 nodes in the input layer, 10 nodes in the hidden layer and 3 nodes in the output layer. In addition, 42 groups (70%) of data are randomly used for model training, 9 groups (15%) of data are used for cross-validation and 9 groups (15%) of data are used for model testing.

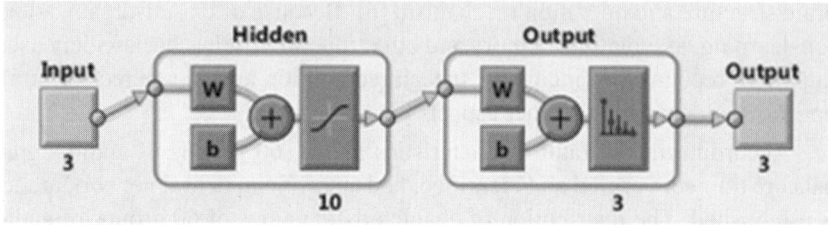

Figure 7: BP neural network structure diagram

6. Model training and result

BP algorithm [9–13] transforms the mapping problem of input and output of the neural network into a nonlinear optimization problem. In this processing, gradient descent algorithm, the most common optimization method, is used [14]. Iterative operation is used to modify the network weight, by which the mean square deviation between the network output and the expected output can realize minimized [15]. As shown in Fig. 8, in the process of network training, the gradient is still in the stage of continuous decline by the 16th iteration, and it is confirmed that the error curve of the sample will not decline for 6 consecutive iterations, also the target error meets the preset requirement, and finally the network iteration stops.

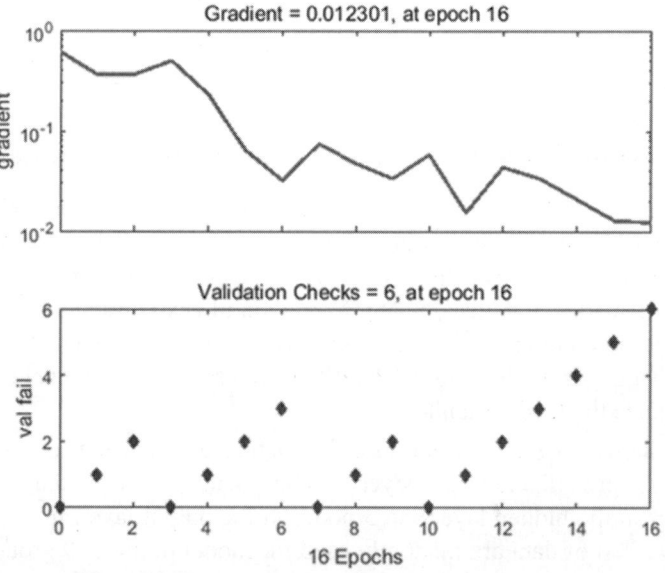

Figure 8: Gradient curve and validation check

In the process of iteration, the effective evaluation of network performance is very important. Cross entropy is one of the indicators to evaluate the network training performance. The smaller the value is, the smaller the loss between the model's recognition value and the real value will be, and the higher the similarity will be. In other words, the model generated by this algorithm is the closest to the optimal scheme. As shown in Fig. 9, in the iterative process of network training, the cross-entropy values of the training curve, the test curve and the verification curve all keep decreasing, and the cross-entropy value of the test curve is lower than the verification curve value, indicating that the test performance of the network is better than the verification performance, and the test result of the network is effective. In the figure the iteration number of times corresponding to the minimum entropy value is 10, and this point is the optimal network performance point, and the weight corresponding to this point is used to uniquely determine the parameter values of the network model.

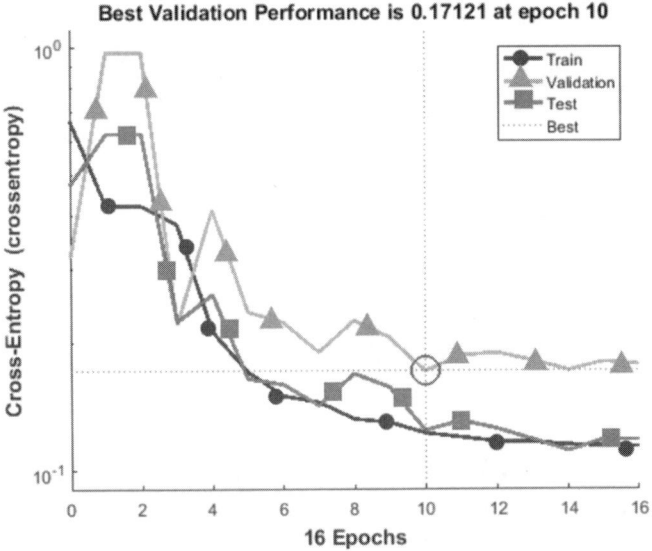

Figure 9: Model training performance diagram

Above, various parameters of the model are determined through multiple iterations of the minimum gradient method, and the accuracy of the model in recognizing sample data is an important criterion for judging if the model is good and reliable. For the analysis of recognition accuracy, the confusion matrix method is commonly used, which is the comprehensive embodiment of algorithm performance and test results in the recognition process through a specific matrix. As shown in Fig. 10 is the confusion matrix of this model. It can be obtained from the figure that the recognition accuracy of iron in the training matrix, verification matrix and test matrix are respectively 84.6%, 50% and 80%. Among the 20 samples, 1 sample is wrongly identified as wood and 3 samples are wrongly identified as coal. However, the identification results of wood and coal are relatively similar, 6 of the wood samples are wrongly identified as coal, while 5 of the coal samples are wrongly identified as wood. Based on the overall confusion matrix, the overall recognition accuracy of the model for iron, wood and coal reaches 80%, 65% and 60%, respectively.

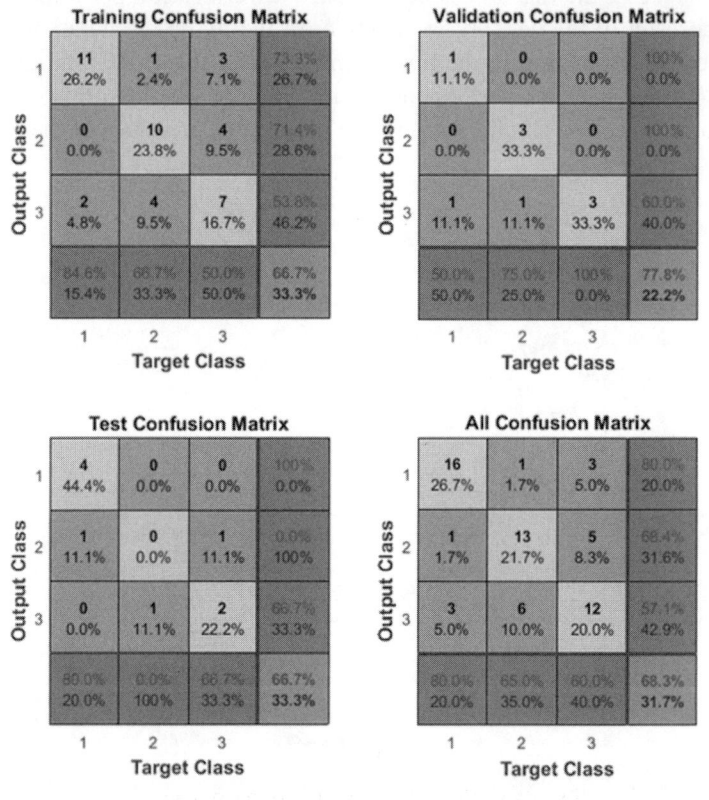

Figure 10: Model confusion matrix diagram

7. Conclusion

(1) To some extent, the result indicates that it is feasible to recognize the iron through the difference of audio signals preliminarily, and the selection of the audio signal characteristic quantity is reasonable.

However, due to the high sensitivity of audio signal and poor anti-interference ability to other external factors, the accuracy of iron recognition by audio signal still exist some errors, and other signals such as vibration signal can be added for auxiliary recognition to improve the accuracy and reliability of iron recognition.

(2) The recognition accuracy of wood and coal is obviously lower than that of iron, and misclassification between each other exists, which shows that the similarity of the two kinds of audio signal features is much higher, and the degree of differentiation is much lower. Therefore, it is necessary to study the differences of the audio signals between coal and wood, and optimize the method of the characteristic extracting in order to improve the recognition accuracy for coal and wood.

(3) For other coal preparation equipment, such as vibrating screen and cyclone, the audio recognition method in this paper can also be used to establish the corresponding artificial neural network model to monitor the abnormal sound signal of the equipment in the process of operation. Combined with the big data technology, the fault can be diagnosed online. And then the control system can intelligently adjust the rotation speed, positive and negative rotation, start and stop of the equipment, through which a closed loop control system of intelligent monitoring and diagnosis can be established. The system can not only ensure the safe operation of the equipment to the greatest extent, but also make great significance for the efficient production of the mine.

(4) In addition to the research content of this paper, there are still many factors to be studied in depth to adapt to the practical application in the industrial field, such as the expansion of data collection, the optimization of recognition characteristic quantity, the mixed feeding condition of multi-particles and the change of material diversity.

8. References

[1] Dong Wei. Research on the application of feature extraction and feature optimization in vehicle acoustic classification. North University of China, 2010.

[2] Liu Bo. Research on vehicle acoustic characteristics analysis and vehicle recognition. Wuhan University of Technology, 2007.

[3] Wei Hongfeng. The parameter model method of picking up the automobile audio signal information. Northeast Normal University, 2005.

[4] Chen Jiayan, Chen Dongjiao, Zhang Daxiang. Collecting and processing of sound signal with Matlab. Jisuanji Yu Xiandaihua, 2005(06):91–92+96.

[5] Song Zhiyong. Matlab voice signal analysis and synthesis (Second Edition). Beijing: Beihang University Press, 2018:170, 21–22.

[6] Shen Shiyi. Neural network system theory and its application. Beijing: Science Press, 2000, 1–20.

[7] Lou Shuntian, Shi Yang. System analysis and design based on MATLAB-neural network. Xian: Xidian University Press, 1998.

[8] Werbos P J. The roots of backpropagation. NY: John Wiley & Sons, 1994.

[9] Ku Xiangchen, Guo Yuefei, Duan Mingde, Cao Beibei. Predicting tool wear by vibration frequency spectrum. Machinery Design and Manufacture, 2017(10):113–116.

[10] Zhang Kaifeng, Yuan Huiqun, Nie Peng. Prediction of tool wear based on generalized dimensions and optimized BP neural network. Journal of Northeastern University, 2013, 34(9):1292–1295.

[11] Chen Chao, Xu Jianlin, Huang Jianlong. Tool condition monitoring system based on artificial neural network. Chinese Journal of Mechanical Engineering, 2002, 38(8):135–138.

[12] Gao Hongli, Xu Mingheng, Fu Pan. Tool wear monitoring based on integrated neural networks. Journal of Southwest Jiaotong University, 2005, 40(5):641–653.

[13] Ghosh N, Ravi Y B, Patra A. Estimation of tool wear during CNC milling using neural network-based sensor fusion. Mechanical Systems and Signal Processing, 2007, 21(1):466–479.

[14] Hecht-Nielsen R. Theory of the Backpropagation. Proceedings of the IEEE International Joint Conference on Neural Networks, 1989, 1:593–605.

[15] Zhang Rui. Research on the mechanical fault diagnosis technology based on artificial neural network theory. Northeast Forestry University, 2001.

30

Building information modeling and integrated practice in JSC SUEK by the example of the Tugnuy CHPP construction

Aleksandr Khvan

SibNiiUgleobogashenie LLC/SUEK JSC, Moscow, Russia

Abstract: Integrated practice in the construction industry is identified as one of the solutions that could be used to minimize the challenges associated with the construction process. Any construction project has to be profitable and reasonable at each stage of life cycle. New technologies in building information modeling (BIM) open new rules in the project management, construction and operation, as well as allow to evaluate the capital costs effectively. Several benefits of implementing BIM are also discussed below, including the visualization and collaboration benefits, synchronization of design and construction planning, and conflict detection and cost reduction. The advantages and limitations of BIM integrated practice in designing and construction Coal Handling and Preparation Plant are discussed in this paper. The conclusion of this paper is that the BIM could provide an ideal platform for integrated practice since it is capable of integrating and linking the project information to the capital construction.

Keywords: BIM modeling, coal preparation plant, construction, design, The Fourth Industrial Revolution, SUEK, SibNiicoal

1. Introduction

The word building information modeling (BIM) came to Russia recently. Though many companies are trying to use BIM in design and construction (digitize business) process. None of them actually implement BIM in each life cycle of the facility being constructed.

A life circle usually contains eight stages (Fig. 1) [1].

The practise of BIM integration in the design of Coal Handling and Preparation Plant (CHPP) is described below.

2. Project description

SUEK is one of the world's largest coal companies and Russia's leading coal, heat, and electricity producer. Key assets of SUEK include (http://www.suek.com/):

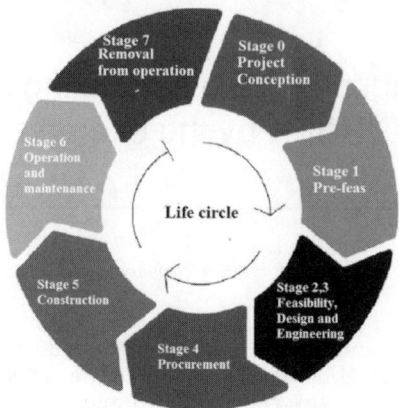

Figure 1: Life circle of a facility under capital construction

- 18 open pits,
- 8 mines,
- 9 washing plants and processing facilities,
- 3 ports,
- 24 combined heat and power plants,
- Global sales and distribution network,
- Research institute,
- 66,383 employees.

The Tugnuy CHPP is located in Buryatia republic with the short working season (from April to October). The territory represents a special environmental zone (near Baikal lake) with the high seismic activity. The proposed Tugnuy Coal Handling and Preparation Plant will be constructed next to the existing CHPP (capacity 1200 t/h) that is producing a single product, which is thermal coal, which is being washed to the required customer specifications available at both Russian and export markets. The capacity of the CHPP after upgrade is 2000 t/h (the ultimate designed value is 2500 t/h).

The main capital facility:

- The second coal preparation module (capacity 1000–1250 t/h),
- The raw, reject, and product handlings

3. The concept of BIM

BIM as a modeling technology and associated set of processes is designed to produce, communicate, and build models. BIM is the acronym of "Building

Information Modeling," reflecting and emphasizing the process aspects, and not of "Building Information Model." The objects of BIM processes are construction models, or BIM models.

Since the UK Government BIM have adopted the concept of "BIM Levels," the following chart provided for the four levels (Level 0–Level 3) have become a widely adopted definition of the criteria for a project to be deemed BIM-compliant (Fig. 2) [2].

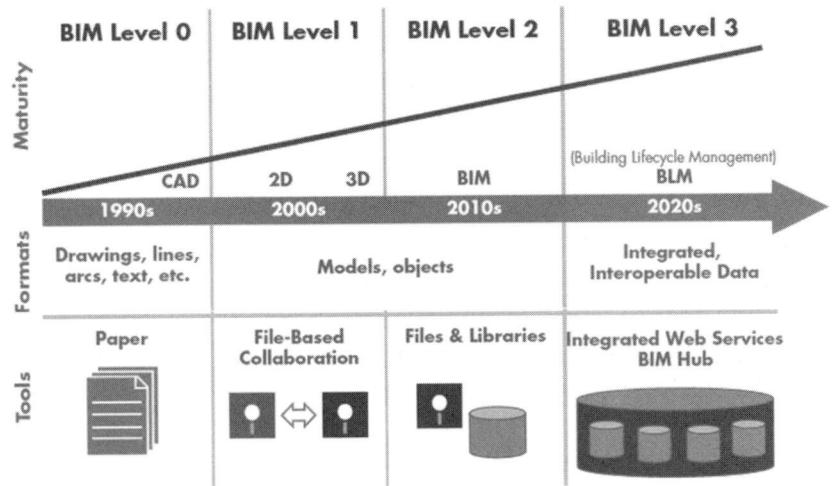

Figure 2: Levels of BIM

3.1 Level 0 BIM

This level is defined as unmanaged computer-aided design (CAD). This is likely to be 2D, with information being shared by traditional paper drawings or in some instances, digitally via PDF, essentially separate sources of information covering basic asset information. The majority of the industry is already well ahead of this now.

3.2 Level 1 BIM

This is the level at which many companies are currently operating. This typically comprises a mixture of 3D CAD for concept work, and 2D for drafting of statutory approval documentation and Production Information. CAD standards are managed to BS 1192:2007, and electronic sharing of data is carried out from a common data environment (CDE), often managed by the contractor. Models are not shared between project team members.

3.3 Level 2 BIM

This is distinguished by collaborative working—all parties use their own 3D models, but they are not working on a single, shared model. The collaboration comes in the form of how the information is exchanged between different parties—and is the crucial aspect of this level. Design information is shared through a common file format, which enables any organization to combine that data with their own in order to make a federated BIM model, and to carry out interrogative checks on it. Hence any CAD software that each party uses must be capable of exporting to a common file format. This is the method of working that has been set as a minimum target by the UK government for all work on public-sector work, by 2016.

3.4 Level 3 BIM

This level represents full collaboration between all disciplines by means of using a single, shared project model that is held in a centralized repository (normally an object database in cloud storage). All parties can access and modify that same model, and the benefit is that it removes the final layer of risk for conflicting information. This is known as "Open BIM."

4. First step of BIM

Usually in SUEK a big project was designed with 0–1 level BIM. In this project SUEK decided to design with 2 level BIM (shared 3D model, link with 3D and 2D, data extraction from 3D).

The main disciplines designed with 2 level with CAD Program:
- Process (LIMN, AUTOCAD Plant 3D),
- Construction (Bentley ProStell, AutoCAD Advance steel, Bentley ProConcrete, Strand7),
- Mechanical (Inventor),
- Piping (AUTOCAD Plant 3D, AutoCAD P&ID, Start),
- General layout (Civil 3D),
- Architecture (Revit Architecture).

The model viewer software was Navisworks (read annotation, property of engineering objects, clash-checking). Other disciplines designed with the same level, but used the shared 3D model.

Figure 3: 3D model (left) and construction of CHPP (right)

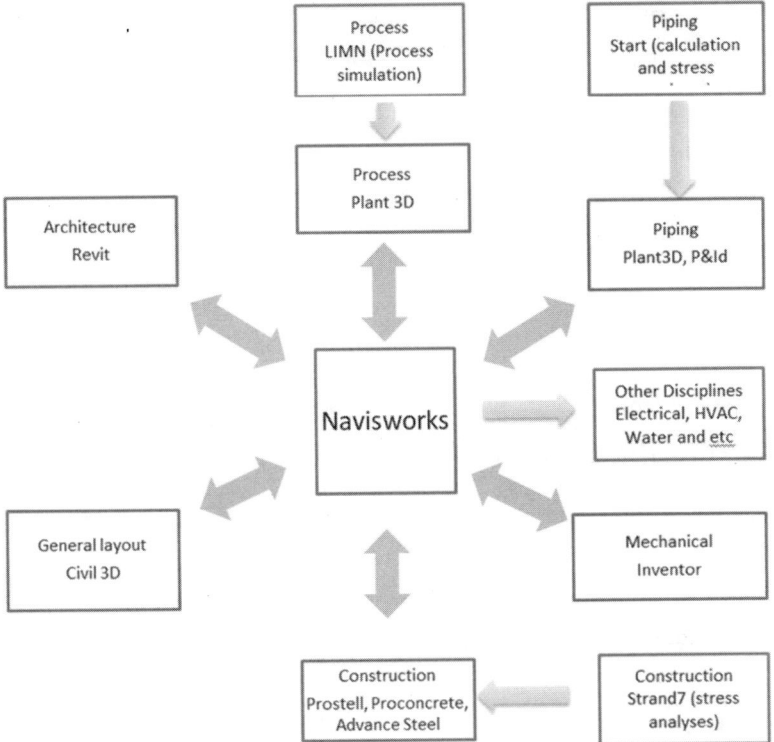

Figure 4: Block diagram of connection between disciplines

5. Advantages and disadvantages

Though not all of the disciplines designed at the level 2 of BIM, all of them felt the "BIM effect." The task description between all disciplines became much faster. Each change along with subsequent influence was detected and tracked through the shared 3D Model.

After a global change, the correct specification of material was issued by means of the data extraction from 3D Model, and provided to the Procurement department. During mechanical construction it was detected that the dimensions of equipment (five screens and motor of conveyor) differ from the approved drawing supplied by the manufacturer. Within a three-day period we managed to detect all of the clashes, and have issued the correct drawing with all the necessary changes to be introduced.

The main disadvantage was the longer period required to issue the drawings for construction. Unlike 2D designs, it takes more time to create a 3D library, to issue specifications, and to train people.

It also takes more time to work with 3D model; more discussions are required between all disciplines, facility managers, and plant operation. It is a good benefit in terms of capital costs due to the ability to time influence and cost more before issued drawing than after (Fig. 5) [3].

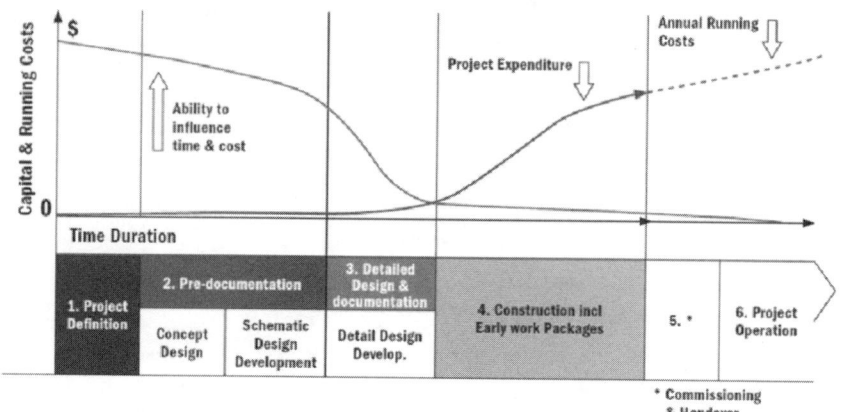

Figure 5: Ability to influence cost during a project's life cycle

6. Conclussion

For the purpose of assisting integrated practice in construction, BIM could provide a good platform regardless of whether it is taken as tools or as a process and concept. BIM provides for a good influence on design, procurement, and construction. In the next project SUEK will design at BIM level 2 in any disciplines with a little integration with BIM level 3.

7. References

[1] A little book about BIM, BSI Group, 2018.

[2] BIM handbook: a guide to building information modeling for owners, designers, engineers, contractor, and facility managers by Rafael Sacks, Charles Eastman, Chang Lee, Paul Teicholz, 2018.

[3] Building information modeling in integrated practice, Haron, AT, Marshall-Pontiing, AJ and Aouad, CF, 2009.

31

Innovative ways to develop coal production wastes after bioleaching

Joanna Calus Moszko, Agnieszka Klupa, Aleksander Wrana,
Robert Fraczek, Rafal Przystas

Główny Instytut Górnictwa, Katowice, Poland
TAURON Wydobycie S.A., Jaworzno, Poland

Abstract: The article presents the activity of coal mining plants in the field of waste management using TAURON Wydobycie S.A. as a case study. Good practices in coal production wastes and coal combustion by-products management are presented as an element of the synergy of mining and energy sectors. In addition, the article presents research on new ways of managing coal production wastes following bioleaching to eliminate acid-generating potential. These activities were carried out as part of the RFCS-funded CEReS project "Co-Processing of Coal Mine and Electronic Wastes: Novel Resources for a Sustainable Future". The aim of the work was to convert environmentally harmful coal production wastes to stabilised material, in order to re-use it. These studies develop new directions of application of stabilised coal production wastes (after the desulphurisation process) obtained as a result of the biological leaching process. On the basis of stabilised waste, full-value commercial products were developed in the form of ceramic products, granulates, additives for cements and polymer concrete products. The work carried out contributes to increasing the amount of waste recovered, which limits the amount of waste deposited.

Keywords: Coal mine waste, electronic waste, bioleaching, ceramic product

1. Introduction

In conducting their basic activities, coal mines produce huge amounts of mining waste. Their characteristics and physico-chemical composition are very diverse. In 2015, 142 million tons of waste were generated in Poland, of which only 8% was municipal waste. The remaining part, approx. 53%, is waste from mining and quarrying, about 16% waste from electricity generation and supply, and 21% from industrial processing. The total amount of waste stored (accumulated) in landfills and waste disposal facilities (heaps, sedimentation basins) has remained at the level of 1.7 billion tons in the last decade (Central Statistical Office – Warsaw, 2016).

The exploitation of coal is directly related to the production of electricity and heat, resulting in the generation of combustion by-products (UPS) – ashes, slags and gypsums. As a rule, the majority of both mining and energy wastes are treated as non-hazardous wastes. According to the waste catalogue and the Regulation of the Minister of Environment from 9th of December 2014 on the waste catalogue, most belongs to the group: 01 – wastes generated during exploration and extraction, physical and chemical processing of ores and other minerals and 10 – waste from thermal processes (Regulation of the Minister of the Environment, 2014).

The energy sector is strongly pressured to constantly improve the quality of manufacturing processes and to reduce emissions of pollutants in waste gases. Therefore, it is important to obtain stable, high-quality fuels. Additional problems are by-products produced during the combustion process.

The TAURON Group owns three mines that supply fuel to power and heat and power plants. The coal mines are also receivers of UPSs produced in the process of combustion. TAURON Group is one of the largest business entities in Poland, employing about 25,000 people and owns about 29% of Polish hard coal resources. TAURON Wydobycie S.A. is responsible for coal mining, processing and sale. The structure includes three mining plants: Sobieski Coal Mine in Jaworzno, Janina Coal Mine in Libiąż and Brzeszcze Coal Mine in Brzeszcze (Website: www.tauron-wydobycie.pl). The extraction and processing of coal in the processing plants is accompanied by waste rocks and materials classified as mining waste. According to the waste catalogue they are classified: 01 04 12 wastes from washing and cleaning of minerals other than those mentioned in 01 04 07 and 01 04 11 – mainly in the form of waste rock, less frequently coal sludge produced in mechanical processing plants and water and mud management plants, wastes of code 01 01 02 – wastes from mineral extraction other than metal ores – wastes generating in preparatory and mining operations, 19 13 06 – sludges from groundwater treatment other than those mentioned in 19 13 05 – sludges accumulated in the sedimentation basins (ACT of 10 July 2008 on mining waste, ACT of December 14, 2012 on waste).

The types of waste differ not only in terms of grain size or consistency, but above all in their mineral composition, flammable and physicochemical properties (Fraś and Przystaś, 2015). TAURON Wydobycie S.A. has been working on successively reducing the growing costs of mining waste management for years. Organisational activities were also followed by the implementation of investments supporting the processing of mining waste into products with specific applications. Waste rock is used for production of aggregate in road engineering and engineering works, construction of railway embankments, construction and strengthening of flood embankments, reclamation, revitalisation and levelling of land, building earth objects, sealing of municipal waste dumps (Fraś and Przystaś, 2015; Cała et al., 2016). On the other hand, coal sludge has been converted into a low-calorific granular fuel or sealing material as an additive to waste rock for the construction of above-ground structures.

The high content of clay substances and water in coal sludge from TAURON Wydobycie S.A. causes problems with using them in building process of earth objects because it causes the effect of the so-called flowing. In the case of use of sludge in combustion processes in power plants, high moisture and doughy consistency causes problems with their dosing, transport or mixing with other fuels. In order to reduce the amount of generated waste, measures have been taken to turn sludge into useful products. As with much coal production waste, an elevated sulfide content (mostly pyrite) can limit options for reuse, especially in civil engineering. Bioleaching can be used as a method of desulphurisation, removing the acid-generating potential and thus stabilising the waste prior to reuse. These activities were carried out under the CEReS project "Co-Processing of Coal Mine and Electronic Wastes: Novel Resources for a Sustainable Future", financed by the Research Fund for Coal and Steel: RFCS-2015.

2. Results

The aim of the research was to transform environmentally harmful post-mining waste into a stabilised material for reuse. Research allowed the development of new directions for the utilisation of stabilised hard coal mine waste obtained as a result of a biological leaching process (desulphurisation). On the basis of stabilised waste, full value commercial products in the form of ceramic products, aggregates, additives to cements and polymer–concrete products were developed.

2.1 Ceramic products

A lot of non-plastic raw materials are used in the building ceramics industry, which are correction admixtures. The main purpose is to reduce the plasticity of clays, reduce the susceptibility of masses to drying, dry masses with too much moisture, speed up the firing process, obtain greater porosity and reduce the volume weight and reduce the tendency to form a ring structure. The use of admixtures containing flammable components (coal sludge) should be considered mainly as an ingredient facilitating the sintering of the ceramic mass, improving the colour. The condition for introducing this ingredient into the masses is only a technological assessment of the clay raw material, including mainly the behaviour of the material during firing (melting point, sinter interval, formation of eutectic alloys) and also the characteristics that the ceramic material should achieve after firing, such as mechanical strength, porosity and absorbability. The brick production technology was based on mechanical processing and shaping (screw-vacuum press). The bricks were fired under the same conditions (Fig. 1). The addition of coal sludge contributed to a less intense colour. The bricks were examined for compressive strength tests (Table 1).

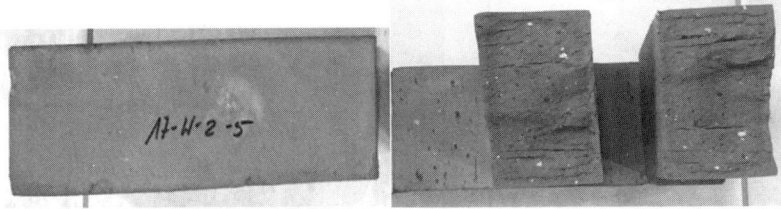

Figure 1: Brick with coal sludge in a wet state (a) and brick with coal sludge – breakthrough reduction cores visible at the brick (b)

Table 1: Strength tests

CERADBUD brick standard	Sample	*fb* (MPa)
	1/1	30.8
	1/1	33.0
	1/2	30.7
	1/3	37.0
	Mean	**32.8**
	Std.	2.9
CERADBUD brick class 20, sludge 8%	2/1	46.8
	2/2	45.2
	2/3	40.1
	2/4	45.0
	Mean	**44.3**
	Std.	2.9
CERADBUD brick class 20, sludge 15%	3/1	43.6
	3/2	32.3
	3/3	52.1
	3/4	52.6
	Mean	**45.2**
	Std.	9.5

On the basis of the laboratory tests it was found that the highest compressive strength of 44 MPa and 45.2 MPa was obtained for products with 8% and 15% content of coal sludge. Tests carried out on averaged samples of ceramic products confirmed that a small standard deviation guarantees stability of compression strength of ceramic products with hard coal sludge content. On the basis of comparative tests (Table 2) it can be concluded that other parameters, especially water absorption do not differ from the level achieved in standard products with a slight tendency to decrease open porosity.

Table 2: Results of research on bricks

Content		Mean				Density	Absorb. Z	Absorb. G	Absorption coefficient	Compressive strength	Standard strength
		Length	Width	Height	Mass						
Examination of bricks with coal sludge											
Brick – with coal sludge 8%	17-W-1	248.3	19.7	64.5	3.64	1898	10.7%	12.6%	0.85	38	30.8
Brick – with coal sludge 15%	17-W-2	248.9	20.0	64.2	3.46	1805	13.1	14.6%	0.90	32.3	26.1
Examination of bricks without admixture of coal sludge											
Brick – facing	17-1	249.9	20.8	64.2	3.76	1955	11.6%	12.6%	0.92	46.3	37.5
Brick – class 20	17-2	249.9	20.6	61.5	3.67	1986	9.9%	11.1%	0.89	52.5	42.5
Brick – facing	17-3	250.0	20.9	63.4	3.72	1942	12.1%	13.1%	0.92	40.3	32.6
Brick – class 20	17-4	247.8	19.8	62.7	3.71	1993	10.6%	11.5%	0.92		0.0

2.2 Additives to cement

The aim of the study was to determine the effect of the addition of post-mining waste after the stabilisation (bioleaching) process on the strength of cement mortar. Four types of cement mortar were prepared in accordance with the PN-EN 196-1 standard. The samples were prepared as follows: cement 42. 5R from the Górażdże cement plant, sand and water in the ratio of 450:1350:250, where 5% of sand was replaced by waste, 10% of sand was replaced by waste and 15% of sand was replaced by waste. Other components – cement and water – were not different from the reference sample. Strength tests (compression) of products after 28 and 90 days (Fig. 2) were carried out in a MTS-810 testing machine. Compression strength tests for cement were carried out on the basis of PN-EN 196-1: 2006 Cement testing methods. Part 1: Determination of strength. The test method is to load the lateral surface of the half of the beam with a compression force increasing steadily (in accordance with the requirements of the standard) until the destructive force is reached, and to calculate the compressive strength R_c, in MPa, according to the formula (Table 3):

$$R_c = F_c/1600$$

where R_c is the compressive strength in MPa, F_c is the highest load during the crushing of the sample in N and 1600 surface of tiles (40 mm × 40 mm), mm^2.

Figure 2: View of products after 28 and 90 days

On the basis of laboratory tests it was found that the addition of 15% of waste did not cause any deterioration of the compressive strength of the obtained concrete.

Table 3: Strength tests after 28 and 90 days

Sample	Number of sample	R_c (MPa)	Sample	Number of sample	R_c (MPa)
Strength tests after 28 days			Strength tests after 90 days		
0% waste	1	44.1	0% waste	1	40.1
	2	45.4		2	40.69
	3	45.1		3	42.01
	Mean	44.9		Mean	40.93
	Std.	0.7		Std.	0.98
5% waste	4	46	5% waste	4	43.9
	5	46.5		5	45.4
	6	45.1		6	44.95
	Mean	45.9		Mean	44.75
	Std.	0.7		Std.	0.77
10% waste	7	52.1	10% waste	7	49.93
	8	51.4		8	50.29
	9	50.8		9	47.02
	Mean	51.4		Mean	49.08
	Std.	0.7		Std.	1.79
15% waste	10	51.4	15% waste	10	49.97
	11	51.9		11	45.21
	12	52		12	49.05
	Mean	51.7		Mean	48.08
	Std.	0.3		Std.	2.53

2.3 Product of granulates

Industrial tests were carried out in order to obtain four types of granules. Coal sludge from the Janina coal mine was used in granulation process. The product in the form of granulate consisted of 2% ash with coal waste, 5% ash with coal waste, 2% lime with coal waste and 5% lime with coal waste. The analysis of water extract (Table 4) from the examined granules did not exceed the admissible values of pollutants determined for granules used in underground mining techniques. After the biostabilisation process, the material can be used to produce granules that can be safely used in underground mining techniques (Fig. 3).

Figure 3: Industrial test production of granulates in TAURON Wydobycie
S.A. and view of granulates

Table 4: List of results of analysis of water extract from granulate with maximum
permissible values of impurities specified for granulates used in underground mining
techniques (TAURON Wydobycie S.A.)

Pollution indicator	Concentration of ingredients in the water extract (mg/l)	Maximum permissible values of pollution indicators determined for granulates used in underground mining techniques	
		Filling solidified and caulking gobs	Hydraulic filling
Chlorides	71.2	1000	1000
Sulfur	324.2	500	500
Sodium (Na)	77	n.n.	800
Arsenic (As)	<0.001	0.2	n.n.
Lead (Pb)	0.033	0.5	0.5
Cadmium	<0.004	0.1	0.1
Copper (Cu)	<0.001	0.5	0.5
Mercury (Hg)	0.01	0.02	n.n.
Chromium (Cr)	<0.004	0.5	0.5
Nickel (Ni)	0.055	n.n.	2
Zinc (Zn)		n.n.	2
pH	8.9	6–12	6–9

n.n., non-standard value; <, value below the lower limit of quantification of the applied
measurement method

2.4 Production of polymer concrete

On the basis of stabilised mining wastes, a product was developed that can be used as underground roadway lagging beams (Fig. 4). Beams were made of polymer concrete with the addition of stabilised mining waste. These elements are currently made of cement concrete and used mainly in long lasting underground roadway. From that perspective the high resistance to corrosion is important parameter.

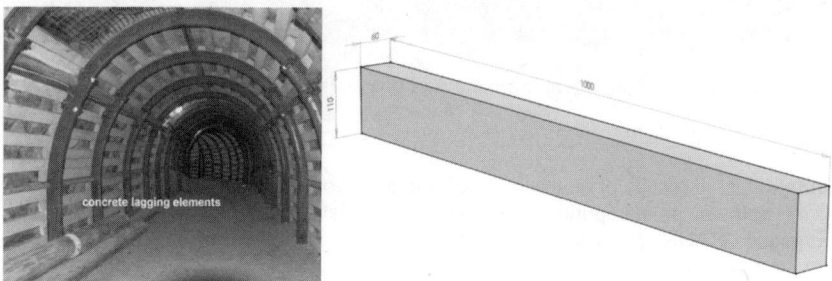

Figure 4: Concrete lagging beams: installed in roadway (left); single concrete beam (right)

Unfortunately due to the limited amount of stabilised mine wastes it was impossible to conduct the test on real scale elements. That is why small scale samples (40 mm × 40 mm × 160 mm) were prepared and tested. The set for samples preparation and samples before the tests are shown in Fig. 5. The three-point bending test was conducted in accordance with EN-196-1:2016 standard (Methods of testing cement. Determination of strength).

Figure 5: Set for samples preparation (left) and polymer concrete samples (right)

Two different compositions of polymer concrete mixture were prepared. Three samples for each mixture were tested. The list of samples together with the summary of three-point bending tests results are presented in Table 5. Also the reference sample was prepared where the polymer concrete was based on sand and fly ash. The strength of the sample was 25 MPa. The bonding time of the reference sample was about 45 min.

Table 5: Characterisation of polymer concrete samples with the results of three-point bending test

Sample no.	Polymer concrete mixture composition				Bending strength (MPa)	Average value (MPa)	Standard deviation (MPa)
	Polyester resin (g)	Hardener (g)	Accelerator (ml)	Waste (g)			
BL-0.1	208	4	0.42	1500	0.76	0.59	0.15
BL-0.2					0.49		
BL-0.3					0.52		
BL-1.1	600	12	1.2	1500	14.62	15.96	1.39
BL-1.2					17.39		
BL-1.3					15.86		

In the two mixtures prepared, the content of stabilised mining wastes was the same (1500 g) for comparable results. First, the composite polymer mixture with lower content of resin (208 g) was prepared for possible cost reduction of final product. Unfortunately, the bending test result showed that the strength is insufficient (Table 1 – samples BL-0.1–BL-0.3).

In the next stage the resin content in the mixture was 600 g. The bending test results were lower than the strength of reference sample but in some cases sufficient (Table 1 – samples BL-1.1–BL-1.3). The biggest problem of these samples was the long bonding time. After 24 h the samples were not properly bonded so they were stored for another day in elevated temperature (80°C). It allowed the bonding of the resin and to conduct the strength test. The long bonding time was probably caused by acidity of the wastes and its reaction with the hardener.

To summarise it can be sated that it is possible to use the stabilised mining wastes to produce polymer concrete elements as roadway lagging beams. Nevertheless, the problems with long bonding that require the storage of the concrete in elevated temperature or additional waste treatment. The consequence of it could be low economic efficiency of the production process.

3. Conclusion

The conducted research indicated new directions of application of stabilised coal mine wastes (after the desulphurisation process) obtained as a result of the bioleaching process. On the basis of stabilised wastes, full-value commercial products were developed in the form of ceramic products, granules, additives for cements and polymer–concrete products. In addition, work carried out contributes to the increase in the amount of reutilised wastes, which limits the amount of deposited wastes.

Tests carried out for various silty raw materials with an addition of coal sludge allowed for preliminary understanding of the parameters and characteristics of tested ceramic products. Based on the results of testing and their interpretation, it can be stated that coal production waste is suitable for commercial use in the ceramic industry after the desulphurisation process. On the basis of laboratory tests of concrete additives, it was found that the addition of even 15% of waste does not cause any deterioration in the compressive strength of concrete. The material after the biostabilisation process can also be used for the production of granules, which can be safely used in underground mining techniques, as indicated by analyses of the water extract. Also stabilised mining wastes can be used to produce polymer–concrete elements as roadway lagging beams. Also stabilised mining wastes can be used to produce polymer–concrete elements as roadway lagging beams.

4. Acknowledgement

This work was carried out as part of the CEReS (Co-Processing of Coal Mine and Electronic Wastes: Novel Resources for a Sustainable Future) project. This project has received funding from the Research Fund for Coal and Steel under grant agreement No 709868.

5. References

[1] Cała M., Tajduś A., Pomykała R., Przystaś R., Adamczyk J., Stopkowicz A., Kępys W. "Energy and mining wastes as components of products for mining, construction and geoengineering", Materials of the XXIII International Conference "Ashes from energy", Zakopane October 2016.

[2] Central Statistical Office – Warsaw 2016 Environmental protection 2016 Environment.

[3] Fraś A., Przystaś R. "Ecological and economic aspects of mining waste management and UPS in TAURON Wydobycie S.A." Materials of the XXII International Conference "Ashes from energy" Krynica Zdrój 21-23.10.2015.

[4] Regulation of the Minister of the Environment of 9 December 2014 on the Waste Catalog.

32

Utilization and disposal of coal washery rejects: An environmental challenge

Amarjeet Singh, D.K. Sah, P.C. Jha, Pushkar

Central Mine Planning and Design Institute Ltd, Dhanbad, India

Abstract: Coal beneficiation is required for supplying coking coal to the steel industry as well as non-coking coal with ash content of <34% to the coal based thermal power plants located at a distance of 500 km or more from the pit head. Coal rejects, produced during coal washing process, either remain unutilized or used for some small scale industry like brick kilns or power generation in fluidized bed combustion boiler. Washing of coal increases its efficiency and quality, therefore increasing its price. It can also help to reduce emissions from burning. The environmental impacts of burning of coal are required to be kept in mind. The mineral matters of the coal should be reduced during its processing such that, the emissions of sulfur dioxide (SO_2), carbon dioxide (CO_2). Particulate matters are minimized during burning in washed coal. The unutilized washery rejects undergoes interaction with water resulting in formation of leachates. The leachates may pollute the water bodies by increasing the concentration of total suspended solids (TSS), chemical oxygen demand (COD), trace metals, etc. Leachability study of reject samples obtained from coking and non-coking coal washeries after analysis show that some elements like iron (Fe), manganese (Mn), and copper (Cu) may leach from the reject dumps and contaminate the water bodies. Regular monitoring of surface and ground water and comparison with baseline condition is required for assessing the long term impact of dumping washery reject in mine void or on the surface.

Keywords: Coal washery, coking and non-coking coal, leachability study, environmental challenge, phytoremediation

1. Introduction

Coal beneficiation is required for supplying coking coal to the steel industry as well as non-coking coal with ash content of <34% to the coal based thermal power plants located at a distance of 500 km or more from the pit head. Coal rejects, produced during coal washing process, either remain unutilized or used for some small scale industry like brick kilns or power generation in fluidized bed combustion reactor. Interaction of reject with water results in formation of leachates. The leachates may pollute the water bodies by increasing the concentration of total suspended solids (TSS), chemical oxygen demand (COD), trace metals, etc. In this study, rejects obtained from coking and non-coking coal washeries have been subjected to Toxicity Characteristics Leaching Procedure (TCLP) Test (USEPA, 1986) to simulate leaching of trace metals. The synthetic leachates were analyzed for various trace elements and compared with the primary drinking water standards (PDWS) i.e. IS-10500 and corresponding TCLP limits. The TCLP limit for an element has been taken as 100 times the corresponding value of PDWS. The TCLP tests, conducted on coking coal washery rejects of Moonidih, Madhuband (BCCL), Kathara and Swang (CCL), show that only iron (Fe) concentration in leachates (prepared at pH 2.88 ± 0.05) is more than the TCLP limits in all samples, whereas the manganese (Mn) concentration in leachate (prepared at pH 2.88 ± 0.05) exceeds the TCLP limits for Swang washery reject only. The analysis results indicate possibility of water pollution owing to higher concentration of iron and manganese in the leachate of washery rejects. The concentration of the rest of the trace elements (arsenic, selenium, mercury, copper, zinc, nickel, cadmium, lead, chromium, and boron) are below the corresponding TCLP limits in leachate samples of all coking coal washery rejects. In case of Piparwar Washery (non-coking coal), TCLP test shows that only iron (Fe) and copper (Cu) concentration in leachates (prepared at pH 2.88 ± 0.05) is more than the corresponding TCLP limits, whereas the concentration of rest of the elements remain below TCLP limits.

2. Collection and preparation of samples

The coal reject samples were collected from Moonidih and Madhubandh Washeries of Bharat Coking Coal Limited (BCCL), Kathara, Sawang, and Piparwar Washeries of Central Coalfields Limited (CCL). Piparwar washery is a non-coking coal washery, whereas the rest of them are coking coal washeries.

The coal reject samples were subjected to coning and quartering and finally 1 kg of coal sample was carefully ground to 72 mesh for conducting various tests.

3. Analysis of samples

Coal reject samples were mixed with water in 1:5 ratio (solid:water) and analyzed for pH and conductivity. Synthetic leachates prepared from the rejects as per Test Method 1311 of USEPA were analyzed for trace elements in the environment laboratory.

4. Interpretation of TCLP test results

A close examination of the analysis results of the leachates prepared from washery rejects are as follows:

(i) The pH of the suspension formed from reject samples (solid:water, 1:5) is in the range of 6.73–7.47. The pH value for the suspension obtained from Moonidih washery reject is weakly acidic whereas the values for the rest of the samples have been observed in slightly alkaline range. The pH indicates that the washery rejects contain mostly neutral ingredients or the substances which in contact with water almost produce neutral solutions.

(ii) The conductivity of the suspension formed from reject samples (solid:water, 1:5) have been found in the range of 35–202 **µS/cm. The older rejects of coking coal washeries** of BCCL i.e. Moonidih and Madhuband have conductivity values i.e. 35 and 95.7 **µS/cm,** respectively in comparison to coal washeries of CCL i.e. Piparwar, Swang, and Kathara, showing conductivity values of 81.1, 137, and 202 **µS/cm,** respectively. Higher conductivity values indicate higher concentration of total dissolved solids. This indicates that with ageing/weathering, the soluble contents get removed from the rejects by the action of water.

(iii) Two types of extraction fluids are used for preparation of leachates. One at pH 2.88 ± 0.05 and another at pH 4.93 ± 0.05. Generally higher concentration of trace metals were observed in leachates prepared from extraction fluid having lower pH.

(iv) The concentration of trace elements like arsenic (As), selenium (Se), and mercury (Hg) has been found to be below detection limits (BDL) in leachates prepared from both types of extraction fluids for all four non-coking coal washery reject samples and one number of non-coking coal washery (Piparwar) reject sample.

(v) The average concentration of trace elements in leachates are in the following order:

Fe > Mn > Zn > Cu > B > Ni > Cr > Pb > Cd (Coking coal)

Fe > Cu > Zn > Mn > Pb > B > Ni > Cr > Cd (Non-coking coal)

(vi) The concentration of iron (Fe) and manganese (Mn) in leachates samples of coking coal is high (mean value Fe: 43.9 mg/l and Mn: 7.1 mg/l). It indicates high concentration of water soluble iron and manganese salts in the rejects. The threshold limit for Fe and Mn is 0.3 and 0.1 mg/l, respectively as per Drinking Water Standards (IS-

10500). The corresponding TCLP limit is 30 and 10 mg/l, respectively for Fe and Mn. TCLP tests indicates that Fe concentration in leachates prepared at pH 2.88 ± 0.05 is more than the TCLP limits for all samples, whereas the Mn concentration exceed the TCLP limits for Swang washery reject only. The analysis of Piparwar coal washery rejects indicates the concentration of iron is 86.02 mg/l which is 2.87 times higher than the TCLP limits, whereas the concentration of Mn is 2.09 mg/l which is below the TCLP limits of 10 mg/l/.

(vii) The threshold limit for copper (Cu) is 0.05 mg/l (IS-10500). The concentration of Cu is equal to or higher than the permissible Indian Standards for drinking water in 7 out of 8 leachate samples (87.5%) prepared from coking coal washery rejects. The maximum value of copper in leachate sample is 1.86 mg/l, which is much below the TCLP limits. The concentration of Cu in leachates prepared from Piparwar Coal washery reject is 14.36 mg/l. It is 2.87 times higher than the corresponding TCLP limits.

(viii) The threshold limit for zinc (Zn) is 5.0 mg/l (IS-10500 standards). The concentration of zinc in all leachate samples prepared from coking coal washery rejects have been found to be below the threshold value as per drinking water standards. The maximum concentration of zinc in leachate prepared from Piparwar washery has been observed as 5.4 mg/l.

(ix) The concentration of nickel (Ni) is more than the permissible value for drinking water in all leachate samples (IS-10500) prepared from coking and non-coking coal washery rejects. The maximum value of nickel in leachate sample is 0.29 mg/l, which is quite below the TCLP limit of 2 mg/l.

(x) Cadmium (Cd), lead (Pb), and chromium (Cr) are considered as highly toxic elements. The permissible limits as per IS-10500 are 0.003, 0.01, and 0.05 mg/l, respectively. The concentration of Cd, Pb, and Cr are equal to or higher than the threshold values in 3 out of 8 (37.5%), 8 out of 8 (100%), and 4 out of 8 (50%) samples, respectively for leachates prepared from coking coal washery rejects. The maximum observed concentration of Cd, Pb, and Cr in the leachates are 0.0058, 0.055, and 0.21 mg/l, respectively for coking coal washery rejects. These values are lower than the corresponding TCLP limits of 0.3, 1.0, and 5 mg/l. In case of Piparwar washery, the observed concentration of Cd, Pb, and Cr are 0.0036, 0.376, and 0.07 mg/l, respectively.

(xi) The permissible limits for boron in drinking water is 0.5 mg/l as per IS-10500. The concentration of boron is equal to or higher than the threshold value in 4 out of 8 (50%) samples for leachates prepared from coking coal washery rejects The maximum concentration of boron in leachate was observed as 0.7 mg/l, which is quite below the TCLP limit of 50 mg/l. In case of Piparwar washery, the maximum observed concentration of boron has been found as 0.28 mg/l.

The potential of various trace metals to pollute ground water due to leaching from washery rejects of coking coal washery has been presented by taking the ratio of maximum concentration of an element in the synthetic leachate and corresponding TCLP limits and the same has been depicted in Fig. 1. If the ratio is more than 1, it indicates probability of ground water contamination.

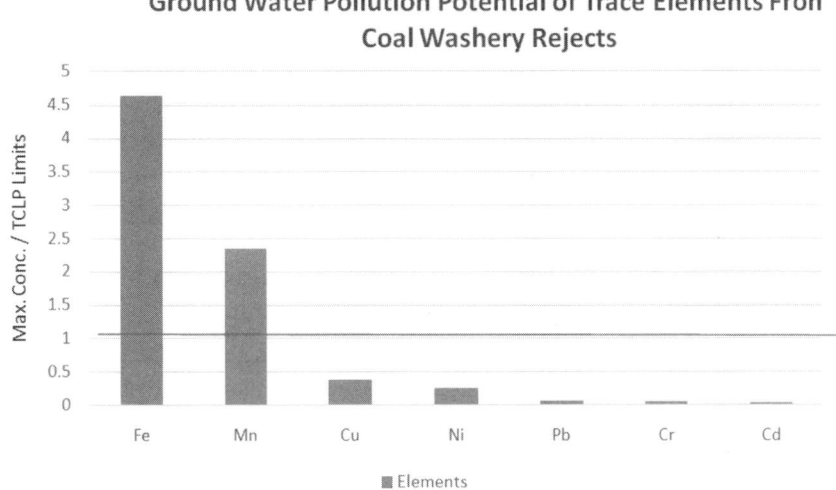

Figure 1 : Ground water pollution potential of trace elements from coal washery rejects

5. Impact of keeping the washery rejects on the land

Washery rejects, kept on the land, interacts with air and water. Air contact may induce spontaneous combustion resulting in generation of various air pollutants. Trace metals in coal are either associated with organic or inorganic matrix. Metals, associated with organic matters, are generally not leached when they come in contact with water whereas inorganic portion gets leached relatively easily by action of water. Spontaneous combustion in reject heaps results in destruction of complex organic matrix and increase in *insitu* temperature of the dump and it finally leads to faster leaching of trace metals. Leachates containing iron and manganese may enter the surface water bodies including nearby abandoned mine voids. Iron and manganese exist in variable oxidation states i.e. Fe^{2+}, Fe^{3+}, Mn^{2+}, and Mn^{4+}. Dissolved oxygen content is generally lower (<1 mg/l) in the bottom of the water bodies and ground water. Anaerobic condition facilitates conversion of these metals to lower oxidation states. Both iron and manganese compounds have higher solubility in water if dissolved oxygen content is less than 1 mg/l. Seepage from the water bodies may increase the concentration of these metals in the ground water. It may be prevented by taking suitable measures which may include the followings:

(i)   Reduction in volume of reject generation by washing non coking coal with more than 34% ash content depending upon its feasibility. It will also reduce the carbon contents in the rejects and may help in controlling spontaneous combustion. This will help in controlling air and water pollution arising out of washery reject disposal.

(ii) Use of suitable chemical inhibitors for controlling the spontaneous combustion.

(iii) Providing a suitable cover over the reject heap both during progressive dumping and on final closure of the site.

(iv) Run-off from the reject dumps should be passed through a settling pond for arresting TSS. A garland drain around the dump may be provided for this purpose. The trace metal contents in the output of the settling tank may be removed through phytoremediation.

(v)   Reject heaps may be covered with inert materials and top soil for biological reclamation using native varieties of plant species. Suitable species of plants may be grown for phytoremediation.

6. Impact of keeping the washery rejects in mine void

Washery rejects, if kept in a mine void, may interact with water and generate leachates having Fe, Mn, Cu, and other metals. High concentration of Fe and Mn in leachate is expected due to prevailing anaerobic condition. The contamination of ground water may be controlled by using suitable liners if required. The actual condition of the ground water has to be assessed by regular monitoring and comparison with baseline data.

7. Conclusion

On the basis of the leachate test carried out for washery rejects for some samples, the following conclusions may be drawn:

(i) Leachates generated from coking coal washery rejects show that only iron (Fe) concentration in leachates prepared at pH 2.88 ± 0.05 is more than the TCLP limits for all samples, whereas the manganese (Mn) concentration exceed the TCLP limits for Swang washery reject only.

(ii) Leachates generated from non-coking coal washery show that the iron (Fe) and copper(Cu) concentration in leachate prepared at pH 2.88 ± 0.05 is more than the corresponding TCLP limits.

(iii) The physio-chemical analysis of washery rejects along with the leachate (TCLP) tests should be carried out before deciding about their mode of disposal with adequate control measures as coal is a heterogeneous material and reject generated from each washery may exhibit different characteristics. The actual condition of the ground water and surface water bodies around the reject dump has to be assessed by regular monitoring and comparison with baseline data.

8. Acknowledgement

The authors are thankful to Central Mine Planning and Design Institute Ltd., Ranchi for giving permission for publishing the data in the present paper. The views expressed in the present paper are of author/authors not of CMPDI/ Coal India Limited.

9. References

[1] Assessment of Leachability of Trace Heavy Metals from Ash Ponds of Ground Water, Central Pollution Control Board, May 2007.

[2] Indian Standard, Drinking Water Specification (Second Revision), IS-10500:2012.

[3] New South Wales, Australia, Environment Protection Agency, and the Coal Washery Rejects Order 2014.

[4] Standard Methods for Examination of Water and Waste water 22nd Edition, APHA Publication, 2012.

[5] Toxicity Characteristics Leaching Procedure (TCLP), Method 1311 of United States Environment Protection Agency, Revision 0 (July 1992).

33

Waste Management in Opencast Coal Mines

Arvind Kr Singh, Sr Manager (Mining), Ranjeet Prasad, DGM (Mining)

VC Dubey, AGM (Mining), Partha Mazumder, ED (Coal Mining)

Abstract: This paper provides an overview of types of waste, sources of waste, the volume of waste, issues in waste management, etc., during the mining of coal and its usage. Challenges in waste management in view of increasing coal production from Open cast mines with higher stripping ratio. Problems related to Environment, safety, land acquisition and R&R involved in Waste Management. Safe, economical and environmentally acceptable management of coal waste involves consideration of geology, soil and rock mechanics, hydrology, geochemistry, etc. Scientific and meticulous planning (for coal block boundary and sequence of mining) and execution (for waste management) is critical for sustainable, environment friendly and safe coal production.

Keywords: Overburden, Coal reserve, Stripping Ratio, External dump, Internal Dump, Slope Stability, Out of seam dilution, Coal Processing waste, Fly Ash.

1. Introduction

India is poised to become US $5 trillion economy by 2025 as envisioned by the Government of India, which requires securing 9 percent average annual economic growth rate. The long-term growth perspective of the Indian economy is positive due to various reasons including accelerated reforms in almost all sectors.

Affordable energy is basic and fundamental to any growth. India is the world's third largest producer and third largest consumer of electricity. The Government of India's program *"Power for All"* is intended to ensure continuous and uninterrupted electricity supply to all households, industries and commercial establishments. Though, there is a greater push towards investment in renewables but India›s electricity sector is dominated by fossil fuels, and in particular, coal, which during the Fiscal year 2017-18 produced more than 65% of the country›s electricity. Coal is the most important and abundant fossil fuel in India.

To fuel the growth of the country and meeting the Government of India's program *"Power for All"*, total coal production from the country will keep increasing. By greater thrust on mechanisation, application of modern technologies and start of coal production from captive coal blocks, it has been possible to produce 730.354 million tonnes of coal in FY 2018-19 which is positive growth of 8.1%from FY 2017-18 (675.40 million tonnes). More than 80 percent of total coal production comes from Opencast Mines, which requires removal of a large volume of waste (Overburden) to produce coal e.g last year Northern Coal Fields produced 101.50 Million te of Coal and that required removal of 318.22 Million Cum of Overburden. Coal is likely to remain the main source of energy in India in the near future due to the availability of large geological and mineable reserve. At the same time, the share of coal production from opencast mines is also likely to increase which will require large volumes of solid waste (Overburden) to be handled. The depth of opencast coal mines is likely to increase in the times to come and that will require more waste to be handled to get more coal lying at larger depth. This increase coincides with a new awareness in which environmental concerns have become a growing challenge. The mining process generates a large quantity of waste that must be strategically and safely managed to combine economic efficiency with demands for environmental sustainability. The social demand has increased for the sustainable development of all of the activities related to mining, particularly the adequate management of waste products during each phase of the mining process, including development, extraction, transport, washing, etc.

We have attempted to cover the topic in following sub-topics:

  (i)   Present status of coal reserve in India
  (ii) Present level of coal production and corresponding volume of waste (Overburden) removal
  (iii) Types of waste in mining and use of coal
  (iv) Sources of Waste in Mining of coal
  (v)   Estimated volume of waste from coal blocks of NTPC and related issues
  (vi) Issues in Waste Management and their Mitigation
  (vii) Suggestions and Recommendation

Lets have a look at the coal reserves our country have and present status of waste generation to produce present level of coal production:

2. Present status of coal reserves in India

The Coal resources of India are available in older Gondwana Formations of peninsular India and younger Tertiary formations of north-eastern region. Exploration is being carried out by the GSI, MECL, CMPDI, SCCL, etc up to the maximum depth of 1200m and different types of the reserve, as on 01.04.2018, are tabulated below

Inventory as on	Proved/ Measured	Indicated	Inferred	Total (Million te)
1.4.2018	148787	139164	31069	319020
1.4.2017	143058	139311	32780	315149
1.4.2016	138087	139151	31564	308802
1.4.2015	131614	143241	31740	306596
1.4.2014	125909	142506	33149	301564

(Source: Web site of Ministry of Coal, www.coal.nic.in)

India has a reserve of 319 Billion te of Coal and It can be observed that total as well as Proved reserve is increasing due to the increased exploration activities and the trend is likely to continue.

Detailed Exploration in selected blocks, where boreholes are less than 400 meter apart, upgrades the resources into more reliable 'Proved/Measured' category. Out of 319 Billion tonnes, 282.9 Billion te is non-coking or coal which is mainly used for power generation. Type wise coal reserve is as below:

Type of Coal	Proved/ Measured	Indicated	Inferred	Total(M te)
(A) Coking :-				
-Prime Coking	4649	664	0	5313
-Medium Coking	13914	11709	1879	27502
-Semi-Coking	519	995	193	1708
Sub-Total Coking	19082	13368	2073	34522
(B) Non-Coking	129112	125697	28102	**282910**
(C) Tertiary Coal	594	99	895	1588
Grand Total	**148787**	**139164**	**31069**	**319020**

3. Present extent/volume of waste handling (solid waste) in coal mining

The volume of Overburden being removed vis a vis coal production from opencast coal mines of subsidiaries of Coal India Ltd and overall CIL is tabulated below to have an idea of the volume of waste(Overburden) being handled today. Handling such large volume and having a look at the challenges being faced today will help us in preparing ourselves to be ready to address in a better way all the challenges of increased coal production.

Subsidiary	Year	2018-19	2017-18	2016-17	2015-16
NCL	Coal production	101.50	93.02	*Coal production in Million te and OB removal in Million cum.*	
	Opencast Coal	**101.50**	**93.02**		
	OB(M Cum)	**318.22**	**316.95**	*Subsidiaries of CIL are producing coal from such opencast mines which are*	
SECL	Coal production	157.35	144.71		
	Opencast Coal	142.58	130.25	*economically favourable or in other words*	
	OB(M Cum)	**183.44**	**205.018**	*having less stripping ratio (Volume of OB to*	
CCL	Coal production	68.72	63.41	*produce one te of coal). Most of the captive coal*	
	Opencast Coal	68.41	63	*blokes allocated recently*	
	OB(M Cum)	**100.49**	**95.622**	*for captive consumption are having high stripping*	
MCL	Coal production	144.15	143.06	*ratio means much larger volume of Overburden*	
	Opencast Coal	143.28	142.02	*needs to be handled to*	
	OB(M Cum)	**130**	**138.179**	*produce coal.*	
WCL	Coal production	53.18	46.22		
	Opencast Coal	**48.62**	**41.27**		
	OB(M Cum)	**192.03**	**185.287**		
CIL	Coal production	606.89	567.365	554.140	538.754
	Opencast Coal (M te)	**576.40**	**536.823**	522.663	504.968
	OB(M Cum)	**1161.99**	**1178.12**	**1156.377**	**1148.908**

(Source: Annual report of Coal India Limited for FY 2018-19 available at website www. coalindia.nic.in)

4. Type of wastes related to mining and use of coal

Any materials which are thrown away because of no value are called waste. There is a solid waste (Overburden and interburden), Carbonaceous waste (shale), Liquid waste from workshop area, Mine water, Chemical waste, tailings or middlings during washing of coal, fly ash & Bottom ash during the burning of coal for power generation, etc. The various waste during different stage are depicted below:

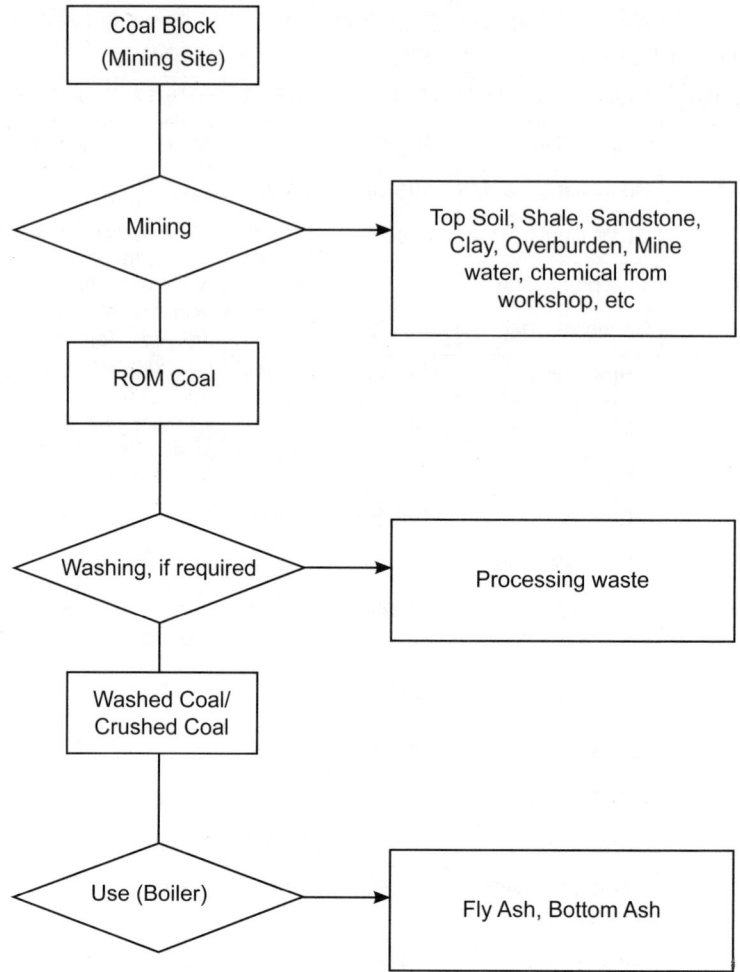

5. Sources of waste in complete cycle of coal (mining of coal to the burning of coal) and some related terms

Strategies for effective management or handling of waste can be framed better if we know or understand the sources of waste.

(i) **Soil and rock on the top of the coal seam (Overburden):** In case of opencast mining, the mine waste materials ("Overburden") which consist of soil and rock on the top of the coal seam is required to be removed to get access to the coal and its extraction. This waste removal activity is known as 'Stripping'. In opencast mines, the mining company has to incur such expenses over the life of the mine. In the case of Coal deposits, they generally consist of top soil, weathered coal, shale, dirt band, thin coal bands, sandy dirt, and sandstone. It is removed by cutting, drilling, blasting, loading and hauling to the Dump. A study of geochemical characteristics of Overburden may determine the contamination potential of the material and would help in deciding about the future course of action for environment friendly mining.

Dump (internal/backfilling and external OB dump): Dump is the place where the loosed or fragmented Overburden is placed. That should be nearby and non-coal bearing area. If decoaled (all coal extracted and no coal is lying beneath) space is available inside coal mine then that is filled with Overburden (internal dump or backfilling). The Overburden which is placed or dumped outside of the pit area that is called an external dump. As the volume is generally very large, the Dump is designed(shape, height, etc) scientifically ensuring safety, minimum damage to environment and compliance to all regulations. Internal dumping is economically preferable as external dumps require the additional acquisition of a large area of land and expenditure in transporting the Overburden to the external area.

(ii) **Carbonaceous material (part of Interburden and Overburden):** After detailed exploration, geological Report is prepared which provides details about quantity (Reserve) and Quality (Grade, GCV, etc) of coal along with geo-mining constraints. Generally, the following norms are followed for delineation of coal seams:

Coal   : Ash+moisture upto 40%
Shaly coal   : Ash+moisture>40% to 55%
Carbonaceous shale : *Ash+moisture>55% to 75%*
Obvious bands   : *Ash + Moisture >75%*

Coal and shaly coal are grouped together to form a coal seam. Carbonaceous shale band is considered as combustible dirt band while shale and sandstone are treated as obvious bands.

While computing effective thickness of the seam, all the carbonaceous shale bands, and obvious bands as well as contiguous dirt bands are excluded/included as detailed below.

BCS: Excluding all carb shale & obvious bands irrespective of thickness

I 30 : Excluding all carb shale & obvious bands of more than 0.30m thickness

I 100: Excluding all carb shale & obvious bands of more than 1.00m thickness

Ip: Including all carb shale and obvious bands irrespective of Thickness

The Overburden also includes sandy soil, weathered rocks and dirt bands >1m thickness (I100 basis of reserve estimation). While computing inseam burden, dirt-bands of >1m generally being excluded to arrive at the effective thickness of the seam.

Carbonaceous shale with Ash+Moisture % more than 55% are considered as waste (Overburden). Coal seam of thickness less than 1m is also considered as part of Waste (Overburden).

(iii) **Middlings and rejects during washing of coal:**

Tailings are the waste products generated during the recovery of the washed coal. Middlings are by-products of the coal washing/ beneficiation process which is a fraction of feed (crushed ROM coal). This can be used for power generation and also used by brick manufacturing units, cement plants, industrial plants, etc.

Since the increased mechanization of opencast coal mining, the proportion of out-of-seam dilution (OSD) mining from the immediate roof and floor strata has increased and the dilution shall be separated during washing and will be rejected as waste.

Rejects are the products of coal beneficiation process after separation of clean and / or middling, as a fraction of feed raw coal. This may be used for road repairs, briquette (domestic fuel) making, land filling, etc.

(iv) **Fly ash and bottom ash during the burning of coal in boiler:** Fly ash is a by-product of power generation with coal. Ash that falls to the bottom of the boiler's combustion chamber is called bottom ash. Fly ash is generally captured by electrostatic precipitators **or other particle filtration equipment before the flue gases reach the chimneys. Depending upon the source and composition of the coal being burned, the components of fly ash vary considerably, but all fly ash includes substantial amounts of** oxides of silicon, aluminium, and calcium which are the main mineral components in coal-bearing strata. Fly ash management is itself a big subject in thermal power generation and hence further detailing has not been done.

6. Estimated volume of approximate waste removal from coal blocks allocated to NTPC

NTPC is a leading power generation company in India and as a backward integration it has entered into Coal Mining. NTPC currently has 10 coal blocks i.e. Pakri Barwadih, Chatti Bariatu (including Chatti Bariatu-South), Kerandari, Dulanga, Talaipalli, Bhalumuda, Banai, Mandakini-B, Banhardih and Badam Coal block with geological reserves of over 7.3 Billion Tonnes & production potential of about 113 Million Tonnes Per Annum.

NTPC has made substantial progress in coal mining and approx. 7.31 MMT coal was extracted from Pakri-Barwadih and Dulanga mines in the last financial year. NTPC has targeted to produce 10.4 Million MT of coal in this financial year. The estimated volume of waste (Overburden) removal from its own blocks is tabulated below. This topic assumes more importance considering great thrust on the environment, safety, and problems in land acquisition and R&R.

Coal Block	Approx Mineable reserve (Million te)	Approx Overburden (Solid waste) (M Cum)	Approx External Dump (m Cum)	External Dump height (m)	Approx Land Requirement (Ha)	Maximum Depth of Mine(m)
Pakri-barwadih	500	2100	600	90	900	300
Chhatti-Bariatu	150	260	10	30	38	150
Kerandari	140	620	90	120	370	400
Dulanga	150	400	40	60	110	250
Talaipalli	840	3700	260	90	Rehandling back into decoaled area after initial 9 Year onwards.	400
Banai	530 (GR)	Mining Plan not yet approved. Due to non-availability of land for dumping, it may be proposed to consider both the blocks as one and dump the OB of Banai on Bhalumuda and later external OB shall be rehandled back into void of Banai.				
Bhalumuda	680(GR)					

7. Waste management strategies

7.1 Waste avoidance

Waste avoidance is the first hierarchical step in reducing the amount of waste produced. e,g going for underground mining, underground coal gasification, etc

7.2 Waste separation

Waste, where practicable and taking into account health and safety issues, to be segregated and collected on-site at the point of generation. e.g Separation of top soil, carbonaceous material, sandstone, and other waste.

7.3 Use of waste

The appropriate management and storage of wastes will prevent on-site and off-site pollution and enhance opportunities for reuse like the use of sandstone in road making, embankment, sale of carbonaceous material, etc

7.4 Waste Disposal

Wastes generated to be disposed of, considering rock-mechanics, hydrogeology, geo-chemical properties, etc in a way that causes the least harm to the environment.

8. Issues in waste management and their mitigation

(i) **The requirement of non coal bearing area land for dumping Large Volume of Overburden**: As the depth of coal is increasing and coal reserve lying at shallow depth are depleting at a very fast rate. Stripping ratio of most of the captive coal blocks allocated recently are high hence requiring handling of a large volume of Overburden. Availability of non-coal bearing area land in the nearby area is prerequisite for external dumping. Availability of non-coal bearing area land in coal fields is not common and acquiring land, to accommodate a large volume of waste of a mine (which is having high stripping ratio) rated for higher production, is a major challenge. This issue may be addressed in the following way

(a) Planning the sequence of Mining in such manner that decoaled area is created at an early stage of mine so that internal dumping can be started.

(b) If decoaling at early stages is not possible, putting the Overburden on coal bearing area within coal block and rehandling the same at a later stage when decoaled area will be available. For that two adjacent blocks may be clubbed together to be treated as one coal block for Mining and Mine Planning purposes. In Talaipalli coal block of NTPC, Overburden during initial years shall be dumped at coal block and will be rehandled later. Waste management is the biggest challenge in Banai and Bhalumuda coal blocks of NTPC due to very high Stripping ratio (1:8) and non-availibility of non-coal bearing area land nearby. Mining of coal will only be possible if Banai and Bhalumuda are combined together as a single coal block and planning will be done for putting OB of Banai on Bhalumuda initially and the same will be rehandled later.

(c) Even if the land is available, there are safety concerns as the height and shape of OB dump to be designed in such manner as to ensure the stability under adverse conditions of the rainy season. The same is true about the stability of internal dumps. The solution is conducting a scientific study and ensuring the compliance of recommendations. More research is required in this area to find out other scientific ways to accommodate more Overburden in less space.

(d) Deciding the optimum depth of mining of coal by opencast method. Two of our coal blocks have been planned for mining of coal by opencast method up to the depth of approximately 400m. Proper consideration to the volume of waste generation must be given by estimating the volume of waste generation vis a vis Coal production for 50 m of depth increase.

(e) Deciding the technology level and method of mining: Underground method of mining requires very less amount of waste to be handled. China is the largest producer and consumer of coal in the world. China Produces more than 3300 Million te of coal per annum and more than 70% of coal production comes from underground method of mining which does not require removal of Overburden to extract coal thereby avoiding the issues involved in the handling of a large volume of Overburden. We should also adopt mechanisation and automation to make underground mining more economical and safe.

(ii) **Waste in the form of carbonaceous material:** Norms for reserve estimation may be changed and the scope of economic mining of Coal bands less than 1m in thickness to be explored. Carbonaceous material (norms as mentioned above) when dumped as part of Overburden (waste), become the source of spontaneous heating and fire in OB dump which is a danger to the safety of persons and challenge to the environment in form of air and water pollution. The study must be done to find out the washability of these carbonaceous material.

(iii) **Requirement of land for disposal of ash (fly ash+ bottom ash):** As Power generation will increase, which in turn, will increase the generation of Fly Ash which requires a large chunk of land for disposal. Environment friendly and economical use and consumption is the answer. NTPC's Ash is being used as Pozzolana in cement, cement mortar and concrete. The fly ash generated at NTPC stations is ideal for use in the manufacture of cement, concrete, concrete products, cellular concrete products, bricks/blocks/ tiles, etc.

9. Recommendations, suggestions and conclusion

(1) Coal Block should be allocated after making a conceptual plan and ensuring that internal dumping will start at early stages.

(2) Block boundary should be demarcated such that there is space of putting Overburden on non-coal bearing area nearby or on the block which shall be rehandled later.

(3) Gradual shift back to underground mining of thermal coal since economically mineable surface reserves are depleting.

(4) Use of Overburden in other areas is not very well researched. Conventional wisdom suggests that research should be conducted to explore the use of waste materials in the construction sector, road construction, embankment construction, filling, etc.

(5) Selective mining shall be practiced and coal band of thickness less than 1m should be accounted in reserve.

(6) Wash ability study of carbonaceous materials such as shale shall be conducted and if found feasible, washing shall be done to extract the valuable source of energy.

10. References

[1] Chugh, Y P and Behum, Paul T (2014), " Coal waste management practices in the USA: an overview", Int J Coal Sci Technol(2014).

[2] Information available at website of Ministry of Coal (www.coal.nic.in)

[3] Annual Report of Coal India Limited available at website www.coalindia.nic.in.

[4] Information available at website of NTPC Limited(www.ntpc.co.in)

[5] BHP Billiton Waste Management (https://www.bhp.com › coal › bhp-billiton-mitsubishi-alliance › red-hill)

34

Development and creation of environmentally friendly technology and equipment for the processing of coal preparation waste

V. I. Murko[1], V. A. Khyamyalyainen[2], M. A. Volkov[3], M. P. Baranova[4]

[1]Kuzbass State Technical University, CJSC Scientific Production Enterprise «Sibecotechnika», Novokuznetsk, Russia
[2]Kuzbass State Technical University, Kuzbass, Russia
[3]JSC «SUEK Kuzbas», Leninsk-Kuznetsky, Russia
[4]FSBEI "Krasnoyarsk SAU", Krasnoyarsk, Russia

Abstract: A feature of modern coal preparation plants (CPP) of Russia is the use of closed water slurry circuits (without a reset slurry waters outside the plant into external hydro-dumps) and lack of thermal drying of small classes of coal due to their effective mechanical dehydration. As a result, a significant amount (up to 10–12% of the volume of coal processing at the plant) of toxic fine-dispersed coal-enrichment waste (FDCEW) with a particle size of less than 0.5 mm, humidity of 30–45% and an ash content of 25–65% appeared on the output. This product is not in demand in the market, it is very difficult for processing and, as a rule, shipped outside the plant with a breed or separately by road to rock dumps or sludge storage sites. Taking into account the high toxicity of the FDCEW caused by the presence on the surface of flocculants and coagulants particles, which are carcinogenic and mutagenic, used on the CPP, storage of such wastes poses a high risk, which creates significant environmental problems in the region.

To solve the problem of using the FDCEW, a technology and a set of equipment for obtaining on their basis of suspended water coal fuel (WCF) and its effective combustion with capture and utilization of ash and slag wastes generated (ASW) was developed.

Keywords: Fine coal preparation waste, filter-cake, preparation, burning of suspension coal-water fuels

1. Introduction

A feature of technological schemes of modern coal preparation plants (CPP) of Russia is the use of the closed water-slime cycle that allows to liquidate dumping of slurry waters out of limits of plant in external settlers and hydrodumps, and lack of thermal drying of small classes of coal at the expense of their more effective mechanical dehydration in the precipitating-filtering centrifuges – decanters and vacuum filters of various designs. At the same time it was planned to solve two problems: decrease in cost of process of coal preparation and increase in its environmental friendliness. However, actually, as shows operating experience of CPP, at least one of these problems is not solved fully. In fact, at plants nonenriched coal slimes and waste of coal preparation with fineness of particles to 300 (1000) microns are condensed in radial thickeners, whose condensed product goes to tape or chamber filter presses for dehydration. For an intensification of processes of condensation, clarification or filtering flocculants of anion and cationic types, whose expense makes up to, 460 g/t of a solid phase are used. According to this process the considerable share of the dissolved flocculants contains in a liquid phase and on a surface of firm particles of a deposit (filter-cake) of a filter press. Calculations show that the filter-cake contains in 1 ton up to 300 g of extremely toxic flocculants the withdrawal of which outside the factories is invalid.

At the same time, at present filter-cake with a particle size of 0–1000 microns with a content of class 0–100 microns up to 90%, humidity 30–45% and ash content 23–62% cannot be added to the product, is not used as an independent product and goes to the dump together with a large breed. As a result, the environment is significantly polluted and a significant proportion of processed coal is lost (up to 10–12%). In addition, the circulating water is saturated with residual flocculants that disrupt the coal preparation process [1–5].

To solve this problem, SUEK-Kuzbass, LC, decided to create a pilot sample of a technological complex for processing fine-dispersed coal preparation waste, by preparing and burning suspension coal fuel based on the filter-cake of enrichment plants of SUEK-Kuzbass, LC.

This decision was based on the conducted research on the preparation and combustion of coal suspension fuel, obtained on the basis of fine coal preparation waste made on the experimental stand of Kuzbass State Technical University [5–7]. This paper presents the results of completed studies and shows the prospects for solving the existing problem. Thus, the aim of the work was to prepare the initial data for the creation of a pilot sample of the technological complex.

Objectives of the study:

- determination of the possibility of preparing a suspension of coal–water fuel based on the FDCEW of CPP SUEK-Kuzbass LC with the choice of the optimal composition of the complex plasticizer and equipment;
- determination of the possibility of burning obtained experimental batches of WCF at the experimental stand of KuzSTU;

- determination of the composition and amount of harmful emissions in the flue gases generated during the combustion of fuel-containing fuel;
- assessment of the prospects for utilization of industrial volumes of FDCEW received at the CPP SUEK-Kuzbass, LC at the nearby Belovskaya SDPP.

2. Characteristics of raw materials

To study the possibility of preparing suspension coal–water fuel based on fine coal preparation waste from the CPP «Komsomolets» and «named after S.M. Kirov» mines were delivered filter cake samples weighing 2000 kg each. Delivered samples were analyzed in the coal chemical laboratory. The qualitative characteristics of the studied samples are given in Table 1.

Table 1: Sample characteristics

Indicator name	CPP (Komsomolets)		CPP (named after S.M. Kirov)	
	Sample no. 1	Sample no. 2	Sample no. 3	Sample no. 4
Total moisture (%)	35.2	34.8	40.3	40.8
Ash content (on dry fuel condition) (%)	30.7	26.8	32.8	48.4
Volatile yield (dry ash-free fuel) (%)	43.0	43.1	42.4	41.4
Sulfur common (on the dry state of the fuel) (%)	00.52	0.48	0.15	0.13
Higher calorific value (dry state of fuel) (MJ)	33.70	33.04	33.08	33.00
Lower calorific value of the working fuel (MJ)	13.75	14.3	11.8	8.7
Granulometric composition (mm)				
0.250–3.0	5.7	8.4	4.5	6.3
0.071–0.250	15.1	18.7	26.8	17.3
−0.071	79.2	72.9	68.7	76.4
Total	100.0	100.0	100.0	100.0

Analysis of the data in Table 1 showed that the humidity of the filter cake samples submitted for research was consistently high $W_t^r = 35.2$–40.8%. In contrast to humidity, the ash content of the filter-cake depends on the quality of the feedstock and changes simultaneously with it and can vary both in a narrow range of values (=30.7%; 26.8%) and in a wide range of deviations up to 15, 6% (=32.8%, 48.4%) of the filter cake with CPP «named after S.M. Kirov» mine. The particle size of the filter cake from CPP «Komsomolets» and «named after S.M. Kirov» mines included size classes up to 3.0 mm.

3. Research methodology for the preparation of suspension coal fuel

The resulting filter-cake is actually a semi-finished product for the production of water–coal fuel (WCF) with characteristics that allow it to be effectively burned in a boiler with a vortex furnace or co-burning with conventional fuel in pulverized coal-fired boiler units.

Evaluation of the possibility of preparing coal–water fuel from coal preparation waste (filter-cake of the «Komsomolets» and «named after S.M. Kirov» mines) and the selection of the optimal plasticizing additive were carried out under laboratory conditions on a universal vibrostand (IED).

In the process of research, laboratory samples of suspension coal fuel were prepared from a mixture of filter-cake and an aqueous solution of a plasticizer reagent. Taking into account the initial size of the filter-cake, for the preparation of samples of WCF, a mixing or grinding chamber of a periodic action of a universal vibrating stand was used. Dosing of the starting components was carried out in manual mode. The purpose of laboratory research was to select the optimal variant of the additive based on the analysis of the values of the main structural and rheological characteristics of the prepared WCF.

Samples of WCF were analyzed for the mass fraction of the solid phase, particle size distribution and viscosity. Static stability was determined by the presence of sediment and dewatering during storage of the sample under static conditions. The mass fraction of the solid phase was determined by the standard drying method according to GOST 27314-91, or according to GOST 11014-2001, the particle size distribution – by wet sieving on sieve of 0.355 mm; 0.250 mm and 0.071 mm according to GOST 2093-82, ash content according to GOST 11022-95. The lower calorific value was calculated using measured values of the mass fraction and ash content of the solid phase. Recalculation of the results of the analysis for different states of the fuel was carried out according to GOST 27313-95.

Viscosity measurements were carried out on a rotational viscometer "RHEOTEST" in the range of shear rates from 1.0 to 437.4 s^{-1} with a standard cylinder system S2. The measurement temperature was 20 ± 5°C [5–10].

Further, on the experimental stand (in semi-industrial conditions), the preparation of the pilot batches of WCF and their burning at the boiler facility of the experimental stand was carried out. At the same time, the composition and amount of harmful emissions generated during combustion were determined on the stand.

As a result of studies on the preparation of WCF on a shaker table in two ways: mixing and grinding determined the optimal type of plasticizer (high molecular weight inorganic compound), its consumption (0.3%) and the mass fraction of the solid phase in the prepared fuel (57.0%).

Then, on the basis of the obtained results, experimental batches of WCF were prepared on the experimental stand according to the technological scheme presented in Fig. 1.

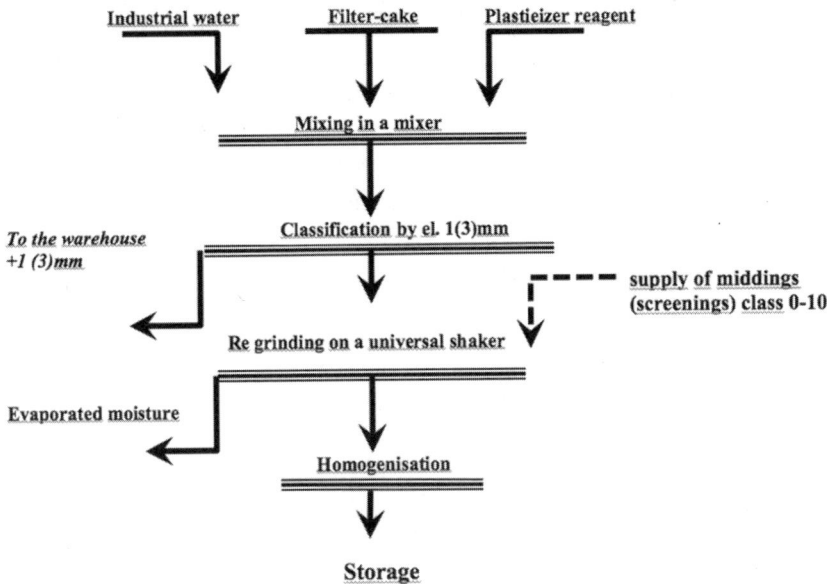

Figure 1: Technological scheme of the CWF preparation from the filter-cake

According to the technological scheme, the initial filter-cake and an aqueous solution of the plasticizer reagent were fed to a batch mixer, then the resulting coal–water slurry was dosed to a universal vibration unit, where regrinding and additional mixing of the resulting fuel occurred. The universal vibration unit is a bicamera vibration mill consisting of cylindrical chambers loaded with balls and interconnected by channels. The initial suspension enters the internal chamber of the vibromill, moves downwards and through the channels enters the external chamber. In the outer chamber, the movement of the suspension is carried up. Unloading the crushed material through the external threshold of the outer chamber. This principle of operation of the vibromill allows providing low energy consumption for grinding particles in the mixture at the required size of large particles in the finished fuel. Prepared WCF pumped into storage tanks [11].

To stabilize the quality of WCF by ash, the possibility of supplying a vibromill with a screening or crushed middlings with a particle size from 0 to 10 mm was provided.

The influence of mechanical activation on the structural and rheological characteristics of finely dispersed coal suspensions, obtained on the basis of filter-cakes in the created experimental sample of a special pump-activator, was determined. In the activator pump, WCF was processed under shear deformation in a gap of $0.2 \div 0.3$ mm two disks with a relative angular frequency of rotation of the movable disk 2900 rev/min. As a result, on the peripheral part of the disk, the shear rate reached $2100 \div 3160$ s^{-1}, i.e. the conditions corresponding to the utmost destruction of the structure were realized. Passing the resulting fuel through an activator pump made it possible to reduce the viscosity of the suspension by 30% and increase the static stability of the fuel.

Table 2 shows the structural–rheological and thermophysical indicators of pilot batches of WCF.

Table 2: Characteristics of the prepared experimental batches of WCF

Ash content A^d (%)	Output class over 0.25 mm (%)	Mass fraction of solid phase S_t (%)	Effective viscosity at shear rate 81 s^{-1}, η (MPa s)	Stability (day)	Net calorific value, Q_i^r (MJ/kg)
CPP filter-cake of the «Komsomolets» mine					
26.8	1.9	56.9	178	15	12.22
CPP filter-cake of the «named after S.M. Kirov» mine					
48.4	1.4	56.6	148	15	8.24

It was established experimentally that the performance of the universal unit for the initial suspension varies in the range of 0.155–0.217 t/h, depending on the particle size of the initial suspension. In this mode of operation of the plant, the yield of class +0.250 mm in the finished suspension did not exceed the limit required by the combustion conditions (R250 ≤ 5%) and was 1.4–1.9%.

Considering that in the initial product – filter-cake, the content of micron size classes, as a rule, more than 70% (Table 1), it is advisable to install a rod mill with a grinding mill in the regrinding operation in the pilot process complex. The use of a core-grinding load in a drum mill makes it possible to obtain a more uniform particle size of the solid phase, and the presence of micron particles in the initial product contributes even more to this fact. Thus, it becomes possible to ensure that the particle size distribution of the solid phase particles is close to bimodal in the finished ground product. As a result, the solid content in the finished fuel increases by 2–3% while maintaining the structural and rheological characteristics, which, in turn, leads to an increase in the heat of combustion of the fuel. To implement this condition, the design of a rod mill (Fig. 2) was developed, the principle of operation of which is similar to the mechanism of operation of the test tube bicameral ball mill.

The vibrating mill consists of a body 1, divided by inserts 2 into a central 3 and peripheral 4 cavities, which are hydraulically interconnected by slots 5 formed between the lower edges of the inserts and the bottom 6 of the body. On the top cover 7 of the body installed feed hopper 8. The cavity is filled with grinding medium 9 (rods, balls). The design of rectangular cavities allows the use of rods as a grinding medium. Outside the peripheral cavities on their outer side walls 10 are fixed drain chute 11, connected to the drain nozzles 12. The body is mounted on the frame 13 through the springs 14. On the bottom of the body the exciter 15 is fixed.

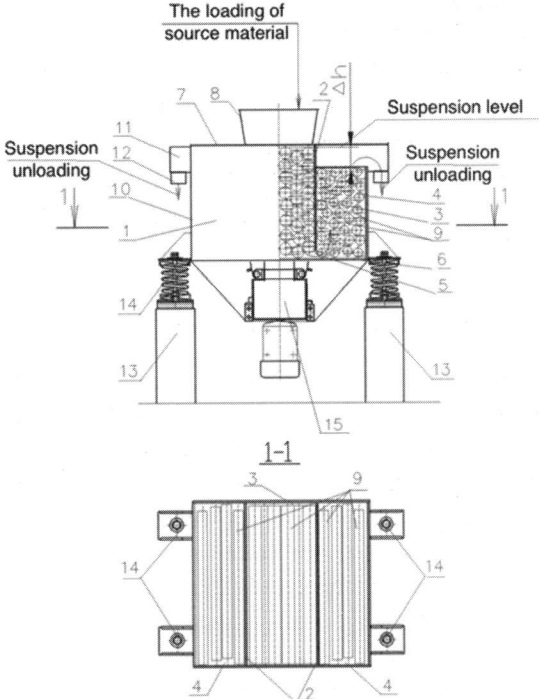

Figure 2: The design of the developed vibrating mill

Vibrating mill works as follows. The source material and the liquid phase (water) through the feed funnel 8 are fed into the central cavity 3 of the vibrating mill chamber, filled with a grinding medium (rods or balls). Due to the impact of the vibrating grinding medium provided by the vibration exciter 15, the incoming products are mixed and the wet grinding of particles of solid material is carried out. The resulting slurry through the slots, equipped in inserts 2, enters the outer peripheral cavities 4 of the chamber, also filled with a grinding medium (rods or balls), under the vibration action of which the wet grinding of solid particles continues. At the same time, the movement of the grinding medium is carried out from the bottom up with a speed substantially lower than in the central cavity, which ensures high grinding efficiency with less energy consumption. At the same time, the speed of movement of the slurry in the cavities is regulated by changing the difference in the height of the slurry in the central and peripheral cavities by changing the height of the outer sidewalls of the peripheral cavities. Since the size of solid particles in the internal and external cavities is different, the size of the grinding elements in the external cavities is smaller than in the internal. The crushed product from the outer peripheral cavities through the inner wall of the drain grooves 11 through the pipes 12 is discharged into the receiving tank (not shown in the figure). Thus, due to the constructive execution of the camera according to the proposed embodiment, it is possible to load both balls and rods as grinding elements. This ensures the implementation of a more efficient method of wet grinding of the material in a vibromill with the possibility of obtaining a particle size, close to bimodal, during the processing of a mixture of materials of different size [11].

4. Burning of WCF on an experimental stand

Combustion of pilot batch of WCF was carried out at the boiler plant, consisting of a boiler with a thermal capacity of 0.63 MW, fuel supply system, ash collection system, heat removal heater, and equipment. The created boiler consists of a firebox – a vortex combustion chamber located in a water-cooled case and an economizer for removing heat from the hot flue gases formed in the firebox. The WCF is fed into the vortex furnace through a burner with a pneumatic mechanical nozzle tangentially the inner cylindrical surface of the combustion chamber. Also tangentially blown air is fed into the combustion chamber. Spray fuel produced compressed air supplied to the nozzle. The fuel supply is regulated by varying the engine speed of the fuel pump. The combustion chamber of the firebox is equipped with a water-cooled pinch, which makes it possible to hold burning coal particles and sprayed WCF droplets the necessary time for their complete burning out.

The dust collection system is two-step, and consists of a battery cyclone unit and a fabric filter that provides a high degree of purification of flue gases from dust. Table 3 presents the results of burning pilot lots of WCF.

Table 3: The results of burning pilot batches of coal–water fuel

Parameter	The numerical value for CPP mine	
	«named after S.M. Kirov»	«Komsomolets»
Boiler heating capacity (MW)	0.47–0.57	0.52-0.66
Temperature condition in a fire chamber (°C)	980–1050	980–1050
WCF consumption (kg/h)	220–250	170–210
Pressure WCF (MPa)	0.20	0.19
Compressed air pressure (MPa)	0.21	0.20
Flue gas temperature (°C)	250–260	250–260
Efficiency (%)	0.83	0.85

5. Determination of the composition and amount of harmful emissions in the flue gases generated during the combustion of fuel-containing fuel

For analysis on the chemical composition, as well as on the complete burnout, dust and waste ash were taken from the flue gases and the grate, located at the bottom of the combustion chamber, respectively. All batches of fuel differed little in the magnitude of underburning – for WCF from FDCEW CPP «named after S.M. Kirov» the content of combustible in the ash and slag waste dump (carbon in fly ash) was 4.8%, and for fuel CPP «Komsomolets» mine – 4.7%, i.e. the value of carbon in fly ash did not exceed 5.0%. During the tests, gas emissions were measured using the Testo 300 XXL gas analyzer [12–16]. The process of sampling to determine the composition and amount of harmful emissions in the flue gases generated during fuel combustion was carried out with a stable boiler.

The composition and amount of harmful emissions from the combustion of prepared fuel samples are given in Table 4. As the data in the tables show, the obtained values of harmful emissions are significantly less than permissible values for coal-fired boilers of such power when using high-ash fuel.

Table 4 : The composition and amount of harmful emissions in flue gases

Name indicator	Maximal permissible concentrationa	WCF of CPP «Komsomolets» mine	WCF of CPP «named after S.M. Kirov» mine
Dust (mg/m³)	250	No more than 170	No more than 200
CO (mg/m³)	375	No more than 75	No more than 75
NOx (mg/m³)	750	No more than 250	No more than 230
SO$_2$ (mg/m³)	1200	No more than 200	No more than 200
PAC (benzo[a] pyrene) (mg/m³)	0.1×10^{-3}	Less than 0.1×10^{-3}	

[a] Standards for specific emissions of particulate matter, carbon monoxide, nitrogen and sulfur oxides, benzo(a)pyrene (GOST P50831-95)

Based on the results obtained, a pilot technological complex was designed for using filter-cake, the parameters of which are presented in Table 5.

Table 5: Technical and economic indicators of the pilot technological complex

Usage	Heat energy production
Operation mode	All-day
Production capacity of the WCF preparation (t/h)	2.0
Heating capacity of the boiler site (MW t/h)	0.58
Fuel consumption (WCF), nominal (t/h)	0.2
Compressed air consumption, maximum (m³/min)	1.0

Table 6 presents the calculated technical and economic indicators for the partial replacement of WCF coal fuel at boilers PK-40 of the Belovskaya SDPP (the SDPP is located 40 km from the industrial sites of the «Komsomolets» and «named after S.M. Kirov» mines).

Table 6: Estimated technical and economic indicators for the partial replacement of coal in boilers PK-40 of the Belovskaya SDPP WCF on the basis of coal slimes

Annual demand WCF (thousand tons)	1000.0
The share of WCF in the fuel balance of the boiler unit (%)	Up to 25
The lowest heat of combustion of fuel (Gcal/ton)	2.8
The cost of WCF at SDPP (rubles/ton)	433
The economic effect of the replacement of 1 ton of coal on WCF (rubles/ton)	763
Cost reduction for 1 Gcal of thermal energy (rubles/Gcal)	144
Payback period (years)	No more than 1.5

6. Conclusion

The technology and equipment have been developed for the preparation and incineration of low- and medium-capacity boilers of suspension coal–water fuel obtained on the basis of fine coal preparation waste (filter-cakes) of the CPP «Komsomolets» and «named after S.M. Kirov» mines. It is shown that on the basis of these wastes, it is possible to prepare a suspension coal–water fuel with a solids content of 56–60%, with the required structural and rheological characteristics and a lower calorific value of up to 13 MJ/kg.

A boiler with a thermal capacity of 0.63 MW with a vortex combustion system, operating efficiently on fuel from coal enrichment wastes, was developed, manufactured and tested. The results of the boiler operation on this fuel showed its high efficiency (efficiency is 83–86%) with a level of harmful emissions in flue gases significantly lower than permissible values. High levels of carbon in fly ash and incomplete combustion of fuel were achieved (respectively, no more than 5% and 80 mg/m^3, which is significantly less than the permissible values).

According to the results of the research, a working draft has been developed for the creation of a pilot technological complex for the processing of coal preparation waste with the production of suspension water–coal fuel and its combustion at a boiler plant. An assessment was made of the use of the total volume of FDCEW CPP SUEK Kuzbass at the nearby Belovskaya SDPP, which showed high economic and environmental efficiency of the proposed project.

7. References

[1] E. I. Wan, M. D. Fraser, C. N. Logan, Proceedings of the IEA-CLM, 22, 00879 (1993).

[2] R. A. Ashworth, T. A. Melick, D. K. Morrison, J. J. Battista, Twenty Third International Technical Conference on Coal Utilization & Fuel Systems (Coal & Slurry Technology Association and ASME-FACT, Clearwater, 1998).

[3] J. D. Morrison, A. W. Scaroni, J. J. Battista, XIII International Coal Preparation Congress (ICPC, Brisbane, 2012).

[4] A. M. Musalam, A. Fattah, A. Qaraman, International Journal of Energy and Environmental Research, 4(3), 27 (2010).

[5] V. I. Murko, V. I. Karpenok, Yu. A. Senchurova, V. A. Khyamyalyainen, O. V. Tailakov, Coal in the 21st Century: Mining, Processing and Safety, 297 (2016).

[6] V. Murko, V. Hamalainen, E3S Web of Conferences, 21, 01029 (2016).

[7] M. Baranova, Energy and Resource-Saving Sources of Energy in Small Power Engineering of Siberia. Founder of the Second International Innovative Mining

Symposium (T. F. Gorbachev Kuzbass State Technical University, Kemerovo, 2017), P. 02001.

[8] V. I. Murko, V. I. Fedyaev, V. I. Karpenok, I. M. Zasypkin, Yu. A. Senchurova, A. Riesterer Investigation of the spraying mechanism and combustion of the suspended coal fuel. Thermal Science, T. 19, No. 1, pp. 243–251 (2015).

[9] V. I. Murko, V. I. Karpenok, V. I. Fedyaev, E. M. Puzyryov, M. P. Baranova, XVIII International Coal Preparation Congress Conference proceedings, 345 (2016).

[10] V. I. Murko, V. I. Fedyaev, H. L. Aynetdinov, M. P. Baranova, Environmentally Clean Technology of Fine Waste Coal Utilization (XVII International coal preparation congress, Istanbul, Turkey), (2013).

[11] Patent No. 2145038. M.cl. F 23 Q 5/00. Method of Combustion and Combustion Stabilization of the Water–Coal Fuel in the Settling Chamber (in Russian), No. 97120914/06. Appl. 03.12.97. Published on 27.01.2000, Bulletin No. 3.

[12] M. G. Prudhon, Boiler plant with a circulating liquid phase of 320 MWth, France unification project collieries SODELIF. Helsinki, Finlande (1993).

[13] V. Biletskyy, P. Sergeyev, O. Krut, Fundamentals of highly loaded coal–water slurries. Mining of Mineral Deposits, London (2013), pp. 105–113.

[14] A. Kijo-Kleczkowska, Analysis of cyclic combustion of the coal–water suspension Archives of Thermodynamics, 32(1), pp. 45–75 (2011).

[15] A. Kijo-Kleczkowska, Combustion of coal–water suspensions, Fuel, 90(2), pp. 865–877 (2011).

[16] V. Murko, V. Hamalainen, M. Baranova, Use of ash-and-slag wastes after burning of fine-dispersed coal-washing wastes: E3S Web of Conferences Electronic edition, (2018).

35

Oil contaminated soil remediation through oil/coal agglomeration

Talal Omar, Vincent De-Paul Osaji

Norman B. Keevil Institute of Mining Engineering, University of British Columbia, Vancouver, BC, Canada

Abstract: Large quantities of hydrocarbon/oil contaminated soils are produced as a result of various industrial and municipal activities such as active oilfields, refineries or urban gas station facilities. This contamination causes adverse effects to human and animal activities and needs to be remediated. Coal mining production produces large amounts of fine coal tailings as a result of coal mining and processing activities. The proposed agglomeration technique takes advantage of coal/oil agglomeration properties to remediate contaminated soil, while at the same time reducing the ash content of the fine waste coal. In this study, kerosene contaminated soils and coal flotation tailings are used to test the effectiveness of the proposed process at different operating conditions. The test conditions are tailored to promote coal agglomerates' growth while attempting to clean coal and soil at the same time. The remediation, or proposed separation process is tested at different conditions to determine it' effectiveness.

Keywords: Coal, oil, contamination, agglomeration, upgrading, cleaning

1. Introduction

Following over 200 years of industrialization, soil contamination is a widespread problem in Europe. The most frequent contaminants are heavy metals and mineral oil. According to estimates by the European Environment Agency (EEA), the number of sites where potential polluting activities have been carried out in the EU is approximately three million and, of these, an estimated 250,000 sites may need urgent remediation (European Commission, 2013). Soils are consistently contaminated by industrial human activities. Oil extraction, processing, transport, and consumption are major causes of oil/soil contamination. Tanker spills, pipeline leaks, and other forms of mishandled hydrocarbons and oils have adverse effects on soil quality, directly affecting surrounding flora and fauna (Mariana et al., 2017). Soil remediation is a major industry today, and currently relies on the use of different surfactant remediation techniques and bioremediation methods. Although some methods have proven very effective, the cost and ability to treat soil in remote locations has posed large challenges.

One of the industrial sectors with the largest impact on the environment is the oil and gas industry (Gossen and Velichkina, 2006). The proposed solution is the use of tailings coal to decontaminate the soil from hydrocarbons using coal/oil agglomeration. This process, if effective, will be able to create a favorable economic avenue for soil remediation, can be applied directly to the contaminated site, and can additionally help clean up the fine coal tailings. The adverse impacts to ecosystems and the long-term effects of environmental pollution by oil spillage call for an urgent need to develop remediation processes for cleaning up oil from oil impacted areas

Coal agglomeration depends on the differences between the wettability of clean coal and the mineral matter and ability of the agglomerating or binding agent, such as hydrocarbons and various types of oils to aggregate coal particles. Agglomeration is possible when the coal is preferentially wetted by oil, instead of water. This means that all coals are suitable for agglomeration unless some specific types which do not have enough carbonaceous matter in their content (Özer et al., 2017). Multiple factors affect the agglomeration process, such as the petrographic composition of the coal, chemical structure, rank, surface oxidation, as well as the concentration of the agglomerating agent. It is possible to enhance the coal surface using modification agents to improve the preferential wetting capability of the coal to improve the agglomeration.

Figure 1 shows a flowsheet of an oil agglomeration process that includes units: (a) slurry preparation and conditioning unit; (b) oil heating and emulsification unit; (c) oil agglomeration unit; (d) agglomerate recovery unit; and (e) agglomerate dewatering unit. It is also possible to recover the oil from the agglomerate by incorporating other units. In the conditioning vessel, the coal pulp is prepared. In the next unit, the agglomerating agent is dosed, and in turn allows the formation of coal agglomerates. The slurry is then mixed well to improve the probability of collisions, which will result in the growth of the agglomerates. The agglomerate is then recovered through skimming, wet screening, or other methods (Mehrotra et al., 1983). Dewatering of the agglomerates can be done by using sieve bends, centrifuging, filtration, or even stockpiling to allow water to drain Capes et al., 1977). The coal can then be used as is, or in some instances, can be pelletized.

The focus of this paper is to explore the cleaning of fine coal tailings and contaminated soil originating from petroleum or chemical manufacturing contamination, using the principle of coal/oil agglomeration. In the proposed process, coal slurry is mixed with contaminated soil (kerosene) at different concentrations to induce selective agglomeration of coal particles, while leaving the particles of soil dispersed. The agglomerates are wet-screened, leaving an oil free soil and producing clean coal. The proposed process should reduce the ash content of the coal and remediate the contaminated soil by releasing oil from the soil

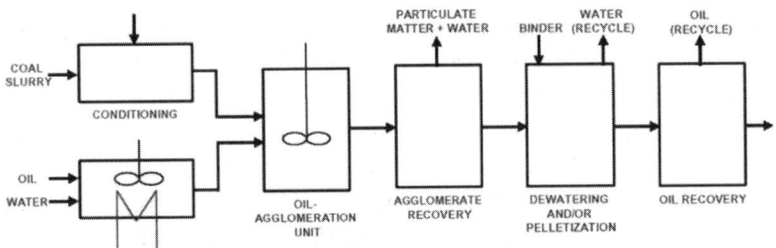

Figure 1: A typical flowsheet of an oil agglomeration process

2. Methodology

2.1 Experimental matrix

Different variables can be manipulated to obtain an understanding of the agglomeration process. The first controlled variable is the concentration of contaminant in the soil. Since the contamination is simulated, the effectiveness of the agglomeration process can be tested with respect to contaminant concentrations. The amount of coal used to treat the soil sample, the pulp density of the slurry, the mixing/conditioning time, and the mixing intensity can all be manipulated to obtain data to characterize the effectiveness of the process. In this study, the controlled variables are: soil contaminant wt% and ratio of coal to soil. The response is the ash content in the product in the form of agglomerates and a visual observation of the quality of agglomerates. For the scope of this study, the experimental matrix is presented in Table 1

Table 1: Experimental matrix

	Run 1	Run 2	Run 3	Run 4	Run 5	Run 6
Pulp density	10%					
Contaminant wt% in soil	30%		20%		10%	
Coal wt% wrt soil	30%	60%	30%	60%	30%	⁼60%

2.2 Experimental procedure and setup

2.2.1 Sample preparation

For the experiment itself, two separate components had to be prepared; the soil and the coal. Potting soil was acquired at a commercial outlet and was prepared by drying and then followed by screening. The soil was screened to −300 µm. This clean, dry soil was then artificially contaminated. Artificial contamination was selected since this method allows for contaminant concentration control. The soil was contaminated with kerosene, a common fuel oil. The contamination was carried out by spraying the kerosene and mixing the soil sample until homogenized. This was conducted on three different soil batches with different contaminant concentrations.

The coal samples used in this study were fine coal tailings obtained from processing coal samples by flotation in the laboratory. These were characterized by an average ash content of 44%. The target coal particle size was also −300 µm. The coal was prepared by drying for 24 h, followed by grinding, to achieve the target size and homogeneity in the sample. The sample was then screened and classified to obtain a top size of 300 µm. The sample was then riffle split to obtain a representative sample to testing, and a portion of that sample was tested using ASTM D7582-15 to determine the ash content.

2.2.2 Agglomeration procedure

The contaminated soil and coal samples were weighed to meet the ratio requirements for each of the experimental runs described in Table 1. The water coal slurry was subsequently prepared using the required amount of distilled water, and then the required amount of coal sample was added. The slurry was then mixed for 5 min at medium speed (1200 rpm). This step ensured the homogeneity of the slurry. The contaminated soil sample was then added to the prepared pulp. The blender was run at low speed (800 rpm) for 2 min. This promotes the formation of coal agglomerates. The blender speed is then raised to medium speed, and run for another 5 min. This stage promotes the growth of the coal agglomerates. Once this is complete, the slurry is then placed in a series of wet screens, with the coarsest being 300 µm. This facilitates the recovery of the formed agglomerates, as all the soil and coal samples had a passing size of 300 µm before processing, therefore any material remaining being 300 µm is considered to have been agglomerated during the process described above. Distilled wash water and a vibrating screen set up were used to recover the formed agglomerates. The samples were collected, filtered and then dried at 60°C for 24 h. Once dry, optical microscopy slides were prepared with the recovered agglomerates (coarse) and tailings (non-agglomerated, mostly fines) samples. The rest of the samples were bagged for ASTM D7582-15 ash analysis. A brief illustration of the experimental setup is shown in Fig. 2.

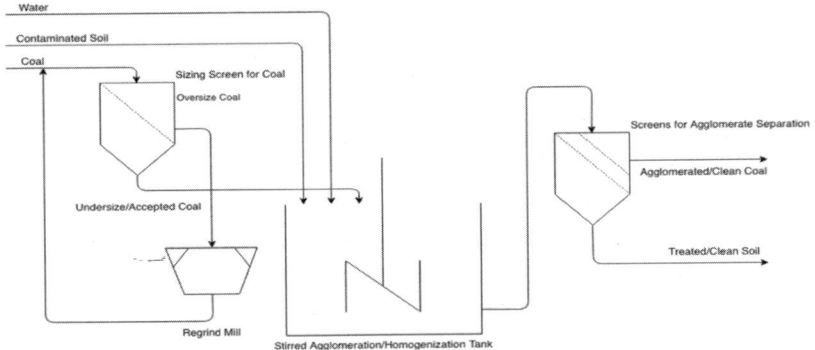

Figure 2: Experimental setup

3. Results and discussion

The soil and coal particles were collected and tested as control specimens before any treatment. The coal and soil samples were subjected to the ASTM D7582-15 procedure to obtain the ash content in all samples. The ash content in coal was found to be 44%, whilst the soil ash content was found to be 61% ash. The ash content in the soil is mostly a result of the presence of sand in the soil. These samples were also examined under the microscope, and the images captured are presented in Fig. 3. The difference in shape and color are quite obvious. Coal particles are more spherical in shape while soil particles have very assorted shapes, mostly being angular and fiber like.

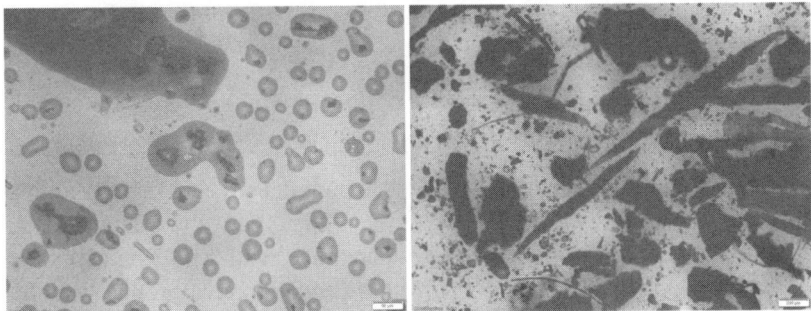

Figure 3: Coal (left) and soil (right) control sample under the microscope

Run 1 and Run 2 both had identical pulp density at 10%, and a contaminant concentration at 30 wt% in soil. The only difference was that Run 1 had less coal present at 30 wt% wrt to soil and Run 2 had 60% coal content wrt to soil. The ash content in the final agglomerated product was 45% for Run 1, and

15% for Run 2. This yielded very promising results, as it shows that there is significant ash reduction in overall ash content after the treatment for coal from Run 2 and some improvement in coal from Run 1 (calculated feed ash for Run1 was 55.9%; feed ash for Run 2 was 50.8%). From these two runs it can be concluded that with a higher content of coal in the slurry, the contaminant (oil) was released from the soil, allowing for more effective agglomeration and this also promoted cleaning of coal tailings.

This could also be due to the higher probability of collisions between coal particles and this allowed the agglomerates formation and growth. When compared to the control ash contents of 44% for coal and 61% for soil and their composite feed ash 55.9% and 50.8% for Run 1 and Run 2, respectively, it is evident that the clean coal product from agglomeration is indeed 'cleaner' than the ash in the feed samples. Through visual observation under the microscope, it was evident that the majority of the agglomerated particles found in the product were mostly coal, with some soil that was trapped within the agglomerates. Figure 4 shows a comparison between the agglomerated product and fine (non-agglomerated coal or soil) product. The presence of coal is evident.

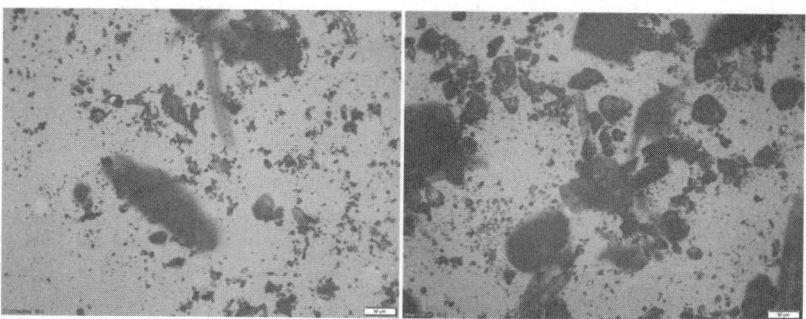

Figure 4: Run 1 agglomerated coal product (right) and fine product (left)

Run 3 and Run 4 also maintained the 10% pulp density, but the contaminant concentration was much lower, and it was 20%, with the amount of coal between Run 3 and Run 4 increasing, where once again Run 3 had 30% coal, and Run 4 had 60%. The ash content in the agglomerated product was 28% and 20% as compared to feed ash of 55.9% and 50.8% for Run 3 and Run 4, respectively. This falls in line with the trend observed earlier. However, when comparing Run 2 and Run 4, which had the higher amount of coal, it can be concluded that with a decrease in contaminant concentration, the effectiveness of agglomeration is reduced. An 8% difference in ash content in the agglomerated product is observed between the two. This can be explained by the lower amount of oil as a bridging agent which will in turn reduce the quantity of agglomerates formed. Similar observations to Run 1 and 2 were made when microscopic images were used to compare particles after the agglomeration treatment in Run 3 and 4.

In Run 5, the agglomerates and tailings had 22% and 73%, respectively, while the feed ash before the treatment was 55.9%, hence from the mass balance only 35% of the feed material was agglomerated while 66% remained in the tailings (non-agglomerated fines). From the optical microscopy observations, it was evident that most of the coal did not agglomerate and was found in the fines and hence passed through the screen. This can be explained by the fact that most of the coal with the majority of the inorganic components in the soil were in the tailings. Run 5 had the lowest amount of coal and contaminant oil of all the runs thus far. The lack of a sufficient amount of bridging liquid, combined with the reduced amount of coal particles, corroborates the previous conclusion that most of the coal remained in the tailings. Run 6 ash results were inconclusive due to the lack of samples for analysis, but it can be inferred, based on the results from Run 5 that the same trend would follow with slightly less ash content in the fines (non-agglomerated fraction).

4. Conclusion

Based on the obtained results, it can be concluded that with higher concentrations of contaminant oil, more successful coal agglomeration occurred. At lower concentrations, although some coal agglomerates were formed, the efficiency of agglomeration was poor. This can imply that only highly contaminated soils (at least when contamination is 20% by weight) could benefit from oil/agglomeration remediation treatment.

Runs 2 (30% contaminant; 60% coal; feed ash 50.8%), 3 (20% contaminant; 30% coal; feed ash 55.9%), and 4 (20% contaminant; 60% coal; feed ash 50.8%), have yielded the most promising agglomeration results. It was observed that the ash content in the agglomerated fraction for all three of those runs, and when compared to the feed ash, resulted in the significantly lower ash content for these three runs as 15%, 28%, 20%, respectively. Hence it can be concluded that the coal was successfully upgraded.Microscopy of the agglomerated and tailings fraction confirmed that coal was predominantly agglomerated, hence was found in the coarser fractions. It is important to note that even in the most promising runs (2, 3, 4), soil material was still recovered in the agglomerated product (Fig. 4, right), and a loss of some coal into the fine product/tailings was noticed (Fig. 4, left). This would indicate that only part of the contaminant oil was released from the soil. However, the fact that contaminant oil was recovered through the agglomeration treatment revealed a promising technique for industrial application but needs additional test work to be carried out to validate these initial results.

This study is considered to be a work-in-progress and further work is required to explore more conditions by manipulating conditioning and mixing times, as well as pulp density and the coal to soil ratio. The results were found to be promising and to have solid grounds for industrial application to remediate soil contaminated with hydrocarbons after further refining and optimization.

5. References

[1] E. Capes, C. et al., 1977. *Method of separating inorganic material from coal*, Patent US4033729 A.

[2] European Commission, 2013. Science for Environment Policy In-depth Report Soil Contamination: Impacts on Human Health, (5).

[3] Gossen, L.P. & Velichkina, L.M., 2006. Environmental problems of the oil-and-gas industry (Review). *Petroleum Chemistry*, 46(2), pp. 67–72. Available at: https://doi.org/10.1134/S0965544106020010.

[4] Mariana, D., Stoian, M.A. & Karademir, A., 2017. Crude Oil Contaminated Sites : Evaluation by Using Risk Assessment Approach. *Sustainability*, 9, 1365.

[5] Mehrotra, V.P., Sastry, K.V.S. & Morey, B.W., 1983. Review of oil agglomeration techniques for processing of fine coals. *International Journal of Mineral Processing*, 11(3), pp. 175–201.

[6] Özer, M., Basha, O.M. & Morsi, B., 2017. Coal-Agglomeration Processes: A Review. *International Journal of Coal Preparation and Utilization*, 37(3), pp. 131–167. Available at: http://dx.doi.org/10.1080/19392699.2016.1142443.

36

Beneficiation of coal washery tailings using conventional and column flotation techniques

U.S. Chattopadhyay[1], T. Gouri Charan[1], R. Venugopal[2], S.C. Maji[1],
Dhanjay Roy[1], Pradeep Kumar Singh[1]

[1]CSIR – Central Institute of Mining & Fuel Research, Dhanbad, India
[2]IIT-Indian School of Mines, Dhanbad, India

Abstract: Investigations were carried out using conventional and column flotation techniques to recover coking coal fines from the tailings generated at flotation plant of one of the operating washeries. A 400 kg/h conventional flotation and a 100 mm diameter bench scale column were used in these studies. Washabilitiy studies indicated that the tailings have the potential to yield 50–55% of clean coal at 16–17% ash level. Flotation studies in cell indicated that a product with 16–17% ash may be obtained at 45–50% yield from these tailings. The coal washery tailings were also tested using a 100 mm column flotation cell and the results of column flotation was compared with the standard flotation cell data. It was observed that the flotation column could recover the ultrafines present in the sample compared to that of the conventional cell. Most of the coal values being rejected presently may be recovered by using the combination of conventional and column flotation techniques. A suitable flow sheet was developed to recover additional clean coal from the tailings of the operating coal washery.

Keywords: LTGK, Pulp, Parameters, Tailings, Washability studies, Proximate, Lagoons, Flotation

1. Introduction

In India, the Central coal washeries with their present beneficiation circuits are producing about 1.6 Mt of coking coal fines per annum [1,2]. These slurries are not processed further in most of the washeries because of varied reasons. One of the reasons is that most of the above washeries were designed to wash coarser coals up to 0.5 mm only with Jig and heavy media circuits to produce metallurgical cleans at 17% ash level with the then available better quality upper seam coals. Thus in those days the coal fines below 0.5 mm were not processed and only it was mixed with clean coals produced from coarse coal washing circuits. With fast depletion of the good quality coals and deterioration in the quality of coal mined, the existing washeries suffer from severe technical constraints in terms of throughput and discharge capacities of individual washing units while treating the fines of 'difficult to wash' lower seam coals. As a result, a huge quantity of unwashed slurry is left as such and discharged as tailings to nearby ponds or lagoons, thereby polluting the surroundings.

The coal fines (below 0.5mm) having better coking propensity is a valuable prerequisite in preparing cleans for metallurgical coke. Though enriched in vitrinite content, the fines cannot be mixed directly with clean coal due to their high ash content (more than 25%) and of high percentage of silica content. The slurry needs to be washed through improved coal cleaning technology for enrichment of coking propensity and for utilization of total cleans as high valued component for coke making. Conventional flotation that uses mechanical cells has been known to be ineffective in processing fine coal mainly due to entrainment of fine gangue minerals in the froth that requires complex circuit arrangements that incorporate several cleaner stages [3]. CSIR – Central Institute of Mining & Fuel Research had developed an improved design of Mini Flotation Plants suitable to treat the inferior coal fines, highlighting the process technology and construction aspects with due considerations to the hydro-dynamics and geometric parameters of the cells [4].

Column flotation has been effective in cleaning fine coal as it has many advantages over conventional flotation owing to its ability to effectively reduce entrainment of fine gangue minerals due to less turbulence in the pulp, having a deep froth bed and using wash water to drain back the entrained gangue. In addition, column flotation is preferred due to its simpler construction, convenience in incorporating automatic control, as well as a having single-stage system which embodies rougher, cleaner, and scavenger [5,6]. Several studies also have emphasized that column flotation is superior to mechanical flotation in handling both coarse and fine fractions and gives a higher recovery with lower ash content [7–9].

Detailed pilot scale flotation studies and column flotation tests were carried out and the results are presented in this paper.

2. Experimental process

A total amount of 10 t samples were collected from settling pond of the operating washery. The collected coal slurry samples were initially air dried by spreading it on the floor. Representative samples were collected for conducting characterization tests with respect to proximate, CSN (crucible swelling number) and LTGK (low temperature Grey-King test), conventional flotation tests, and column flotation tests. The coal fines sample was subjected to screen analysis and the data is shown in Table 1. The washability studies of the coal fines were carried out and the data is shown in Table 2.

Table 1: Screen analysis

Size (mm)	Wt%	Ash%
0.5	3.5	19.1
0.5–0.25	22.5	25.5
0.25–0.125	23.8	29.5
0.125–0.063	18.5	35.1
−0.063	31.7	40.2
	100	32.7

Table 2: Washability data

Sp. Gr	Wt %	Ash %	Cumulative float		Cumulative sink		Ch. Wt %	Mayer's pt. value
			Wt %	Ash %	Wt %	Ash %		
<1.30	22.1	4.6	22.1	4.6	77.9	39.7	11.1	1.0
1.30–1.40	20.2	16.4	42.3	10.2	57.7	47.9	32.2	4.3
1.40–1.50	19.1	27.9	61.4	15.7	38.6	57.7	51.9	9.7
1.50–1.60	10.3	45.3	71.7	20.0	28.3	62.3	66.6	14.3
1.60–1.70	11.0	55.5	82.7	24.7	17.3	66.6	77.2	20.4
1.70–1.80	8.0	62.3	90.7	28.0	9.3	70.3	86.7	25.4
>1.80	9.3	70.2	100.0	31.9			95.4	32.0
	100.0							

2.1 Conventional Flotation tests

The systematic laboratory flotation tests on the coal fines was carried out earlier [10] and the studies showed that clean coal with appreciable quantity and quality may be recovered through physic chemical methods. Further, research initiatives in the institute led to the development of improved flotation plant for treating the inferior grade slurry [4]. The conventional pilot scale flotation plant as shown in Fig. 1 was used for conducting the tests. The coal slurry was first conditioned in the conditioner with diesel oil at the rate of 1.5 kg/t as collector reagent. The conditioned slurry was then fed to the bank of flotation cells where the frother MIBC (methylisobutylcarbinol) was added at the rate of 0.5 kg/t. The froth was collected as cleans and dewatered in the drum filter, while the tailings were sent to the tailings sump. Samples of clean coal from different flotation cells and combined tailings were collected at regular intervals and analyzed for its ash content and the data is shown in Table 3. The data is also depicted in Fig. 2.

Table 3: Pilot plant flotation data

Cell no	Concentrate	
	Wt%	Ash%
C1	28.5	13.8
C2	13.8	15.1
C3	8.1	17.9
C4	4.1	19.8
Total cleans	54.5	15.2
Tailings	45.5	53.8

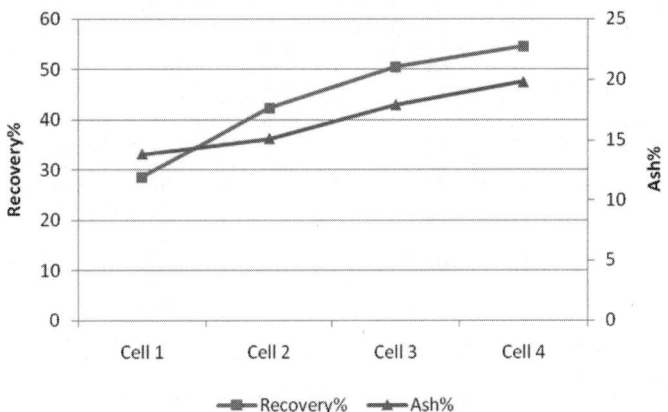

Figure 1: Cumulative recovery% versus ash% with number of cells

Figure 2: Pilot scale flotation cells

Figure 3: Column flotation test rig

2.2 Column flotation

The flotation column of 100 mm diameter, was used in the present study to know the effect of its operating parameters on the process. The schematic details of laboratory column were shown in Fig. 3. Electronically controlled metering pumps were used to feed and discharge the slurry. Slurry/froth interface was maintained using differential pressure transmitter (DPT). The output signal of the DP cell was looped with the stroke controller of the tailings pump so that the pumping rate could be automatically varied to maintain the interface level at a fixed froth depth. Sintered ceramic tube air sparger was used. Purge rotameter with a differential pressure regulator was used to control the flow of air to the column from a compressor. The column was filled with water and stabilized at required froth depth. The coal slurry at 10% solids by weight was conditioned initially with diesel (collector) in a conditioner for 3 min. In the second stage, MIBC (frother) was added and conditioned for another 3 min. This slurry was pumped to the column. The column was allowed to run for a minimum period of 3–4 residence times. Samples were drawn under near-steady state conditions. Both the process parameters and column operating conditions were recorded before collecting the samples. Samples were analyzed for ash content by standard method.

Table 4: Column flotation data

Sl. no.	Collector (kg/t)	Frother (kg/t)	Details	Wt%	Ash%
1	1.25	0.5	Conc.	45.7	14.1
			Tails	54.3	48.5
			Feed	100.0	32.8
2	1.5	0.5	Conc.	56.5	15.8
			Tails	43.5	54.5
			Feed	100.0	32.6
3	1.75	0.5	Conc.	59.3	17.5
			Tails	40.7	54.2
			Feed	100.0	32.4
4	1.5	0.25	Conc.	40.2	13.5
			Tails	59.8	45.8
			Feed	100.0	32.8
5	1.5	0.75	Conc.	61.5	19.1
			Tails	38.5	55.3
			Feed	100.0	33.0
6	1.5	1.0	Conc.	65.5	20.9
			Tails	34.5	56.5
			Feed	100.0	33.2

3. Results and discussion

Characterization tests of raw coal revealed that ash percentage of coal is 32.7% and moisture percentage (on as received) of the coal is 1.3%. The sample tested is medium volatile in nature, the VM% being 20.5.1. The coking propensities of the coal fines seem to be good in quality. The swelling index (CSN) is 3.0 which may be termed as good coking coal and LTGK is C. The petrographic analysis shows the vitrinite content is 39.8% and the inertinite is 29.2%. The reflectance of the coal fines is 1.0% which indicates the maturity of coal.

The screen analysis on the coal fines collected from washery was carried out in the laboratory at sizes 500, 250, 125, and 63 micron. It may be observed from the screen analysis data that the quantity of ultrafines (−63 micron) is 31.7% and ash content being 40.2%. It may be observed from Table 2 that more than 40% material is lying in less than 1.40 specific gravity fraction and the theoretical yield at 15.7% ash content is 61.4%.

The conventional flotation pilot plant data as shown in Table 3 shows that more than 40% of the clean coal is reporting from the cell 1 and cell 2, while the overall recovery was observed to be 54.5% at 15.2% ash content and the total tailings is 45.5% and the ash content being 53.8%. The cumulative recovery% versus ash% with number of cells showed that the maximum clean coal particles were recovered from the 1st and 2nd cells.

In the column flotation studies it could be observed that as the diesel dosage is increased from 1.25 kg/t to 1.75 kg/t, the yield of the froth increased from 45.7% to 59.3% with ash in it increasing from 14.1% to 17.5%. This could be attributed to the non-selectivity of diesel at higher dosages and physical entrainment/entrapment of ash forming minerals in the froth resulting in higher yield and ash content in the froth. The dosage of frother plays vital role as it affects the bubble size distribution and mean bubble size. In flotation, the bubble is the driving force and it determines the separation efficiency. The flotation results at various MIBC dosages and at a constant diesel dosage are presented in Table 4. The results clearly indicate that MIBC plays very important role in coal flotation. At constant diesel dosage, flotation concentrate yield increased as the MIBC dosage increased correspondingly the ash content also increased.

4. Acknowledgement

Authors are thankful to the Director, Central Institute of Mining & Fuel Research, Dhanbad for giving permission to publish the paper. The authors are also thankful to all the staff members of Coal Preparation Division, CIMFR (Digwadih Campus) for their kind support.

5. References

[1] Sen K. et.al., Abatement of coal fines pollution by the concept of slurry flotation plants, National Seminar on Mining & Environment (ME-98): Seminar: Vol 1, IICM, Ranchi, Bihar (India).

[2] Sinha K.M.K., Chattopadhay U.S., and Sen. K., Recycling and treatment of settling pond fines – A challenge to industry and environment, International Symposium: Processing of Fines (2), Nov. 2000, National Metallurgical Institute, Jamshedpur, India: 831007, page no. 416–422.

[3] Sastri S.R.S., Reddy P.S.R., Bhattacharyya K.K., Kumar S.G., and K.S. Narasimhan, Recovery of coal fines using column flotation, Minerals Engineering 1 (1988) 359–363.

[4] Kumar A., and Sen K., Design and construction aspects of mini coal flotation plant, International Symposium: Processing of Fines (2), Nov. 2000, National Metallurgical Institute, Jamshedpur, India: 831007, page no. 388–402.

[5] Ityokumbul M.T., Salama A.I.A., and Taweel A.M.A., Estimation of bubble size in flotation columns, Minerals Engineering 8 (1995) 77–89.

[6] Altun N.E., Xiao C., and Hwang J.Y., Separation of unburned carbon from fly ash using a concurrent flotation column, Fuel Processing Technology 90 (2009) 1464–1470.

[7] Hacifazlioglu H., and Sutcu H., Optimization of some parameters in column flotation and a comparison of conventional cell and column cell in terms of flotation performance, Journal of the Chinese Institute of Chemical Engineers 38 (2007) 287–293.

[8] Barraza J., Portilla A., and Piñeres J., Direct liquefaction of vitrinite concentrates obtained by column flotation, Fuel Processing Technology 92 (2011) 776–779.

[9] Jena M.S., Biswal S.K., Das S.P., and Reddy P.S.R., Comparative study of the performance of conventional and column flotation when treating coking coal fines, Fuel Processing Technology 89 (2008) 1409–1415.

[10] Chattopadhyay U.S., Kalyani V.K., Venugopal R., and Gouri Charan T., Application of Response Surface Methodology in Effective Recovery of Settling Pond Coal Fines by Froth Flotation, International Journal of Coal Preparation and Utilization, 35:4, 206–215, DOI:1080/19392699.2015.1011326.

37

Discard waste amelioration approaches at Greenside colliery

David Power[1], Christian Swanepoel[1], Michael Phali, Maynard Lombard

Anglo American Coal, Johannesburg, South Africa
Greenside Colliery, Emalahleni, South Africa
Khwezela Colliery, Emalahleni, South Africa

Abstract: Greenside Colliery is a South African coal mining operation that has been in operation since the 1940s; it is located close to Emalahleni (formerly known as Witbank) and has at various stages in its lifetime mined and processed the Witbank Nos. 1, 2, 4 and 5 seams. In the past, the plant has had a smorgasbord of unit operations in terms of baths, cyclones, spiral and fines dense medium processing to name but a few to its current configuration of cyclone, 3-product cyclone and flotation methodology for addressing the processing requirements of the coal. The refuse generated from these processes was in the form of a typical South African co-disposal method; i.e. a compacted discard encapsulating a slimes dam which requires careful management of the mass balance in terms of total refuse generation and the management of the discard/slimes ratio in the Mineral Residue Deposit (MRD).

Late in the first decade of this century, it was identified that there was a looming constraint in terms of placement of the final refuse. A simple solution (if available) would be to expand the facility but as in most such conundrums, this was not necessarily an easy first choice; although it was a consideration in the planning process until 2018. The eventual solution(s) adopted did not start out as an integrated approach but as a series of abatement measures but when eventually the momentum gained, the approach became an integrated plan managed by the coal processing team at Greenside. The resultant solutions presented in the paper focus on some of the processing initiatives which include the implementation of a flotation and tailings filtration solution, conversion from a conventional dense medium cyclone to a 3-product cyclone operation and discard retreatment.

Keywords: Amelioration, discards, slimes, processing, solution, MRD

1. Introduction

Anglo American acquired Greenside Colliery through a sale agreement with Goldfields in 1998. The plant had washed a combination of Witbank No 1, 2 and 4 seams in several dense medium module plants with spiral concentrators. At one stage a fine dense medium module was in operation during Goldfields tenure, but this was discontinued. In the final stages of their tenure, Goldfields installed a modular design plant (3 modules) to treat Witbank 4 Seam Coal – known as the 4 Seam Plant. On acquiring the mine, Anglo American quickly decided to rationalise operations and focussed on the mining and treatment of the Witbank No 4 Seam through the DRA designed modular plant. In 2000, a decision was made to open a five seam reserve and to commission a purpose-built plant to supply a sized metallurgical coal product to Highveld Steel (then an Anglo American affiliate), this plant is known as the 5 Seam Plant. In 2004, a flotation module was added to the 4 Seam Plant based on the Anglo Multicell technology utilising solid bowl centrifuges for product recovery. The decision making process for this configuration was faulty as it was based on work carried out at another operation which did not support this configuration based on the centrifuge technology available to Anglo American at that time. This resulted in very poor reconciliation of flotation recovery, which was a problem that bedevilled such installations in the Witbank Coalfield over the past 30 years – there have only been three sustainably successful flotation operations there to date.

Towards the end of the 2000s, Greenside had to face yet another challenge regarding the airspace for the placement of discard and slimes. The discard and slimes are placed in a facility constructed with discard encapsulating an inner dam of slimes. In this case the facility is limited by the N12 highway which is the main artery between South Africa and Mozambique as shown in Figs. 1 and 2.

Figure 1: Aerial view of the Greenside co-disposal facility

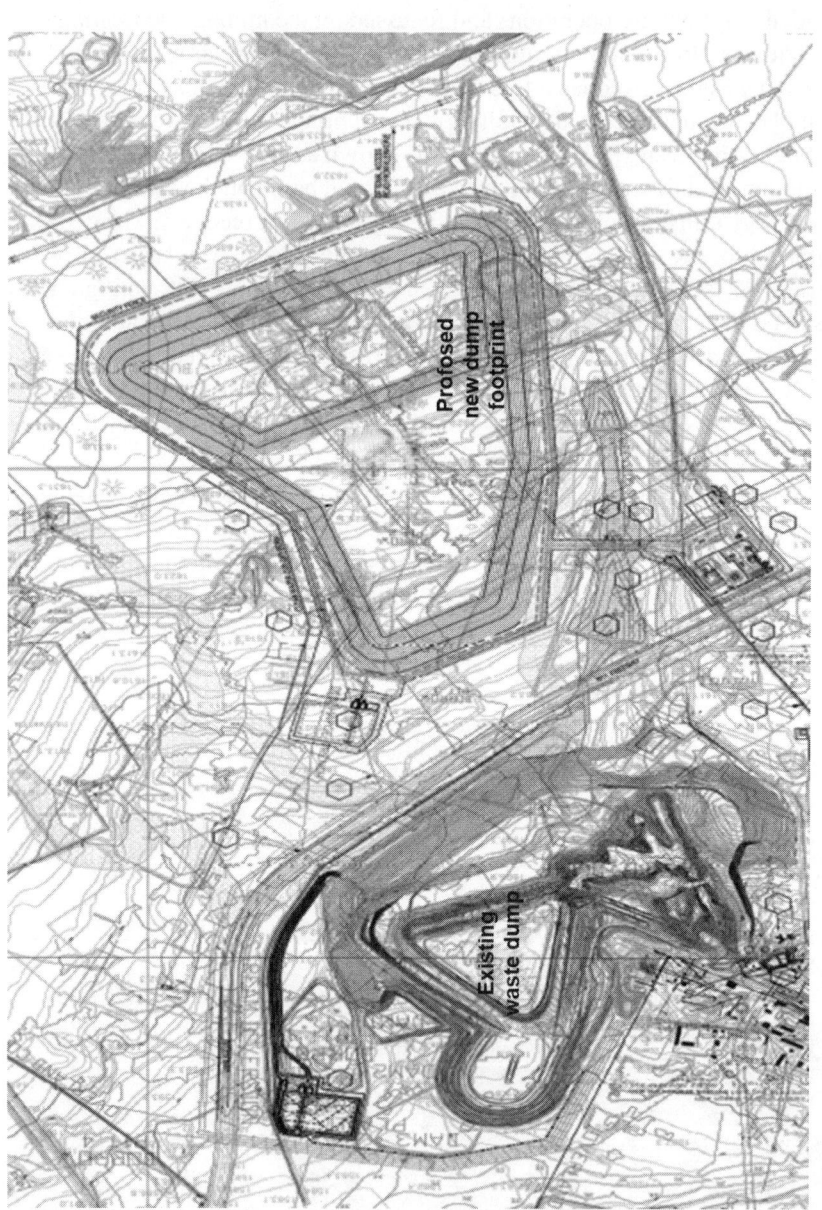

Figure 2: Plan view of existing facility with proposed location for new facility

The practicality of installing such a new facility on the other side of the highway was rather daunting but not impossible; there is a service bridge over the highway which could have served as a conduit for a conveyor, slimes line or a trucking option. The expense and likely licencing challenges led to other solutions being investigated.

2. Amelioration solutions

The following illustrates some of the processing solutions devised by the Greenside team over the course of several years from 2012 onwards to address the issue:

- Reconfiguration of the flotation circuit with tailings filtration which eventually led to the elimination of a wet tailings dam.
- Installation of 3-product cyclones in the 4 Seam Plant.
- Re-configuration of the 5 Seam Plant to enable the small-scale mining and retreatment of the discard dump.
- Large scale mining of the discard dump.

Mass balancing techniques at the Anglo American mines is based on the methodology shown by (Power, 2006) to determine the slimes discard ratio which enabled forecasts to be developed to understand the criticality of the airspace constraint. This guided the prioritisation of the project pipeline.

2.1 Reconfiguration of the flotation plant with tailings filtration

As earlier described, the initial layout of the flotation plant had many flaws both with the technology (original Multicell) and the configuration of the ancillary equipment. A decision was taken with mine personnel to go back to the drawing board and address the shortfalls. The Multicell technology is an Anglo American concept described by (Esterhuizen, 2008). An upgrade of the cell shown in Fig. 3 designed by Enprotec which greatly enhanced the contact effectiveness of particle–bubble interface was chosen.

In partnership with Enprotec a cost-effective filtration solution was investigated in China and a decision was made to utilise the plate and frame technology manufactured by Jinjing which is capable of filtering high throughputs of material with a fine particle size distribution (ca. 60% – 44 micron). The plant was designed to recover coal with a calorific value of ca. 27–27.2 MJ/kg with a recovery of ca. 50–60% and this has been attained consistently since commissioning. The filtration plant enabled proper mass balancing to be carried out and it could be clearly demonstrated on a techno-commercial basis for capital closeout purposes and a clear reduction of refuse material being delivered to the dump.

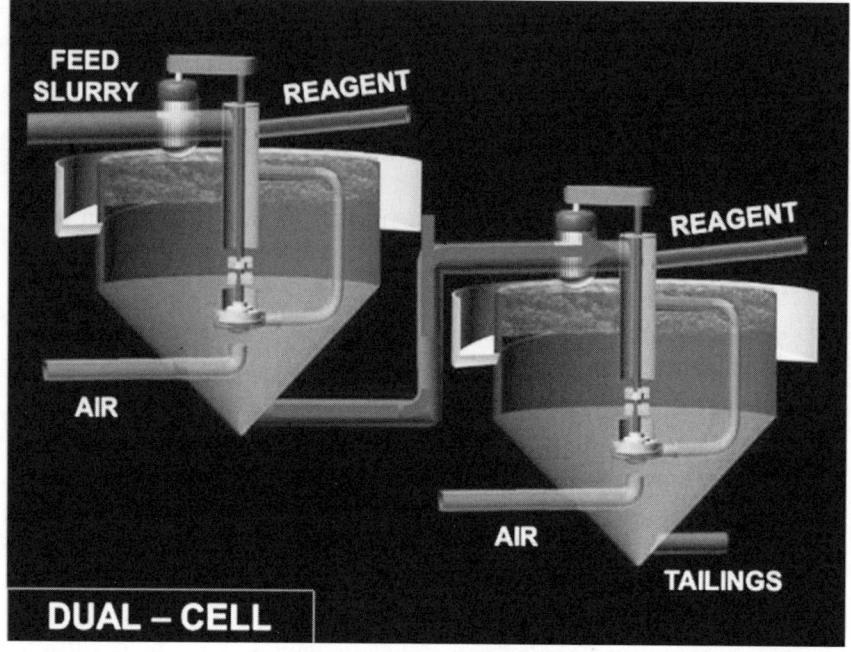

Figure 3: Layout of dual cell flotation cell

After this configuration had proven successful, a decision was made to close the water circuitry of both 4 and 5 Seam plants with the installation of a tailings filtration plant using the Jinjing technology. A complete dry mass balance was enabled following this installation which provided more certainty to the life of mine forecasts for the operation.

2.2 Installation of 3-product cyclones

The 3-product cyclone has been utilised extensively in the Chinese coal industry and its application in South Africa described by (de Korte, 2014) in terms to the amenability for the technology to treat a typical Witbank No. 4 Seam coal as shown in Table 1.

Table 1: Typical washability of a 4 Seam reject washed at a specific gravity of 1.50

Float	Mass (kg)	Yield (%)	Cum yield (%)	IM (%)	Cum IM (%)	Vol (%)	Cum vol (%)	Ash (%)	Cum ash (%)	FC (%)	CV (MJ/kg)	Cum CV (MJ/kg)	S (%)	Cum S (%)
F@ – 1.300	0	0	0	*		*		*		*	*		*	
1.300–1.350	0	0	0	*		*		*		*	*		*	
1.350–1.400	0.141	0.35	0.35	2.8	2.8	27.4	27.4	9.5	9.5	60.4	29.6	29.6	0.57	0.57
1.400–1.450	0.16	0.4	0.75	2.9	2.8	23.2	25.2	12.4	11	61.5	28.28	28.9	0.48	0.52
1.450–1.500	0.382	0.95	1.7	2.9	2.9	21.5	23.1	17.7	14.7	58	26.16	27.36	0.67	0.6
1.500–1.550	1.155	2.88	4.58	3	2.9	21	21.8	19.7	17.9	56.3	24.77	25.73	0.44	0.5
1.550–1.600	2.989	7.45	12.03	2.8	2.8	21.2	21.4	24.2	21.8	51.9	23.1	24.1	0.62	0.58
1.600–1.650	4.517	11.26	23.28	2.6	2.7	21.4	21.4	28.1	24.9	47.8	21.18	22.69	0.98	0.77
1.650–1.700	4.01	9.99	33.28	2.3	2.6	20.3	21.1	33.4	27.4	44	19.35	21.69	0.79	0.78
1.700–1.750	3.152	7.85	41.13	2.3	2.6	19.8	20.8	35.4	29	42.5	18.77	21.13	1.76	0.97
1.750–1.800	2.571	6.41	47.54	2.1	2.5	19.4	20.6	40.3	30.5	38.2	17.24	20.6	2.63	1.19
1.800–1.900	5.355	13.34	60.88	2.2	2.4	19.6	20.4	43.2	33.3	35	15.1	19.4	3.65	1.73
1.900–2.000	2.329	5.8	66.68	2.1	2.4	20.6	20.4	44.8	34.3	32.6	13.63	18.9	5.65	2.07
2.000 – @S	13.371	33.32	100	1.3	2	18.6	19.8	60.4	43	19.7	9.17	15.65	9.97	4.7

For this washability, at a target CV of 27.2 MJ/kg there is little material left to be recovered but if one peruses the data at a cut of 1.70, a middling product of ca. 21.5 MJ/kg with a sulphur less than 1.0% is recoverable. The 3-product cyclone as shown in Fig. 4 at Umlalazi had proven very successful in this application on 4 Seam Coals and based on this work there, a decision was taken to retrofit the units to replace the existing DSM cyclones in the 4 Seam Plant.

Figure 4: 3-Product cyclone installation at Umlalazi

The principle of the operation relies on the pressure differential arising within the barrel unit of the first stage being the driving force for the second stage cyclone. In the second stage, a separation takes place at approximately 20 points higher than the first stage separation with absolutely no control of this cutpoint and resultant quality of the middlings product (it must be noted that subsequently the control algorithm has been adjusted to allow for dilution water to be added at the second stage to enable some improvement of product quality by impacting the cutpoint on the second stage although at a cost of a higher magnetite consumption). The Umlalazi installation was also a retrofit and provided learning experience for the Greenside installation in terms of design (mass flow, medium flow, screen capacity and wear characteristic due to the abrasive nature of South African coals). In Fig. 5, the layout for two of the modules is shown and the flows from the second stage to the re-configured sinks screen which is split to allow for recovery of middlings and final discards on the one unit.

A further complication for this intervention concerned the placement of the resultant middlings product within the environs of the plant as there was no real estate to install a separate stockpile. A novel approach was to fluidise this product and pump the mix across the plant complex to an area adjacent to the 5 Seam Plant where a new screen was installed. The effluent is classified (underflow to a dewatering screen and overflow to a thickener) to limit recirculating fines loading. This route is approximately 800 m long (with a minimum velocity of 2 m/s to prevent settlement in the line) and was designed by Paterson and Cooke, the pipeline layout is illustrated in Fig. 6 (du Toit, 2012).

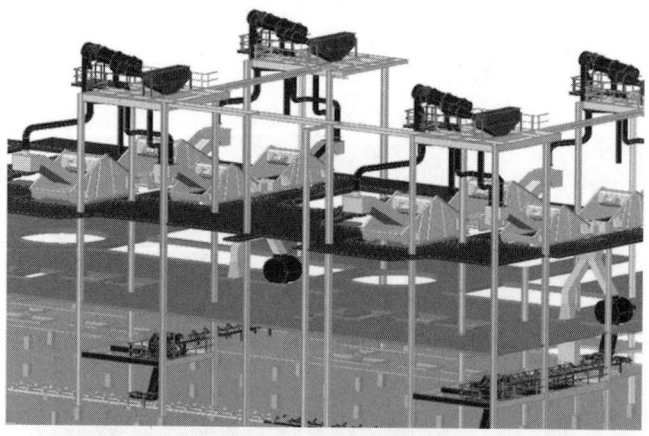

Figure 5: Retrofit of 3-product cyclone at 4 Seam Plant

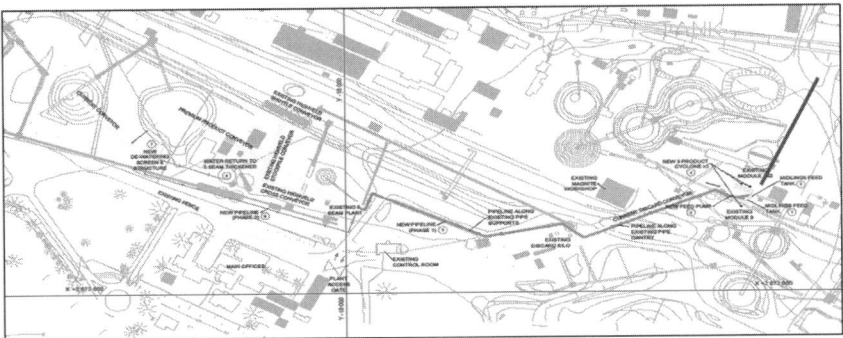

Figure 6: Route showing the middlings pipeline between 4 Seam and 5 Seam Plants

Again, this intervention led to a ca. 3.5–4% recovery of the feed to plant which no longer reported to the discard facility thus relieving pressure on the available airspace.

2.3 Re-configuration of the 5 Seam Plant to enable small scale mining and retreatment of the discard dump

The 5 Seam resource was exhausted in 2012, which left the 5 Seam Plant potentially idle. As the airspace issue was ever present in the minds of the technical team at Greenside, an examination of the circuit, yield and size envelopes ensued to establish if the dump could be selectively mined to create usable airspace on the dump. To put the challenge of utilising this capacity into perspective, one has to understand that the coals being treated in the 5 Seam and 4 Seam Plants were washed at similar specific gravities but the 5 Seam Plant produced a sized washed product (−100 mm) through a combination of dense medium bath/cyclones whereas the 4 Seam Plant utilises cyclones only (−50 mm); the constraint for treating 4 Seam Discard in the 5 Seam Plant is highlighted in Fig. 7, where it can be seen that the 4 Seam Discard is finer than the 5 Seam feed.

In terms of a typical near density, the discard was not found to be any more difficult to treat than a typical South African coal as shown in Fig. 8 but the yield characteristic was somewhat lower than the typical for a typical 5 Seam coal. Because of the finer particle size distribution and lower yield characteristic and inherent constraints in the 5 Seam Plant (drum, cyclone and sinks screen capacity), the split feeding the drum had to be made quite fine to maintain the feed rate to the plant at 150 tph.

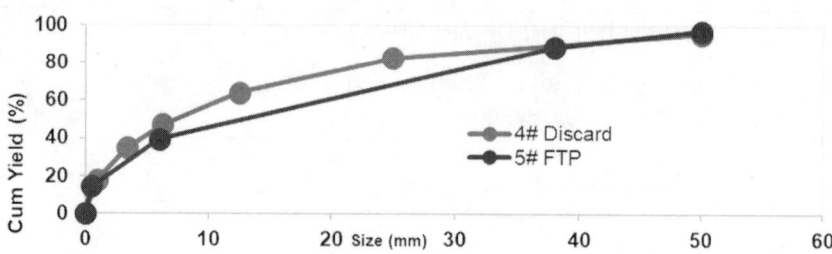

Figure 7: Particle size distribution of 4 Seam arising discard and 5 Seam feed

During 2018, as the airspace projections for the dump became even more exacerbated, it was decided to abandon the dump extension due to both the life of mine economics and technical considerations. Further circuit changes were conceived for the 5 Seam Plant to be carried out in two distinct phases; firstly, to increase the head feed to plant from 150 tph to 180 tph by increasing the DSM Cyclone size from 610 mm to 710 mm with some associated ancillary equipment changes. This was implemented successfully at the end of 2018. The second phase involves the removal of the drum and 710 mm DSM cyclone replacing them with a 900 mm DSM cyclone and associated ancillary equipment which will enable a further increase in the head feed to plant to 250 tph. This will be a substantive change to the plant and requires alterations to filtration and thickener circuitry, conveyor and centrifuge capacity which will be commissioned in mid-2019.

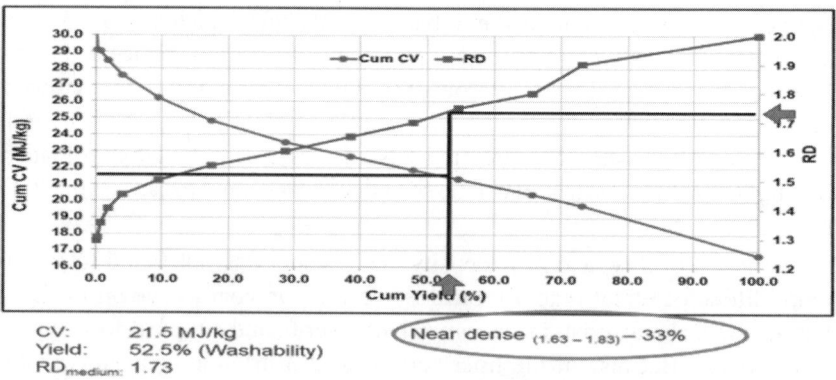

CV: 21.5 MJ/kg
Yield: 52.5% (Washability)
RD_{medium}: 1.73

Near dense $_{(1.63 - 1.83)}$ − 33%

Figure 8: Washability characteristics of 4 Seam arising discard sample

2.4 Large scale mining of the dump

Since 2003, several initiatives had been examined to commence the mining of the dump with a Black Economic Empowerment (BEE) entity for supply to the national power supply utility Eskom to no avail. However; the work concluded entailed the characterisation of the dump reserve to a measured status, the evaluation of transportation routes for mining and product removal and the establishment of a dense medium plant in the very limited real estate at the plant complex. Despite the lack of success to full implementation of these initiatives but because of the decision not to build a dump extension, a decision was made to avail of the opportunity of available toll washing capacity situated close by to mine the dump. This operation commenced in mid-2018 and is expected to peak at a rate of 1 million tons per annum from 2019 onwards. In Figs. 9 and 10 the construction of the roadway from the MRD to the main road is shown.

Figure 9: View of the roadway of the MRD

Figure 10: View of the roadway to the main road

2.5 Summary

The accompanying Table 2 summarises all the initiatives taken including the processing initiatives taken above.

Table 2: Amelioration actions implemented to alleviate MRD air space constraint at Greenside

Amelioration actions implemented	Implementation date	Impact on MRD air space
Relocation of the Greenside East Village	2011	Created 1300 kt additional placement capacity
As-arising rejects from the 4 Seam Plant treated in the 5 Seam Plant	2012	Increased MRD capacity by nominal 1.5% of total head feed for the period May 2012 to Dec 2014 and generated additional revenue
Construction and commissioning of flotation and filtration plant	2012	Increased MRD capacity by nominal 3% of total head feed for Life of Mine and generates additional revenue
Construction and commissioning of tailings filtration plant	2014	Ensured "dry" disposal thereby allowing for MRD to be reclaimed as resource and reduced water consumption per feed tonne washed
Product change from 6000 kcal/kg to 5850 kcal/kg	2014	Increased MRD capacity by nominal 8% of total head feed for Life of Mine and generates additional revenue
Installation and commissioning of 3-product cyclone in the 4 Seam Plant	2014	Increased MRD capacity by nominal 3.5% of total head feed for Life of Mine and creates additional revenue
Construction of a new pollution control dam	2014	Created 8300 kt additional placement capacity
Discard retreatment in the 5 Seam Plant from the MRD facility	2015	Increased the MRD capacity by 1315 kt over a three-year period and generated additional revenue
Reclamation and raw discard sale of the MRD	2018	Will increase MRD capacity by nominal 1000 kt/year and generate additional revenue
Upgrade of the 5 Seam Plant capacity from 150 tph to 180 tph head feed	2018	Increased the MRD capacity by 50 kt in 6 months and generated additional revenue
Upgrade of the 5 Seam Plant capacity from 180 tph to 250 tph head feed	2019	Will increase MRD capacity by nominal 200 kt/year and generated additional revenue

3. Conclusion

Hindsight is the perfect art, when one must deal with a number of legacy issues, logistic challenges and the daunting challenge of a physical constraint within an environment where commercial and legislative issues (both environmental and economic) dictate further conditions for your license to operate; it is difficult to arrive at a perfect solution i.e. one that is made in the most economic and timeous manner. In this case, the solutions embarked upon were correct and made sense at the time, but all were dictated by having a clear understanding of the mass balancing and a fluidity in interpretation of the mass balancing when undergoing the various scenario planning situations when performing life of mine studies accompanied by some well thought out solutions by the various processing teams at the operation over a period of eight years.

4. Acknowledgement

With any conundrum of this nature, it takes considerable patience and faith from a management team and with this uppermost in our minds we would like to thank the successive managements at Anglo American for working with us on this journey. As this work has proceeded over several years and with several iterations in approach, we would like to acknowledge the contributions of our technical colleagues at Anglo American and the various plant teams at Greenside Colliery. We would also like to acknowledge and thank the following engineering houses DRA, Enprotec, Fraser Alexander, Paterson and Cooke & Royal Haskoning DHV. In some cases, their respective work was implemented and sometimes not but in the latter the quality of input framed our thoughts as to how to proceed. A final acknowledgement is given to Coaltech through the support from Johan de Korte regarding the performance evaluation of the 3-product cyclones.

5. References

[1] de Korte GJ., 2014, The Current Focus for R&D within the South African Environment, 15th Australian Coal Preparation Conference, Gold Coast, pp 2–14.

[2] du Toit F., 2012, Private Correspondence on Greenside 4 Seam Plant 3-Product Cyclone Upgrade.

[3] Esterhuizen JJ., 2008, Coal Flotation – Goedehoop Colliery, 12th Australian Coal Preparation Conference, Sydney, pp 268–279.

[4] Power D., 2006, Metallurgical accounting for residues from coal preparation plants, XV International Coal Preparation Congress, Beijing, pp 707–717.

38

Effect of feed characterization on solid bowl centrifuge performance for dewatering of fine-coal tailings

Majid Ejtemaei, Andrew Doi, Jianlong Wang, Anh V Nguyen, Dave Osborne

School of Chemical Engineering, The University of Queensland, St Lucia, Queensland, Australia

Somerset International, Brisbane, Australia

Abstract: Solid bowl centrifuge applications in fine-coal tailings dewatering have been increasing in recent years. However, there are still challenges to make the technology work well enough to be economically viable for clay-rich tailings. The degree of difficulty associated with fine-coal tailings dewatering differs from mine to mine as the process water and surface properties of tailings can change significantly. These dewatering and handling problems are caused by the complex surface properties and gelation due to the swelling characteristics of smectite-type clays, which can result in high yield stress, high dosages of dewatering aids, low settling rates, and poor supernatant clarity. In this paper, the results obtained for dewatering fine-coal tailings containing different characteristics are presented. Specifically, pilot-scale experiments were conducted on samples received from different coal preparation plants in the Bowen Basin, Australia using a pilot-scale solid bowl centrifuge. Pilot-scale trials on tailings thickener underflow were conducted at different pool depths and with different differential rates to maximize capture of ultra-fine clay mineral particles. The results showed that in the presence of clay tailings containing more negative surface charge and swelling types, pool depth in the clarification zone of the centrifuge must be increased to recover an acceptable clear centrate with some sacrifice in cake moisture content. It was also discovered that process water chemistry plays a critical role in the dewatering performance through neutralization of the particles surface charge.

Keywords: Coal tailings, solid bowl decanter, feed properties

1. Introduction

Centrifugal dewatering is a process that applies the g-force, gravitational force, from rapid rotation of a bowl, to solid–liquid separation. There are several forms of centrifuges used for solid/liquid separation with different operation procedure. The solid bowl centrifuge operation is based on the settling rate of the particles. In this case, all that is necessary is that the solids are heavier than the liquid phase so that the solids will settle out quickly in the machine by centrifugal force. The settled particles are removed by a scroll conveyor, and liquid is discharged over weirs. The bowl is cylindro-conical-shaped that facilitates the transfer of solids out of the liquid and into a drainage zone allowing them to dry on an inclined surface before being discharged (Das et al., 1996; Klima et al., 2011). However, in the screen bowl centrifuge, there is an additional screen section of the bowl, after the conical section, which improves the dewatering through filtration. The potential loss of ultrafine particles during filtration in the screen section is a disadvantage of the screen bowl centrifuge. The ultrafine particles can be effectively captured in a solid bowl but with some sacrifice in moisture content (Parekh, 2009; Klima et al., 2011), if some form of chemical treatment is applied to capture these particles within the formed cake. Therefore, it may be possible to effectively use a solid bowl in dewatering tailings with ultrafine clay particles.

Optimum performance of a solid bowl centrifuge is determined by measuring the feed solids, discharged cake consistency, and centrate solids content generated from tests which vary feed rate, g-force (bowl speed), pool depth (weir height), and solids retention time (conveyor speed). The efficiency of a solids/liquid separation is also dependent upon the physical and chemical nature of the materials and process water. Particle size distribution, shape, porosity, compressibility, concentrations of the dissolved salts in process water, and particles surface properties all affect the separation process. Fine particles, porous or plate-like solids produce a cake product higher in residual moisture than hard, coarse, non-compressible substances. Viscous slurries are more difficult to handle than water like slurries (Bochkov and Zarubin, 1967; Parekh, 2009; Meiring, 2015).

In this paper, the effect of the particles surface charge, size distribution, and process water chemistry on the solid bowl dewatering performance were studied. The optimum pool depth and differential rate were also defined using tailings feed with different properties.

2. Methods and materials

2.1 Pilot-scale sold bowl centrifuge

Pilot-scale dewatering trials were carried out on the tailings thickener underflow samples with around 35% w/w solids received from two Bowen basin CHPPs (sample A and B). A 150 mm (6″) pilot-scale, solid bowl centrifuge was used to dewater the tailings and assess the effect of the pool depth on the performance of the dewatering process. Applied parameters regarding the operation of the centrifuge are given in Table 1.

Samples of cake and effluent (centrate) water were collected for a given duration at each condition after getting a consistent cake discharge. As feed rate was known, collected samples were analyzed for moisture content and total dissolved solids (TDS) calculations based on the mass balance. TDS was used because filtration of ultrafine suspended solids proved to be unreliable. A baseline value for dissolved solids was calculated from a filtered sample. Each experiment was repeated at least three times, and the standard deviations were calculated.

Table 1: Applied Hz of scroll motor at 40 Hz (2000 × g-force) of bowl motor at a different differential rate

Differential rate	14	20	24
Hz of scroll motor	40	35.5	32.5

2.2 Zeta potential measurements

To measure the zeta potential, 0.01 g of coal tailings paste was dispersed in 10 mL of 10^{-6} M KCl background electrolyte at neutral (unadjusted) pH and wait for 10 min to syphon off the colloidal particles in the supernatant as a representative sample. A ZetaPlus instrument (Brookhaven Instruments Corp., Holtsville, NY) was used to measure the zeta potential of the particles. The Smoluchowski equation was applied to calculate the zeta potential from the electrophoretic mobility of the particles measured by the instrument. The zeta potential mean values with standard errors were reported for five runs.

2.3 Slurry water analysis

ICP-OES (Inductively coupled plasma optical emission spectrometry) was used to measure the concentration of the elements in the process water. The technique can analyze water samples for all elements apart from C, H, N, O, the halogen, and gaseous elements.

2.4 Particle size distribution

The size distribution was statistically characterized using a Malvern Mastersizer instrument (Malvern Instruments Ltd., Worcestershire, UK). Size measurements of the particles were carried out using the experimental procedure was as follows. First, a 100 mL of coal tailings suspension was stirred at 500 rpm for 10 min. Then, the sample was taken from bottom to the top of the suspension during the mixing to make sure that it is a representative sample. The collected sample was added into the 1 L of deionized water. The suspension was then stirred gently using an overhead impeller at 800 rpm (in order to keep the particles in the suspension during the measurements) and fed to the measurement cell of the Malvern instrument.

3. Results and discussions

Flotation tailings thickener underflow were diluted with process water to reach around 15% solids slurry as a feed sample to the centrifuge. To find out the optimum operation condition, the effect of conveyor speed, or differential speed (lower differential rate means less difference between the bowl speed and the conveyor speed, and hence higher residence time), and pool depth were studied. As the main aim of the test work was to identify the effect of the pool depth and differential rate on the dewatering behavior of the tailings samples with different properties, the feed rate and bowl speed (g-force) were kept constant at 0.22 m³/h and 2000 × g-force, respectively.

3.1 Cake moisture and centrate TDS studies

The differential rate is the speed difference between the scroll and main bowl. It should be noted that scroll speed should be slower or faster than the bowl speed depending on the configuration of the scroll drive mechanism to ensure consistency in the discharged cake. In this test programme, this differential is a constraint which limits the residence time at different bowl speeds, i.e., with increasing bowl speed the minimum achievable differential rate decreases. Pool depth also plays a critical role in increasing centrate clarity. It is accepted that increasing the pool depth will result in a high solid recovery or centrate with low TDS%, but increase in the cake moisture will have an adverse effect. Therefore, it was essential to find the optimum condition of residence time and pool depth to achieve the optimum operating condition. Figure 1 shows the effect of differential rate and pool depth on sample A cake moisture at 40 Hz bowl motor speed (equal to 2000 × g-force). As can be seen, with decreasing differential rate (increasing residence time) at a low pool depth condition, the final cake moisture dropped from ~33% to 30%. Changing the pool depth also confirmed that increasing pool depth showed an increase in the cake moisture. This effect was more highlighted at a low differential rate of 14.

TDS percentages in the centrates at different running conditions of sample A are shown in Fig. 2. TDS in the centrate decreased with increasing differential rate, indicating increased residence time allowed enough time to ultrafine particles to overflow the effluent weirs. As seen, with increasing the pool depth, TDS in the centrate decreased from 0.31% to 0.25% at a differential rate of 24. It should be noted that the conductivity of the process water at the site ranges from approximately 2000–7000 μS/cm. This equates to a TDS of approximately 1300–4500 mg/l (or 0.13–0.45%).

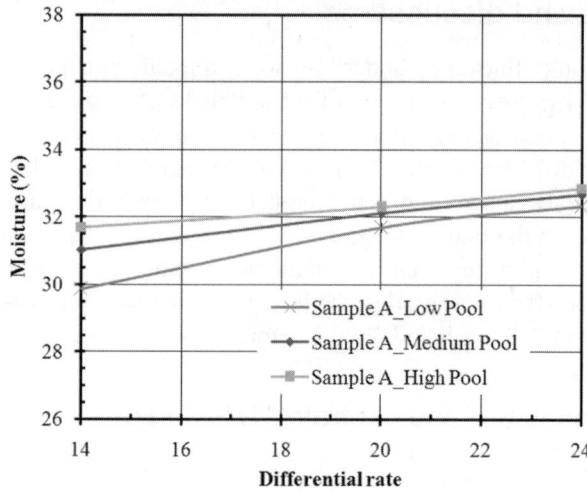

Figure 1: The effect of differential rate and pool depth on sample A coal tailings cake moisture using pilot-scale solid bowl decanter

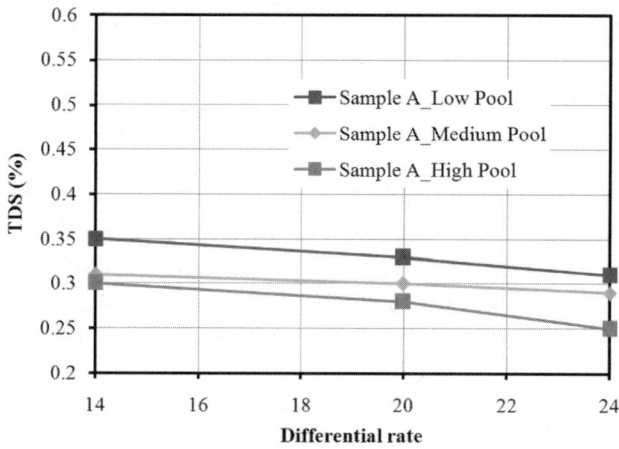

Figure 2: The effect of differential rate and pool depth on sample centrate TDS using pilot-scale solid bowl decanter

The results on sample B cake moisture have been plotted in Fig. 3. Applied different pool depth and differential rates showed a similar trend on the cake moisture. However, obtained cake for samples B trials contained more moisture than sample B in a specific operation condition. TDS of the generated centrates with sample B trials are shown in Fig. 4. Similar to the sample A results, centrate quality was dropped with increasing differential rate. The best TDS of 0.52% for sample B centrate was achieved at high pool depth and differential rate of 24, but it was higher than the sample A centrate at the same condition. These results display that different feed characterization could be the only reason for the significant difference in the results.

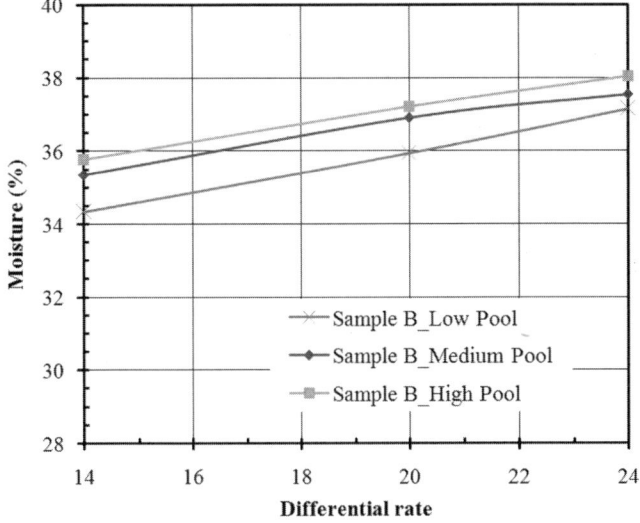

Figure 3: The effect of differential rate and pool depth on sample B coal tailings cake moisture using pilot-scale solid bowl decanter

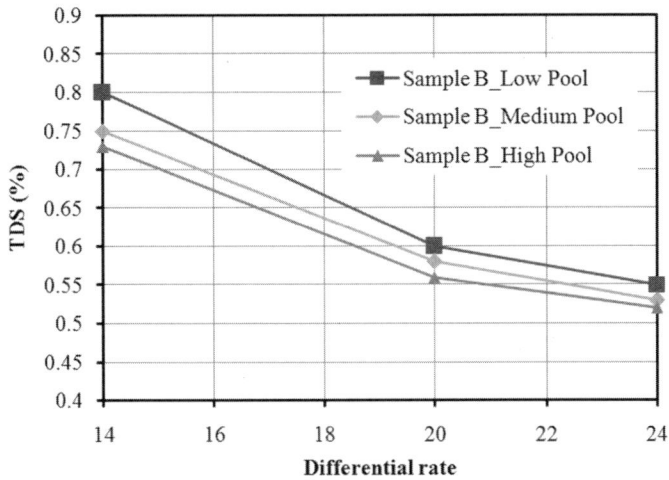

Figure 4: The effect of differential rate and pool depth on sample centrate TDS using pilot-scale solid bowl decanter

3.2 Characterization analysis

Characterization of the samples was carried out to understand the different behavior of the samples A and B in solid bowl centrifugal dewatering. Particle size distribution is an important factor that could affect the dewatering process. Particle size distribution data for samples A and B are shown in Table 2. The results exposed that 90% of the particles in the sample A are below ~60 microns while it is 37 microns for particles in sample B. However, 50% of the particles in sample A are below 9.6 microns which is much smaller than the particle size in sample B. Figure 5 displays the samples particle size volume-based distribution. As seen, sample A contained a mixture of fine and coarse particles with a wide range of particle size distribution. However, sample B included more fine particles with a uniform size distribution that could also result in a poor dewatering performance in comparison with sample A.

Table 2: Samples A and B particle size distribution data (microns)

	d(0.1)	d(0.5)	d(0.9)
Sample A	2.933	9.660	59.958
Sample B	3.445	10.670	37.371

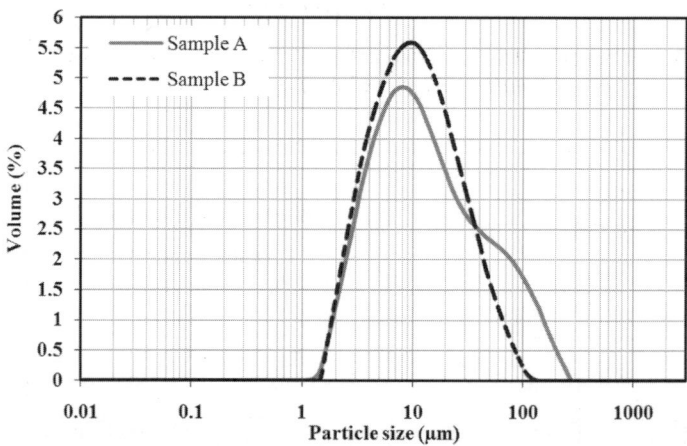

Figure 5: Particle size volume-based distribution of samples A and B

Particle dispersion in the suspension that is governed by surface charge is another factor affecting the dewatering performance. Zeta potential results of the samples are presented in Table 3. As can be seen, zeta potential for sample B was more negative than sample A, which could possibly be due to the presence of smectite-type clay minerals in the sample B. However, the mineralogical studies showed sample A contained slightly more smectite-type clays than sample B which therefore suggest that this is not the reason for the high difference in the zeta potential outcomes. On the other hand, the amount of the coal in sample B was higher than sample B. It may be that this sample contains coal which was strongly oxidized and therefor hydrophilic. It should be noted that oxidized coal comprises more carboxyl groups (COOH) on the particles surface which results in a more negative zeta potential value. Therefore, dispersion of the particles in the suspension could happen as a result of the repulsion force between the intensely negatively charged particles.

Table 3: Coal tailings samples zeta potential measurements

	Sample A	Sample B
Zeta potential (mV)	-8.7 ± 1.1	-45 ± 1.5

Particle zeta potential is also related to the concentration of the salts in the process water, and the process water from samples A and B was also analyzed to attempt identify the reason for the significant difference in the zeta potential. The results indicated that sample A had much higher concentration of Ca^{2+} and Mg^{2+} than sample B. The high concentration of Ca^{2+} and Mg^{2+} in the process water can neutralize the surface charge of the particles and minimize their dispersion force. It seems the presence of the multivalent cations in the process water governs the zeta potential behavior of the samples (Table 4).

Table 4 Sample A and B slurry water analysis

Element (mg/l)	Sample A	Sample B
Ca^{2+}	111	31.8
Mg^{2+}	212	36.4
Na^+	1638	1618
K^+	33	43
S^{2-}	377	227

4. Conclusion

Solid bowl centrifugal dewatering of the coal tailings showed promising results in reducing moisture content, but these levels can be further enhanced by neutralizing surface charge of the particles by salts. Minimums of 30% and 34% moistures were attained for samples A and B, respectively, at low pool depth and 14 differential rate condition. At this operational condition, TDS of the samples A and B were 0.3% and 0.55%, respectively.

The results showed that increasing the pool depth yielded an improvement in the centrate quality, but there was an adverse effect on the moisture content of the cake.

The characterization studies for the samples revealed the importance of the particle surface charge on centrifugal dewatering performance. It was found that repulsion forces for highly negatively charged particles are strong so their settling rate could be very slow. The presence of multivalent cations in the process water could neutralize the surface charge on the particles and facilitate their dewatering process.

The major deliverable of this research has therefore been that such a tailored approach is needed to achieve the consistency in tailings cake and centrate quality and that the solid bowl centrifuge is a suitable and efficient dewatering technology for coal tailings treatment, particularly where the goal is to dispose of combined CHPP discards in a haul truck handleable form.

5. Acknowledgment

The authors acknowledge the financial support from ACARP (project C25012), the feedback and inputs from the industrial monitors, and the significant input and assistance provided by engineers from Somerset International Australia, Mrs. Pamela White (now with BHP), who was instrumental to the start of the project.

6. References

[1] Attia, Y. A. & Yu, S. 1991. Flocculation and Filtration Dewatering of Coal Slurries Aided by a Hydrophobic Polymeric Flocculant. Separation Science and Technology, 26, 803–818.

[2] Besra, L., Sengupta, D. K. & Roy, S. K. 1998. Flocculant and Surfactant Aided Dewatering of Fine Particle Suspensions: A Review. Mineral Processing and Extractive Metallurgy Review, 18, 67–103.

[3] Bochkov, Y. N. & Zarubin, L. S. 1967. Influence of Proportion of Finely-Dispersed Particles on Dewatering of Slack Coal in Screw-Type Thickening Centrifuges. Coke & Chemistry USSR, 11-&.

[4] Das, S., Mohanty, B. & Murty, J. S. 1996. Performance of solid bowl centrifuge in de-watering coal water slurry. Solid Liquid Separation in Mineral and Metallurgical Industries, 205–211.

[5] Gochin, R. J., Lekili, M. & Shergold, H. L. 1985. The Mechanism of Flocculation of Coal Particles by Polyethyleneoxide. Coal Preparation, 2, 19–33.

[6] Klima, M. S., Dehart, I. & Coffman, R. 2011. Baseline Testing of a Filter Press and Solid-Bowl Centrifuge for Dewatering Coal Thickener Underflow Slurry. International Journal of Coal Preparation and Utilization, 31, 258–272.

[7] Meiring, S. 2015. Thickeners versus centrifuges – a coal tailings technical comparison. In: (EDS), I. R. J. A. F. (ed.) Proceedings of the 18th International Seminar on Paste and Thickened Tailings. Australian Centre for Geomechanics, Perth.

[8] Parekh, B. K. 2009. Dewatering of fine coal and refuse slurries-problems and possibilities. Proceedings of the International Conference on Mining Science & Technology (Icmst 2009), 1, 621–626.

[9] Singh, B. P., Besra, L., Reddy, P. S. R. & Sengupta, D. K. 1998. Use of surfactants to aid the dewatering of fine clean coal. Fuel, 77, 1349–1356.

39

Coal tailings treatment through advanced tailings separation process

Rainer Raberger[1] and Robert Stantish[2]

[1]*ANDRITZ AG, Graz, Austria*
[2]*ANDRITZ Separation, Kuthambakkam, Tamil Nadu, India*

Abstract: Traditionally, tailings have been discharged to ponds, where they constitute a long-lasting environmental burden. In addition, valuable process water is lost together with the solids, which is a major disadvantage particularly in arid regions.

Environmental concerns demand actions to tackle the huge amounts of tailings that are already stored in dams. They need be treated to clean up existing environmental damage and to avoid accidents like dam breakage, which may cause high collateral damage in residential areas.

Moreover the liquid can be used as process water, the waste volume is minimized and the hazardous impact is reduced. Separation of the tailings into product and waste as well as into a liquid and a solid fraction is an essential process to generate additional profits.

By dewatering the fine tailings, process water can be recovered and the cake, with final moisture content reduced to a minimum, can be deposited safely. Compared to paste thickening, the filtration process results in lower residual moisture. Also, a reduction in tailings volume is achieved. Filtration is, therefore, a viable technology for dealing with this task.

Examples are the treatment or filtration, respectively, of coal, iron ore, and copper tailings, and even red mud in alumina refineries. A plant concept for coal slurry pond treatment is presented where coal tailings are recycled. Approximately 50% of the solids content in the tailings is converted into a valuable product meeting all the required specifications for ash content and residual moisture.

Keywords: Tailings treatment, heavy duty belt press, dewatering, filtration

1. Introduction

Liquid–solid separation units are an important part of a beneficiation plant. Filters, belt presses, thickeners, centrifuges, and dryers are used for liquid–solid separation. Selection of the equipment for the possible operating range is an important task for process engineers and plant designers [1]. Beneficiation plants increase the valuable fraction of the mined material. During this process, tailings are inevitably produced as a waste material [2]. There is a general trend towards processing increasingly lower-grade materials that require highly sophisticated beneficiation processes and result in larger amounts of tailings. Traditionally, tailings have been discharged to ponds, where they constitute a long-lasting environmental burden with a serious impact for residential areas if the pond dams break. In addition, valuable process water is discharged together with the solids, which is a major disadvantage in arid regions. By dewatering the fine tailings, process water can be recovered and the cake can be deposited safely. This minimizes the risk for surrounding areas. Thickening and filtration are viable technologies to deal with this task [3].

2. Coal tailing dewatering [4]

Coal tailings behave quite differently compared to other tailings. Filtration is particularly difficult when the ash content of coal tailings is high [5]. As a result, heavy-duty belt presses are often given preference over other filtration equipment. Unlike copper and red mud tailings, coal tailings require substantially less flocculating agent when using a heavy-duty belt press. Thus, it is feasible to use a heavy-duty belt press in this case and also attractive due to the comparably lower investment costs.

2.1 Selection of dewatering equipment

Dewatering properties depend on particle size distribution, the surface properties of the particles, and the mineralogical composition of the suspension. For coal applications, the mineralogical composition is defined by the ash content, which is significantly higher in tailings than it is in concentrates. The specific permeability of the material is a value calculated as a guide in equipment selection for feasible dewatering or filtration units.

The specific permeability of a coal suspension K depends on the fine grain fraction below 75 mm and the ash content in this fraction:

$$K = A \cdot \sqrt{W} \tag{1}$$

where K is the specific permeability (filtration coefficient), A is the ash content in grain sizes up to 75 mm in % by weight, and W is the grain class 0–0.075 mm in % by weight.

The ash content forms the fraction for the clay and potter's earth contained in the material. The equation shows that the influence of this ash content is high compared to the influence of the particle size distribution [6]. Suspensions with low specific permeability are easy to dewater, and satisfactory performance will be achieved with vacuum filtration.

In coal beneficiation plants, dry solids content of about 30–40% by weight are common in the feed material to the filters (in the event of raw slurry).

A renowned institute for coal beneficiation in Katowice, Poland, developed this formula further in 1975 and identified distinct classes of specific permeability linked to their filtration behavior. Subsequently, they suggested types of dewatering equipment that can be considered as the optimum for these specific classes. During the 1980s, this classification was adopted and refined by ANDRITZ to enable proper preselection of equipment for hard coal slurry dewatering as shown in Table 1 [6].

Table 1: Preferred dewatering equipment

Specific permeability K	Filtration properties	Preferred dewatering equipment
<100	Good	Vacuum filter or centrifuge
100–200	Average	Hyperbaric filter
200–300	Rather poor	Hyperbaric filtration, chamber filter press or belt press
>300	Bad – very bad	Chamber filter press or belt press

ANDRITZ offers many different types of dewatering equipment, such as belt presses, centrifuges, vacuum disc filters, and hyperbaric filters. In Russian coal plants, ANDRITZ investigated and verified this method of equipment preselection. The result of this study provides an overview of six different examples where different dewatering technologies are used. The specific permeability K was used to confirm right or wrong selection of the dewatering equipment. The operating experience in the various plants verified the right or wrong selection. This rather simple preselection guideline is very helpful, but nonetheless dewatering test work is recommended for final selection to enable proper equipment selection and process guarantees.

3. Coal tailings pond treatment

The objective of the plant shown in Fig. 1 is to recover the coal from an existing coal tailings pond that has accumulated over years.

Treatment of coal tailings to extract valuable coal starts at the tailings pond, where a screen (1) prevents too large pieces or particles from entering the process. Due to inefficient beneficiation processes in the past, these ponds often contain a reasonable fraction of valuable material that allows economically beneficial recovery. After the screen, the material is collected in a sump (2), and cyclones (3) split the stream into a fine and a coarse fraction. The underflow (fine fraction) is treated with Humphrey's spirals (4).

After some screening, the overflow from the Humphrey's spiral coarse fraction is again disposed of as tailings, while the underflow is combined with the underflow from the fine fraction screens. The overflow from the

Humphrey's spiral fine fraction screen is treated with centrifuges (5) to obtain fine concentrate material.

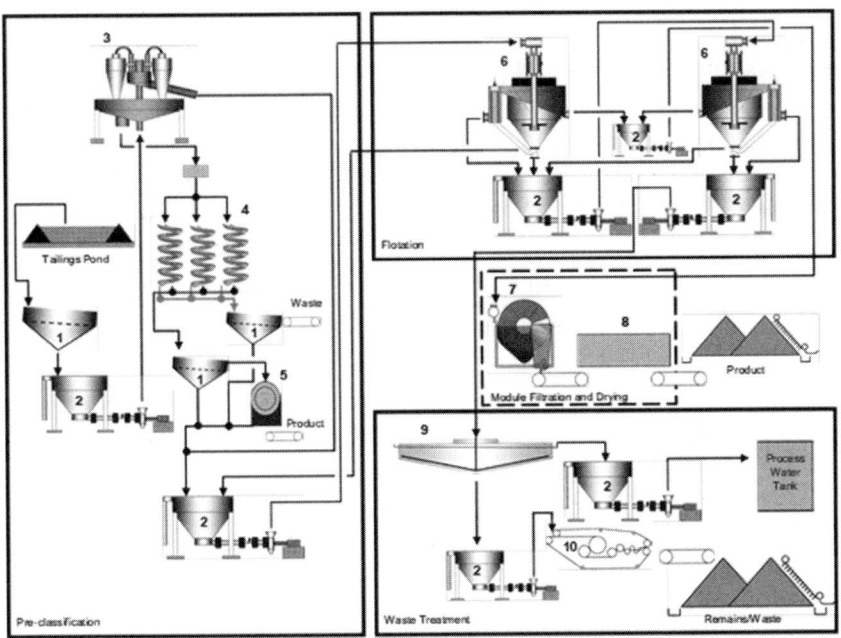

Figure 1: Simplified flow sheet of pond slurry treatment concept

The underflow from the Humphrey's spiral screens (fine and coarse fraction) and the centrate from the centrifuges are floated (6) to separate the tailings from the flotation concentrate, where the overflow forms the flotation concentrate and the underflow remains as tailings. The tailings fraction is thickened (9) and dewatered on a heavy duty belt press (10) specially designed for this application. The remaining tailings are disposed of as flotation tailings, while the overflow from the thickener (9) is recycled as process water. The flotation concentrate is filtered in vacuum disc filters (7) and dried on fluid bed dryers (8) to extract the final, valuable fine coal product.

3.1 Filtration and drying

The vacuum disc filters (equipment no. 7) are used for many different applications, and the technology has become well established over the years. The filters are generally used in heavy-duty applications such as dewatering of iron ore, coal, aluminum hydrate, copper concentrate, pyrite flotation concentrates, and other beneficiation processes.

Fluid bed dryers are used for thermal drying after filtration. The fluid bed is generated by blowing fluidization gas uniformly over the dryer cross-section. In the dryer, the free-flowing granulate begins to float, and it is mixed thoroughly at the same time. A fluid bed is characterized by movement of the granules, achieved by a gas stream passing through the product layer.

The filtration and drying module marked in Fig. 1 consists of vacuum filtration and thermal drying. This module can be replaced, particularly by hyperbaric filtration (HBF), which depends upon the necessary residual moisture (Fig. 2). This has to be evaluated by test work in a case study.

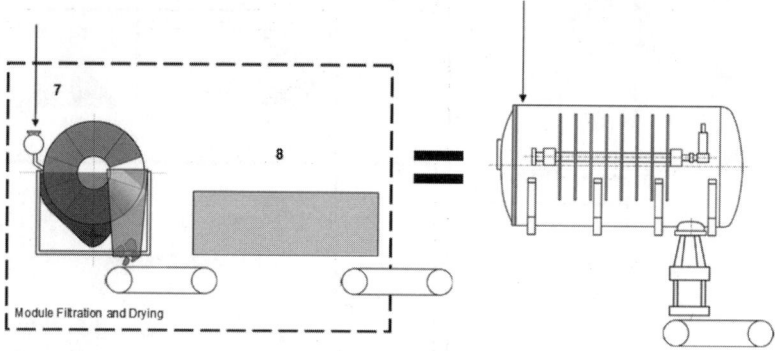

Figure 2: Hyperbaric filtration instead of vacuum filtration and thermal drying

The HBF (Fig. 2) consists of a set of filter discs mounted inside a pressure vessel. The main difference between conventional vacuum disc filtration and the HBF is the positive pressure applied to the filtration surface. In theory, vacuum filtration pressure is limited to 1 bar. This is never achieved, however, and a differential pressure of up to 0.7 bar is applied in practice. Hyperbaric filtration is unlimited in theory, but mining applications have a practical operating range of 2–6 bar, which results in three to eleven times the differential.

4. Dewatering of coal tailings

The tailings from an existing tailings pond, which are processed as described above, or from a coal beneficiation plant are dewatered before landfill disposal in order to avoid any hazardous impact on the environment.

But again, a tailings stream needs to be dealt with, as is inevitable in beneficiation plants. To avoid tailing ponds and recover process water, this tailing stream enters a thickener stage, which is used for separating particles from the liquid by gravity. Solids that sink to the bottom of the thickener are captured and liquid is removed from the surface. The thickened fraction is further dewatered by a filter unit – in this case a heavy-duty belt press – and the waste product that is produced has a dryness that allows uncomplicated landfill disposal (Fig. 3).

ANDRITZ introduced a new belt press with optimized roller geometry to obtain the best dewatering result for tailings treatment.

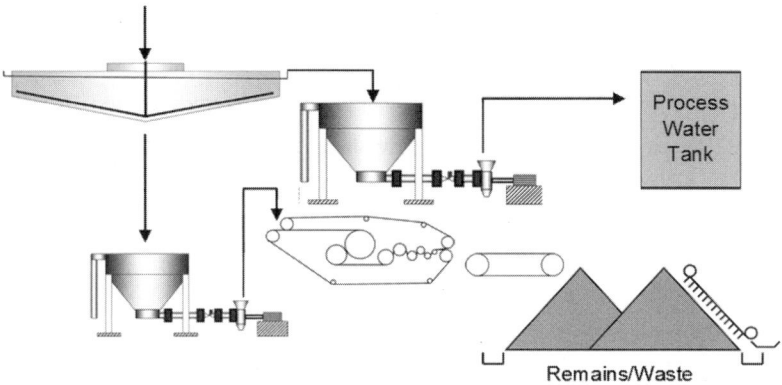

Figure 3: Simplified flow sheet of a tailings treatment stage

5. Conclusion

Coal tailings pond treatment is based upon advanced liquid–solid separation equipment that is readily available on the market. Further optimization is possible by replacing the energy-intensive thermal dryer by hyperbaric filtration. Again, hyperbaric filtration technology is widely utilized, especially for large throughputs. The total tailings volume is drastically reduced by such treatment and the risk of groundwater pollution is thereby minimized because the waste product is quite dry. The solid waste quantity is reduced by a factor of 0.5 overall.

Coal tailings are inevitably produced during traditional coal beneficiation. Settling tanks followed by a heavy-duty belt press form a viable concept for treating these tailings, not only to lessen the sometimes devastating environmental impact, but also to recover valuable process water and, in some cases, also extract additional valuable material.

6. References

[1] Raberger, R., Ziaja, D., Godwin, P. (2010) Environmental and economic success through advanced tailing separation process, Mine Waste 2010, Proceedings of the

1st Seminar on the Reduction of Risk in the Management of Tailings and Mine Waste, 29th Sept.–1st Oct. 2010, (eds) Fourie, A. & Jewell R., Australian Centre for Geomechanics, Perth, Australia.

[2] Nguyen, Q. D., Boger, D. V., Application of rheology to solving tailings disposal problems, Int. J. Miner. Process. 54 (1998) 217–233.

[3] Raberger, R., Krammer, G. (2009) How to meet the challenge of tailings filtration, First International Seminar on Environmental Issues in the Mining Industry, enfiromine2009, 30.9.–2.10.2009, Santiago, Chile, paper (chapter 4, #37) published in proceedings CD eds J. Wiertz & C. Moran.

[4] Raberger, R., Krammer, G., Schnabel, G., Podratzky, K. (2013) Efficient Coal Tailings Dewatering for Coal Benification Plants, XVII ICPC, Istanbul.

[5] De Kertser, R., Scales, P. J., Boger, D. V., Improving Clay-Based Tailings Disposal: Case Study on Coal Tailings, AIChE J., Vol 43, No. 7, 1894–1903 (1997).

[6] Reuter, J., Stand und Probleme der mechanischen Feinstkorn-Entwässerung in der Steinkohle-Aufbereitung; Chemie-Ing.-Techn. 34 Jahrg. 1962/Nr. 4, Page 299.

40

Environmental protection of dormant historical slurry dams: Case study

Nandipha Ziphethe, Collen Mabada

Ifalethu Colliery, South32 Energy Coal, Middelburg, South Africa

Abstract: In coal beneficiation, a product at a specific yield is achieved with the balance of the run of mine coal being waste in the form of coarse discard and slurry, legally referred to as mine residue. In the newly promulgated National Environmental Management: Waste Act 59 of 2008 (NEMWA) classifies Mining Residue as a Type 3 waste and therefore its disposal is regulated and forms part of a colliery's legal environmental authorisations which have an impact on the maintenance of the license to operate.

Due to market constraints and limitations of commercially viable processing/handling solutions of slurry, this material is currently classified as a waste stream which is an environmental liability. Rehabilitation of these facilities is paramount and undertaking of investigations to determine feasible options such as re-mining or processing the material further for valuable product.

This paper will explore the legislative framework governing the disposal of mine residue (that is, discard and slurry) using a case study of dormant slurry dams that exists in a mine that has been subject to spontaneous combustion. The basic state of spontaneous combustion of the slurry dam, the zone of influence should the integrity of the dam be compromised as well as the environmental risk and impact assessment of the facility will be reviewed. Exploration of immediate remedial actions and mitigation controls to prevent further burning and stabilisation of the existing facility that had been compromised. Finally, the long-term options for future protection of such facilities and rehabilitation including the closure of the facilities will be discussed.

Keywords: Mine residue, run of mine, spontaneous combustion

1. Introduction

Reliance on coal mining for South Africa's energy demand will remain for the rest of the century despite migration to a hybrid energy mix introducing green energy. Building the world's fourth-largest coal-fired power plants in South Africa will increase the demand for coal mining and production with the consequence of increasing environmental impact, i.e. land disturbance, waste management, water and air pollution.

The issue of the environmental impact in South Africa is a combination of historic areas which were not subjected to strict legislation and future mining that is highly regulated with direct impact on the economics of mine development. The promulgation of the National Environmental Management Act and changes in legislation has driven development costs more than 10-fold and failure to comply can lead to prosecution and closure of operations.

Coal producers mine, beneficiate and sell coal to both domestic and export market. The run of mine (ROM) coal quality is upgraded through coal beneficiation stages to achieve the required market quality specification. During the beneficiation process a specific yield product is achieved depending on the quality of ROM coal. This yield results in the remainder ROM to be under specification and is considered a waste in the form of coarse discard and slurry, also legally referred to as mine residue.

This paper will thus focus on one of the coal producers, namely South32 South African Energy Coal (SAEC). SAEC has four operations all located in Mpumalanga namely Khutala Colliery, Klipspruit Colliery, Wolvekrans and Ifalethu Colliery with associated coal processing plants. The selected operation is Ifalethu Colliery which is an opencast mine located in Middelburg.

The objective will be to highlight issues faced by the industry, using selected operation as a case study. The question being how to deal with historical tailings facilities without further impact to the environment and compliance with legislation. These facilities' life is attributed to the history of coal beneficiation in the country which started in early 1930s. Coal smaller than 10 mm was discarded at the time, but with improvement with washing techniques, coal greater than 150 µm had value.

Historically, coarse and fine coal discards were disposed separately. Coarse discards were formed by end tipping, levelling and compacting while the fine coal discards were impounded in dams constructed with earth walls.

With future mining being dealt with, a challenge remains with the historical disposal facilities that were filled which lay dormant and have not yet been rehabilitated. Slurry was disposed in voids and in the latter years in surface facilities. These surface facilities mostly were open ponds and in the last decade co-disposal facilities are utilised. All these facilities had been licensed as per environmental authorisations. Typical problems associated with the old dumps are as follows:

- Spontaneous combustion;
- Contaminated runoff and seepage;
- Gaseous emissions and dust;
- Visual pollution;
- Dam wall instability;
- Difficult and expensive to rehabilitate.

The National Environmental Management: Waste Act (NEMWA) classifies Mining Residue as a Type 3 waste and therefore needs to be disposed of with specified lining. Coarse discard is considered a mine residue and currently being disposed of in several ways for the different coal operations. Changes in environmental legislation over the years led to the mining industry adapting its disposal methods to comply with the legal requirements for disposal. The current disposal methods are increasing in costs and create both capacity and legal issues that need to be dealt with upfront.

Approximately 290 million tonnes of mining residue will be generated by SAEC's operations over the next 26 years according to the Life-Of-Operations-Plan (LOOP). This equates to 38% of total ROM that will need to be managed appropriately. Future mining has clear strategies in order to deal with the mine residue, however SAEC has 17 million tons of waste in the form of slurry or coal fines disposed in historic tailings facilities across WMC.

2. Applicable Republic of South Africa legal requirements

Below are the legal requirements taken into consideration when evaluating compliance to authorisations (Table 1).

Table 1: Applicable legislation requirements

Applicable South African legislation to the mine
1.   The Constitution of the Republic of South Africa, 1996 (Act No. 108 of 1996)
2.   National Environmental Management: Waste Act, 2008 (Act No. 59 of 2008) (NEM:WA)
3.   National Environmental Management Act, 1998 (Act No. 107 of 1998)
4.   National Water Act, 1998 (Act No. 36 of 1998) (NWA)
5.   National Environmental Management: Air Quality Act, 2004 (Act No. 39 of 2004) (NEM:AQA)
6.   The Mine Health and Safety Act, 1996 (Act No. 29 of 1996)

3. Ifalethu colliery overview

Ifalethu Colliery mines five opencast active pits and a dump which are the main sources of coal. There are three processing facilities namely North Plant, South Eskom Plant and VDD Plant. North and VDD plants have both disposed slurry historically and have operating plants.

4. Ifalethu colliery mine residue situation: Slurry ponds

The Middelburg opencast mine began operations, producing export thermal and domestic coal for Duvha in 1982 through the North Plant. The North Plant historically dumped coal fines in voids followed by surface facility slurry dams. There are three old facilities (old slurry dam, dam 45 and dam 46) and one current co-disposal facility (Fig. 1). The current co-disposal facility has been in operation since September 2011 and is a fourth-generation ring-dyke dam with a pumped decant system. It accommodates the fine tailings generated from the plant in the basin, and the coal discard in the outer walls.

Figure 1: Ifalethu colliery north section coal tailings disposal facilities

4.1 Ifalethu colliery: North plant old slurry dam

The dam, located close to the entrance of the mine, comprises the original slurry dam for the mine. The dam was originally compartmentalised and is currently burning but is devoid of any ponding other than occasional rainwater. It has not been rehabilitated, as an opportunity to re-mine the deposit was being explored. The North plant old slurry dam has approximately 800 kt (dry) of slurry at concentration of solids (CV) of 19.2 MJ/kg.

There are basically two major reasons for preserving these tailings facilities, namely that of the potential value of the deposit and the legal implications of polluting. Pollution is a result of the products of the combustion of the coal such as carbon dioxide, sulphur dioxide and sulphuric acid. These are liberated into the atmosphere when coal self-ignites. The challenge with the current facility is that it is now burning, thus any future economic value is nil. This poses a potential contravention of the legislation if no action is taken.

In response to the observation that this dormant slurry facility was subject to spontaneous combustion, a basic assessment was conducted to determine extent of issue.

Golder Associates Africa (Pty) Ltd (Golder) was appointed by South32 SA Coal Holdings (Pty) Ltd (South32) to carry out a basic assessment of the spontaneous combustion of the dormant slurry dams 1 and 2 at Ifalethu Colliery in Mpumalanga, Republic of South Africa.

This paper outlines the observations made and provides basic designs for two short-term alternative options identified for the remediation of the dam walls that were compromised because of the spontaneous combustion. The report also provides a high level basic design for the long-term remediation measure.

The topography of the site generally falls in the direction from south west to north east with an average fall of about 2.5–5% (1V:40H to 1V:20H).

The infrastructure surrounding the slurry dam are located downstream of the facility (on the east side) and potentially within the zone of influence (in the event of a wall failure or a catastrophic facility failure). The rest of the infrastructure is as follows:

- Water dams are present on the northern side;
- Access roads are present around the slurry dam complex;
- Solution trenches are present around the north, east and south sides;
- Storm water bunds are present around the west, south and east sides;
- A storm water trench and bund wall is in place on the north-eastern side which diverts dirty water from the facility towards the pollution control dams.
- Telephone lines are present on the west and south sides;
- Power lines (11 kV) are present on the east and south side;
- Rail offices and workshops are present on the east side approximately 70 m from the south-east corner of the slurry dam complex;
- Conveyors lines traverse the site passing approximately 40 m from the south-east corner of the slurry dam complex running in the south west to north east direction;

- Substations and associated works are present on the south and east sides. The closest distance from the slurry dam complex to these infrastructure is approximately 290 m on the east side and 160 m on the south side;
- Railway lines traverse the site passing approximately 100 m from the south-east corner of the slurry dam complex running in the south west to north east direction; and
- Most gate and security offices are present approximately 165 m from west side of the slurry dam complex.

4.1.1 State of burning and resulting instability of outer walls (Fig. 2)

Figure 2: Google earth picture burning slurry pond

It is evident that most of the slurry mass has burnt out and the remainder is in a state of burning. The cause of the burning is mainly due to spontaneous combustion. The burning of the slurry material created a state of significant instability of the outer wall. It was observed that circular failures have occurred caving into the inner slope of the dam wall and significant cracking has developed into the outer zone of the wall. The failures of the wall lead to significant deformations, including, settlement, thinning and cracking of the wall.

Significant emissions of CO and SO_2 were observed and no standing water within the facility was observed, as the facility is dormant. The general condition of the outer embankment appears to be unsafe with the wall crest width reduced to approximately less than 1 m in places.

Figure 3 shows a typical section of the slurry dam complex outer wall and illustrates the sequence leading to the deformations and thinning of the outer wall due to spontaneous combustion.

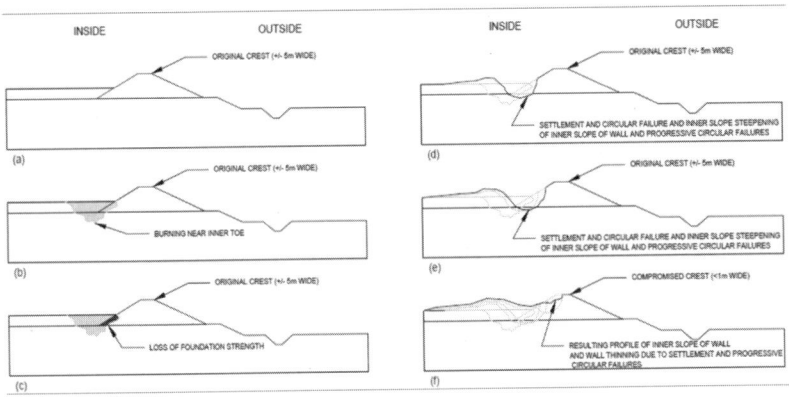

Figure 3: Typical cross section of the outer wall showing wall deformation and wall thinning due to spontaneous combustion

A total of 24 cross sections of the slurry dam walls were extracted from the 3D model of the slurry dam complex to examine the current geometry of the walls and any signs of failures of the dam walls. For each side of the TSF (East, North, West and South sides) a total of six cross sections were taken. Outer walls are present around the entire slurry dam complex. The geometry of the walls at the cross sections considered is summarised as follows:

- The maximum wall height is 13.9 m on the east side (at section B in Fig. 4);
- The minimum wall height is 1.7 m on the west side (at section Q in Fig. 4);
- The average crest width is 5 m;
- The minimum crest width is >1 m in some places due to settlement of inner portion of the wall because of spontaneous combustion. The crest width is minimal at sections A, C and F on the east side; sections G, H and I on the north side; sections Q and R on the west side and section W on the south side;
- The minimum outer slope (overall) is 1V:1.7H on the north side (at sections J and K in Fig. 4 and 1V:1.8H on the west side (at section P in Fig. 4) and 1V:1.7H on the south side (at section S in Fig. 4;
- The minimum inner slope (overall) is 1V:1.5H on the north side (at sections H in Fig. 4) and 1V:1.8H on the east side (at section B in Fig. 4) and 1V:1.9H on the south side (at section V in Fig. 4).

4.1.2 Critical problem zones around the outer wall

A high-level assessment of the critical problematic zones around the facility was done using the survey information and by analysing the cross sections in the figures above. A more detailed assessment will need to be done on site prior to implementation phase.

Figure 4 is the layout of the slurry dam complex and highlights the areas identified as critical problem zones. These are the zones in the wall where the crest width has reduced significantly. It should be noted that as the TSF height is maximum at the east side and minimum on the west side, the high-risk areas are located towards the east side where the wall height is maximum and that these areas should be the priority during the implementation phase.

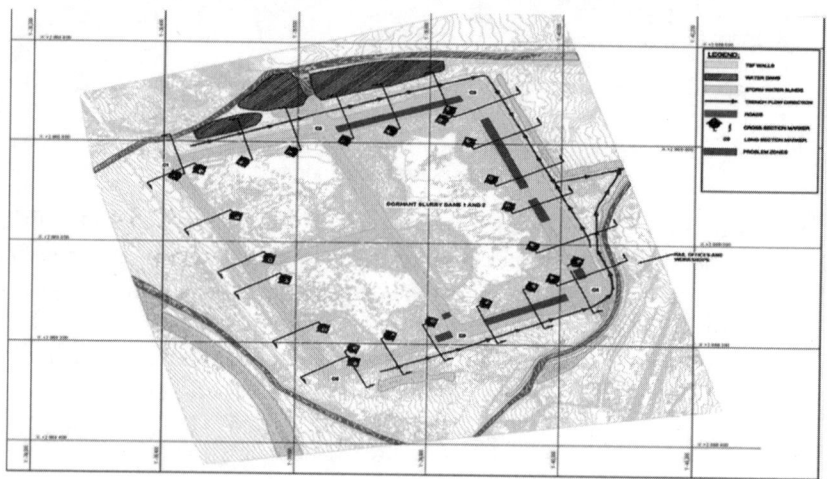

Figure 4: Layout of the slurry dam showing the problem zones

4.1.3 Zones of influence

The guidelines in SANS 10286 were applied to estimate the zone of influence of the slurry dam complex. The zone of influence is the area that would be impacted by the flow of the slurry and water if the outer walls fail of the slurry dam were to fail releasing the contained slurry and any storm water contained.

For estimating the zone of influence, the worst-case scenario has been assumed, namely that the slurry dams are full of storm water and can flow as with hydraulically placed residue. This is a conservative approach. According to SANS 10286, the following guidelines were used to determine the boundary of the zone of influence:

- Upstream of any point on the perimeter, the lesser of a distance of 5h from the toe (where h is the height of deposit at the point under consideration); and the distance to the point where the ground level exceeds $h/2$ above the elevation of the toe at the point on the perimeter;
- On sides parallel to the ground slope – a distance of 10h from the toe; and
- Downstream of the lowest point on the perimeter, a distance of 100h.

Figure 5 is a Google layout of the slurry dam site wherein the zone of influence of the slurry dam complex is shown.

Figure 5: Zone of influence of the slurry dam

It is evident that with the worst-case scenario adopted here, the zone of influence is located to the east and north east of the slurry dams and would impact the rail offices and workshops and all other infrastructure located on the zone of influence.

It should be noted that the zone of influence may be altered to favour South32 by for example constructing bund walls high enough to divert the potential flow path away from the mine workings.

5. Options for short term mitigation measures

The identified and evaluated number of short-term mitigation measures that may be implemented in the short term to reduce the risks of failure of the dam and safety to people and property downstream of the slurry dam complex. The sections below outline the short-term mitigation options.

5.1 Option 1 – Buttressing of the problem areas

In this option, localised buttresses are constructed at the specific problem areas starting with the most critical areas closest to the rail offices and workshop areas (i.e. near the north-eastern corner of the slurry dam complex) and progressing to the lower risk areas.

Figure 6 is a layout of the slurry dam complex showing the localised buttresses that would be constructed as per option 1.

It should be noted that as the buttress may interfere with the existing solution trenches, storm water trenches and bunds, there may need to be modifications to the trenches and bunds to maintain the flows in the trenches and the storm water diversion effect of the bunds. It has been assumed that storm water pipes will be installed to replace the trenches at the affected areas where the localised buttress is constructed over the trenches.

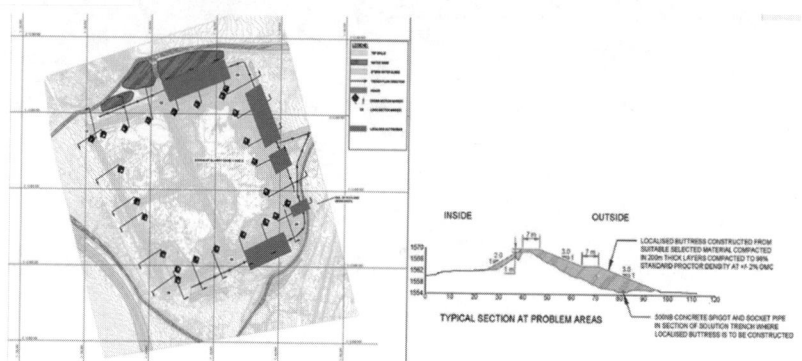

Figure 6: Layout of slurry dams 1 and 2 showing short term mitigation measures option 1 (localised buttresses)

5.2 Option 2 – Ramp and end tipping

In this option, an access ramp is constructed at a suitable location to access the crest of the dam wall and end tipping of suitable material is done into the inner slope of the dam wall with machines driving along the crest of the wall. Reworking of the existing crest will first be done by ripping and re-compacting before new material is placed and compacted on the crest of the wall. The ramp would be constructed to tie into the most critical area closest to the rail offices and workshop areas.

Figure 7 is a layout of the slurry dam complex showing the ramp and end tipping areas that would be constructed as per option 2.

It should be noted that as the ramp may interfere with the existing solution trenches, storm water trenches and bunds, there may need to be modifications to the trenches and bunds to maintain the flows in the trenches and the storm water diversion effect of the bunds. It has been assumed that storm water pipes will be installed to replace the trenches at the affected areas where the ramp is constructed over the trenches.

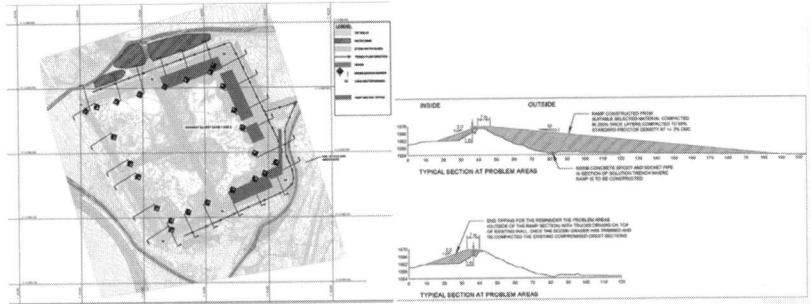

Figure 7: Layout of slurry dams 1 and 2 showing short term mitigation measures option 2 (ramp and end tipping)

6. Long term mitigation measures

The mitigation measures for the long term will commence once the short-term measures are completed or will be done in parallel with the short-term measures, depending on the selected option.

The expert recommendation is option 2 as the preferred short-term mitigation measure with an immediate follow up with the long-term solution.

The long-term mitigation measure will involve continuation with remediation of the outer wall as well as placing and compacting of a capping layer of suitable soil over the top of the slurry dams and to subsequently place topsoil over this material. The capping layer may be obtained from local borrow sources or from the mine stockpiles of overburden materials (if available).

The current outer wall crest elevations (and width) vary significantly (possibly due to failures and settlements caused by spontaneous combustion). The average highest elevation of the crest of the outer walls is approximately 1568.76 mamsl while in some places the crest elevations are as low as 1564.31 mamsl. As part of the long-term measures, it has been assumed that the crest of the outer wall will first be remediated first to have approximately the same elevation (1568.76 mamsl) all around the facility and with a minimum width of 5 m to create sufficient capacity to contain the 1:50 year 24-h storm event with a minimum freeboard of 800 mm, as required by Regulation 704 of the Water Act. A detailed calculation will need to be done at the detailed design phase. Reworking of the existing crest will first to be done by ripping and re-compacting before new material is placed and compacted on the crest of the wall.

For this basic assessment, it has been assumed that the minimum as-built thickness of the capping layer will be 1000 mm compacted and that the thickness of the topsoil will be 300 mm. The borrowed soil capping will be constructed in 200 mm thick layers compacted to 95% Proctor density at ±2% of optimum moisture content.

A detailed design will be required to determine the required levels of the capping layer over the top of the facility based on the topographical survey of the facility. The placing and compaction of the capping layer will be such that the levels of the top of the facility is designed to have grader berms and trenches that will catch and safely route the storm water falling on top of the facility to an outlet structure to be provided at a suitable location on the dam wall crest which will safely lead the storm water to the return water dams at the toe of the facility, via the solution trenches.

Figure 8 shows a typical cross section of the outer dam wall and illustrates the basic design of the long-term mitigation measure whereby a capping layer and topsoil are provided over the top of the facility.

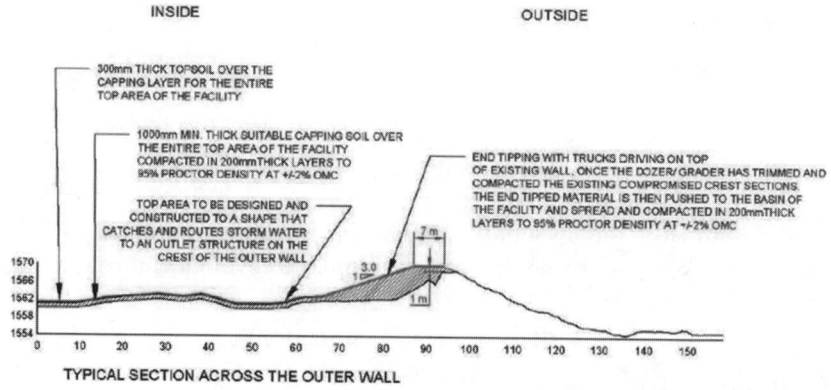

Figure 8: Typical cross section of outer wall showing long term mitigation measures

7. Conclusion and recommendations

A basic assessment of the spontaneous combustion of the dormant slurry dams 1 and 2 has been conducted.

A detailed safety and constructability risk assessment on both short-term and long-term mitigation measures to confirm the feasibility to execute and then select the preferred option needs to be conducted.

A detailed geotechnical investigation and design work including a bill of quantities will be required to underpin the implementation program.

Based on the basic assessment conducted, it is evident that rehabilitation of this facility must be expedited. Any burning spoils or discard within the mine shall be stopped immediately to prevent any air quality impact which would be in violation of environmental legislation.

8. References

[1] Mashamba T, (2017), South32 Eastern Complex Environmental Legal Compliance audit.

[2] Wimberley F et al, (2007), Best Practice Guideline A2: Water Management for Mine Residue Deposits.

[3] Department: Water Affairs and Forestry, (2007), Best Practice Guideline A4: Pollution control dams.

[4] Ifalethu Colliery Internal Audits, Inspections and observations of the current facilities, (2018).

[5] Tshabalala et al, (2018), South32 SA Coal Holding Basic Assessment of the Spontaneous Combustion of the Dormant Slurry Dams 1 and 2, Report 1789271, Golder Associates Tailings Engineer.

[6] Kothuis B et al, (2007), A cleaner production assessment of the ultra-fine coal waste generated in South Africa, The Journal of The Southern African Institute of Mining and Metallurgy; Volume 107.

[7] Van Rooyen K C et al, (2002), An Integrated Method of Coal Discard and Slurry Disposal to Reduce the Environmental Impact from Coal Residue.

[8] Position Statement on Preventing Catastrophic Failure of Tailings Storage Facilities; International Council on Mining & Metals, December 2016.

[9] Wates J et al, (2016), Review of Tailings Management Guidelines and Recommendations for Improvement, International Council on Mining & Metals.

[10] Lizette van der Walt, (2016), Key Changes Promulgate during 2014/2015, Environmental Legal Update Seminar.

41

The importance of high efficiency low emissions technologies for sustainable coal utilisation

Dr Andrew J Minchener OBE

IEA Clean Coal Centre, London, UK

Abstract: Coal utilisation for power generation is facing many challenges including the need to limit greenhouse gas and non-greenhouse gas emissions, minimise water use, and ensure flexible operation. However, it is a key energy choice for developing countries because it is low cost and provides a reliable source of grid-based energy. There is a strong emphasis on its use in Asia and, increasingly, this is based on the introduction of high efficiency low emissions coal power systems, for which there continues to be significant ongoing deployment. At the same time there is ongoing innovative development to further improve overall performance and operational flexibility in order that intermittent renewable sources can be maintained on the grid. For the future, further transformational developments are underway that offer the prospect of maintaining robust coal-based units while also offering the prospect of low cost further reduction of greenhouse gas emissions.

Keywords: Clean coal utilisation; HELE technology; sustainability

1. Introduction

Coal utilisation for power generation is facing many challenges including the need to limit greenhouse gas and non-greenhouse gas emissions, minimise water use and ensure flexible operation. However, it is a key energy choice for developing countries because it is low cost and provides a reliable source of grid-based energy. There is a strong emphasis on its use in Asia and, increasingly, this is based on the introduction of high efficiency low emissions coal power systems, for which there continues to be significant ongoing deployment. At the same time there is ongoing innovative development to further improve overall performance and operational flexibility in order that intermittent renewable sources can be maintained on the grid. For the future, further transformational developments are underway that offer the prospect of maintaining robust coal-based units while also offering the prospect of low cost further reduction of greenhouse gas emissions.

There are three key issues to be considered, namely security of energy supply, economic competitiveness and environmental issues including climate concerns. This trilemma represents an energy selection compromise as it is not possible to maximise all three criteria. In Western Europe, most countries have established anti-coal political positions and are seeking to close coal plants and replace them primarily with intermittent renewable energy resources such as solar and wind power. This approach is very expensive and requires reliable back-up power generation options to avoid major perturbations in grid stability. Such support is typically supplied by coal power plants (Wiatros-Motyka, 2019)

In contrast, in the developing regions, especially Asia, there is a very strong recognition that sustainability is not just about climate issues. It also includes having access to robust, reliable and affordable energy sources that can have a key role in helping to lift people out of poverty. Coal fulfils all these criteria, which is why it is the energy source of choice in the majority of such countries.

From a sustainability perspective, coal utilisation impacts on many issues, as shown in Fig. 1 (United Nations, 2019).

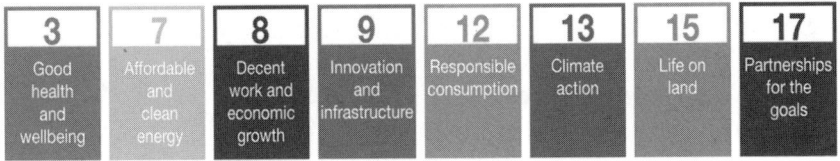

3	7	8	9	12	13	15	17
Good health and wellbeing	Affordable and clean energy	Decent work and economic growth	Innovation and infrastructure	Responsible consumption	Climate action	Life on land	Partnerships for the goals

Figure 1: United Nations sustainable development goals and their applicability to coal production and utilisation

In the developing world, coal use is already very significant and continuing to grow. Most of these coal power plants are young, are at least part owned by the governments and will continue to be used for decades to come. Many more such plants are being constructed (Barnes, 2018). There are also worldwide development programmes to establish high efficiency, low emissions, low water use, coal power units, which can be operated on a flexible basis to meet rapid changes in demand.

2. The case for sustainable coal utilisation

Global energy needs are continuing to rise, which means that we will need all energy sources that are available to us, with maximum benefit being obtained through interconnection. At the same time, there is increasing urbanisation in developing countries and industrialising economies, with some one billion people without access to any power sources and a similar number who only have limited use. It is important to remember that, in such countries, energy use per capita is much lower than in the developed world. The provision of reliable grid-based power will help lift such people out of poverty, with consequent benefits such as improved education and job opportunities. Currently, such power can only be provided effectively by fossil fuel-based systems.

Coal is readily available worldwide, low cost without the price volatility of oil and gas, and can be used for power generation, industrial applications such as cement and steel manufacture, as well as converted to high amenity products such as future fuels and high value chemicals (WEC, 2018). Coal is the second source of primary energy in the world at some 30%, behind oil and ahead of gas, and the leading fuel for power generation at some 40%. However, it has a high carbon content, which raises concern about its potential contribution to global warming (Baruya, 2014).

A great many developing countries have indicated that they intend to continue to use coal, since most cannot afford significant quantities of imported liquefied natural gas and do not recognise any major strategic grid-based benefits from the use of intermittent renewables. A realistic way forward is to encourage them to introduce high efficiency coal power options since these will require less coal per unit of power produced, with a corresponding decrease in CO_2 emissions. This can be achieved through the deployment of high efficiency low emissions (HELE) coal power plant and in due course, when market conditions are right, the application of carbon capture utilisation and storage (CCUS). Such an approach will be effective and allow us to limit future carbon emissions at a much lower cost while ensuring that the advantages of coal use are maintained. It is also possible to ensure levels of conventional pollutants such as particulate matter (PM), sulphur oxides (SOx) and nitrogen oxides (NOx) will meet ever tighter regulations as they are easily removed using state of the art low-cost technologies (Zhu, 2016). It is important to recognise that technology suppliers and users certainly see the benefits of coal use technologies and there is a wealth of transformational development underway to achieve greater efficiencies while ensuring emissions of non-greenhouse gases remain lower than those that can be achieved by gas-fired units. The current best efficiency level is some 47% (lower heating value (LHV), net basis), the exact values depending in part on plant location. As well as establishing units with higher steam pressures and temperatures, very innovative improvements are being taken forward to boost efficiencies to over 50%.

With regard to CCUS, the inclusion of such techniques will allow near-zero emissions to be achieved from a coal plant, while the captured CO_2 can be used in several applications, especially in enhanced oil recovery. However, there is a need for government drivers to ensure commercial-scale units will be established.

3. Current HELE coal fired power generation and beyond

A HELE coal power system comprises the essential components employed in all coal power systems, while operating at higher steam temperatures and pressures than conventional units, Fig. 2. In particular, steam generated in the boiler is carried to a steam turbine that comprises a high-pressure (HP) turbine, an intermediate-pressure (IP) turbine and one or more low-pressure (LP) turbines. Steam passes from one to the next in sequence. Further efficiency gains can be achieved by reheating the steam between the HP and IP turbines. This recycle can take place either once or twice, known as single and double reheat respectively. The latter provides a more efficient system but has a higher capital cost requirement.

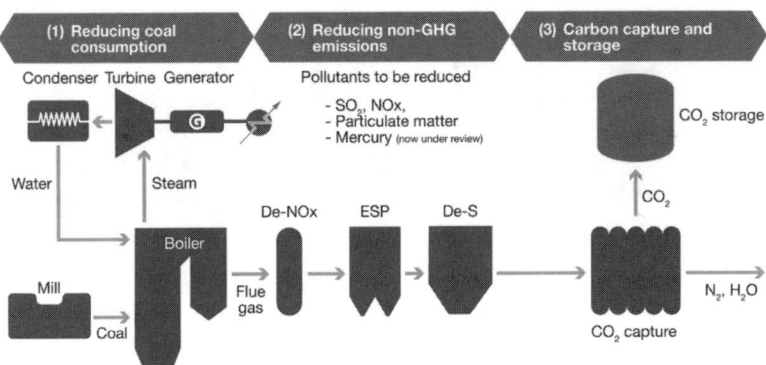

Figure 2: Schematic of a HELE coal power plant

3.1 Lower CO_2 emissions through higher efficiencies

A subcritical plant typically operates at steam temperatures up to 540°C and has a thermal efficiency of between 30% and 39% (net, LHV basis), depending on the unit size, coal quality and local conditions. In order to achieve higher efficiencies, supercritical and, most importantly, ultra-supercritical (USC) coal-fired technologies have been developed. On large-scale units (6601000 MWe) with USC steam conditions, a design thermal efficiency level of 45–47% has regularly been achieved depending on the exact conditions (Fig. 3). The latest designs can achieve efficiencies close to 48% with some units under construction expected to achieve over 49% (net, LHV basis). This represents a >30% reduction in CO_2 emissions compared to the global average.

The driver to increase the thermal efficiency of power generation, that is increase the amount of energy in the coal that is converted to electricity, is essentially an economic decision for which there is a trade-off between the capital and operating costs involved, and the risks involved. Thus, USC plants have higher capital costs than the conventional units because of the higher requirements of the steel needed to withstand the higher pressure and temperature. However, this is offset by costs savings due to the higher efficiency of the process. These include reduced coal use for a given electricity output, while the plant has a smaller footprint with respect to size of coal handling and emission control systems.

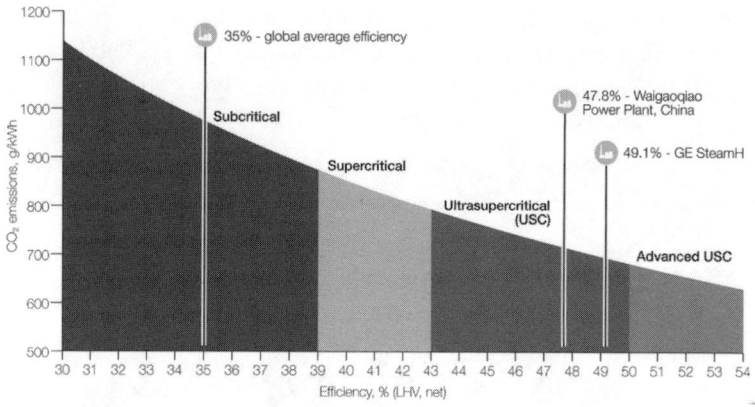

Figure 3: HELE efficiency and CO_2 emissions (Baruya, 2018)

In Asia, for example, the overall cost for an USC coal power plant is lower than for a unit with conventional (subcritical) steam conditions (IEA, 2017). Although the capital investment is greater, the coal cost is lower as less fuel is needed. That said, such a financial balance is sensitive to the cost of capital (Fig. 4).

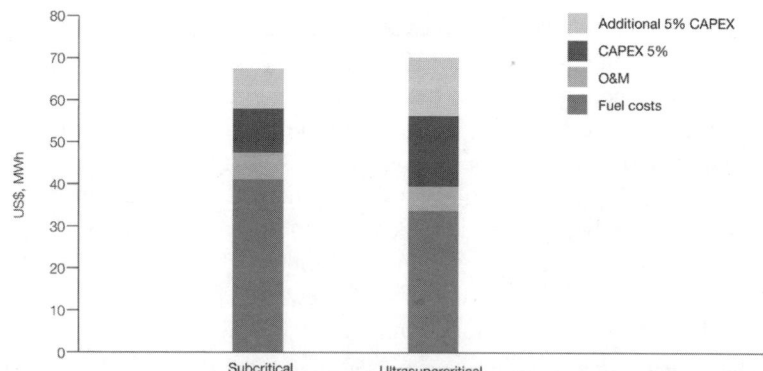

Figure 4: Full cost comparison of a subcritical and ultra-supercritical coal power plant in Asia (IEA, 2017)

3.2 Market penetration for HELE coal power

HELE coal power technology refers to the commercially available widely deployed units with USC steam conditions. These are being deployed in some 17 countries worldwide, especially in China, Germany and Japan, which are also the countries most involved with technology exports worldwide. Table 1 shows the number and distribution of such plants either in operation or under construction in June 2018 (Platts, 2018).

Table 1: Distribution of HELE coal power plants with USC steam conditions (Platts, 2018)

Region	In operation (MWe)	Under construction (MWe)
Asia	224,203	88,228
Europe	19,208	4,970
Middle East	0	2,400
Eurasia	300	0
North America	665	0

3.3 Means to improve the environmental performance of coal power plants

Improvements in environmental performance are driven by the legal requirement to meet emission standards. The tightest are those set in China, as shown in Fig. 5. Coal power plants in Eastern China were required to meet the ultra-low emission standards by 2017 and in Central China by 2018, while coal power plants in Western China are encouraged to achieve emissions that meet or are close to ultra-low emission levels. There are some exemptions for circulating fluidised bed combustion units that burn low-grade fuels and wastes and down-fired W flame boilers that burn low volatile coals. These do not have to meet the ultra-low emissions standards but must meet the emission standards that came into force from 2012.

For the conventional pollutants (IEA, 2014), the inclusion of appropriate well-proven flue gas cleaning units can meet all current requirements reliably and economically, such as electrostatic precipitators (ESPs) or bag filters for fine particulates removal, flue gas desulphurisation (FGD) for SO_2 control, together with combustion modifications and/or catalytic reduction systems for control of NOx, Fig. 6.

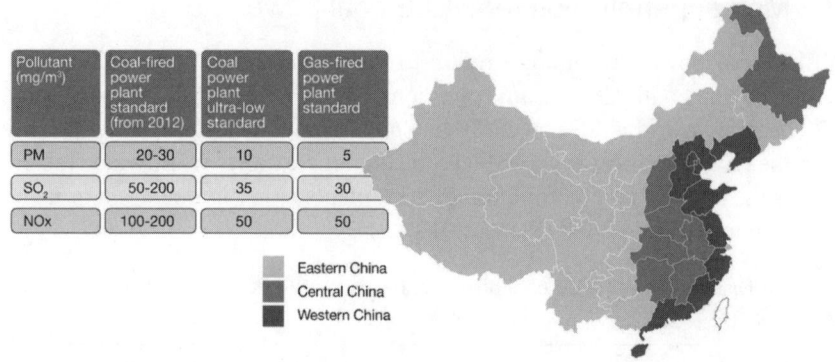

Figure 5: Coal power emission standards in China (Zhu, 2016)

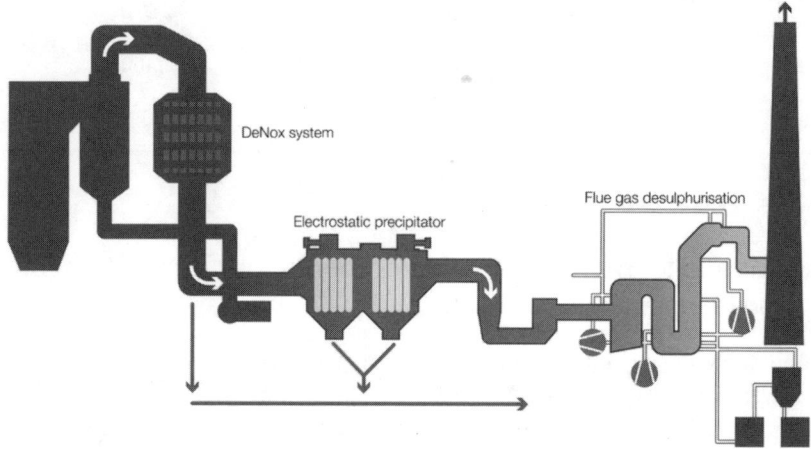

Figure 6: Schematic of the layout of conventional pollutant control systems (Feng, 2015)

Fine particulates and NOx emissions control systems have a relatively small effect on the overall thermal efficiency of the power plant, while the inclusion of FGD can result in a one percentage point loss in thermal efficiency (Minchener, 2012). The capital cost of these three measures can represent about one third of the cost of the unit when meeting the more stringent current standards.

Multipollutant technology offers the scope to combine all individual devices in to a single integrated system. This offers the significant potential of an efficient single solution that will reduce the land footprint and reduce the capital cost, thereby providing another step towards zero pollutant emissions.

These systems are currently under development, particularly in Japan (METI, 2013).

3.4 Ongoing technology developments and demonstrations

For the future, new higher temperature alloys are being developed, with R&D underway in China, Japan. India, Europe and the USA (Table 2). The aim is to achieve steam temperatures of 700-760°C, which would mean that coal power plants could reach net thermal efficiencies of 50–55%, although a considerable amount of work remains to be done, with timelines to implementation for demonstration projects over the period of 2021–2025.

Table 2: Materials development programmes for advanced USC coal power plants (IEACCC, 2017)

Programme	Steam temperature (°C)	Target efficiency (%, LHV, net)	Programme start date	Demonstration plant date and size (MWe)
EU	700	50	1998	2021 (500)
USA	760	45–47 (HHV)	2000	2021 (600)
Japan	700	>50	2008	2021 (600)
China	700	46–50	2011	2021 (660)
India	700	>50	2011	– (800)

While the aim of these demonstration programmes is to prove the performance of nickel alloy components for use with 700°C steam conditions, an alternative approach is also being pursued. With more rapid advances in martensitic steels, it is possible that the steam temperature limits for advanced USC plant will be closer to 650°C, since the materials burden is lower, while the shortfall in efficiency has been limited through careful design and component integration. In this technology variant, GE is taking forward advanced USC technology, with steam conditions of 33 MPa/650°C/670°C, which is linked into their digital optimisation control system. The design cycle efficiency is 49.1% (net, LHV basis). Materials of construction include martensitic steels for most components with high nickel alloys in areas of critical importance such as steam pipes and the steam turbine inlet. The technology was launched in October 2017, with projects underway for advanced coal power plants in Turkey and China (GE, 2018).

The third strand of this global R&D programme is to determine and implement overall design changes for the power plant, through the deployment of optimised individual components and their tighter integration. A key example is being championed by Prof Feng Weizhong, formerly employed by Shenergy but now operating in an independent capacity. He is leading the development and demonstration of an advanced USC technology, which incorporates all the improvements that he made on the Waigaoqiao No. 3 units, each of 1000 MWe capacity, together with additional innovative components (Zhang, 2018). This is being built at the Pingshan Phase 2 site in Anhui Province. It comprises a 1350 MWe double reheat USC with an adapted steam turbine layout (Fig. 7). In this arrangement the turbines are split into two trains. The front train comprises the high-pressure turbine (HP) and intermediate pressure turbine (IP1) coaxial with one generator as the front unit. This is mounted on top of a two-pass boiler near the outlets of the tower type boiler steam headers, which is around 80–85 m above ground level. The rear train, which consists of the IP2 and the two LP turbines coaxial with another generator as the rear unit, remains in the conventional position, some 17 m above ground level. This approach minimises the lengths of the main steam pipe, cold reheat steam pipe, hot reheat pipes and the cold reheat steam pipe. The shorter pipework represents a significant cost saving and reduces the pressure drop and temperature loss of steam from the boiler, which increases efficiency. There is close cooperation with Siemens, who supplied the adapted turbine, GE and the East China Electric Power Design Institute (Table 3).

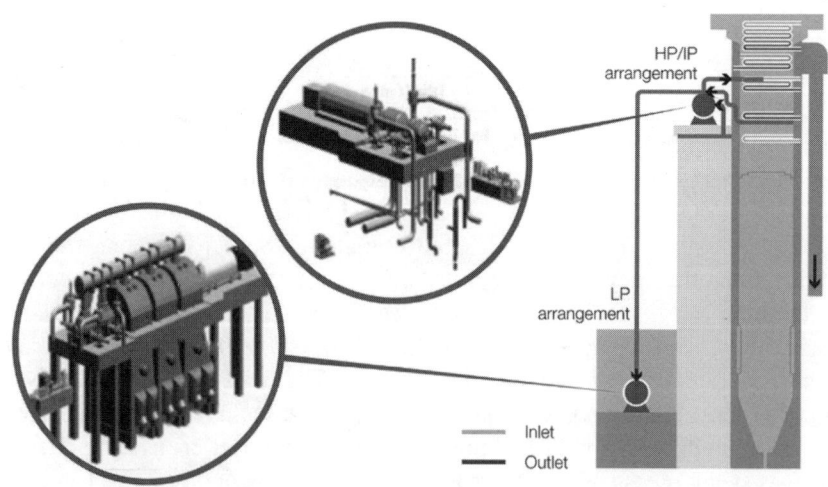

Figure 7: Schematic of the split turbine approach (Zhang, 2018)

Table 3: Design conditions for the Pingshan 2 1350 MWe double reheat USC coal power plant with adapted steam turbine layout (Zhang, 2018)

Design condition	Expected output
Rated output (MWe)	1350
Rated main steam flow (t/h)	3229
Max main steam flow (t/h)	3416
Main steam pressure/reheat steam I pressure/reheat steam II pressure (MPa)	30/9.17/2.25
Main steam temperature/reheat steam I temperature/reheat steam II temperature (°C)	600/610/620
Cooling water temperature (°C)	19

The other key requirement for coal and gas-fired power plants is to provide stability to the grid systems which is reduced by the introduction of intermittent supply from wind and solar sources. A rising proportion of these renewables is reducing the inertia of grids, presenting challenges for stability and security of supplies. Consequently, coal and gas-fired power plants have to operate at highly variable loads and turn on and off at short notice to keep grids stable. Thus, the fossil fuel plants have to operate with greater resilience for high availability, have the capability for fast start-up and rapid ramp rates, the means to meet a wide load range with a lower minimum output, and be able to minimise the need for sudden shut-downs as these can be damaging and expensive. All this has to be done while maintaining efficiency as far as possible and ensuring emissions compliance. This operational approach was not envisaged when such coal plants were designed but retrospective changes and improvements have ensured that it can be achieved, albeit at some cost (Henderson, 2014; Wiatros-Motyka, 2019).

3.5 The role of CCS/CCUS in lowering carbon emissions

Carbon capture, utilisation, and storage (CCUS) is a combination of techniques that can remove CO_2 from the flue gas of a power plant or other industrial process and in principle from the atmosphere, followed by processing of the CO_2 either for utilisation or for transportation and subsequent storage in secure geological formations (AIChE, 2018). As was indicated above, the introduction of HELE technologies can make significant reductions in CO_2 emissions; however, CCUS/CCS can achieve over 90% reductions. Various projections suggest that despite the adoption of alternative energy sources and energy efficient systems to reduce the rate of CO_2 emissions, if the aim of the Paris Agreement is to be achieved then CCUS/CCS should be included in the mitigations measures, otherwise the overall cost will be some 138% higher and it may prove impossible to achieve the reduction targets (IEA, 2018) There is also no other cost-effective technology solution capable of delivering the deep emissions reductions needed across key industrial processes including steel, cement and chemicals manufacturing, and oil and gas production, all of which will remain vital building blocks of modern society.

While the first-generation CO_2 capture techniques, which include a solvent-based process, pipeline transportation of CO_2 and its use for enhanced oil recovery from declining wells, have been shown to work well, albeit at limited scale, the consequence is an operational efficiency penalty and a major increase in capital costs for the power plant or other industrial process. This raises the cost of energy and while CO_2 enhanced oil recovery will provide a valuable income stream, some form of financial subsidy will be necessary especially when CO_2 storage rather than utilisation is undertaken. This suggests that the technology is unlikely to be introduced at commercial scale without regulatory pressure. Equally important, while the first-generation options are being taken forward, it is also critical to address various scientific, economic and societal aspects to ensure successful development and implementation of further CCUS technology options that are not yet ready for demonstration, let alone commercial deployment, but which appear to have potential advantages (Zhu, 2018; IChemE, 2018; Lockwood 2018).

4. A way forward

Globally, the need for a continued supply of secure and affordable energy will increase as demand continues to grow. As such, all available energy sources will need to be used to meet demand, with newer options such as renewables supplementing the traditional fossil fuels of coal, gas and oil (Crooks, 2019). At the same time, societal drivers require an increasing emphasis on environmental sustainability such as clean air and reduced carbon emissions, as is reflected in the United Nations Sustainable Development Goals. The production and use of coal can be linked to each of these challenges.

There is no one-size-fits-all solution to the environmental sustainability challenges. Many OECD (Organization for Economic Cooperation and Development) countries have added intermittent renewables to the overall energy mix, often at the expense of coal. In contrast, while some renewables will be used, developing countries will continue to use coal and are unlikely to include CCUS under current commercial conditions. Consequently, to start to limit carbon emissions, there is a need to establish a staged approach. This is based on encouraging and supporting developing countries to introduce HELE coal power technology either as retrofit or new build options, preferably based on USC technology and in due course advanced USC as the development activities reach maturity. This can be followed by the addition of CCUS/CCS on such units when the business case is right.

Many nations see coal as a key part of their ongoing strategic energy mix and this is reflected in their National Determined Contributions (NDC) to reducing carbon emissions as signatories of the historic Paris Agreement (UNFCCC, 2019). Thus, Afghanistan, Bangladesh, Bosnia and Herzegovina, China, Egypt, Georgia, Ghana, India, Indonesia, Japan, Kazakhstan, Kenya, Mongolia, Montenegro, Myanmar, Nigeria, North Korea, Pakistan, Philippines, Republic of Macedonia, South Africa, Turkey, United Arab Emirates and Vietnam all identified a role for HELE coal power technology in their NDCs. Between them these countries account for over half of global coal power carbon emissions. In addition, nations including Bahrain, Canada, China, Egypt, European Union (currently representing 28 countries), Iran, Iraq, Malawi, Montenegro, Norway, Saudi Arabia, South Africa, United Arab Emirates and USA made either direct or indirect reference to CCUS/CCS in their NDCs.

5. Conclusion

Even with the drive to push the introduction of intermittent energy sources such as solar and wind power, coal currently provides 41% of the world's electricity and is an essential raw material in the production of 70% of the world's steel and 90% of the world's cement. It is set to remain a significant and integral part of the global energy mix for well into the future.

To move towards near-zero coal power, new plant should be based on the deployment of ever improving HELE technologies with the scope to deploy CCUS in due course.

National governments should support the deployment of HELE technologies as part of an emissions reduction strategy while also determining a coherent business model to ensure deployment of CCUS.

With the rise of intermittent renewable power, coal fired plants now have to be ever more flexible, with the need for fast ramp response times and limited operation at maximum load.

In many parts of the world, water utilisation for power production is becoming an issue, with countries introducing legislation that requires utilities and other industrial users to minimise water usage, maximise water recycling in order to meet new environmental regulations, which will impact to varying degrees on operational efficiencies.

The technology development is moving rapidly, to take forward the advanced USC concepts while beyond that are some interesting and innovative alternative systems for which development and planned demonstration are underway.

The IEA Clean Coal Centre will continue to maintain a watching brief on all of these issues, which will be reported to all stakeholders through its assessment study reports, its various dissemination activities, and its outreach programme. As noted above, this is based on ensuring complementarity with the United Nations Sustainable Development Goals (IEA Clean Coal Centre, 2018).

6. References

[1] AIChE (2018) Carbon capture utilization and storage. Available from: www.aiche. org/ccusnetwork/what-ccus (2018).

[2] Barnes I (2018) HELE perspectives for selected Asian countries. CCC/287, London, UK, IEA Clean Coal Centre (June 2018).

[3] Baruya P (2014) Coal reserves in a carbon constrained future. CCC/233, London, UK, IEA Clean Coal Centre (March 2014).

[4] Baruya P (2018) Personal communication (September 2018).

[5] Feng W (2015) Developing green, highly efficient coal-fired power technologies. ASME 2015 Power Conference, San Diego, CA, USA, June 28–July 2, 2015.

[6] GE (2018) Smarter. Cleaner. Steam power. Available from: www.ge.com/content/ dam/gepower-steam/global/en_US/documents/2018-Steam-Power-Product-Catalogue.pdf (2018).

[7] Henderson C (2014) Increasing the flexibility of coal-fired power plants. CCC/242, London, UK, IEA Clean Coal Centre (September 2014).

[8] IChemE (2018) Update on the Demonstration of the Allam cycle. Available from: https://www.google.co.uk/search?q=Liekly+ efficiency+for+coal+fuelled+Allam+ Cycle&rlz=1C1GCEA_enGB745GB745&oq=Liekly+efficiency+for+coal+fuelle d+Allam+Cycle&aqs=chrome69i57.17363j1j7&sourceid=chrome&ie=UTF-8 (11 April 2018).

[9] IEA (2014) Emissions reduction through upgrade of coal-fired power plants. Available from: www.iea.org/publications/freepublications/publication/Partner Country Series Emissions Reductionthrough Upgradeof CoalFired Power Plants.pdf (2014).

[10] IEA (2017) Full cost comparison of a subcritical and ultra-supercritical coal power plant in Asia (2017).

[11] IEA (2018) IEA and UK kick-start a new global era for CCUS. Available from: www.iea.org/newsroom/news/2018/november/iea-and-uk-kick-start-a-new-global-era-for-ccus.html (November 2018).

[12] IEACCC (2017) Compiled from data presented at the 3rd AUSC Workshop, Rome, Italy, (December 2017).

[13] Lockwood T (2018) Overcoming barriers to carbon capture and storage through international collaboration. CCC/284, London, UK, IEA Clean Coal Centre (March 2018).

[14] METI (2013) Advancing highly efficient technology and environmental performance. Available from: www.meti.go.jp/english/publications/pdf/journal2013_10a.pdf Ministry of Economy, Trade and Industry of Japan (2013).

[15] Minchener (2012) Non-greenhouse gas emissions from coal-fired power plants in China. CCC/196, London, UK, IEA Clean Coal Centre (April 2012).

[16] Platts (2018) World electric power plants database. Available from:www.spglobal. com/platts/en/products-services/electric-power/world-electric-power-plants-database (June 2018).

[17] UNFCCC (2019) The Paris agreement and NDCs. Available from: https://unfccc. int/process/the-paris-agreement/nationally-determined-contributions/ndc-registry (January 2019).

[18] United Nations (2019) About the sustainable development goals. Available from: www.un.org/sustainabledevelopment/sustainable-development-goals/   (January 2019).

[19] WEC (2018) Coal. Available from: https://www.worldenergy.org/data/resources/ resource/coal/ (World Energy Council 2018).

[20] Wiatros-Motyka M (2019) Power plant design and management for unit cycling and load fluctuation, London, UK, IEA Clean Coal Centre (report in preparation) (January 2019).

[21] World Economic Forum (2017) More people live inside this circle than outside it –and other demographic data you should know. Available from: www.weforum. org/agenda/2017/07/more-people-live-inside-this-egg-than-outside-of-it-and-other-overpopulation-data/ (July 2017).

[22] ZhangX (2018) Modern power systems, Vol 38, No 12, page 28, 30 December 2018.

[23] Zhu Q (2016) China – policies, HELE technologies and CO2 reduction. CCC/269, London, UK, IEA Clean Coal Centre (September 2016).

[24] Zhu Q (2018) Developments in CO2 utilisation technologies. CCC/290, London, UK, IEA Clean Coal Centre (October 2018).

42

Determination of rational parameters of mechanoactivated coal briquets

N. Nikolaeva[1], A. Afanasova[1], A. Aleksandrov[2]

[1]Saint-Petersburg Mining University, Sankt-Peterburg, Russia
[2]Graduate School of Technology and Energy, Saint Petersburg State University of Industrial
Technologies and Design, Saint Petersburg, Russia

Abstract: Coal is unstable during the storage and disposed to self-ignition. A great quantity of fine leads to the difficulties of transportation and the impossibility of burning in standard fire grate furnaces. The solution of a problem of underutilization of coal resources as a result of direct fuel utilization of run-of-mine coal is briquetting. New briquette composition including binder and filler was developed. Technological approaches include sizing of run-of-mine coal and briquetting of coal fine with the refinery waste and anthropogenic carbon-containing waste products of hydrolytic industry were designed. Technological and granulometric characteristics of brown coal and hydrolytic lignin were researched. Optimal parameters of coal briquetting were established. Influences of moisture content, binder proportion, and briquetting pressure on briquette compressive strength were studied. Fundamental flow-sheet of fine coal briquetting with the addition of technical hydrolytic lignin as a fuel briquette filler and binder was developed. The technology was adopted for one of the deposits. It can be varied to fit another objects.

Keywords: Coal briquetting, briquette composition, coal briquetting parameters, moisture content, binder proportion

1. Introduction

The production of fine coal in mines and coal preparation plants is increasing. A fraction of this coal can be blended with the larger-size clean coal and shipped to the user. However, a convenient means to handle, store, transport, and use the balance must be devised. Coal reconstitution, encompassing briquetting, disk pelleting, extrusion pelletization, and roller-and-die pelletization, is a means frequently proposed for this service (Nicholas Conklea and Ragavana, 1992).

Reasonability of fine coal briquetting is caused by its fine-disperse state and the difficulty of transportation, by the impossibility of burning in standard fire grate furnaces. Coefficient of efficiency of utilization of fuel briquettes is 75% in comparison with 46.7% for run-of-mine coal (Petrova and Bychev,

2001; Aleksandrova and Shabarov, 2016; Rasskazova and Alexandrova, 2016; Rasskazova et al., 2013).

In developed countries, there is an increasing interest in the combustion of coal and biomass mixtures. There is also an increase in the biomass cultivated for energy. The technologies utilizing renewable energy sources have been known well enough, and some were tested in developed countries (Borowski, 2007).

The coal processing industry usually discards fine-size (-150 µm) coal because of its high-moisture content and handling problems. Compacting and briquetting of the fines with and without adding of biomass can be solution of the problem (Patil et al., 2009). One of the most effective additives to use in a briquetting is technical hydrolytic lignin (THL) – large tonnage waste hydrolysis industry. At the moment when the sawmill is formed about 40% of waste, which have the highest energy parameters and can be used to generate energy. Forest raw waste may be part of fuel briquettes, but they may also be briquetted without binders, since they contain lignin in their composition, which is plasticized during pressing and is a binder. Earlier studies (Aleksandrova and Rasskazova, 2016; Koga, 2013; Flynn and Wall, 1966) showed that the mechanical activation of technical hydrolytic lignin can improve the strength characteristics of fuel briquettes. In this case, the analysis of the fractional composition of the obtained components of fuel briquettes becomes important.

2. Materials and methods

The object of research is coal of one of deposits (Rasskazova et al., 2013; Aleksandrova and Korchevenkov, 2017). The goal of research is substantiation of technological parameters of brown coal briquetting. Mechanical activation (MA) THL was carried out using a high-speed planetary mill E_{max}, the view of which is shown in Fig. 1(A) (in general form). The maximum speed of 2000 rpm, which has no analogues in its class, as well as the innovative shape of the grinding bowl make it possible to achieve high grinding efficiency as a result of impact and friction forces.

The combination of shape and circular movements of the grinding bowl significantly improves the mixing efficiency of the sample, the final fineness, and also narrows the range of size distribution compared to other ball mills. Analysis of the fractional composition and measurement of its specific surface area was carried out using the laser diffraction method on a Mastersizer 2000.

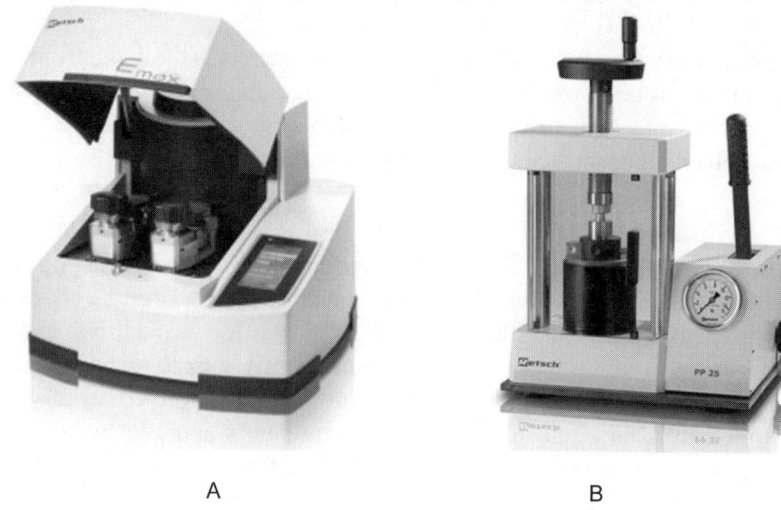

A   B

Figure 1: Used equipment. (A) Planetary mill and (B) tablet press

Thermogravimetric analysis (TGA) were performed on the analyzer SDT Q-600 TA Instruments company (USA), which allows to record simultaneously the sample weight change (TGA) and the process is accompanied by thermal liberation or absorption (differential scanning calorimetry and thermal analysis).

In conducting the research, following methods and approaches were used:

- Determination of the calorific value of the fuel in the calorimeter type Berthelot solid fuel produced according GOST147-2013.
- Organic carbon content was studied using the TOC-V analyzer (Shimadzu). Low ash and high heat of combustion are also factors that positively affect the quality characteristics of the fuel briquette.
- Briquetting was carried out using a tablet press PP-25 (Fig. 1B) and specially made dies for varying the size of the briquette.

2.1 Technological characteristics of the coal

Technological indicators of coal are given in Table 1, where W is the coal moisture, V is the volatile substances output, A is the coal ash content (r, a, daf indicates run-of-mine, analytical, and dry ash-free state, respectively), S_{tot}, N are the sulfur and nitrogen content, respectively, and Q^a_s is the high heat value.

Size test analysis revealed that initial coal samples are characterized by fine dust (−0.15 mm) content at the average of 8% and varies from 3% up to 14%.

High output of fine fractions predetermines the reasonability of coal utilization for briquette production. It is considered that in the case of co-combustion of biomass whose content is higher than 5%, there are necessary technological lines that would provide loading of this fuel to the boiler in a way independent of coal. In such a way, it is possible to increase the content of biomass from 10% to 15% of the calorific value of a fuel flux. In such briquettes, biomass made approx. 20% by weight (Borowski, 2007).

Lignin is researched as briquette filler in this study. Lignin is a waste product of hydrolytic factories. It has a porous structure that improves combustion kinetics of fuel briquette.

Low ash content and high calorific value (Table 1) are also factors improving quality characteristics of fuel briquette.

Table 1: Technological characteristic of coal briquette components

Material	W^a (%)	V^a (%)	A^a (%)	S^{tot} (%)	N (%)	Q^a_s (MJ/kg)
Coal	24.32	29.37	17.59	0.37	0.43	23.39
Hydrolytic lignin	61.8	29.5	0.65	1.1	0.57	16.6

Sulfuric acid content is 1.78 kg/ton (0.17%). Consequently lignin dumps are potential source of sulfuric acid and it is necessary to organize the processing of the waste. Lignin dumps are prone for spontaneous ignition in hot and dry weather.

3. Results and discussion

The kinetics of thermal decomposition of THL was determined on the basis of an analysis of experimentally obtained data of thermoanalytical curves, whose thermoanalytical signal can be expressed by total enthalpy for differential scanning calorimetry (DSC) and total mass loss for thermogravimetry (TG). The most popular methods of kinetic analysis are: Friedmann differential analysis, integral-kinetic analysis of Ozawa-Flin-Wall, Kissinger method. In this work, the Friedman differential method was used. According to this method, the Friedman kinetic model for the n-th order of the reaction has the form:

$$y = \ln A + n\ln(1 - x) + \left(\frac{-E_a}{R}\right)z \qquad (1)$$

where $y = \ln(dx/dt)$, $y = 1/T$.

The dependence of the logarithm of the rate conversion dx/dt from $1/T$ is a straight line tangent angle of inclination which is:

$$m = \frac{E_a}{R} \tag{2}$$

A graphic visualization of the results before and after the mechanical activation is shown in Fig. 2.

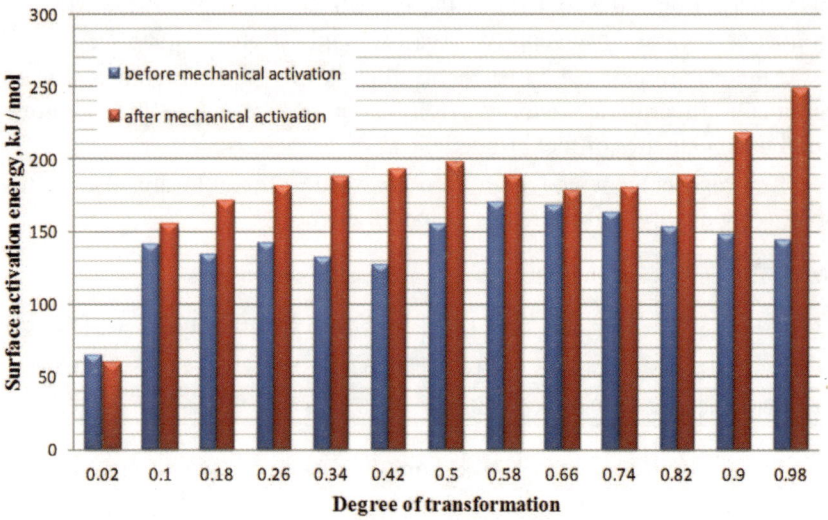

Figure 2: Comparison of surface activation energy before and after mechanoactivation of THL

The research results showed that the average value of the surface activation energy after mechanical activation (180.7 kJ/mol) is greater than before mechanical activation (141.92 kJ/mol). Analysis of the data shows an increase in the activation energy of the sample over the entire range of conversion degrees. The activation coefficient K_a can be defined as the ratio of the sample activation energy after mechanical activation to the activation energy of the original sample. The average value of K_a was 1.25 with an activation time of 5 min, which makes it possible to conclude that the potential interaction energy of fuel briquettes increased by 25%.

The value of the fractional composition is determined by the influence of the total contact surface of the grains, the number and size of voids in the structural frame of briquettes, the content of acute-angular grains, the relief of a solid surface, and the presence of dust particles.

The kinetics of the MA process was investigated in the range from 0 to 20 min of mechanical action. Control measuring point for determining the specific surface at intervals of 5 min. It was established that the maximum increase of the specific surface of the particles is observed after 5 min of mechanical action.

The specific surface of the mixture grains is an important indicator in briquetting, since it determines the thin-layer distribution and structurization of the components, as well as the proportion of adsorption contacts. The higher the number of grains, the more active centers – surface elements, in which atoms with unoccupied valences are concentrated. Packing density is closely related to grain size. Small grains are more ribbed, and the heat of their wetting is four times higher than that of large ones. The high content of large grains (more than 6 mm) adversely affects the strength of briquettes. When pressed, such particles easily crack, new surfaces appear, uncoated with a binder. The presence of dust particles increases the surface area and, consequently, the ascending flow binder that promotes densification briquettes resulting active filling voids. On briquettes packing density significantly affects the structure porosity. How would thoroughly solid granules were packed into bricks, between them there is always the pores. The briquettes of fine-grained particles have small pores and they are mostly filled with a binder. Defects in the form of voids are few, the strength of briquettes is great. Briquettes with a predominance of large grains have a large number of defects, a bulk layer of binder for filling voids in them is not enough, so these briquettes have low strength. To increase the strength of the packaging, it is recommended to inject the mixture into briquette fines which readily penetrate into the voids.

Irregularities and roughness of the material have a positive effect on the mechanical fastening of the binder on it, increasing the strength of the briquettes. The strength of the briquettes is lower than homogeneous sieve composition. The homogeneous mixture cannot provide proper packing density, grain stacked with a large number of voids in the frame, the compacting pressure is unevenly distributed in the volume of the system, the briquettes are easily deformed.

In order to identify the joint effect of mechanical activation time and THL share a part of the preform is applied experimental design method. The method of conducting an experiment using a scheduling matrix allows obtaining statistical mathematical models of processes using factor planning, regression analysis, and gradient motion. For this purpose, a planning matrix was constructed corresponding to a fractional-factorial experiment (Kono's plan on the Ko cube).

Applied scheduling matrix which is close to D-optimal has properties and uniformity rotatable, has a small number of experiments. A smaller number of experiments compared to the matrices of the rotatable central compositional experiment (RCCE) is achieved by reducing the number of experiments that have equal dispersion of the output parameter. Selection of the plan was due to its efficiency and good statistical properties.

As a parameter optimization Y considered the calorific value of the briquette, depending on the content of THL and its activation time chosen for the study factors and ranges of their variation, are shown in Table 2.

Table 2: Factors and ranges of their levels

Selected factor	Designation	Factor levels		
		−1	0	1
Exposure time	X_1	0	5	10
Lignin consumption (g/t)	X_2	0	15	30
Pressing pressure (kgf/cm²)	X_0	600	600	600

The analysis of data was obtained from regression equation for predicting the calorific value of briquettes from lignin content and the time of mechanical activation:

$$Q_{br} = 16.17 + 0.41 \cdot t_1 - 0.90 \cdot q_1 \cdot t_1 - 1.90 \cdot q_1^2 - 0.21 \cdot t_1^2$$

where is the calorific value, mJ/kg, q_1 is the THL consumption, %; t_1 is the mechanical activation time, min. Graphic visualization of the prognostic model is shown in Fig. 3.

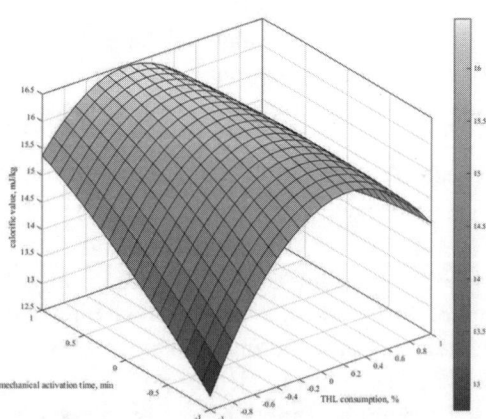

Figure 3: Visualization of the dependence of the calorific value of a briquette on the consumption of lignin and the time of its mechanical activation

4. Conclusion

The developed algorithm and methods for evaluating the results of the analysis of thermographic studies allow us to predict the calorific value of fuel briquettes. The results of the dependence of the activation energy on the composition and time of mechanical activation show the non-additive contribution of each component. This can be explained by the positive effect of the THL addition, which helps to improve the uniformity of the briquette structure during thermoplasticization of lignin. It can be assumed that non-additiveness is a consequence of the effect, the physical meaning of which is as follows: in the process of interaction between the components of the briquette, united into the system, they are synchronized under the influence of both external and internal factors and the technological behavior of each individual component acquires a coordinated direction. The resulting effect of such a coherent collective action is obtained other than a simple sum of the effects of the actions of each component separately.

Thus, based on the interpretation of laser diffraction data and thermograviometric data, mechanoactivation criteria have been proposed. The developed methods and criteria for evaluating the results of laser diffraction analysis and thermographic studies allow us to predict the efficiency of mechanical activation and the calorific value of THL. At the same time, by varying the composition and technological parameters of briquetting, it is possible to predict the production of fuel briquettes of required quality from wood waste.

5. Acknowledgement

The work is carried out under financial support of the Ministry of Education and Science of the Russian Federation, the project RFMEFI57417X0168.

6. References

[1] Aleksandrova, T.N., Korchevenkov, S.A. Ecological and technological aspects of ash and slag wastes utilization. (2017) Journal of Ecological Engineering, 18 (4), pp. 15–24.

[2] Aleksandrova, T.N., Shabarov, A.N. Key issues for improving of carbonaceous ore beneficiation processes for the extraction of valuable components. (2016) XVIII International Coal Preparation Congress: 28 June–01 July 2016 Saint-Petersburg, Russia, pp. 1179–1183.

[3] Patil, D.P., Taulbee, D., Parekh, B. K. & Honaker, R. Briquetting of coal fines and sawdust – effect of particle-size distribution. (2009) International Journal of Coal Preparation and Utilization, 29 (5), pp. 251–264.

[4] Flynn, J.H., Wall, L.A. A quick, direct method for the determination of activation energy from thermogravimetric data. (1966) Polymer Letters, 4 (5), pp. 323–328.

[5] Friedman, H.L. New methods for evaluating kinetic parameters from thermal analysis data. (1969) Journal of Polymer Science: Polymer Letters, 7 (1), pp. 41–46.

[6] Borowski, G. The possibility of utilizing coal briquettes with a biomass. (2007) Environment Protection Engineering, 33 (2), pp. 79–87.

[7] Nicholas Conklea, H., Ragavana, J.K. Reconstitution of fine coal. (1992) Coal Preparation, 11 (1–2), pp. 67–76.

[8] Koga, N. Ozawa's kinetic method for analyzing thermoanalitical curves. (2013) Journal of Thermal Analysis and Calorimetry, 113, pp. 1527–1541.

[9] Petrova, G.I., Bychev. Electrochemical processing of brown coal. (2001) Yakutsk, p. 168.

[10] Rasskazova, A.V., Aleksandrova, T.N., Lavrik, N.A. Rational use of brown coals of the South of Far East. (2013) Gornyi Zhurnal, (10), pp. 42–44.

[11] Rasskazova, A.V., Alexandrova, T.N. Determination of rational parameters of brown coal briquetting. (2016) XVIII International Coal Preparation Congress: 28 June–01 July 2016 Saint-Petersburg, Russia, pp. 683–687.

[12] Sadardinov, I.V. Fuel – energy complex of the Far East in the system of FEC of Russian Federation. (2006) Moscow: 294 c.

43

Complex processing of ash wastes to recover of valuable components and environmental protection

T. Aleksandrova, S. Korshevenkov, N. Nikolaeva

Saint-Petersburg Mining University, Sankt-Peterburg, Russia

Abstract: The article is devoted to the study of the possibility of iron concentrates obtaining from fine grain of ash and slag waste from one of the Russian Federation thermal power plant (TPP) using a laboratory pulsed magnetic field separator. Comparison of technological indicators of ash separation on various types of magnetic separators was made. The characteristics of the laboratory pulsed magnetic field separator, and a scheme of experiments are given. Separation products are analyzed and separation indicators are calculated. A scheme for the comprehensive processing of ash and slag dumps with the production of several types of commodity products has been proposed and substantiated.

Keywords: Fly ash, waste, dry magnetic separation, pulsed magnetic field, iron

1. Introduction

One of the most important objects of complex processing are ash and slag waste (ASW), which are often accumulated within the boundaries of settlements and consist of fly ash and slag. The volumes of ash and slag waste of thermal power plants are about 1.3 billion tons and increase annually, according to various sources, by 40–50 million tons (Ryabov et al., 2013).

Fly ash (hereinafter – ash) is a fine material consisting of particles ranging in size from microns parts to 0.14 mm. Ash is formed as a result of burning solid fuel at thermal power plants (TPPs), and is captured by electrostatic precipitators. Then it is collected in a dry state using an ash collector for production needs, or together with water and slag is sent to the ash dump (Melent'yev, 1985). The structure and composition of ash depends on the whole complex of simultaneously acting factors: the type and morphological features of the fuel burned, the fineness of grinding during its preparation, the ash content in the fuel, the chemical composition of the fuel mineral part, the temperature in the combustion zone, the residence time of the particles in this zone, etc. (Melent'yev, 1985).

By chemical composition, the ASW material mainly refers to acidic particles. The basic mass (96–98%) of ash and slag waste is the sum of oxides: silicon oxide, 45–60%; calcium oxide, 2.5–9.6%; magnesium oxide,

0.5–4.8%; iron oxide, 4.1–10.6%; aluminum oxide, 10.1–21.8%; and sulfur trioxide, 0.03–2.7% (Ahmaruzzaman, 2010).

The ash extracted from the filters of TPP by the dry method must be processed by dry processes in order to extract valuable components, since the contact of various ash grades with water leads to surface hydration and loss of commercial properties for the construction materials industry (Wang et al., 2014; Loya and Rawani, 2014; Chanturiya et al., 2014).

The complex technologies of wet magnetic or gravitational separation of ash–slag waste materials known today make it possible to process wet ash waste already accumulated by a large number of TPPs, but this does not solve the problems of ash and slag incoming, which should be dry and processed separately (Delitsyn et al., 2013; Prokopieyev et al., 2015; Aleksandrova et al., 2015). Modern dry processing technologies can be divided into dry magnetic and dry gravity beneficiation. Both methods are difficult to implement for fine grain; one of the possible methods is separation in a pulsed magnetic field (PMF) or electrodynamic separation in a low field. The parameters of this field change simultaneously in three ways: space, time, and along the inductor plane (Hirajima et al., 2010; Arsentiyev et al., 2015; Nifantov et al., 2006).

The transition of the TPP technology to the system of dry ash and slag removal in the future will improve the technical discipline and efficiency of the stations themselves, stabilize the burning of coal, and orient them to the specified indicators of fly ash. In Germany, each unit is certified for the production of ash. The quality of ASW and the amount of its sales began to determine the efficiency of the station's technology and management (Putilin and Tsvetkov, 2003; Cherepanov and Kardash, 2009; Kalachev, 2014; Korchevenkov and Aleksandrova, 2017).

One of the new directions in the beneficiation and processing of low-quality and non-traditional raw materials is microwave processing. The features of the electromagnetic effects on substances include uniform heating throughout the volume, high speed and low inertia of heating, as well as the possibility of selective heating.

Thus, obtaining additional competitive raw materials for ferrous and non-ferrous metallurgy from ash and slag will reduce the production resource intensity, reduce emissions, comply with environmental standards and introduce progressive low- and non-waste technologies (Sizyakov et al., 2016).

2. Chemical and material composition

The purpose of the study is to develop a comprehensive processing scheme using pre-reduction using electromagnetic radiation and to assess the fundamental possibility of obtaining conditioned iron concentrate using an electrodynamic separator with a travelling magnetic field and analyze the qualitative and quantitative characteristics of the separation.

A sample of ash and slag waste from one of the Russian Federation TPPs was collected as an object of research. The particle size distribution and the averaged chemical composition are given in Tables 1 and 2, respectively.

Table 1: Particle size distribution of ASW sample

Grain size (mm)	+2	-2+1	-1+0.5	-0.5+0.2	-0.2+0.1	-0.1+0
Yield (%)	8.26	16.15	8.2	28.26	18.24	20.29

Table 2: Averaged chemical composition of ASW sample

Name of compound	Concentration (%)
SiO_2	54.79
TiO_2	0.76
$Al2O_3$	10.71
Fe_2O_3	**13.10**
MnO	0.06
CaO	3.90
MgO	1.11
C	12.57
П рочие	3
Итого	100

For a more complete study on the possibility of valuable components extraction from ash and slag waste, mineralogical analysis was carried out in two grain sizes (+2 and −2 mm). Carbon shales with magnetite interlayers were found in the +2 mm class, and unburned carbon – brown, rarely black coal with inclusions of lenses, sockets, and magnetite impregnation; carbon–clay–magnetite–quartz slag cake (grains and spheres); scrap. Unburned carbon: brown coal is fissured, dull, and less often shiny (burnt), sometimes with an tar or bitumen influxes, often with a relict woody texture. Unburned carbon sharply prevails in the nonmagnetic fraction; carbon shale – black and dark gray, the texture is lamellar and massively layered; clay – rose–gray–white, less often rusty-shaded and reddish-brown; fine grains – quartz + clay, slag + coal + magnetite; fine-grained cake – quartz + clay + coal; asbestos; and asbestos wool.

In the grain size of 2 mm, the ASW was divided into magnetic and non-magnetic fraction, and then analyzed. *Magnetic fraction* ≈ 17%: granular magnetite, less often spheres; coal with magnetite inclusions; magnetite carbon shale; clay slag, light gray with magnetite inclusions; fine-grained, quartz-clay-magnetite cake; and scrap.

Nonmagnetic fraction: unburned carbon ≈ 40% – brown coal and carbonaceous shale; quartz ≈ 25%; clay slag ≈ 25%; quartz-clay cake ≈ 7%; black slag, glassy ≈ 3%; and columnar, flattened columnar asbestos grains and single mica flakes.

The results of the carbon content determination in different grain sizes are presented in Table 3.

Table 3: The results of the carbon content determination by ash grain size

Проба № 1	Выход класса крупности (%)	Распределение углерода по классам крупности (%)
+2	8.26	23.76
-2+1	16.15	8.29
-1+0.5	8.8	9.36
-0.5+0.2	28.26	23.58
-0.2+0.1	18.24	21.12
-0.1+0	20.29	13.89
	100	100.00

3. Results and discussion

The initial raw material is prepared by sieving into two grain sizes: coarse and fine. Classification by size allows to select a coarse class containing a small amount of magnetic and carbon-containing components. And a fine class that contains a significantly larger number of magnetic and carbon-containing components, which allows it to be processed according to the technology below. A coarse class goes to the flotation extraction of the coal fraction. It was found that the tailings of coal flotation contain noble metals (Fig. 1).

The fine class is processed by microwave radiation with a frequency of 2000–3000 MHz, with radiation power from 400 to 800 W, with an exposure time from 2 to 3 min, and then separated in a low-intensity magnetic field or in a magnetic field. This produces a magnetic fraction, which is sent for metallurgical processing, and a non-magnetic fraction, sent for further processing, for example, as a raw material for the construction industry.

Small class microwave processing leads to intense heating of weakly magnetic components. In this case, the reaction of their reduction with carbon present in the material to highly magnetic forms takes place. The iron oxide Fe_2O_3 is reduced to the magnetic form C, by reaction (1):

$$3Fe_2O_3 + 2C + (O_2) = 2Fe_3O_4 + CO_2 + 111 \text{ kJ/kg Fe} \tag{1}$$

Figure 1: Microphotography of coal flotation tailings

After microwave treatment, the material is sent for beneficiation in a magnetic field. Since weakly magnetic particles are restored to strongly magnetic during the microwave treatment, they need a rather low-intensity magnetic field to isolate them.

The frequency of the microwave field below 2000 MHz does not allow fully convert weakly magnetic particles into strongly magnetic ones. A frequency above 3000 MHz also leads to a decrease in the degree of conversion of weakly magnetic particles into strongly magnetic ones.

The power of the microwave field below 400 W significantly reduces the number of particles converted to highly magnetic, respectively, significantly reduced technological performance. A power above 800 W does not improve technological performance, leading to unjustified power consumption.

Time of treatment by microwave field of less than 2 min does not allow fully convert weakly magnetic particles into strongly magnetic ones, while technological indicators significantly fall. The processing time of more than 3 min almost does not increase the technological performance and leads to unnecessary energy consumption and decrease in the performance of the process.

The use of a travelling field can significantly improve the quality of the magnetic product. In the laboratory of Mineral Processing Department of Mining University, a PMF separator was developed; it is easiest to imagine it as a winding of a linear asynchronous motor with a non-uniform alternating magnetic field.

It is known that all inductors of an alternating electromagnetic (traveling) field create a field varying according to the law:

$$H = H_0 K \cos \omega t \tag{2}$$

where, H_0 is the field strength amplitude; K is the gate factor that has a variable value; t is the actual time; $\omega = 2\pi f$; and f is the current frequency.

Considering the graph of this function, it can be noted that the intensity on the inductor surface monotonously increases, reaches a maximum and gradually decreases to zero, changes polarity and reaches the minimum value of the opposite polarity.

The mathematical description of the intensity variation law in this case will be:

$$\vec{H}(t) = \vec{H}_0 \frac{3.1}{\pi} K \left(\cos \overset{\rightarrow}{\omega t} - \frac{1}{3}\cos 3 \overset{\rightarrow}{\omega t} + \frac{1}{5}\cos 5 \overset{\rightarrow}{\omega t} - \frac{1}{7}\cos 7 \overset{\rightarrow}{\omega t} + ... \right) \tag{3}$$

In the local current range (1–8 A), the change in the induction of a traveling magnetic field at a standard industrial frequency of 50 Hz has been investigated. During experiments performance, it was established that the particles of the magnetic fraction move along the surface on which the separation takes place, at different speeds from 2 to 6.15 cm/s.

Modernized optimality criterion of Hancock–Luyken was taken as effectiveness of separation, namely the extraction of iron in the magnetic fraction minus the output of the magnetic fraction.

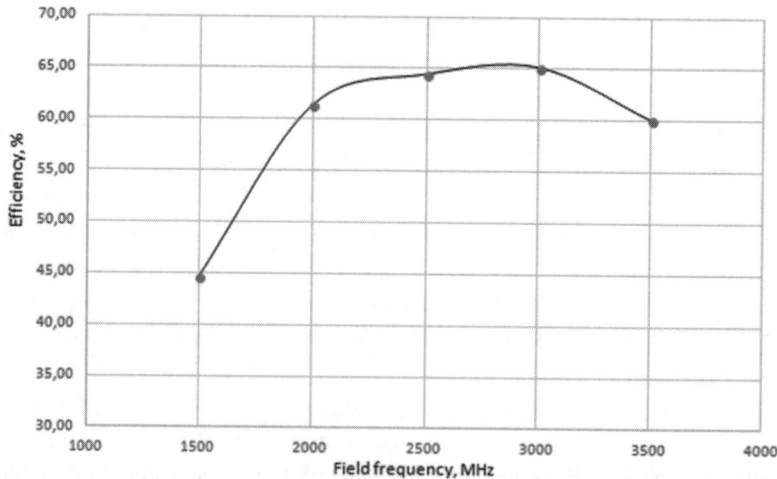

Figure 2: Relationship of separation efficiency on field frequency

Relationships of separation efficiency as a function of the field frequency and field power are presented in Figs. 2 and 3, respectively. The relationship of separation efficiency on microwave processing time with a field power of 600 W and a field frequency of 2500 MHz is presented in Fig. 4.

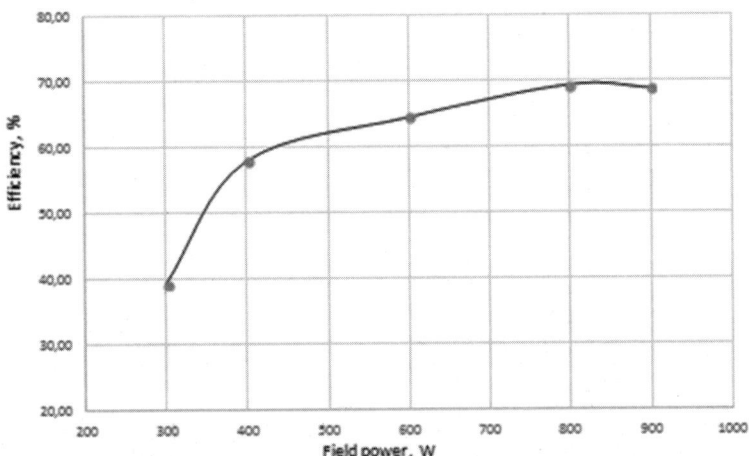

Figure 3: Relationship of separation efficiency on field power

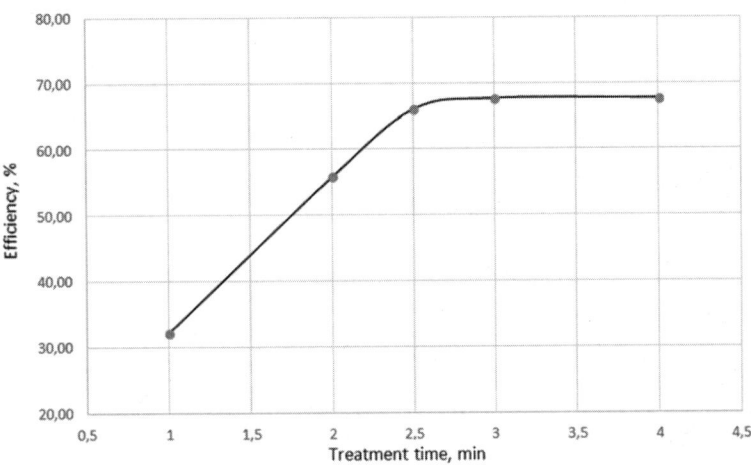

Figure 4: Relationship of separation efficiency on processing time

The best separation results with a field power of 600 W and its frequency of 2500 MHz and a processing time of 2.5 min in a traveling magnetic field (field induction 50 mTs, frequency 50 Hz) are shown in Table 4.

Table 4: Results of separation in a travelling magnetic field

Product name	Yield (%)	Grade Fe (%)	Recovery Fe (%)	Separation efficiency (%)
Coarse class	32.8	3.54	14.13	
Fine class	67.2	10.50	85.87	
Total: initial material	100	8.22	100.00	
Magnet fraction	9.4	67.82	77.56	68.16
Non-magnet fraction	57.8	1.18	8.32	
Total: fine class	67.2	10.50	85.87	

Figures 5 and 6 show microphotographs of the separation products in traveling magnet field.

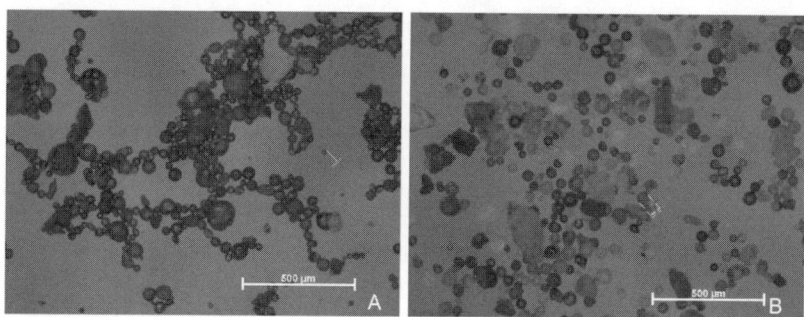

Figure 5: (A) Magnet fraction and (B) non-magnet fraction

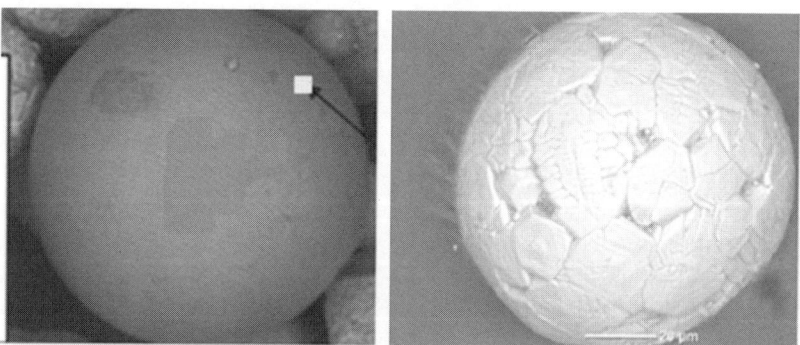

Figure 6: Microphotographs of micro spheres

The results of previous studies on the extraction of the magnetic fraction from a similar material of ash and slag waste using the Outotec Slon 100 wet high-gradient magnetic separator were compared with data obtained on a EBM 32/20 wet electromagnetic separator (Aleksandrova et al., 2015).

This high gradient magnetic separator Outotec Slon 100 has several advantages: high efficiency in beneficiation of fine material due to the vertical structure of the separator, the mechanism for creating a pulsation and the core matrix of the separator; the entrapment of non-magnetic particles is minimized due to the arrangement of the matrix and the pulsation mechanism (Dobbins et al., 2010).

As a result of the conducted research and comparison of the obtained results, a scheme for processing ASW was proposed (Fig. 7).

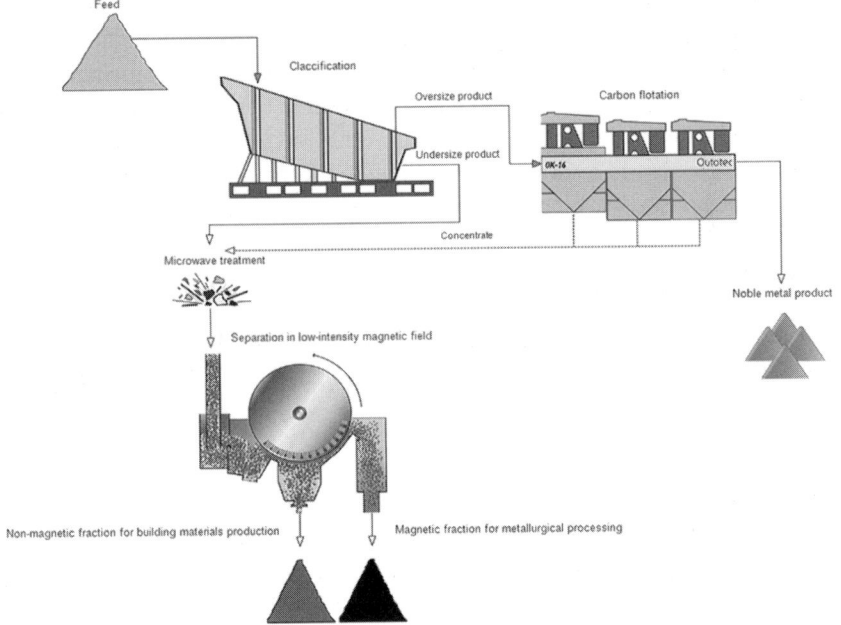

Figure 7: Scheme for processing ASW material

At the same time, the further use of dry magnetic separation tailings is possible in construction and in the production of building materials. Both dry and wet (from the wastes) dispersed ash of TPP is widely used for the manufacture of products from dense or cellular ash concrete (wall blocks and panels, floor slabs). In China, they are developing technologies for the deep processing of ash and slag waste, namely the production of water purification products – zeolites. It is also possible to extract aluminum containing components, unburnt coal and particles of rare and noble metals (Prokhorov, 2015; Shabarov et al., 2016).

4. Conclusion

Production of iron concentrates which are suitable for smelting steel from fly ash from a thermal power plant is possible by a dry process in a traveling magnetic field. In one stage, a concentrate is obtained with Fe content in general at the level of 40%, further refinement allows to obtain an ordinary concentrate with Fe content in general not less than 60%, with high recovery rates. Dry separation of ash takes place at a low magnetic field of about 50 mT to obtain indicators comparable to the separation on a wet high gradient Slon 100 separator with an induction of 20 mT; running field to resource-saving and energy-efficient technologies. Higher extraction and grade indicators are possible to obtain due to a deeper study of the processes occurring on the surface of the inductor, after which it is possible to optimize its operation for obtaining higher technological extraction rates. It is also possible to use the separation process in a travelling field on the electro-filters of the TPP, or to develop separators working in conjunction with cyclones.

5. Acknowledgement

The work is carried out under financial support of the Ministry of Education and Science of the Russian Federation, the project RFMEFI57417X0168.

6. References

[1] Ahmaruzzaman M. A review on the utilization of fly ash. Progress in Energy and Combustion Science. Volume 36, Issue 3, June 2010, Pages 327–363.

[2] Aleksandrova T.N., Prokhorov K.V., Lvov V.V. Recovery of magnetic fraction from coal combustion waste by high-gradient magnetic separation. Gornyi Zhurnal. Volume 2015, Issue 12, 1 January 2015, pp. 4–8.

[3] Arsentiyev V.A., Dmitriyev S.V., Mezenin A.O., Kotova E.L. Veshchestvennyy sostav i tekhnologiya sukhoy pererabotki zoly TETS. Zhurnal «Obogashcheniye rud». 2015 No 4, pp. 49–53.

[4] Chanturiya V.A., Vaysberg L.A., Kozlov A.P. Prioritetnyye napravleniya issledovaniy v oblasti pererabotki mineral'nogo syr'ya. Obogashcheniye rud. 2014, No 2, pp. 3–9.

[5] Cherepanov A.A., Kardash V.T. Kompleksnaya pererabotka zoloshlakovykh otkhodov TETS. Geologiya i poleznyye iskopayemyye Mirovogo okeana. 2009, No 2, pp. 100–117.

[6] Delitsyn L.M. et al. Kompleksnoye obogashcheniye i pererabotka zoly ugol'nykh elektrostantsiy v RF. IX Kongress obogatiteley stran SNG: sbornik materialov. M.: Moskovskiy institut stali i splavov. 2013, Volume I, pp. 225–230.

[7] Dobbins M., Dann P., Sherrell J. Posledniye dostizheniya v oblasti proyektirovaniya i primeneniya magnitnykh separatorov, «Tsvetnyye metally». 2010, No 2, pp. 48–54.

[8] Hirajima T. et al. Recovery of cenospheres from coal fly ash using a dry separation process: Separation estimation and potential application. International Journal of Mineral Processing. Volume 95, Issue 1–4, 1 July 2010, Pages 18–24.

[9] Kalachev A.I. The ash and slag waste market of Russia through the eyes of trader. Phoenix consortium. Proceedings of the V Conference «Ashes from TPPs: removal, transport, processing, storage», Moscow, April 24–25, 2014.

[10] Korchevenkov, S.A., Aleksandrova, T.N. Preparation of standard iron concentrates from non-traditional forms of raw material using a pulsed magnetic field. Metallurgist. 2017 61(5–6), pp. 375–381.

[11] Loya M.I.M., Rawani A.M. A review: promising applications for utilization of fly ash. International Journal of Advanced Technology in Engineering and Science. Volume No. 02, Issue No. 07, July 2014, pages 143–149.

[12] Melent'yev V.A. Sostav i svoystva zoly i shlaka TES. Spravochnoye posobiye. L.: Energoatomizdat, 1985, p. 288.

[13] Nifantov B.F., Zaostrovskiy A.N., Zanina O.P. Issledovaniye opyta promyshlennoy pererabotki zoly sposobom magnitnoy separatsii na yuzhno – Kuzbasskoy GRES. Vestnik Kuzbasskogo gosudarstvennogo tekhnicheskogo universiteta. 2006, No 5, pp. 84–90.

[14] Prokopieyev S.A. et al. Issledovaniya po izucheniyu vozmozhnosti polucheniya kachestvennogo zhelezosoderzhashchego produkta i drugikh tsennykh komponentov iz zoly unosa TETS-9 OAO «Irkutskenergo». X Kongress obogatiteley stran SNG: sbornik materialov. M.: Moskovskiy institut stali i splavov. 2015, Volume I, pp. 256–261.

[15] Putilin E.I., Tsvetkov V.S. Obzornaya informatsiya otechestvennogo i zarubezhnogo opyta primeneniya otkhodov ot szhiganiya tverdogo topliva na TES. Soyuzdornii. M. 2003, p. 60.

[16] Ryabov L.M., Delitsyn A.S. Vlasov U.N., Golubev, U.V. Polucheniye magnitnykh produktov iz letuchey zoly Kashirskoy GRES. Obogashcheniye rud. 2013, No 6, pp. 41–45.

[17] Shabarov A.N., Alexandrova T.N., Nikolaeva N.V. To the question of complex use of ashes and slag waste CHP plant. XVIII International Coal Preparation Congress Proceeding, Volume 1, 28 June–01 July 2016, Pages 603–607.

[18] Sizyakov V.M., Vlasov A.A., Bazhin V.U. Strategicheskiye zadachi metallurgicheskogo kompleksa Rossii. Tsvetnyye metally. 2016, No 1, pp. 32–37.

[19] Wang X.H. et al. The optimization of sintering process for alumina extraction from fly ash. Advanced Materials Research. Volume 878, 2014, Pages 264–270, MPEI Printing House.

44

Evolution of coal processing practices at Tata Steel

Mr. Bhargav Dhavala, Mr. Kunal Mathanker, Mr. Debaprasad Chakraborty, Dr. Suman Sit

Tata Steel Limited, Jamshedpur, Jharkhand, India

Abstract: In1951/52, Tata Steel became a pioneer in Asia by setting up its first coal washery in West Bokaro & Jharia using Chance Cone process. Fine coal (-6mm) was used directly without beneficiation. Chance Cone process was eventually replaced by Dense Media Cyclones and mechanical flotation cells in 1982.1990s witnessed the replacement of DSM cyclones with scrolled evolute followed by the introduction of pump fed cyclones and replacement of flat-bottom flotation cells with U-bottom ones.

Technological improvements between 2000-2018 touch-based most of the critical unit operations: Introduction of Sizers, Replacement of Elliptical screens with Banana screens, Introduction of new-generation mixing mechanism in flotation & substitution of Diesel with green reagent and Replacement of Screen bowl centrifuges with Vacuum Belt Filter. The focus was also put on automation of the technological processes: PGNAA based real-time ash monitoring system to ensure consistency in the product quality.

Increase in demand for coking coal and the deteriorating raw coal quality demand that we continuously scan the world & adopt the best technologies to become a technology leader in the business. Down the line, Tata Steel aims to have: (a) Intermediate size beneficiation circuit, (b) Superior technologies for fine coal beneficiation and (c) Advanced measurement and control systems in place.

Keywords: Dense Media Cyclone, Flotation, Banana screen, Vaccum Belt Filter, Intermediate size beneficiation

1. Introduction

Tata Steel meets its coal requirement for coke making from two captive coal mines: Jharia (underground) and West Bokaro (open-cast) located in the state of Jharkhand. Indian coals are high in ash content (35%) owing to their drift origin and come under the 'difficult-to-wash' category due to high Near Gravity Material (40-50%). Hence, coal produced from the captive mines cannot be used directly for coke making. To make it suitable for coke making, it is beneficiated in a washery and the washed coal is then used for coke making. Tata Steel became a pioneer in Asia by setting up its first coal washery in West Bokaro & Jharia in 1951 & 1952, respectively. The journey of coal beneficiation practices at Tata Steel, in a nutshell, is shown in Figure1 and the story behind each step is explained in detail subsequently.

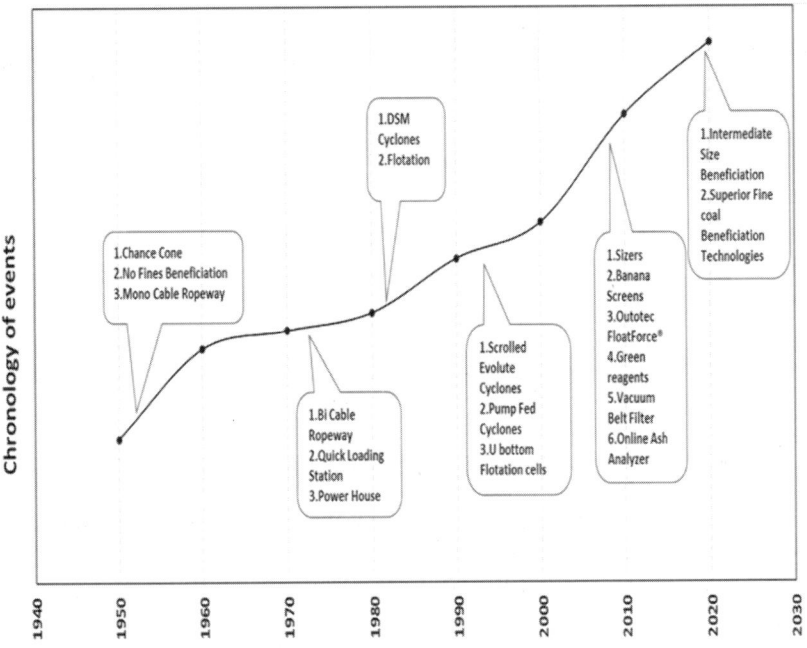

Figure 1: Technological Improvements – chronology of events

2. Technological improvements in coal beneficiation at Tata Steel

2.1 1950s – 1970s (Chance Cone Process)

Raw coal having an ash content of 20-23% was crushed and reduced to -75 mm size followed by screening at 6mm.75*6mm size fraction of coal was beneficiated in Chance Cone process (a dense media separator utilizing sand as media) whereas the finer fraction (-6 mm) was directly added without beneficiation to the washed coal and used (Figure 2). The composite clean coal having ash of 19% was transported to the railway siding via a mono-cable ropeway. The yield of clean coal varied from 65-90% and most of the equipments were manually controlled. Until the early 1970s, only top seams were worked and fed to the washery. In the late 1970s, the middle and lower seams having comparatively higher ash were also mined, a bi-cable ropeway for transportation of washed coal from the washery to the railway siding, a quick loading system at the railway siding and a powerhouse were also added.

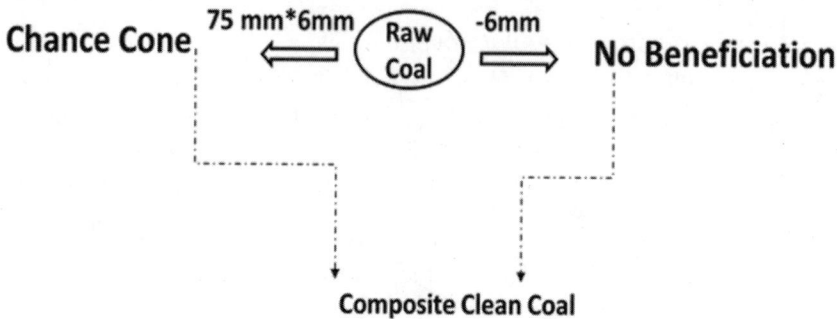

Figure 2: Flow sheet in 1950-1970s

2.2 1980s (Dense Media Cyclones and Froth Flotation)

With the advent of new technologies, continuous deterioration in the mined raw coal quality supplemented by the poor performance of the Chance Cone process was witnessed. Earlier, the coal washeries did not have any fines treatment circuits; they were simply blended with the washed coarser coals owing to the very good quality. However, with time, the quality of fines deteriorated significantly and it was practically impossible to maintain the quality of washed coal by direct mixing. Detailed studies were carried out to arrive at the correct feed top size to the washery to optimize the clean coal yield at the desired ash. Eventually, the Chance Cone process was replaced with Dense Media Cyclones for processing coarser raw coal fraction:13 to 0.5mm and mechanical flotation cells for processing finer raw coal fraction: -0.5mm (Figure 3).

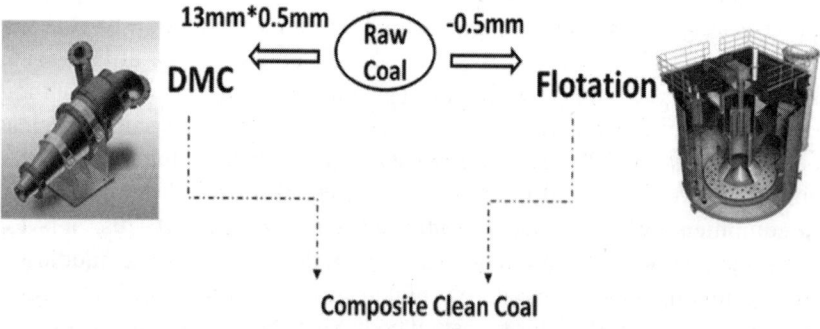

Figure 3: Flow sheet in 1980s

2.3 1990s

2.3.1 Low Ep scrolled evolute cyclones

In the year 1998-99, Washeries switched over from Dutch State Mines (DSM) design cyclones to scrolled evolute design cyclones. Process performance improved as the Ep obtained was 0.02-0.025 as against 0.035-0.040 in the DSM cyclones. These cyclones are also of larger capacity thereby helping in increasing the plant throughput.

2.3.2 Introduction of pump fed cyclones & high – low concept in cyclone circuit

At West Bokaro Washery-3, parallel modular streams in cyclone circuit along with pump-fed cyclones were introduced. The pump feeding system has its own share of pros and cons. Gravity feeding requires a higher footprint but achieves a more consistent flow, less pump wear and feed degradation. There are two stages of cyclone washing in which the primary cyclones are operated at higher specific gravity to discard the rejects. The second stage cyclones are operated at lower specific gravity for separation of clean coal & middlings. These cyclones have V/f converters for maintaining the required inlet pressure thereby reducing the capital costs in comparison to a conventional gravity fed cyclone.

2.3.3 Replacement of flat-bottom flotation cells with U-bottom ones

U-bottom cells minimize the sanding/silting phenomenon. Sanding is high in flat-bottom cells due to lack of velocity in the un-agitated zones thereby allowing the larger particles to settle down.

2.4 2000-2018

2.4.1 Introduction of sizers to improve liberation

Crushing and liberating coal to the correct size is of prime importance, as it would improve the performance of the coal washery. Sizers were introduced in place of roll crushers to get optimum liberation at reduced noise and dust.

2.4.2 Replacement of elliptical screens with banana screens to improve the desliming efficiency

Earlier, elliptical screens were used for desliming raw coal at 0.5mm. Screening efficiency of these screens was found to be poor- it was observed that a substantial quantity of undersize i.e. (-) 0.5mm reported to the screen oversize. These finer coal particles create difficulties in maintaining the cut density inside the cyclone thereby impacting the efficiency. These screens were eventually replaced with robust Banana Screens.

2.4.3 Introduction of 'Advanced new-generation mixing mechanism' in Flotation cells

At West Bokaro Washery-3, significantly high (20-25% by wt.) proportion of plus 0.5 mm size coal particles report in the flotation cell feed. This was always a concern as most of these particles report to the flotation tailings. To improve the fines circuit performance, Tata Steel incorporated a new generation mixing mechanism- Outotec FloatForce® in one flotation bank to start with. This mechanism creates more turbulent energy and generates finer bubbles as it has separate chambers for air and slurry in the rotor assembly of the flotation cell (Figure 4). In the conventional mixing mechanism, the design of rotor-stator assembly is such that there is a single passage for both slurry and air and hence, air mixing with slurry is not fully effective. Outotec FloatForce® has also been found to be more effective in floating relatively coarser particles and pushing more coal slurry as compared to the conventional mixing mechanism.

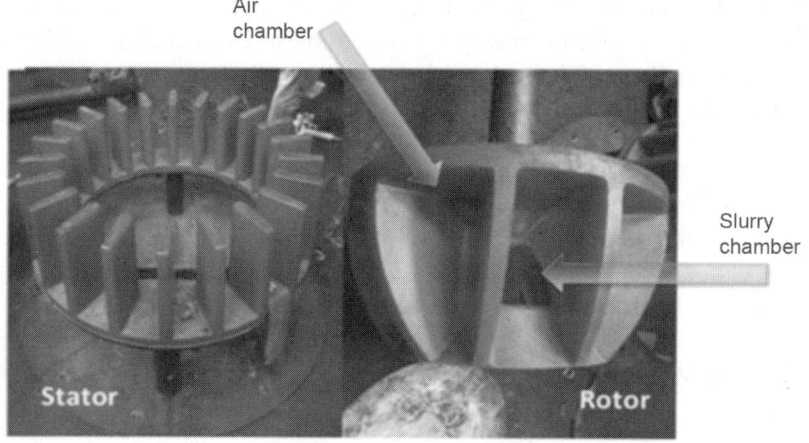

Figure 4: New generation mixing mechanism outotec float force®

2.4.4 Substitution of Diesel with green reagents in Flotation

For ages, Diesel is being used as a collector in coal flotation. With the stringent environmental regulations and policies, replacement of diesel with a reagent that is environmental friendly as well as technically & economically competent became inevitable. Diesel is also highly inflammable and prone to pilferage and hence, poses safety as well as environmental hazards. A joint improvement initiative was taken with Nalco Chemicals for the development of a synthetic collector, which not only gives technically competent results but also is economical, safe and environment friendly.

2.4.5 Introduction of vacuum Belt Filter for dewatering fine clean coal

Initially, Screen Bowl Centrifuges were used for dewatering fine clean coal: <0. 5 mm. However, it was observed that ultra-fine coal particles were getting lost with the centrifuge effluents. To capture such low ash ultra-fine clean coal, Horizontal Vacuum Belt Filters (HVBF) were installed. The belt filter installed at West Bokaro washery#3 is also the world's largest HVBF with an effective filtration area of 145 m² for coal slurry.

2.4.6 On Line Ash analyzers for consistency in product quality

Taking representative samples from the conveyor belt and analyses for effective quality monitoring & control was time-consuming. As a result, corrective actions could not be taken timely resulting in variations in the clean coal ash. To overcome the mentioned problems, an online ash analyzer based on the Prompt Gamma Neutron Activation concept was introduced. The use of an online analyzer to monitor ash has resulted in the minimization of shift wise standard deviation in the clean coal ash.

2.5 2018+ (Some already established, some need to be established)

2.5.1 Intermediate size beneficiation

It has been observed through process audits that recovery of 0.5-0.25/0.15mm size fraction is the lowest of the lot in froth flotation process and 0.5mm is not the ideal top size for flotation. To improve the recovery of 0.5-0.25/0.15mm size fraction, intermediate size beneficiation in a Reflux Classifier (RC) has already been implemented at Jamadoba washery and would be replicated across the remaining coal washeries of Tata Steel in the near future (Figure 5).

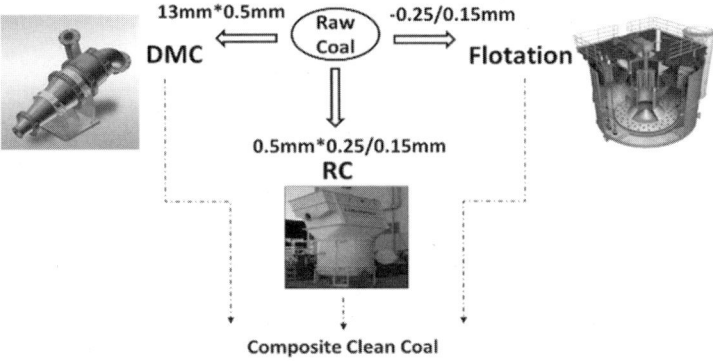

Figure 5: Futuristic Flowsheet

2.5.2 Superior technologies for fine coal beneficiation/difficult-to-float coals

Flotation is a complex process controlled by factors, which can be divided into three facets: coal characteristics, chemistry and machine characteristics. Factors within the coal and chemistry areas are dynamic and hence, need to be dealt with by personnel on an ongoing basis in normal plant operations. However, the most important characteristic of any flotation technology is air bubble generation and the size of air bubbles produced as this controls flotation kinetics and dictates the carrying capacity of the machine. Several technologies have come up for fine coal beneficiation such as Column cells, Jameson cells and Jet flotation cells which have been found superior to the conventional mechanical flotation cells.

2.5.3 Step-by-step approach to a fully automated plant

In the age of Digitalization, IoT and Industry 4.0, it is equally important to have automation – basic & advanced measurement systems & control systems in the technological processes to achieve higher yields at the same grade and to become a technology leader in the business as shown in Figure 6.

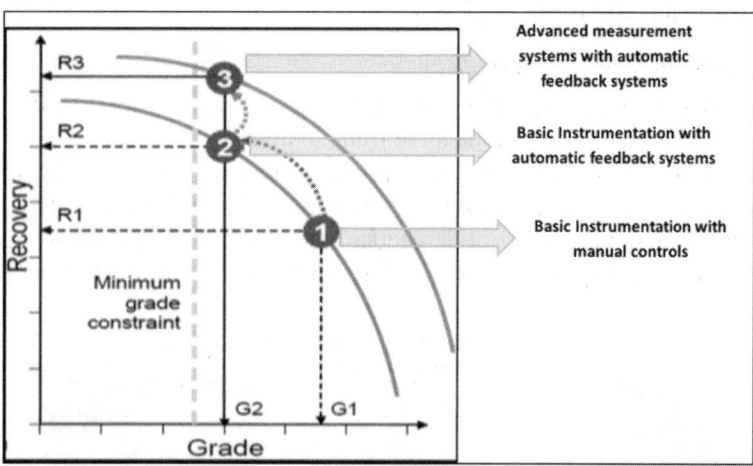

Figure 6: Automation in technological processes

3. Conclusion & way forward

Increase in demand for coking coal and the deteriorating raw coal quality demand that we continuously scan the world & adopt the best technologies to become a technology leader in the business.

4. References

[1] P S Rao, P Khattri, C G Krishna, and S M Rao ,(2012) "Technological advancements in coal processing at Tata Steel ", IMPC2012, Delhi, India

[2] P Khattri, C G Krishna, P Srinivasa Rao, P Pratim Banerjee and S Mohan Rao , (2012) "Gammametry-a technological intervention for optimization & process control at coal washeries " ,IMPC 2012, Delhi, India

[3] I Prasad, S. M. Safyi , (1985) " Some aspects of coal flotation at washeries of Tata Steel",Eprints NML

[4] Dr. T. C. Rao, L. Prasad, Kripal Singh ,(1978) " Some aspects of beneficiation of coal slimes in India " ,Paper presented at Seminar on coal washing organised by the Bureau of Public Enterprises, New Delhi Oct '78

[5] Chakraborty DP, Sen S, Ray PR, Kumar A ,(2015) " Augmenting supplies of metallurgical coal to steel mills by upgrading low and medium coking coals ",CPSI Journal, 2015, 66–72

[6] Noor Ahmed , (1989) "The New Washery at Tata Steel's West Bokaro Colliery ", Tata Tech. Vol. 5

45

"Coal blending – Finding the right mix for powering thermal power plants in India" – A literature review

Ajay Kumar[1], Neeraj Gautam[2]

[1]*AGM (Coal Mining), EOC, NTPC Ltd.*
[2]*Sr. Manager (CM&CW), EOC, NTPC Ltd.*

Abstract: Coal blending in power stations is being adopted to reduce the cost of generation and increase the use of indigenous or more readily available coal. Low-grade (high ash) coal can be mixed with higher-grade low ash coal without deterioration in thermal performance of the boiler, thus reducing the cost of generation. When indigenous coal becomes less available, are of lower quality or are more expensive to mine in some regions, blending of imported coals becomes necessary. It can be challenging to ensure that the resulting blend will maintain plant output without damaging the boiler.

In some cases, coal blending is also used as a tool for pollution control, such as the combination of inexpensive high sulphur coal with more costly low sulphur coal to ensure compliance with sulphur emission limits. It is even possible to blend different coal types to maximise mercury reduction.

Many methods of coal blending are being used. Coal can be blended at the coal mine, at the preparation plant, trans-shipment point, or at the power station. The method selected depends upon the site conditions, the level of blending required, the quantity to be stored and blended, the accuracy required, and the end use of the blended coal.

Keywords: Coal blending, High ash coal, Coal-fired power plants

1. Introduction

Coal-fired power plants are designed to burn coal with defined characteristics, commonly indigenous coal. As noted by Anderson and Nowling (2014), coal characteristics affect nearly every operational aspect of a power plant, including forced outage rate, maintenance costs, auxiliary power requirements, net plant heat rate, emissions, and the ability to meet the full load.

However, over time, the accessibility and affordability of coal may change and the coal plants must, therefore, be ready to adapt to take the coal that does not match the characteristics for which they were originally designed. These new coals may perform differently to the design coal in such a way that blending is required to reduce detrimental effects on the plant.

It is likely that, globally, at least 20% of power plants, probably significantly more, cannot achieve design output due to difficulties in sourcing coals which consistently meet boiler requirements (Petrocom, 2014). This could be resulting in a reduction of 10% or more in potential output from the plants and may be causing a loss of 2% in total output from the power sector as a whole. By optimising blending to provide compliant and consistent fuel stock, plants can increase their power output while reducing negative effects on the plant (such as corrosion and fouling) and potentially reducing emissions of pollutants of concern.

As emission standards tighten globally, the number of coal which can meet these standards drops. This pushes up the demand for compliant coals, and this generally results in a price increase. Blending allows the use of lower quality, non-compliant coals, thus increasing coal reserves and ensuring that all coal can be fully utilised. However, blending coal requires that two major questions be answered:

- How can the characteristics of the final blend be predicted/guaranteed?
- Where and how is the coal to be blended?

2. Reasons for coal blending

1. **Cost:** The fundamental aim of a power plant is to produce power at the lowest possible cost. Once a power plant is constructed, the cost of the coal used to fire the plant is the usually greatest variable affecting the plant economics. Blending allows plants to fire less expensive coal as even poor quality coal which, on its own, would cause detrimental effects on the plant can be blended with higher quality to achieve the desired quality of coal to efficiently fire the plant.

2. **Plant Specifications:** The priority of coal plants is to provide electricity/energy on demand. This demand can vary over time and so the operation of the plant is to be adjusted accordingly. This often requires changes in fuel – high Btu/kJ blends can be used to reach peak load at any given unit and lower Btu/kJ blends can be used during lower load periods (Campbell, 2014). Plant operators should know which coal suits their plants and should make coal selections based on meeting minimum specifications. When all these specifications cannot be met, the plant operator must make a decision as to which parameters matter the most. Coal can then be bought, and/or blended accordingly.

When predicting the behaviour of coal blends, it is accepted that values for proximate, ultimate and calorific contents are additive – that is, the value of the blend will be the average value of the coals within the blend, proportionally. This is not the case for some of the other coal characteristics. In a review of coal blending, Wall and others (2001) summarised those properties that are not additive – that is, the blend property is not the weighted average of the properties of the individual coals.

Table 1: Additive and non-additive coal characteristics (After Wall and others, 2001)

Additive Coal Characteristics	Non-Additive Coal Characteristics
Calorific Value	Free-Swelling Index
Fixed Carbon	Grindability
Hydrogen	**Unsure Coal Characteristics**
Carbon	Ash
Chlorine	Nitrogen (some comes from combustion air)
Sulphur	Ash Fusion Temperature
Oxygen	Volatile Matter & Moisture

3. **Emission Legislation:** As legislation on pollution control tightens internationally, many plant operators are finding that they need to reconsider the coal they use in order to comply with emission limits or reduction targets.

    As a carbon intensive fuel, coal is well established as a source of CO_2 emissions to the atmosphere. Some plants have co-fired other materials such as biomass to reduce the overall CO_2 emissions.

3. Coal blending – Finding the right mix

Plant Managers seek blends, which will allow them to produce the maximum amount of energy from their plant with minimal detrimental effects on plant performance and equipment, simultaneously meeting any relevant emission legislation. Determining how to produce the best blend to meet all these requirements can involve a significant amount of work to determine the characteristics of individual coals and how these coals will behave when they are part of a blend.

In order to ensure that the characteristics of a coal or blend meet requirements, sampling and analysis is required at several steps during the coal delivery chain.

1. **Sampling:** The objective of coal sampling is to obtain a small amount of coal for detailed analysis, which will be assumed to be representative of all the coal in that batch or shipment. Ideally the sample should reflect the overall variability within a coal batch. In coal handling and blending facilities, samples are commonly taken from the conveyors, as the coal is taken from the bins or stockpiles to the final transit point before combustion. For materials such as coal, representative sampling can be a challenge.

   Traditionally used Cross Belt Sampler and Falling Stream Sampler are used to obtain samples for analysis. Various studies have demonstrated that sampling system bias was as high as 35.8% in some cases. The studies showed that over 40% of the cross-belt systems were biased as compared to 27.9% of the falling-stream samplers. It is likely that this is due to many sampling systems not being adequately inspected and maintained (Robinson and others, 2012).

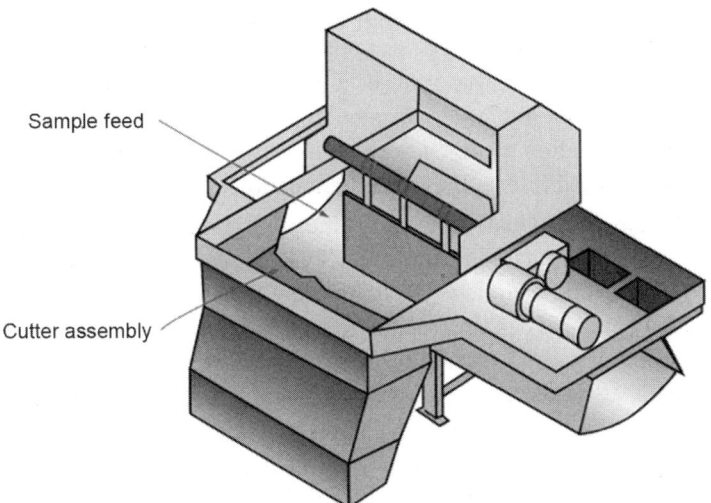

Sample feed

Cutter assembly

Figure 1: True Belt Sampling System

A 'true-belt' sampling system is also in use that throws material off the conveyor belt by a moving plough (along the length of the conveyor) and then a portion of this thrown material is taken by a fixed cutter. As the material moves along the conveyor belt, samples from a long length of the conveyor are pushed off towards a set of bins at the side, as shown in Figure 1. Selecting bins at random should provide a representative sample.

2. **Analysis:** Once a sample has been obtained it must then be analysed to establish the physical and chemical characteristics, as required. There are numerous methods for analysing coal and many of these function in a real-time online basis.

   Some real-time analysers (prompt gamma neutron activation analysis (PGNAA)) can also be used to provide data to automated coal blending systems and to provide quality assurance data on the final blends. Unlike coal analysis performed in labs, analyses at coal prep plants are done in large quantities to provide better data for large volumes – large volumes of the sample will give a better indication of the average qualities of the coal than smaller samples which may reflect minor variations.

3. **Blending Control:** Programmable logic controllers (PLCs) can be installed at coal handling facilities to take the data from sampling and analysis and use these to control blending ratios on a real-time basis.

   For example, a coal mine in West Virginia, USA, uses an online analyser to monitor coal as it passes through the preparation plant. Data from the analyser are used to sort the coal. Any coal with ash <16% is sent directly to the stockpiles; coal with ash >16% is sent to the preparation plant. Once the coal is washed, the analyser evaluates the coal and directs it to separate locations, according to quality.

   At the Arch Coal Catenary Mine, the gamma metrics analyser is used in conjunction with the COBOS coal blending system (coal optimisation blending system, discussed under Modelling) to provide coal blends of certain specifications. In addition to the gamma metrics system sampling the coal as it enters the blending system, the Catenary plant also has a gamma metrics CQM (coal quality management) system sampling 5 t/h of the blend as it is produced to confirm blend consistency and feedback to the blending system to adjust feed rates accordingly (Woodward and others, 2004).

3.1 Modelling coal blends

As shown in Table 1, some characteristics of coal are additive within a blend, while others are not. However, the behaviour of coal characteristics also has effects on different aspects of plant performance and some of this is predictable whilst some are not. Those aspects which are not easily predictable may have to be considered using more complex models, based on actual coal studies, in order to make them more predictable in future

Based on the known additive parameters it is possible to predict blend characteristics based on the characteristics of the individual coals in a blend. Calculation tools are available which will allow coal users to feed data on individual coal characteristics and to receive a 'best guess' on the characteristics of the resulting blend. A simple search on the internet will provide online blend prediction tools such as: http://www.seabase.in/pop-blending.html

An example of the user interface is shown below:

BLENDING CALCULATOR

Blend Ratio				%		%	Blend Result
Specifications	Basis	Unit	coal 1		coal 2		
1. Total Moisture	a.r.	%		0			
2. Inherent Moisture	a.d.b.	%		0			
3. Ash	a.d.b.	%		0			
4. Volatile Matter	a.d.b.	%		0			
5. Fixed Carbon	a.d.b.	%		0			
6. Calorific Value	a.r.	kcal/kg		0			
7. Calorific Value	a.d.b.	kcal/kg		0			
8. Totle Sulphur	a.r.	%		0			
9. HGI				0			

4. Coal blending techniques

The blending of coal can be done in different ways to achieve different outcomes. Blending is rather more complicated than just the mixing of two or more coals together. However, homogenisation goes even further than blending for those who need a truly optimised combination of coals.

1. **Stockpile Blending:** Perhaps the most simple and low cost method of coal blending is to place different coals into one single pile.

    The pile grows as layers of different coals are added in horizontal/inclined layers as shown. The different colours indicate the different coals as they are added to the pile. The more layers of coal there are in a stockpile, the greater the blending effect. However, the homogeneity of the mix will still be limited by the thickness of the layers and the subsequent reclamation of the coal from the stockpile and delivery to the boiler. (Isherwood, 2014).

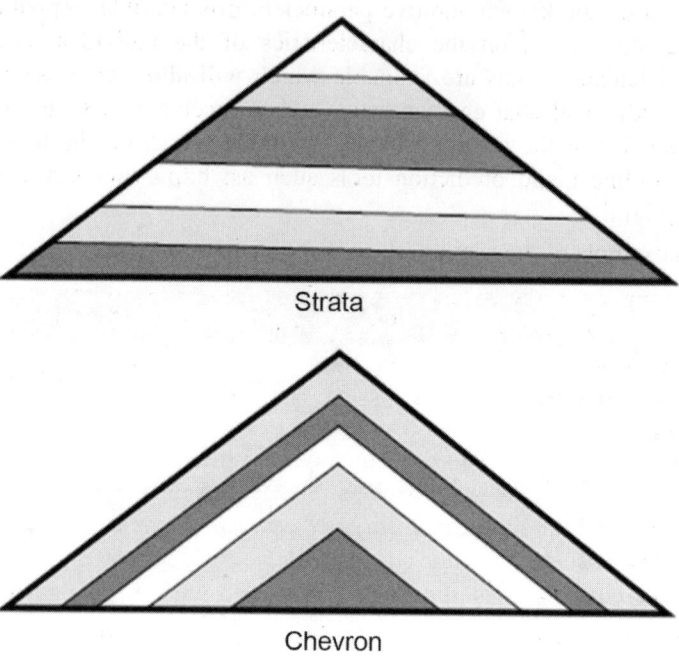

Strata

Chevron

For larger sites, it may be more convenient to keep coal piles separate or in longitudinal piles and then create the blend during reclamation of the coal. Whether the blend is obtained from a pre-mixed stockpile or separate piles, the mode of reclamation is determined by the availability or affordability of equipment. Mixing can be achieved by anything from simple bulldozer manipulation through more complex reclaiming systems which use vibration to ensure an even blend.

Traditional coal blending/reclaiming methods include methods based on extracting the coal from under the stockpile. Figure 2 shows a coal pile with a tunnel running through, equipped with a conveyor belt and scraper. The scraper moves through the tunnel, scraping off layers of the different coals in the stockpile, dropping them together onto the conveyor belt, thus creating a mix. Although this method can achieve significant mixing it is a clumsy method, requiring significant construction investment, high maintenance costs and yet does not ensure the homogeneity of the blend produced (Petrocom, 2014).

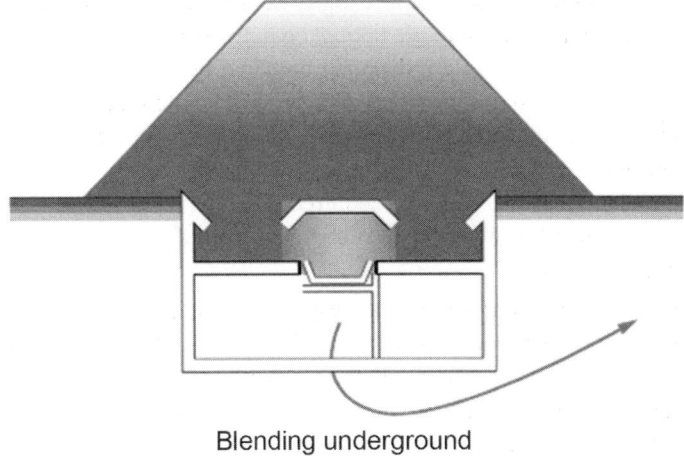

Blending underground

Figure 2

However, the simplest method is probably the bucket-wheel reclaimer which uses buckets in sequence passing through the coal pile to gather the coal from the desired section, usually containing a mix of coals. More advanced systems involve scrapers, which scrape the coal across the face of a pile, giving greater mixing and a more accurate selection.

2. **Bins – Silos & Bunkers:** Blends can be created at sites which have large coal storage units such as coal silos and hoppers. Silos are tall cylindrical storage devices with a single output hole at the bottom whereas bunkers are rectangular systems which may have multiple outlets along the bottom. Both silos and bunkers are often referred to as bins.

   Bins generally hold lower quantities of coals than piles and, of course, have a maximum capacity. Bins can be used to hold pre-blended coals or can be used as part of the blending process. A hopper/feeder can be used to create a pile of known weight under the bin. This will be determined by flow rate and/or by weight. This can then be mixed with a known weight of coal from another bin or pile using either digger-type vehicles or using belt feeders or conveyors. These systems are best for blends from 20–100%. These kinds of systems can create a reasonable blend with down to 5% increments of different coals in the mix. Below 5% can be possible but only with appropriate equipment and may become difficult under some weather conditions (McCartney, 2006).

Belt blending is the combination of coal on a moving conveyor. This can be achieved through taking coal from a stockpile, a blending pile or a bin. Coals are dropped onto the belt either from the crane systems or through the hopper systems. The blending is achieved by controlling the rate at which coal is added to the conveyor. These systems are designed to include a weighing system, to confirm that the blend is being added as required. The conveyor is also commonly the means of moving the blended coal from the blending area to the plant.

3. **Homogenisation:** Homogenisation is not always included as part of the blending process at many coal plants as most boilers can cope with some variation in coal characteristics within the load. However, it can be required at some plants where coal stocks are delivered from several local suppliers into one single stockyard and also at cement plants. Homogenisation could be achieved by layering materials onto feeding conveyors as shown in Figure-3

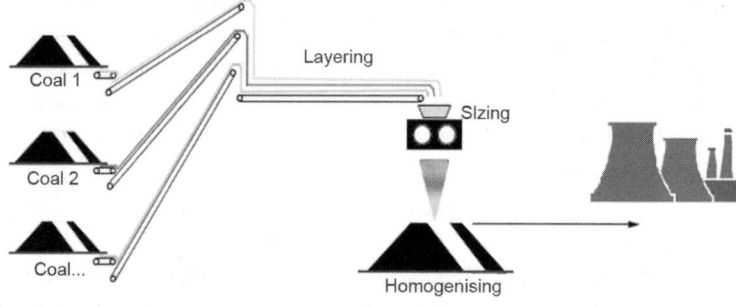

Figure 3

5. Quality control through blending at pakri barwadih coal mines of NTPC Ltd

Coal Fired Power plants of NTPC has been designed for ash from 30 to 40% in general. Pakri Barwadih is a basket mine for NTPC power stations. Coal Blending at Pakri Barwadih is proposed to be done considering the design parameters mentioned above of NTPC plants.

Based on the geological data, mining scheduling is being done to maintain the grade of Coal. Further to optimize, Coal from three mining pit comes through 3 different sizing stations and conveys to transfer house through online Ash Analyzer where coal with 30-35 % Ash is conveyed to railway siding directly.

As there could be difficulties to maintain the quality from mining itself 2 no stock piles are designed to be established symmetrically.

To maintain the quality of coal all time i.e Ash content of 30-35%, there are 2 stackers and 2 reclaimer and four (4) no of piles in two (2) different bays are maintained in the stock yard. Four (4) numbers piles are created in each row with different ash values.

Different Coal stock piles are created with different Ash percentage (No of piles 4 in each bay) in the stock yard are given below:-

- Pile Type A: Less than 31% Ash
- Pile type B: 31 – 35% Ash
- Pile type C: 35 – 38% Ash
- Pile Type D: + 38 % Ash

Coal is reclaimed by reclaimer 1 and reclaimer 2 from two different stock piles of each bay in such a way that Ash content is maintained between 30-35 %, which is analyzed through an Online Ash Analyzer and conveys to the railway siding. Coal is properly blended before it reaches to railway siding as there are 14 no of transfer points in the CHP circuit.

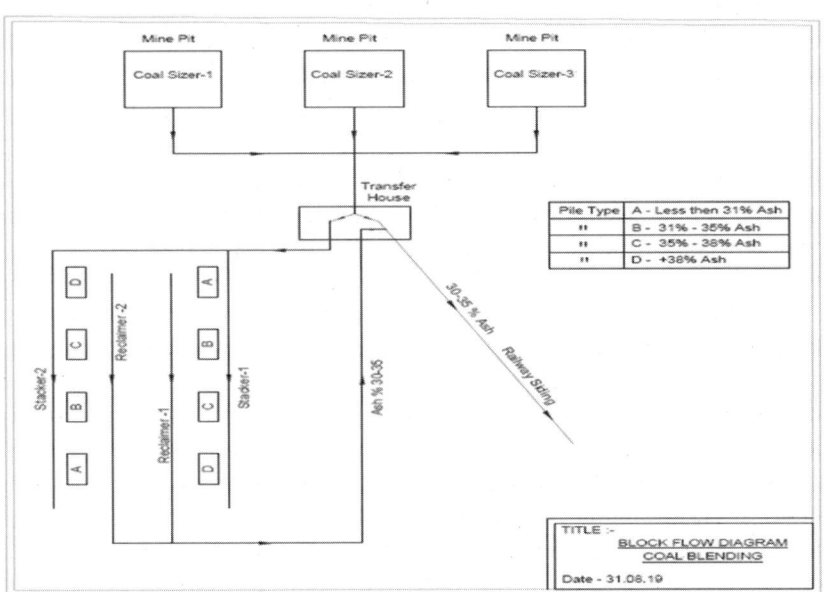

Stacker-1 will feed Bay 1 and Stacker-2 will feed Bay 2. One ash analyser (AS1) will be installed on the Belt Conveyor which analyses the ash content and as per the report received, coal will be stacked on to the different stock pile with specific coal quality. Reclaimers will work independently on two dedicated stockyard rails, which can reclaim from the predefined coal quality stock piles at an average capacity of 2500 TPH. On the reclaiming yard conveyors, two ash analyzers will installed, which give reclaimed coal quality. Mixing of both the coal streams happens on the belt. Quality of this outgoing coal (after mixing) quality will be measured via ash analyzer installed on Belt conveying outgoing coal. Based on feedback from different ash analyzers, the stockpile management team will decide the quantity of reclaiming coal from individual reclaimer to get desired output coal quality.

6. Conclusion

Coal blending adds extra effort, cost and space requirements but gives coal of desired quality suitable to fire at end user plants without detrimental effects to plants at low cost. Thus improves the revenue of coal mines and power plants both.

Coal blending allows the flexibility for plants to alter the blend of coal very quickly. This is ideal for plants that are required to adjust output rapidly to cope with grid demand or for plants that switch coals to ensure that emissions comply with legislated limits. Blending on site also allows plants to manage their own coal purchase and buffer supply as required.

With respect to keeping the coal plants adequately stocked, McCartney (2006) recommends that the relative size of the coal piles should be comparable to the relative burn rates of the coal. The stock should keep up with coal burn rate but not necessarily exceed it in any significant quantity. This is an important consideration when dealing with coal such Indian coal that are prone to spontaneous combustion. The coal should be reclaimed on a first-in first-out basis. Coal delivery rate should also be controlled in the same way. The blending methods used at any site will be, to some extent, determined by the available space and equipment. The stockpiling and blending systems outlined in this paper are all suitable for on-site plant use, depending on the space and equipment available, For example, for smaller plants with little storage space and only one hopper/bin, the coal mix may be dozed into the hopper at the same time, in the required proportional quantities to create the blend. A more accurate method would involve a dozer trap for the first coal type and a conveyor reclaim system for the second coal type. If two hoppers are available then each can contain a separate coal type and the feeder can then be used to withdraw the coal at the required ratio. Larger plants are likely to have more space for storage and a greater number of silos/piles and more hoppers to allow more co-ordinated blending.

The main methods for coal blending used on-site at coal-fired power plants are summarised in Table 2 and can be deployed based upon their suitability to individual plants.

DISCLAIMER: Views expressed in the text belong solely to the author(s) and are not necessarily attributable to the author's employer or organization.

Table 2: Blending Methods – Advantages and Disadvantages (Arora and Banerjee, 2013)

Method	Description	Location of blending	Advantage	Disadvantage
Beds/ Stockpiles	Stacking of two or more type of coal in layers	On the conveyor belt Homogenisation at transfer point	Only one system required. Relatively inexpensive	All coal must be stacked before blending Not possible to change the blending ratio
Silos	Coal in one silo is dropped onto a conveyor below carrying a second type of coal	On the conveyor belt Homogenisation at transfer point	Accurate blending ratio. Ratio can be varied	High capital cost
Ground hopper	First coal is bulldozed into the hopper, second coal is added via wagon or other source	At common transfer point Homogenisation at transfer point	Ground hopper is an additional source / stock for reclaiming	Feeding rate is not accurate
Blending on Moving Belt	Two types of coal are stacked in two yards and gathered by separate stacker/ reclaiming systems	At common transfer point Homogenisation at transfer point	Blending ratio can be changed at any time	All the coal must be stacked first
Blending on Moving Belt (imported coal reclaimed and domestic coal from track hopper)	Imported coal is reclaimed from the coal yard and domestic coal is fed from the track hopper/ wagon	At common transfer point Homogenisation at transfer point	Blending ratio can be changed at any time. Only imported coal needs to be stacked, other can come in as delivered	Lower blending accuracy than silo blending

46

Coal to chemicals: Scope and challenges

Navin Kr. Singh

Central Mine Planning & Design Institute Ltd. (CMPDIL), Ranchi, India

Abstract: Coal is an important source that can play critical role in boosting any countries economy. India is among one of the few countries that has vast coal reserves. Utilizing this resource in the best possible manner can only be achieved by exploring alternative use of coal applying clean coal technologies (CCT). India's commitment towards Paris Climate Treaty to reduce the carbon emissions intensity by 30–35% below the 2005 level by 2030 makes it imperative to adopt alternative use of coal with lesser carbon footprints. Utilizing coal feedstock for chemicals the first stage would be gasification of coal to syngas. Surface coal gasification (SCG) is a flexible, reliable, and commercial clean coal technology that can turn variety of low value feed stocks into high value products, helps in reducing countries dependence on imported oil, natural gas, and various chemicals, can provide alternative source for ammonia/fertilizer, fuels, substitute natural gas (SNG), and many other chemicals. Technologies for converting syngas to various chemicals are well developed and have been in use for a long time. Of these methanol has great potential, it can be directly used as fuel or also be blended with gasoline pool relatively easily converted to di-methyl ether which may be substitute for LPG. It can also be converted into light olefins, ethylene, and propylene (feed stocks for petro chemicals). Recognizing the potential of coal as game changer in the Indian energy sector SCG would be suitable CCT for future and may be proved in the interest of India's energy security, reduce import of various chemicals and oil & gas, and macro economy.

Keywords: Syn-gas, Renewable, Gasification, Ammonia, Petrochemicals

1. Introduction

India is in the early stages of a major transformation, bringing new opportunities to its 1.3 billion people and moving the country to center stage in many areas of international affairs. The energy sector is expanding quickly but is set to face further challenges (such as carbon emissions, alternate use of vast coal resource of India, oil & gas and various chemical import, cost competency, etc.) as India's modernization and its economic growth gather pace, particularly given the policy priority to develop India's manufacturing base.

India has very limited gas and oil reserves and the abundance of coal. At present, the share of coal in power generation is about 80%. However, coal is a major source of CO2 emissions in power generation and therefore it is facing a tough challenge from the alternative source of energy especially fast emerging renewable sources of energy. In addition, there has been a steep fall in the cost of renewable energy. As per India's COP 21 obligations, India's renewable capacity should be 40% of the total capacity by 2030 and our country is already under the process of implementing the 175 GW renewable programme to achieve the 40% criteria. Renewable energy costs have drastically fallen and are forecasted to continue dropping apace. New wind and solar is now cheaper than existing coal-fired generation's average wholesale power price. Hence, in this scenario, it is imperative to use the country's vast coal resource utilization through clean coal technology (coal gasification route) in coal to chemical business with due respect to environment will benefit the country.

Country's vast coal reserves and utilizing the resource in the best possible manner can prove to be highly significant by producing ammonia/urea, chemicals and petrochemicals in addition to conventional use as an energy source either for electricity or heat at present. Currently coal is primarily used for power generation. Considering the current crude and gas price scenario these options especially coal gasification have been found to be challenging in terms of investment required. However, with respect to nation's energy security these are the tried and tested options which cannot be overlooked. Past couple of years, even the government has been promoting coal gasification as the alternate feedstock for the fertilizer industry, methanol, etc. for developing self-reliance, apart from the domestic gas. Globally syngas is also being used to make valuable chemicals at competitive price.

Coal gasification is considered an important strategy towards clean coal technology with a goal of reducing its environmental footprints/low carbon energy development. The product of the coal gasification process is a syngas, which is a mixture of carbon monoxide, hydrogen, methane, carbon dioxide gases, unsaturated carbon, etc. with associated gross calorific value (energy content). The composition of the gas mixture and its calorific values depend on the type of coal fuel, gasifying media, and operating parameters of the gasifiers. It can be utilized just like natural gas in an eco-friendly manner. Gasifiers are designed according to the coal characteristics.

Coal characteristics are the most important factors that influence the gasification process selection and design. Generally, low-rank coals are the preferable feedstocks for the commercially available gasification processes, in comparison to the caking bituminous coals. The main properties of low-rank coals, which impact the choice of gasifier or process application, are the reactivity, the moisture and oxygen content, the volatiles content, the caking properties, the ash characteristics, and the sulfur content.

Utilizing coal feedstock for chemicals the first step would be gasification of coal to syngas. Technologies for converting syngas for various chemicals, are well developed and have been in use for a long time. Of these methanol has great potential, it can be directly utilized as fuel or relatively easily convertible to di-methyl ether (DME) which can be a substitute for LPG. Also it can be converted into light olefins, ethylene, and propylene (feedstock for petrochemicals). However, coal to chemicals and coal to liquid plants require significant capital investment as well as a high level of environment scrutiny.

Gasification is a flexible, reliable and commercial technology that can turn a variety of low value feedstock into high-value products, help reduce a country's dependence on imported oil and natural gas, and can provide a clean alternative source for ammonia/urea, fuels, substitute natural gas (SNG), and many other chemicals e.g. methanol, olefins for petrochemicals, acetic acid, etc. Gasification has the additional benefits of potentially net zero carbon dioxide release to the atmosphere and utilization of readily available low cost fuels.

China has become the global leader in converting coal to chemicals, SNG, fertilizers, and olefins for petrochemicals, and many other industrial chemicals, through coal gasification route. Coal gasification plants are successfully operating in countries like South Africa, China, USA, Netherlands, etc. due to cost competitiveness with that of natural gas feedstock.

2. Coal gasification route for coal exploitation

Productions of various chemicals through gasification route are as follows:

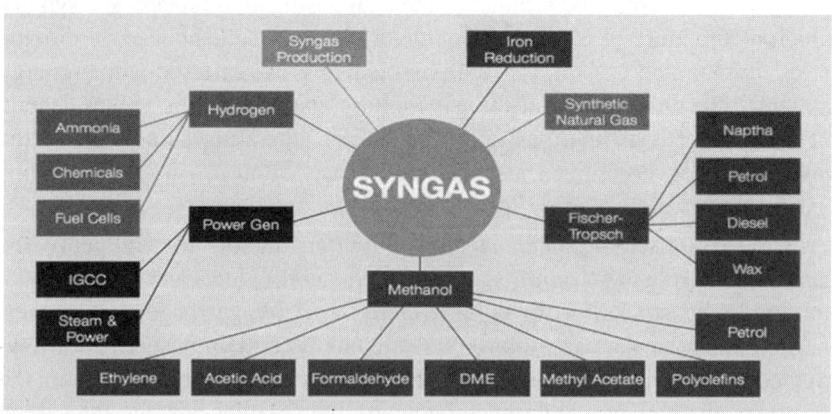

2.1 Global feedstock and application scenario

Primary Feedstocks for Gasification

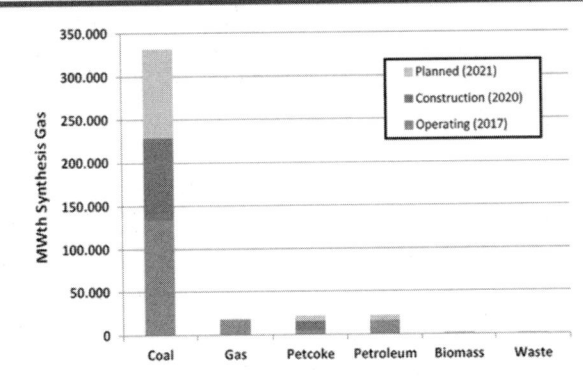

Source: GSTC Database, 2017

End Use Applications of Gasification Syngas

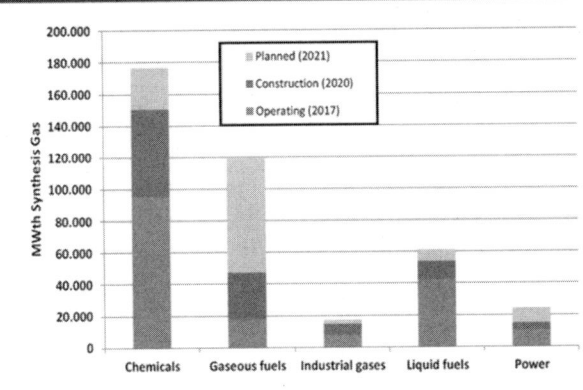

Source: GSTC Database, 2017

Globally there are 938 operating gasifiers and the number will increase to 2020 by 2021 and Gasification Technology provider is led by GE, SHELL followed by ECUST, SEDIN, SIEMENS, LURGI, HT-L, MCSG, and others. In the terms of global cumulative gasification capacity, operating gasification capacity has showed exponential growth since 2010–2016 after that it has been stagnant since year 2016 nearly to 160 MW_{th} synthesis gas but both the construction and planned during 2016–2020 has been increased manifold from 175 MW_{th} synthesis gas to 275 and 400 MW_{th} synthesis gas respectively during the same period.

2.2. Rapid growth of chemical and petrochemical import in India

Major chemical and petrochemical	Production (2016–17)	Import (2016–17)		Net import (2016–17)	Consumption (2016–17)	% Import (2016–17)	% Net import
	Qty (MT)	Qty (MT)	Value (Rs lakh)	Qty (MT)	Qty (MT)		
Acetic acid	158,510	847,808	193,958	834,894	993,410	85	84
Phenol	43,570	282,187	176,601	278,174	321,740	88	86
Methanol	176,960	163,7457	295,762	1,624,833	1,801,790	91	90
LDPE	202,000	391,501	330,125	364,592	566,000	69	64
HDPE	1,520,000	1,000,457	824,364	852,315	2,372,000	42	36
LLDPE	1,318,000	469,075	381,387	415,860	1,734,000	27	24
PVC compound		59,032	45,858	42,131	42,000	141	100
Polypropylene	4,253,000	781,188	611,882	206,954	4,460,000	18	5
PVC	1,462,000	1,702,852	1,015,835	1,696,368	3,158,000	54	54
MEG	1,110,000	1,235,385	604,838	1,173,186	2,284,000	54	51
Styrene		729,627	557,571	724,539	725,000	101	100
VCM	791,000	344,423	177,368	344,423	1,136,000	30	30
Acetone	26,790	135,887	63,055	132,662	159,460	85	83
EVA	0	144,604	141,686	143,110	143,000	101	100
Poly carbonate	0	138,365	204,195	134,375	135,000	102	100

Source: Ministry of Chemicals and Fertilizers

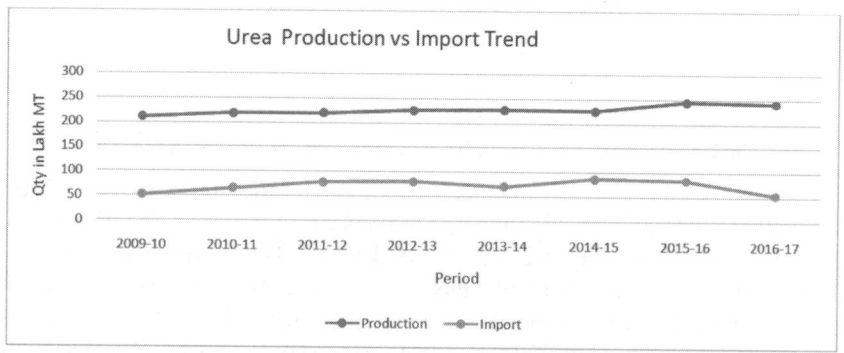

Projected demand–supply gap for urea in India is approximately 8–10 million tonne in between FY 19 and 25. As India is agrarian country so self-reliance in the field of fertilizer is the need.

In India, we are not self-reliant/sufficient in production of chemicals, petro chemicals and fertilizers and the same are being imported to meet the country's demand. Given this backdrop, there is a considerable forex outgo on import of these chemicals. In view of this, there is immense opportunity to set up domestic facility for these chemicals through coal gasification route, which will help in the following:

- India aims to reduce its crude oil import dependency by 10% by 2022, and indigenous methanol production could help achieve this goal.
- India's commitment to Paris Climate Treaty to reduce the carbon emissions intensity by 30–35% below the 2005 level by 2030. Methanol and DME have the potential to contribute towards achieving this target, depending on the production route.
- To keep pace with the government of India target of wiping out urea imports completely by 2022.
- Make in India Programme.

It is estimated that coal gasification can potentially target 90–120 MMT of coal by 2030 for production of methanol, ammonia, dimethyl ether, synthetic natural gas, and many other products. The detailed assessment is presented as below:

Expected coal demand from alternative use, in million tonnes. Coal demand for:	Unit	2019–20	2024–25	2029–30
Methanol	Million tonnes	6–14	8–19	11–25
Dimethyl ether	Million tonnes	44–48	56–62	72–80
Urea	Million tonnes	8–10	9–11	10–12
Total	Million tonnes	58–73	73–92	93–117

Source: Coal Vision 2030 KPMG

2.3 NITI AAYOG's road map for methanol economy comprises

- Production of methanol from Indian high ash coal from *indigenous technology*, in large quantities and adopting regional production strategies and produce methanol in large quantities. India will adopt CO_2 capturing technology to make the use of coal fully environment friendly and our commitments to COP21.
- Utilization of methanol as well as DME in transportation: rail, road, marine, and defense. Industrial boilers, diesel Gen-sets and power generation and mobile towers are other applications.

- Utilization of methanol and DME as domestic cooking fuel-cook stoves. LPG–DME blending program.
- Utilization of methanol in fuel cell applications in marine, Gen-sets and transportation.

The final roadmap for '*Methanol Economy*' being worked out by NITI AAYOG is to substitute 10% of crude imports by 2030, by methanol alone. This requires approximately 30 MMT of methanol. This will also help in an annual reduction of 100 billion dollar by 2030 in crude imports in line with our honorable prime minister's vision. To promote this renewable, alternate fuel a "Methanol Economy Fund" is also being contemplated.

2.4 Government initiative to facilitate to methanol/DME production

- Ministry of Road Transport and Highways has already notified methanol as "Transportation Fuel" and has come out with draft Gazette notifications for blending of 15% methanol with gasoline. In the following proportions viz. 15% (M15), 95% (MD 95), and 100% (M 100) blends to substitute gasoline.
- Pilot project for use of methanol as cooking fuel has been initiated with government of Uttar Pradesh covering 20,000 rural households in the Ganga basin.
- LPG DME blending notification: 20% blending of DME with 80% LPG has been notified by BIS. This program is expected to result in a savings of rupees 6000 crore per annum due to reduced LPG imports and cost savings.

2.5 Commercial scale project through SCG route by Coal India Ltd. (CIL) and its joints ventures (JV)

Commercial Scale Projects have already been taken up in CIL and its JV with respect to coal to chemicals/liquid (methanol) and fertilizer through surface gasification route are given as under:

(1) Talcher Fertilizer Limited (A JV Company of CIL, GAIL, RCF and FCIL):

It is a JV company of CIL, GAIL, Rashtriya Chemicals & Fertilizers (RCF), and Fertilizer Corporation of India Ltd (FCIL) for production of ammonia and urea. The technology on which it is based is Shell-Gasification, Haldor

Topsoe-Ammonia Technology and Saipem's Urea Process. This JV aims to produce 0.726 MMTPA ammonia and 1.27 MMTPA urea after successful completion. The coal requirement for this project will be 2.6–3.3 MMTPA (40–47% ash content). In addition, 0.31 MMTPA pet coke (ash 0.41%) will also be consumed in the plant. Ratio of feed for the gasifier would be 75:25 of coal and pet coke. This prestigious plant has been inaugurated by honorable prime minister on 22.09.2018 and targeted commissioning date of the plant is July 2022.

(2) Dankuni coal complex: Coal to methanol project:

A methanol plant (commercial scale) has been proposed at Dankuni Coal Complex, CIL through Surface Coal Gasification route from Raniganj Coal which is in progress. This *Coal Complex* aims to produce *676500 MTPA methanol* after successful completion. Total coal requirement for DCC will be 1.42 MMTPA (ash content below 25%). Targeted commissioning date is July 2023.

(3) Central Coalfields Limited: Coal based methanol/ammonia/ ammonium nitrate plant:

Poly generation of chemicals (ammonia, ammonium nitrate, and methanol) through coal gasification from North Karanpura coal by CCL is also in progress (pre-feasibility report prepared). This project aims to produce *0.75 MMTPA* methanol after successful completion. Total coal requirement for the project will be approx. 1.39 MMTPA (ash ~ 35%). Project is expected to be completed by March, 2022.

At present, research studies regarding coal to liquid (methanol) have also been undertaken to develop indigenous technology through surface coal gasification for high ash Indian coal by BHEL, IIT Delhi and Thermax and IIT-ISM, Dhanbad.

2.6 Coal availability and outlook: A promising feedstock

The present updated geological coal resource of the country is 3,19,020.33 million tonne as on 01.04.2018 for coal seams of 0.9 m and above in thickness and upto 1200 m depth from surface. Out of the total resources, the Gondwana coalfields account for 3,17,432.65 million tonne while the rest 1,587.68 million tonne from tertiary coalfields of coal resources. This inventory dealt with only the net geological resource assessed, so far and did not take into account the mined out quantity.

2.7 Summary of grade-wise resources of non-coking coal of the country follows (0–1200 m depth) (as on 01.04.2018)

Grade	GCV (kcal/kg)	Resource (million tonne)
G1–G3	6401 and above	2725.99
G4–G5	5801–6400	10044.14
G6	5501–5800	24674.44
G7–G8	4901–5500	42560.98
G9–G14	3101–4900	172964.51
G15–G17	2201–3100	1838.58
Ungraded		28101.55
Total		*282910.19*

2.8 Potential area of non-coking coal resource in Gondwana coalfield (as on 01.04.18) upto 300 m depth for coal gasification (in million tonnes)

Coalfields	State	Measured reserve (G1–G8 – upto 34% – A+M)	Coalfields	State	Measured reserve (G9–G14 – 34–55% – A+M)
Raniganj	West Bengal	8098.13	Talcher	Odisha	19345.10
Wardha-Valley	Maharashtra	2034.45	IB-River	Odisha	10713.89
North Karanpura	Jharkhand	1223.93	North Karanpura	Jharkhand	7377.88
South Karanpura	Jharkhand	1505.06	Mand-Raigarh	Chhattisgarh	6996.71
Sohagpur	Madhya Pradesh	1253.32	Korba	Chhattisgarh	4502.89
Talcher	Odisha	1108.05	Godavari	Andhra Pradesh	0.00
Mand Raigarh	Chhattisgarh	951.62	Singrauli	Madhya Pradesh	3545.13
Kamptee	Maharashtra	847.28	Rajmahal	Jharkhand	3441.94
Pench Kanhan	Chhattisgarh	780.81	Wardha Valley	Maharashtra	2117.94
Bisrampur	Chhattisgarh	799.56	Raniganj	West Bengal	2139.09
Korba	Chhattisgarh	737.1		Jharkhand	867.01
Hasdo-Arand	Chhattisgarh	481.29	South Karanpura	Jharkhand	1735.11
IB-River	Odisha	500.63	Hasdo-Arand	Chhattisgarh	1540.55
Chirimiri	Chhattisgarh	309.34			
Jharia	Jharkhand	655.12 (0–600 m depth)			

3. Conclusion

Looking at the challenges being posed to coal's future, National Energy Security and import of oil & gas, chemicals, fertilizer, petrochemicals and vast reserve of coal in our country, this is a need of the time to use coal as alternative feedstock for chemical, petrochemical, fertilizer, etc. and would be beneficial for Macro & Micro Economy, Employment Generation, saving in foreign exchange, revenue generation-taxes, duties, royalty, favoring energy security of India, low emission of gaseous pollution, etc.

This will also help in (i) PM's ambitious target of reducing 10% import dependence of oil & gas by 2022 from 2014 to 2015 levels, (ii) India's commitment to Paris Climate Treaty to reduce the carbon emissions intensity by 30–35% below the 2005 level by 2030, (iii) to keep pace with the government of India target of wiping out urea imports completely by 2022, and (iv) Make in India Program of government of India.

Ash/slag generated from coal gasification would be required to be disposed properly. Keeping view of this the commercial plant may be planned nearby to the abandoned mine voids and ash disposal have to be done in environmental friendly manner or used for various other purposes. Slag/ash should be utilized in cement manufacturing and as constructional material according to its physical and chemical properties.

Recognizing the potential of coal as game changer in the Indian energy sector, surface coal gasification would be a suitable clean coal technology for future. Pursuing surface coal gasification technology may be proved to be in the interest of our country's energy security and macro economy.

4. References

[1] India's Leapfrog to Methanol Economy by Dr. V.K. Saraswat and Ripunjaya Bansal.

[2] PIB's release on "Methanol Economy": NITI Aayog working on road map for India on World Environment Day, 2018.

[3] Methanol as an Alternative Fuel for India, By Ramya Natarajan.

[4] Indian Coal and Lignite Resources – 2018 – GSI.

[5] Gasification of coal, D. Vamvuka.

Websites consulted

1. http://pib.nic.in/newsite/PrintRelease.aspx?relid=179785.

2. Export/Import data from: http://www.chemicals.nic.in/ and http://fert.nic.in/ shipping-month-port-wise-report/38/2017.

3. https://www.niti.gov.in/.

47

Production of geopolymer materials based on waste from mining and processing of hard coal

Barbara Tora, Tadeusz Olkuski, Stanisław Budzyń, Jolanta Biegańska

AGH University of Science and Technology, Kraków, Poland

Abstract: Geopolymers are inorganic, amorphous, and synthetic aluminosilicate polymers formed from the synthesis of silicon (Si) and aluminum (Al) and obtained geologically from minerals. Their chemical composition is similar to that of zeolite. Geopolymers are usually hard, mechanically resistant solid bodies resembling natural stone or concrete.

The main components of waste from coal mining are illite, quartz, and kaolinite, which contain a large amount of silicon oxide and aluminum oxide. Their chemical composition makes it potentially a good raw material in the process of alkaline activation.

The examples presented in the article show the possibilities of producing geopolymers based on waste, in particular derived from hard coal mining and processing. Materials obtained in the geopolymerization process are characterized not only by the proper mechanical properties, but also by a number of features, i.e. binding of heavy metal elements or refractory.

Keywords: Coal waste, Geopolymers, Activation

1. Introduction

Geopolymers are, by definition, inorganic, amorphous, and synthetic aluminosilicate polymers formed from the synthesis of silicon (Si) and aluminum (Al) and obtained geologically from minerals. Their chemical composition is similar to that of zeolite, but reveals an amorphous microstructure.

Geopolymers are usually hard, mechanically resistant solid bodies resembling natural stone or concrete (Davidovits, 1989).

The main components of waste from coal mining are dolomite, which contain a large amount of silicon oxide and aluminum oxide. Geopolymers are usually hard, mechanically resistant solid bodies resembling natural stone or concrete. The main components of waste from coal mining are illite, quartz, and kaolinite, which contain a large amount of silicon oxide and aluminum oxide (Hycnar, 2015).

Their chemical composition makes it potentially a good raw material in the process of alkaline activation.

However, to ensure adequate reactivity in the alkaline activation process, this raw material requires initial preparation in the mechanical and thermal process. It is possible to use only one of these processes to activate the material, however, optimal properties are obtained by means of mechanical activation – grinding (fine particles show higher reactivity) and thermal activation in the calcination process. It should also be noted that the best material properties are obtained on the basis of waste, which contains a high content of amorphous aluminosilicate. The content of reactive components, in particular active silicon and aluminum, is important (Mikuła, 2014).

The examples presented in the article show the possibilities of producing geopolymers based on waste, in particular derived from hard coal mining and processing. The presented technologies show the possibilities of mineral waste management coming from extraction and processing into materials that can be used in the broadly understood construction. Materials obtained in the geopolymerization process are characterized not only by the proper mechanical properties, but also by a number of features, i.e. binding of heavy metal elements or refractory.

Such properties predestine the obtained product also for applications as so-called special materials. Examples of possible applications are the protection of landfills. Fire-resistant properties suggest that this material can also be used in mining.

2. Geopolymers and alkali-activated materials

Geopolymers consist of long chains (copolymers) of aluminosilicate and aluminum stabilized by metal cations, usually of sodium, potassium, lithium or calcium, and bonded water (Król and Błaszczyński, 2013; Rajczyk et al., 2015). In addition to polymer chains, the material usually contains various mixed phases: silicon oxide, unreacted aluminosilicate substrate, and crystallized zeolite-type aluminosilicates (Rajczyk et al., 2015).

It is a characteristic feature of the geopolymer material that it contains SiO_4 and AIO_4 tetrahedra in its structure (a three-dimensional network). These are alternately bonded by oxygen atoms. The setting occurs in highly alkaline aqueous solutions in which reactive aluminosilicates become dissolved and then, in the process of polycondensation, the tetrahedral, $[SiO_4]_4$ and $[AlO_4]_5$, become linked at corners, forming amorphous or subcrystalline three-dimensional aluminosilicate structures similar to zeolites (Rajczyk et al., 2015; Mikuła, 2014). The presence of monovalent Na+ or K+ cations, or cations of other metals, in chambers compensates for the negative charge of the skeletal structure (Rajczyk et al., 2015; Davidovits, 2015). Considering the whole geopolymerization process (Fig. 2), alkali activation is only the first step in the formation of geopolymer materials (Davidovits, 2015).

Alkali-activated materials do not form three-dimensional networks, and only have a two-dimensional structure, which affects their properties. This structural difference leads to different physico-chemical and performance properties of geopolymer and alkali-activated materials, particularly in terms of their resistance to chemical agents and such performance properties as flame resistance and durability (Davidovits, 2015). It should also be noted that the mechanical properties of alkali-activated materials can exceed those of geopolymers in the short term (Davidovits, 2015).

The raw material and additives used for synthesizing the material have the strongest effect on the process of its formation.

High aluminum content and low calcium content are especially conducive to the formation of geopolymer materials. Please note that bibliographical sources do not specify the exact values of each element or oxide in the base material.

3. The raw materials used in geopolymer production

Geopolymers can be produced on the basis of both natural raw materials and waste material. The most commonly used material in geopolymer synthesis is metakaolin.

The second most common raw material for geopolymers is ash/slag. It should be noted that the level of usefulness of ash depends on the content of active components, which go into solution influenced by an alkali activator (Rajczyk et al., 2015).

What is also important is the ratio of SiO_2 to A_2O_3, as well as the granulation and morphology of particles, including the content of unburnt carbon particles (Rajczyk et al., 2015; Mikuła, 2014).

The synthesis of geopolymers most commonly uses Class F or Class C fly ash (Mer, 2016). Slag from various metallurgical productions is also commonly used in the process of geopolymer formation. Slag is used both as the primary raw material for synthesis and as an additive to other raw materials.

Other potential raw materials include natural clay (e.g. shale clay) and waste containing high amounts of it, including glass and ceramic waste. There have also been attempts to produce geopolymer materials from mineral wool. Glass waste is usually used as an additive to other materials.

Other potential raw materials include the combustion products of sewage sludge and silt. There have been some studies conducted on sediments containing montmorillonite, which was ground and then calcined at 850°C for 6 h. A mixture consisting of 30% ground-granulated blast-furnace slag and 70% sewage sludge powder was activated by mixing it with various alkali solutions of water, sodium hydroxide, and sodium silicate to obtain a geopolymer material.

It is also possible to use paper industry waste for the alkali activation process (Budzyń et al., 2015). The waste incinerated in this process may contain considerable amounts of aluminosilicate ingredients resembling metakaolin, which exhibit good pozzolanic properties (i.e. various forms of calcite and kaolinite clay).

Another type of industrial waste, which can potentially be used as a raw material in the production of geopolymers is petroleum processing waste and oil sands.

Mineral waste, such as mining, extraction, and processing waste, is also used for the production of geopolymers (Pasiowiec et al., 2016).

4. Basic properties of geopolymers

In comparison to traditional structural materials, such as concrete, geopolymers have numerous advantages. They can be used in normal conditions where they represent an environmentally-friendly alternative, but also in extreme conditions, in which traditional materials quickly wear or cannot be used at all (Davidovits, 1989). Additionally, the production of geopolymer composites, when compared to special-purpose materials (for use in difficult conditions), is economically viable, e.g. due to the low energy-intensity of the process (Mikuła, 2014).

An important growth factor in the context of the growing interest in geopolymers is also the increasing public awareness of the need to protect the environment. The production of geopolymers generates considerably lower carbon dioxide emissions than that of traditional construction materials, e.g. Portland cement. It is estimated that six times less carbon dioxide is used to produce geopolymers than to produce cement. At the same time, these materials retain similar or achieve better properties than traditional materials, i.e. cements, including in particular (Davidovits, 1989; Michalíková et al., 2015):

- High initial strength,
- Less shrinkage and low thermal conductivity in comparison to materials on the basis of traditional cements – dimensional stability,

- Good fire-resistance,
- High resistance to various acids and salts,
- Good wear resistance,
- Adhesion to fresh and old concrete bases, steel, glass, and ceramics,
- High capacity to imitate surfaces and reproduce patterns,
- Absence of corrosion of the steel reinforcement in the geopolymer and a high level of adhesion to steel,
- Weather resistance,
- The availability of raw materials and their lower cost, while geopolymer materials can also be produced on the basis of waste material, i.e. ash from power plants or combined heat and power plants or other production waste,
- Lower energy consumption in the manufacturing process – environmentally-friendly,
- The possibility of immobilizing hazardous waste by enclosing it in geopolymer composites, and especially the possibility of immobilizing heavy metals.

Due to the properties of geopolymers, such as chemical and thermal resistance and their excellent mechanical characteristics, their significance in economic use is constantly rising. Currently, the applications of these materials include (Szczygielski et al., 2017):

- For the production of construction materials, including the production of bricks,
- In industry as a versatile material, particularly in areas requiring the final product to be heat- and fire-resistant,
- As a material for protective coatings, including various types of steel,
- In founding, as a component of molding compounds,
- For emergency repairs of such structures as runways,
- As a carrier material for the stabilization of toxic waste, including radioactive substances.

5. Research material

The study examined samples of mining waste and tailings from the polish lead–zinc processing waste mines. The samples had been taken from various locations in the landfills. Coal-associated shales were separated from the sampled material due to their low usefulness in obtaining construction materials.

6. Research methodology

In order to determine the phase and chemical composition, the waste was milled on a ring-roller mill (Längauer and **Čablík**, 2018). The milled material was dried. Loss on ignition in the temperature of 900 °C was also determined for the milled material in order to approximately define the value of elemental carbon in the analyzed waste. The roasted material was subjected to mineralization in a solution of concentrated HNO_3 and HCl acids with the volume ratio of 3:1.

The content of silicate as a residue of mineralization (after repeated roasting) was determined. The obtained aqueous extract and mineralization solution were analyzed using the ASA method for the content of Ca, Mg, Fe, K, Pb, Zn, Na, Ni, Cd, and Cr. The measurements were taken using a Perkin Elmer 370 device.

In order to obtain aggregate for conducting strength tests, the waste material originating from a mining waste landfill site was subjected to mechanical processing. The pulverized material was separated using sieves for the fractions with a granulation of 8–10 mm.

A concrete mix was prepared on the basis of sand and Portland cement. Aggregate prepared in advance was added to it with weight ratios of 1:1 in relation to the weight of concrete, and then an appropriate amount of tap water was added.

7. Research results

Figure 1 presents an example diffractogram of the coal waste sample. The analysis demonstrated the presence of silicate in sample, with additional aluminosilicate phases, iron, calcium and magnesium carbonates. Table 1 presents the content of SiO_2, C (delimited by loss on ignition), Ca, Mg, Fe, K, Na, and Cu. No Cd, Cr, Cu, Ni, Pb, or Zn was found in the analyzed waste. Compositions differed considerably depending on the sampling location.

Table 1: Chemical compound of the waste sample

Chemical compound	(%)
SiO_2	64.8
$Al2O_3$	7.08
Fe_2O_3	4.33
CaO	7.52
MgO	4.06
SO_3	1.66
Na_2O	0.90
K_2O	3.26

Element	Wt %
C	5.99
O	43.38
Mg	0.20
Al	0.43
Si	40.08
P	0.00
S	1.99
Pb	1.08
K	0.00
Ca	0.26
Ti	0.08
Mn	0.48
Fe	1.07
Co	0.31
Zn	4.64

Figure 1: Diffraction pattern of waste sample: A – SiO_2; B – $Al_2(Si_2O_5)(OH)_4$; and C – Ca, $Mg(CO_3)_2$

The obtained aqueous extracts were not found to contain Ni, Cu, Cd, and As within the limit of quantification of the used method. The pH of the analyzed leachates was between 7 and 9. Higher pH values of aqueous leachates were found in samples with lower silicate and higher calcium and magnesium contents.

7.1 Hydrothermal synthesis

After basic analyses (XRD, XRF, and SEM), the coal wastes were homogenized. In 250 ml containers, we take waste, to which we added NaOH at 3 M concentrations. S/l ratio was 1:7.5.

The prepared samples were dried in a laboratory drier and the hydrothermal synthesis was set to proceed for 6–24 h at 90°C, 100°C, 110°C.

Upon completion of the synthesis, the samples were filtered and washed with distilled water to lower the pH value below 10. Samples were homogenized and analyzed on XRD and SEM analysis.

8. Results and discussion

Figure 2 shows the sample of coal waste after geopolymerization using 3 M NaOH. Figure 3 shows the results of determination of a compression force for the sample of geopolymer after 30 days.

Figure 2: Waste after geopolymerization

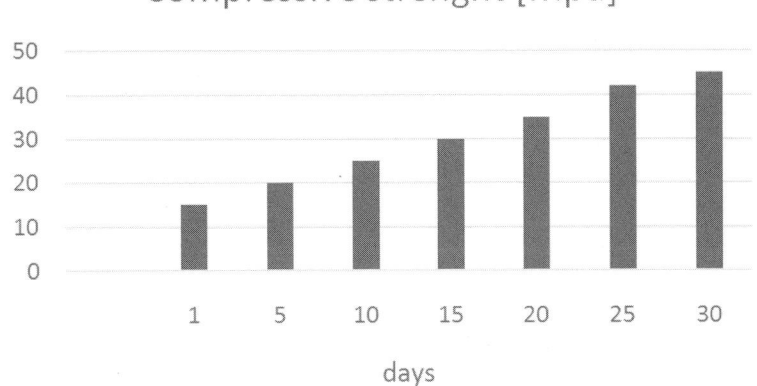

Figure 3: Values of the compression force determined for geopolymer sample

9. Summary and conclusion

The material used in this study was characterized by relatively low elemental carbon content (less than10%). The material did not contain coal-associated shells (black or burnt), which have already been used as a raw material in the production of cement and in road construction. The main crystallographic phases in the analyzed material were silicate, aluminosilicate, and carbonate phases. Due to the phase composition, such material is potentially well-suited to the production of concrete. However, the mechanical properties obtained were only slightly higher (38–43 MPa) when compared to standard concrete (37 MPa).

The principal element limiting the application of mining waste as aggregate for concrete is the heterogeneous composition of the sample taken from one landfill site.

The obtained aqueous extracts were characterized by neutral or weakly alkaline reaction and low sodium, potassium, calcium, and magnesium content. The aqueous extracts were found to contain no heavy metals such as Cr, Cd, Ze or Ni, which is why such aggregates may find applications mainly as a material used for the substructures of roads, bridges, and railway embankments.

The use of aggregates from mining waste has a positive influence on the landscape of mining areas and allows deposited waste to be put into use, while reducing the amount of natural aggregate needed.

10. References

[1] Budzyń S., Andrusikiewicz W., Cablik V., Tora B., Gradoń W., 2015, *Energetic recovery of Cellulose Waste*, Inżynieria Mineralna – Journal of the Polish Mineral Engineering Society, No 2(36), p. 137–141.

[2] Davidovits J., 1989, *Geopolymers and geopolymer new materials*, Journal of Thermal Analysis, vol. 35(2), s. 429–444.

[3] Hycnar J. J., 2015, *Methods of Increasing the Calorific Value of Fine Coal Waste*, Inzynieria Mineralna – Journal of the Polish Mineral Engineering Society, No 1(35), p. 33–55.

[4] Klojzy-Karczmarczyk B., Mazurek J., 2017, *Propozycje rozszerzenia działań celem zagospodarowania materiałów odpadowych z górnictwa węgla kamiennego*, Zeszyty Naukowe Instytutu Gospodarki Surowcami Mineralnymi i Energią Polskiej Akademii Nauk, nr 98, s. 151–166.

[5] Längauer D., Čablík V., 2018, *Preparation of synthetic zeolites from fly ash*, Inżynieria Mineralna z. 1(41), p. 7–12, Polish Mineral Engineering Society, Kraków, DOI:10.29227/IM-2018-01-01.

[6] Michalíková F., Brezáni I., Sisol M., Stehlíková B., Mihok L., 2015, *The use of black fly ash at the production of ceramics materials – Part II*, Inzynieria Mineralna – Journal of the Polish Mineral Engineering Society, No 1(35), p. 103–108.

[7] Mikuła J. (red.), 2014, *Rozwiązania proekologiczne w zakresie produkcji. Nowoczesne materiały kompozytowe przyjazne środowisku*, Wydawnictwo Politechniki Krakowskiej, Kraków, ss. 287.

[8] Pasiowiec P., Hycnar J. J., Tora B., 2016, *Fine coal waste utilisation*, Inżynieria Mineralna – Journal of the Polish Mineral Engineering Society, No 1(37), p. 213–222.

[9] Rajczyk K., Giergiczny E., Szota M., 2015, *Mikrostruktura i własności stwardniałych spoiw geopolimerowych z popiołu lotnego*, Prace Instytutu Ceramiki i Materiałów Budowlanych, nr 23, s. 79–89.

[10] Szczygielski T., Tora B., Kornacki A., Hycnar J. J., 2017, *Fluidal ashes – properties and application*, Inżynieria Mineralna – Journal of the Polish Mineral Engineering Society, No 1(39), p. 207–216.

48

Installation of coal pyrolysis with the improvement of the reactor retort

Alexandra Boytsova[1] and Stanislav Avtamonov[2]

[1]Saint-Petersburg Mining University, Sankt-Peterburg, Russia
[2]Ukhta State Technical University, Ukhta, Russia

Abstract: Russia contains vast coal resources on remote area. They represent a potential energy source for the province. Many countries used coal gasification to produce fuels and chemicals during the last century. The coal-conversion processes involve reacting coal in a large reactor vessel at high temperature and pressure with steam and a limited amount of oxygen to prevent combustion. The chemical bonds in the coal are broken, producing synthetic gas (syngas) with fuel properties and ash residue remaining from the mineral matter in the coal. In addition, one of the perspective methods of coal processing is pyrolysis. In comparison with burning – pyrolysis has less emissions in atmospheric sphere, and, consequently, its pollution decreases. In this paper, it is considered the method of complex processing of coal of various composition and properties. The main stage of conversion is low-temperature pyrolysis, after which the obtained fractions are used for further refining or cracking.

The plant's products are energy resources of a wide range, such as gas, gasoline, kerosene, diesel fuel and heavy residues. The paper describes the development of the design of a combined thermal distillation coal processing plant with the improvement of the reactor retort.

Keywords: Russian coal industry, pyrolysis, gasification, non-waste technology

1. Introduction

At the end of 2017, Russia took sixth place in the world coal production, leading by China, the United States, India, Australia and Indonesia. Last year in the country almost 410 million tons of coals were produced and the Ministry of Russian Energy predicts to increase this volume to 480 million tons. There are 22 coal basins and 129 separate deposits in the country. But in spite of huge coal reservoirs, Russia has a significant problem connected with the remoteness of coal mining centres. This has great influence on logistics costs and in turn on the impact on the final price of coal product. Thus one of the most important things for Russian coal companies is to find solution to process coal near its mining and transport high-valuable products such as syngas that can be used as a feed for methanol, olefins, gasoline, diesel, methane, ammonia, etc. (Fig. 1) [1].

The main reaction for methanol production from syngas is:

$$CO + 2H_2 = CH_3OH$$

Then it could be converted to fuel (mixture of hydrocarbons) as following:

$$CH_3OH = [-CH_2-] + H_2O$$

One of the most promising technologies for syngas production is coal gasification.

Coal gasification is the process of reacting coal with oxygen, steam and carbon dioxide to form a product gas containing hydrogen and carbon monoxide. It is considered that gasification is incomplete combustion. The chemical and physical processes are quite similar; the main difference being the nature of the final products and the amount of oxygen that is added to the coal. In coal gasification, much less oxygen is added to the process (almost 40% in comparison with combustion). From a processing point of view the main operating difference is that gasification consumes heat evolved during combustion [1].

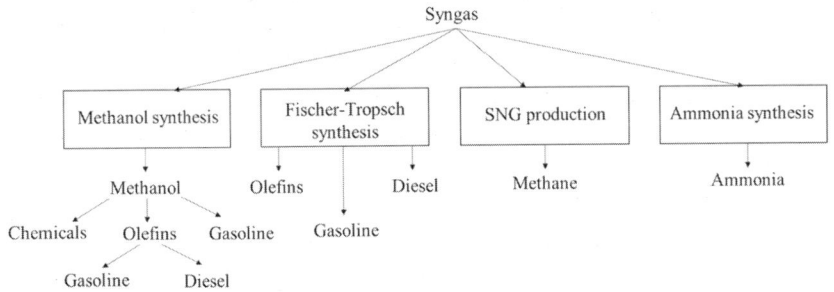

Figure 1: Syngas utilization

These processes were created to convert the bituminous coals into a strong hard coke ideally suited for iron making. It was found that when bituminous coals are heated slowly in the absence of air and then cooled the solid coke is both hard and inherently strong [2].

The first coal gasifier was used by Fontana in 1780 when he directed a flow of water (steam) over red-hot coal that was previously heated in an air blast. This process can be simply described using two chemical reactions: char combustion for producing heat and steam gasification for producing synthesis gas, a combustible mixture of carbon monoxide and hydrogen [3,4].

Coal gasification may be generally represented by reaction:

$$C_mH_n + 0.5mO_2 = mCO + 0.5nH_2$$

where, m and n depends on the composition of coal. The reactions in different stages of the process are as follows:

$$CO + H_2O = CO_2 + H_2$$
$$C + CO_2 = 2CO$$
$$C + H_2O = CO + H_2$$
$$CH_4 + H_2O = CO + 3H_2$$

Coal technologies besides combustion included: pyrolysis, coking, cyclic gas generators and gas producers. All of these processes are heated with insufficient air to convert the coal to the final products of combustion. The products of all of these processes are a solid fuel, condensables and tars and flammable gases [5].

Pyrolysis, or heating coal in the absence of oxygen, was conducted principally for the manufacture of chemicals. This conversion process is geared toward the production of condensable products such as benzene, toluene, naphthalene, phenols, creosote and pitch. The coal beds are thin and temperatures are low and near the melting, or the decomposition, temperature of coal to promote production of liquids and gases [6]. Coking is another process in which coal is heated in the absence of oxygen, the main product is the solid coke. The coal beds are deep and the temperatures are much higher to promote resolidification of the decomposing coal [7].

Gasification has many positive attributes that make it a desirable technology for the production of power, fuels and/or chemicals. Some of those attributes are described below:

- The ability to utilize all carbon-containing feedstocks.
- The ability to produce value-added products.
- The removal of impurities from the reducing gas is easier than from combustion systems.
- Gasification plants can be recombined to reach near zero levels of emissions.
- Gasification is more efficient and eco-friendly technology in comparison with combustion.
- The aim of this study is to develop the design of the combined unit by thermal utilisation of coal with the improvement of the reactor retort.
- In the gasification practice, the basic equipment can be grouped in three main categories:
- Moving bed gasifier (coal bed slowly moves downwards counter currently with respect to air and is gasified. It has the lowest oxygen consumption. It can operate at the lowest temperature, which inhibits the reaction rate and increase the maintenance cost) (Fig. 2).

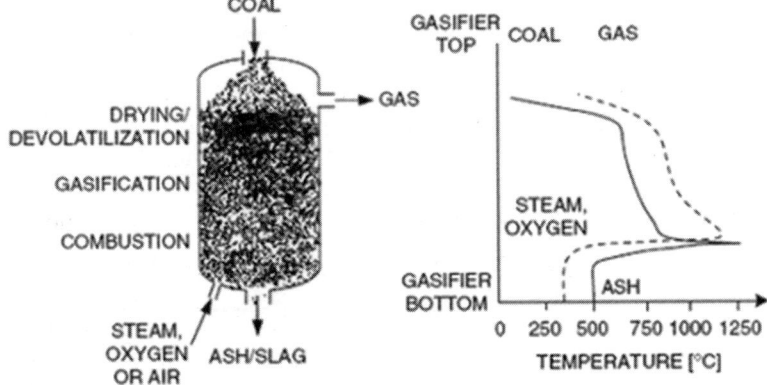

Figure 2: Moving bed gasifier

- Fluid bed gasifier (it has low overall coat and facilitates good mixing. The conversion rate of carbon is lower in comparison with the other two types due to some carbon lost with the ash. It is appropriate for low rank coals like lignite. Equipment is operated only at near atmospheric pressure) (Fig. 3).

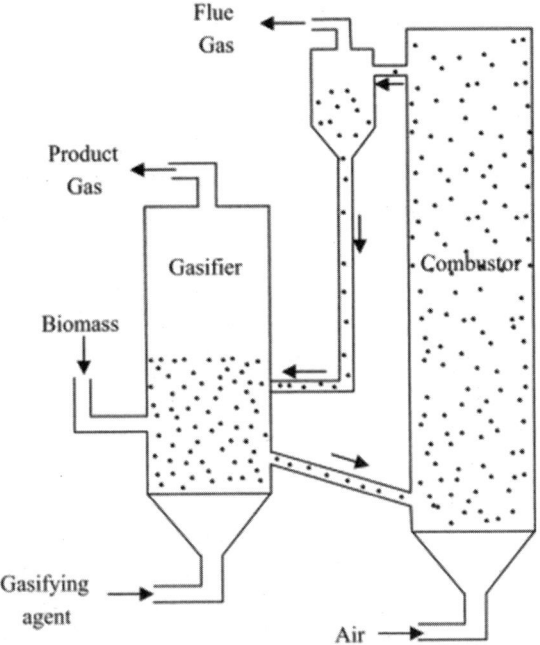

Figure 3: Fluid bed gasifier

- Entrained-flow gasifier (the fine coal particles react with concurrently flowing steam and oxygen Conversion of about 99% is obtained since the gasifier operates at a high temperature. However, it has high oxygen demand and also the high ash content in the sub-bituminous coal would increase the oxygen consumption. It is chosen for high carbon conversion and purity of the resulting syngas) (Fig. 4) [8].

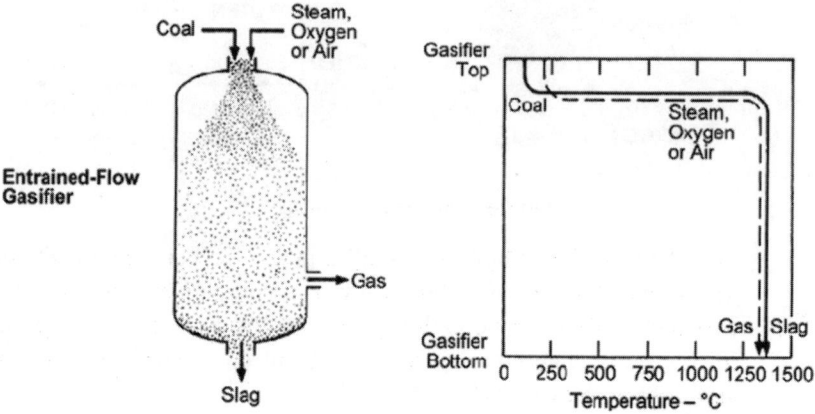

Figure 4: Entrained-flow gasifier

Technical approach

Based on the analysis of coal thermal decomposition methods, equipment operation features and design flaws, an improvement of the reactor retort is considered using a new reactor design, characterised in that a combined thermal decomposition method (pyrolysis and gasification) is used.

The scientific novelty of the proposed solution consists in the implementation of a combined thermal decomposition method consisting simultaneously of the exothermic gas–vapor gasification process and the endothermic low-temperature pyrolysis process with the production of dry mineral residue and energy carrier.

The technical novelty of the solution is that three sealed chambers are combined in one reactor: the lower, middle and upper (Fig. 5).

The lower chamber is designed for the gasification process (700°C) to produce synthesis gas. In the lower chamber there are two screws: upper ("cold") and lower ("hot"). The bottom "hot" screw (Fig. 6) is equipped with an electric frame heater that maintains a constant and controlled operating temperature of 700°C. The upper "cold" screw is a conveyor to the zone of thermal destruction (gasification).

The upper chamber is designed for the process of low-temperature pyrolysis (550°C) to produce pyrolysis gas.

The middle chamber is an intermediate one and is a combustion device working by burning the generated gases generated in the upper (pyrolysis gas) and lower (synthesis gas) chambers.

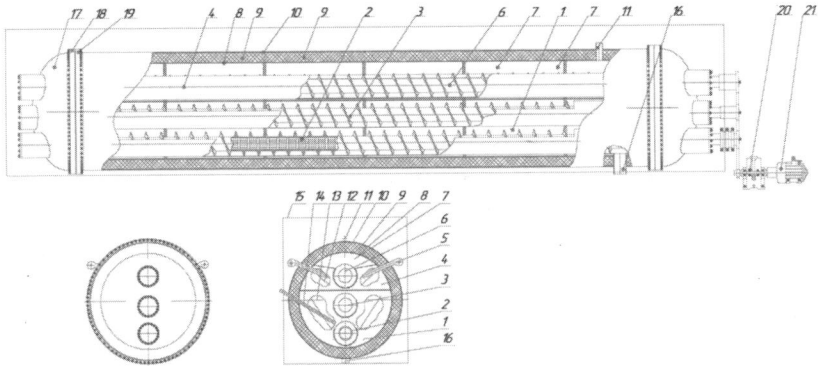

Figure 5: Combined equipment of coal thermal destruction. Labels: 1, gasification chamber; 2, the lower screw with the electric heater; 3, medium screw; 4, combustion chamber; 5, burner; 6, top screw; 7, pyrolysis chamber; 8, sleeve; 9, lining; 10, shell; 11, output of the pyrogas mixture; 12, 13, technological openings; 14, input agent blowing; 15, container; 16, the output of the mineral residue; 17, elliptical bottom; 18, groove construction flange; 19, groove construction flange; 20, reductor; 21, electric motor

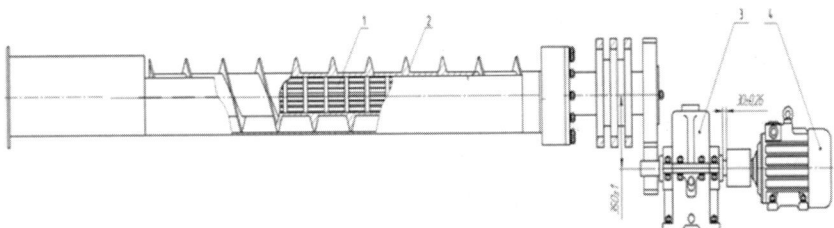

Figure 6: Bottom "hot" screw. Labels: 1, frame electric heater; 2, screw; 3, electric motor; 4, reductor.

The equipment (Fig. 5) is proposed to be placed in sea 40-foot containers (15) with an internal volume of 76 m³ and installed on a mobile base. The installation works as follows: the raw material after grinding to a fraction of 5 mm is fed into the storage, the volume of which is designed for 30 min of the installation. The construction of bunkers and screws (2, 3, 6) is designed in such a way as to minimize the ingress of atmospheric air into the reactor retort (8). After filling the internal volume of the bunkers and part of the screw tubes, by using a vibration device, the internal volume of the reactor retort is purged with an inert gas (nitrogen) to replace air from the system. An electric heater built into the lower screw heats the gasification chamber to an operating temperature of 700°C. The intermediate chamber is heated to a temperature of 500°C due to the burning of the energy carrier (LPG, diesel fuel, etc.) on the burners (5). When the working temperature is reached by the screw, the raw material is fed into the reactor retort (infrared heating zone), superheated steam is simultaneously fed into the gasification chamber, resulting in synthesis gas ($CO + H_2$) with heat generation, and the mineral residue goes through the nozzle to the outside (16). The heat release during the gasification process allows the use of a second screw in the chamber without an electric heater. With the release of a sufficient amount of synthesis gas, the unit goes off-line. Synthesis gas is burned on the furnace intermediate chamber, where infrared heating of the outer side of the screw tube and partitions results in infrared indirect heating, which allows the pyrolysis process to proceed with heat absorption and pyrolysis gas generation, which is subsequently burned on the burner devices. It is advisable to dispose of raw materials with a high carbon content in the gasification chamber, since during pyrolysis a large amount of carbon (soot) is formed, which gradually cements (carburizing) the surface of the internal apparatus, which reduces the mechanical characteristics of the material.

Obtained ash could be used as ingredient in the manufacture of concrete and wallboard or to support roadways. It can lead to positive environmental, economic and performance benefits such as reduced use of virgin resources, lower gas emission, reduced cost of coal ash disposal and improved strength and durability of materials.

However, in this study accurate economic assessment of the equipment is not given while it is necessary to simulate it in modern software, but approximately it is cost of almost one million dollar.

2. Conclusion

Coal gasification is well-proven processes with feed/product flexibility. However, not every gasifier is able to efficiently convert every coal quality. Some theoretical foundations of coal gasification technologies are described. New reactor design – characterised in that a combined thermal decomposition method (pyrolysis and gasification) – is proposed. Equipment consists of three main sections – two of them are pyrolysis and gasification for better and purify separation.

3. References

[1] V. Litvinenko, B. Meyer, Syngas Production: Status and Potential for Implementation in Russian Industry, Springer International Publishing AG, Switzerland, 2018 p. 70.

[2] R. Moreea-Taha, "Modeling and simulation for coal gasification", Report CCC/42, IEA Coal Research, London, UK, December 2000.

[3] Lignite to Methanol: An Engineering Evaluation of Winkler Gasification and ICI Methanol Gasification Route, AP-1592, Electric Power Research Institute, Palo Alto, CA, October 1980.

[4] P. F. H. Rudolph, "The Lurgi Process—The Route to SNG From Coal", Fourth Synthetic Pipeline Gas Symposium, Chicago, IL, 1972.

[5] J. N. Phillips, M. B. Kiszka, U. Mahagaokar, and A. B. Krewinghaus, "Shell Coal Gasification Project Final Report on Eighteen Diverse Feeds", EPRI TR-100687, 1992.

[6] L. G. Massey, in C. Y. Wen and E. S. Lee, eds., Coal Conversion Technology, Addison-Wesley Publishing Co., Reading, MA, 1979, pp. 313–427.

[7] L. Shindman, in H. H. Lowry, ed., Chemistry of Coal Utilization, Volume II, John Wiley & Sons, Inc., New York, 1945, pp. 1252–1286.

[8] B. J. C Van Der Hoeven, in H. H. Lowry, ed., Chemistry of Coal Utilization, Volume II, John Wiley & Sons, Inc., New York, 1945, pp. 1586–1672.

49

Coal gasification -Opportunities and challenges for India

Dipankar Sengupta

Jindal Steel &Power Ltd

Abstract: Coal Gasification is a very environment-friendly chemical process by which low grade, non-coking coal can be converted by reaction with O_2, H_2O and CO_2 into fuel gas called Synthesis Gas or 'Syngas' containing mainly H_2, CO and CH_4. The reaction takes place at high pressure and temperature inside a Reactor called 'Gasifier' in the absence of any conventional catalyst.

The Syngas produced has a huge commercial application potential as it can be directly used as or converted into - industrial clean Fuels, variety of high value Chemicals & Petrochemicals, Fertilizers, also for generating Electric Power as well as Steel (Direct Reduced Iron) making.

Keywords: Gasification, Syngas, Chemicals, DRI.

1. What is Coal Gasification?

Coal Gasification is a very environment-friendly chemical process by which so-called black and dirty low grade, non-coking coal can be converted by reaction with O_2, H_2O and CO_2 into a very clean, high grade fuel gas called Synthesis Gas or 'Syngas' comprising mainly of H_2, CO and CH_4. The reaction takes place at high pressure and temperature inside a Reactor called 'Gasifier' in the absence of any conventional catalyst.

The Syngas produced has an immense commercial application potential as it can be directly used as or converted into - industrial clean Fuels, variety of high value Chemicals & Petrochemicals, Fertilizers, also for generating electricity as well as for making steel via Direct Reduced Iron (DRI) process.

1.1 Chemistry of Coal Gasification

The following are the major reactions that take low quality Coal to high quality products by Coal Gasification:

A. Coal Gasification Reaction :

Coal + with O_2, H_2O, CO_2 $\Longrightarrow$ **Synthesis gas** + CO_2, Tar, Oil, Naphtha, Phenols, P, T $(\mathbf{H_2}, \mathbf{CO}, \mathbf{CH_4})$ **Ammonia, Sulphur, Ash**.

B. Coal to Liquid :

Syngas + H$_2$ $\Longrightarrow$ Fischer – Tropsch $\Longrightarrow$ **Gasoline (Petrol), Diesel, Naphtha**

Reactor with FT Catalyst **LPG, Wax, Lubes**

Syngas $\Longrightarrow$ Catalytic Reacn $\Longrightarrow$ **Ammonia** + CO$_2$ $\Longrightarrow$ **Urea**

$\Longrightarrow$ **Methanol** $\Longrightarrow$ **MTO / MTP** $\Longrightarrow$ **DME, Olefins, Propylenes**

$\Longrightarrow$ **SNG, Acetic acid, Oxo-Alcohols, MEG, MTBE**

Gas Turbine $\Longrightarrow$ **Power (IGCC)**

+ Fe$_2$O$_3$ Pellets $\Longrightarrow$ DRI $\Longrightarrow$ Steel $\Longrightarrow$ Fuel Gas

2. Coal Gasification& Coal-To-Liquid Technologies

Globally, there are 3 broad classes of Gasifiers commercially available which define the types of Coal Gasification technologies for Syngas generation and subsequent application to CTL Technologies. These technologies with a specific type of Gasifiers are currently being offered by various globally reputed Licensors. Some examples are given below. All of these technologies are mostly suitable for Coalfeedstock with 25 – 30 % Ash or less.

2.1 Coal Gasification Technology Licensors

	Gasifier types	**Licensors**
1.	Fixed Bed	Lurgi / AL, SEDIN, BGL / ZEMAG
2.	Fluidized Bed	Uhde / TKIS, Envirotherm, SES / GTI, MHI
3.	Entrained Bed	GE, Shell, CBI / McDermott, Siemens GSP, KBR
		Uhde / TKIS, ECUST - OMB, HT – L, TPRI

2.2 CTL FT Technology Licensors

Lurgi / AL, Shell, Sasol, BP, Axens, ENI, Exxon, Synfuels China

3. Major Benefits of Coal and Petcoke Gasification

It makes a lot of sense for Gasification of low quality coal and Petcoke in our country. The major benefits for India by gasification of Coal/Petcoke and establishing further downstream facilities would be as follows:

 i. Huge potential for Import Savings due to lesser dependence on Import of Crude oil, LNG, Chemicals and Petrochemicals, Fertilizers, etc.

ii. Substantial Monetization of vast resources of unutilized low Grade, High Ash Coal as well as high Sulphur Refinery Distillation Residues.

iii. Adoption of Environment-friendly, "Clean Coal Technology" (low SOx, NOx) with "CCS-ready" CO2 streams

iv. Long term "Energy Security" for the country due to immunity from global politics on Oil & Gas

3.1 Global Gasification Scenario

Coal Gasification is a very well established and widely accepted technology all across the world wherever there are large deposits of Coal available. Soon there will be 350 Coal Gasification plants with about 1000Gasifiers of different operating types. Most of them are in China, USA, South Africa, Mongolia, South Korea, etc.

Total Syngas generation capacity is currently 250,000 MW thermal equivalent likely to go to 310,000 MW thermal by 2019-20. Such huge quantities of Syngas would be mostly converted to Chemicals / Petrochemicals, Liquid Fuels, Gaseous fuels, Fertilizers, and Electric Power.

It is also noteworthy that 25% of the world's total Ammonia and 30% of total Methanol production are through Coal gasification route.

3.2 China CG and CTL Scenario

China has adopted long before and further developed and fine-tuned the Coal Gasification and Coal-to-Liquid technologies due to its vast resources of mostly good quality, low ash Coal all across the country. More importantly, they used this as a Strategic tool to substantially get out of the trap of global politics in oil and gas. They are relatively self-sufficient even if the global oil or gas in-flow gets interrupted due to any political sanction or war scenario.

China produces 80% of its total Ammonia and 90% of total Methanol production through Coal Gasification route. As high as 21 MM TPA of Methanol produced is blended with Gasoline (petrol) to be used as automobile fuel. Various Blend ratios being used are – M15 (15% MeOH + 85% Gaso), M25, M50, M85, M100. They have developed and/or adopted automobile engines to suit these Blends, thus substantially cutting down on Crude imports. They also produce various other useful fuel substitutes, viz. DME (blended with LPG), MTBE, MTG (blended with Gasoline), etc.

The expected CTL production in 2020 is about 178 MM TPA comprising of Methanol, LPG / DME, Ethyl Glycol, Oil products, Ammonia / Urea, etc.

Some of the major companies operating CG / CTL plants in China with more than 1.0 MM TPA capacity are – Shenhua Group, Yitai Ordos, YitaiYili, Jincheng Anthracite, Shanxi Lu'An, etc.

3.3 India CG and CTL Scenario – Scope &Opportunities

India is in an unique advantageous position as it has huge estimated coal resources of about 319 billion tonnes down to 1200 metres depth below the Earth's surface. India is ranked 4th Rank in the world in terms of total coal resources. Proven coal resource is about 148 bn tonnes, indicated resource about 139 bn tonnes and inferred resource is about 32 bn tonnes. Of the total resource, coking coal is about 35 bn tonnes and non-coking coal is about 284 bn tonnes.

But despite such a huge resource, the actual coal production is around 750 million tonnes 2018-19, with projected coal production of one billion tonne in the next couple of years. So there is a huge scope for increasing coal production, which is now limited due to various reasons and also limited usages.

India currently imports chemicals and petrochemicals worth about US$ 45 bn per year. There are hardly any CTL plants today worth mentioning producing these items. It is estimated that that about 50% of the aforementioned Imports can be reduced by going in a big way for CTL plants, contributing to the country's exchequer of about US$ 21 bn per year.

3.4 Current CG and CTL Projects in India

There are at present only a handful of CG / CTL plants in operation and / or in activeProject phase based on Coal, Petcoke and Coal + Petcoke as feedstock.

1. **Jindal Steel & Power Ltd, Angul, Odisha :**
    7 (6+1) nos. Lurgi /AL FBDB type Coal Gasifiers producing Syngas to make DRI (Steel) and Fuel gas (Future Polygen products - Methanol / Ammonia)

2. **Reliance Industries Ltd., Jamnagar, Gujarat :**
    10 (4 + 6) nos. CBI / McDermott e-Gas type Entrained Bed Petcoke(+Coal)Gasifiers to meet refinery hydrogen and captive power demand (future Polygen products – SNG, Chemicals)

3. **Talcher Fertilizers Ltd. , Talcher, Odisha:**
    Planned: 2 (1+1) nos. Shell / Air Products SCGP type Entrained Bed Coal + Petcoke Gasifiers to make Ammonia/ Urea (later Polygen products – Methanol, SNG)

4. **Coal India Ltd., Dankuni, WB :**

Planned: 2 (1+1) nos. Entrained Bed Coal Gasifiers to make Methanol & SNG

There are a few more projects in technology evaluation, pre-feasibility/Feasibility Study or pre-project stage, viz. Adani, GSFC, IOCL, etc.

3.5 Coal Gasification: Challenges in India

There are lots of challenges for the success of Coal Gasification technology in India. Major of the challenges are mentioned below:

Coal related

- Coal available is of poor quality, high ash and low fixed carbon in nature
- Generally, there is the presence of a large number of fines in the feedstock coal. (This leads to problems in operations in Fixed Bed Gasifiers)
- Reutilisation and monetisation of fines is a problem. Not many usages of the fines have been found
- There is a lot of variation in the quality of feedstock coal between the mines. Often the coal available is from different mines, which leads to inconsistency in the quality of the coal
- Dedicated coal mines are not available suitable for Coal Gasification
- The availability and capacity of coal blending and washing facilities are inadequate
- There are frequent Price fluctuations in Domestic as well as Imported Coal. Sometimes large price fluctuations make the operations unviable
- Quite often the plants are located at large distances from the coal source – leading to higher freights for the coal to be transported.

Environment related

- There is a large generation of waste water
- Costly waste water treatment plants have to be put up to meet Pollution Control Board norms
- Costly systems for CO_2, H_2S removal and dust emissions control
- Large areas are required for huge quantities of Ash storage and disposal
- Carbon Dioxide Capturing systems and applications are not yet developed in India

4. Technology and Project Related

- There is a major issue of procuring compatible, proven technology for Indian poor quality coal
- Coal Feed preparation is very critical, complex and costly, but essential for the success of Gasification technologies and sustainable operation of such plants
- Frequent change over of Technology ownership at Global level causes uncertainty
- Very high Project Capex and Coal price significantly affects Syngas& downstream products cost and overall Project viability
- Hardly any Incentives on Coal price or Product price or Interest onLT / ST loans, etc. for viability of such country critical Coal Gasification / CTL Projects
- Large Land mass required for Coal storage, handling, sizing, Blending, long distance conveyers, Ash handling, storage, etc.
- Huge quantity of Water resources should be available nearby
- Costly Cryogenic ASU plants required for Oxygen gas supply
- Sizeable utilities and off sites facilities reqd.
- Inexperienced Vendors vis-à-is fabrication of critical Equipment (mainly Gasifiers& its associated parts/equipment) – to be imported
- Very costly Spares, still have to keep high Inventories
- Experienced EPC / LSTK Contractors not many for Gasification plants
- Experienced O&M Manpower not readily available
- Costly Laboratory Analytical equipment & trained Manpower are required
- No Single Product Plant is viable due to Market demand / price fluctuations only Poly-generation route configuration is viable

5. Success of CG/CTL Projects: Key Factors

i.   Dedicated, nearby Coal Mines to be earmarked for CG/CTL Projects for better Coal Quality consistency & closer Mining & Transportation Cost control

ii. Financial Incentives by the government to reduce very high Capex of CG/CTL Projects is needed to improve Viability of Energy Security Projects

iii. Exemption from applicable cess and duties on coal feedstock prices for the environment-friendly CG/CTL Projects due to Clean Technologies

iv. Price protection of CG/CTL Import substitute Products from the cheaper imported Chemicals/Petrochemicals/ Fuels, viz. Fertilizers, Methanol, Euro – VI Grade Gasoline, Diesel, Jet Fuel, etc.

v. Schemes for Blending Methanol, DME with Gasoline & LPG have to be implemented at the earliest

50

Modelling studies on hydrodynamics of flow inside a solid–liquid fluidizer

Anmol Awasthi, Prasad Kopparthi, A. K. Mukherjee

Tata Steel Limited, Jamshedpur, India

Abstract: Solid–liquid fluidized beds are ubiquitous in industrial applications viz., chemical, pharmaceutical, petroleum, food, and the like. One of the important features of fluidized beds is the ease with which particles mix and/or segregate in them. Thus, they are highly desirable for configurations like good mixing leading to uniform bed properties, subsequently elevated heat and mass transfer rates, convenient solids handling, significantly lower pressure drops leading to reduction in cost, etc. In the present study, the hydrodynamic modelling of fluidized bed using an Eulerian multiphase approach has been carried out. The bed voidage and velocity profiles have been investigated and are evaluated qualitatively. The kinetic theory of solids has been invoked for considering conservation of energy associated with particle fluctuations. The effect of mesh size has been also investigated. The drag has been modelled using the drag law by Pandit and Joshi (1998) which has been derived from first principles. The computational geometry has been considered two-dimensional owing to restricted computational resources.

Keywords: Fluidised, Drag, Purity, Separation, Coking

1. Introduction

Steel production is intrinsically dependent on coal. Nearly 75% of the manufactured steel involves coal, resulting it as an essential ingredient for steel making process. In steel industries, coal is primarily converted to lump coke in a coking oven. Coking coal, a certain grade of coal, is required to make the preferred quality of coke. Consequently, coal preparation is a vital stage that follows mining but precedes its dispatch as a feedstock for steelmaking. Excavated coal consists of a composite mixture of carboniferous material and various non-coal impurities, such as rock, shale, clay, and water. Coal preparation, also referred as beneficiation, is the process where impurities are physically separated from the coal to meet the desired quality criteria. The compositional purity of coal significantly influences the final economic unit value because the presence of mineral impurities inhibits the coal heating value and also mitigates its suitability as metallurgical coke. While the fundamental goal of coal preparation is to improve the value of the raw material, the physical removal of impurities cascades through various sequential stages. Separation of mineral particles by gravity concentration process is one of the oldest techniques, that is widely used in coal beneficiation for its low cost, ease of operation, and eco-friendly nature. Differential settling velocity of the particles in a fluid medium is the basis of such processes. However, as the settling of particles is strongly dependent on particle size, the application of gravity processes is limited to coarse particles, and such practices become ineffective for fine and ultra-fine particles (<1 mm) due to poor settling characteristics. Therefore, in pursuit of developing high throughput advanced gravity separators for fine minerals, several technologies have been developed, which can broadly be classified on the following principles: (a) enhancement of separation efficiency by centrifugal force (e.g., Knelson concentrator), (b) generation of autogenous heavy media (e.g., fluidized bed separator), and (c) combination of centrifugal force and autogenous heavy media separation (e.g., multi-heavy separator).

The reflux classifier is a state-of-the-art fluidized bed separator, which manifests efficient, high capacity fine particle separation on the basis of size and/or density. These classifiers comprise a set of inclined plates atop a fluidized bed that aids the settling of finer and lighter particles and enhances the throughput significantly. Reflux classifiers can be used for size classification as well as gravity concentration. Despite significant advancement of this technology, to best of our knowledge, no previous study clearly demonstrates the impact of process parameters i.e., feed solid concentration and size distribution, physico-chemical properties of fluidization media and flow rate on the separation performance of these processing units for Indian coal. However, such analyses are of paramount importance from the standpoint of its applicability in the Indian context due to differential compositional characteristics of coal than in the locations, where the reflux classifier is developed and demonstrated.

2. Previous work

2.1 Experimental status

A reflux classifier combines a conventional fluidized bed with sets of parallel inclined plates, as shown in Fig. 1 [1]. This design mitigates the effects of particle size, and prompts high separation efficiencies across a feed size distribution [2,3]. Feed slurry enters below the plates while fluidization water is introduced through a distribution plate in the base. The slurry feed moves downwards into the vessel, forming a bed of particles that is fluidized from below. High-density particles settle into the lower portion of the bed, and light as well as fine particles are transported upward, with the majority flow towards the lamellae. The high hydraulic load carries the suspension up into the parallel inclined lamella plates. Here slow settling particles, which are unable to settle against the fluidization water, emerge through the plates and report to the overflow. Faster settling particles drop out of suspension and onto the plates before sliding back to the zone below. At high bed concentrations in the reflux zone, this reject suspension provides an autogenous dense medium, allowing the separation to proceed largely on the basis of density. When the density of the fluidized bed exceeds the set-point value, a valve opens near the base of the unit and discharges some of the denser particles as an underflow stream [4].

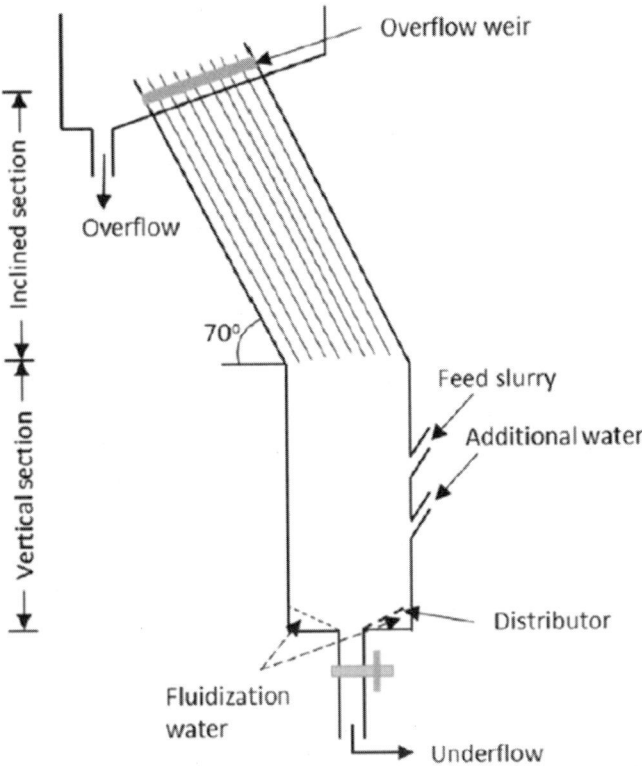

Figure 1: Schematic of a reflux classifier [1]

The inclined channels, the geometry of which is defined by the plate length (l), perpendicular channel spacing (z), and angle of inclination with the horizontal (θ), provide a significant hydraulic advantage over conventional fluidized beds that is consistent with the well-known Boycott effect [5]. Experiments to investigate the effect of inclination angle were carried out by Laskovski et al. [6] at 70°, 60°, and 45°. It was found at an angle of 70°, the separation was much sharper, as lower density particles reported to the overflow. Zhou et al. [7] also proved that over the range $\theta = 60$–$80°$, the performance was optimal, hence all of the work since then has been based on an angle of inclination of nearly 70°. Laskovski et al. [6] investigated the effect of increasing the number of inclined plates in the reflux classifier. This investigation was performed on PVC, ilmenite, and silica sand particles at the same superficial velocity. The results showed that as the number of inclined plates increased, the separation size decreased. Further increase in the number of plates resulted in a significant increase in separation size. Galvin et al. [8] also investigated the use of closely spaced inclined plates in the separation of ultrafine particles (<0.1 mm). The system geometry consisted of 24 inclined plates each 1 mm long with perpendicular spacing of 1.77 mm. It was shown that the use of closely spaced inclined plates promotes a stable laminar flow condition, a shear induced lift force and a local elutriation velocity near the surface of the inclined plates that is directly proportional to the particle diameter. A normal rule of thumb for fine coal beneficiation is that the channel spacing should be a minimum of three times the particle top-size being beneficiated, usually in a spacing range between 2 mm and 6 mm [9].

Orupold et al. [10] studied the lamella high shear rate reflux classifier for −2 mm coal application. It was established that this classifier can process coals at lower cut points with a high efficiency. Thus, low ash clean coals can be easily produced by these classifiers at improved yields. Galvin et al. [11] employed reflux classifier with closely spaced channels for the separation of heavy minerals from minerals sand. The closely spaced channels accomplish improved gravity-based separation through laminar shear mechanism in single stage operation. Research has primarily been performed on the separation of coal from mineral matter such as ash. The results show a high recovery of up to 80% in the size range of 2 mm to 0.25 mm [12]; however, no work has been done on separating pyrite from coal. Recent studies have also shown the reflux classifier to be effective at coal sizes up to 8 mm [13].

Pilot scale testing of the reflux classifier was carried out by Galvin et al. [14] for the beneficiation of −2 mm coal fines. Excellent gravity concentration was reported with a throughput much higher than what can be achieved in a conventional teetered bed separator. The reflux classifier was also shown to offer less variation in separation density with particle size compared to the conventional teeter bed separator. Iveson et al. [15] later studied the use of two 600 mm × 600 mm pilot-scale reflux classifiers in series for the beneficiation of –2 mm coal sample. Both reflux classifiers had close channel inclined plates of 6 mm and 12 mm, respectively. The first stage reflux classifier performed a density separation producing a coal product containing fine high-ash slimes (–38 μm + 75 μm). The second reflux classifier then washed the contaminated stream to produce clean coal product. The result showed that a product ash of 16.5% was achieved from a feed ash of 42%. The reject and slime ash contents were 77% and 64%, respectively.

2.2 Modelling status

Most of the work, hitherto, has been attempted on gas-fluidized beds, with computational fluid dynamics (CFD) explicitly applied to liquid fluidization in only a limited number of cases. The earliest work of Roy and Dudukovic [16] simulates a liquid–solid riser type reactor using a two-fluid Euler–Lagrange model coupled with kinetic theory of granular solids. The drag was modelled using Wen and Yu drag law [17] and the Jackson boundary condition for solids at the wall was applied. Liquid turbulence was handled using the k–ε model. The solids were taken as glass beads of 2.5 mm size. The model was shown capable of predicting the liquid and solids RTD (residence time distribution) in the reactor as well as the solids velocity and holdup pattern. The model was also used to successfully predict the extent of solids back-mixing in the reactor. Doroodchi et al. [18] applied CFD to model the fluidized and inclined sections in a reflux classifier using ballotini particles of two sizes (but same density) and compared the predictions with experimental data and with those of a kinematic model developed by them earlier. Despite assumptions of laminar flow in both the vertical and inclined sections and of neglecting the interparticle interactions, the results were in good agreement with the experimental data and kinematic predictions. Further, Cornelissen et al. [19] studied the effect of drag relationships and the nature of boundary conditions on the simulation predictions and found good qualitative and reasonable quantitative agreement with experimental results. The quantitative agreement was considered with respect to the Richardson and Zaki equation [20]. It was shown that although both the Richardson and Zaki equation and the model underpredicted the experimental data, the model predictions were better than those of the Richardson and Zaki equation and were also within 5% of the experimental data. Reddy and Joshi [21] varied the Reynolds number in creeping, transition, and turbulent flow regimes and predicted the critical velocity at which complete mixing in a binary particle system occurs. They utilized a drag law proposed by Joshi [22] and Pandit and Joshi [23] which was derived using energy balance approach. Very few assumptions were made in the derivation (from first principles) leading to reduction of empiricism in the drag law. The bed voidage predicted showed excellent agreement with the Richardson and Zaki equation. The model also successfully predicted the layer inversion phenomenon occurring in binary mixtures of particles of different size and density. The above work goes to show the potential of commercial or self-developed CFD codes in being able to predict multiphase flow in a fluidized bed accurately and thus warrants the need to develop and test such models for complex real-time fluidization systems with a higher degree of sophistication in terms of computational resources and/or coding capabilities.

3. CFD modelling

3.1 Computational geometry

Table 1 shows the geometry as well as the particle properties chosen for the simulations. The column to particle diameter ratio was taken greater than 25 to eliminate wall effects for the particles. The grid independence study was conducted using mesh sizes of 10 mm, 5 mm, 2 mm, 1 mm, and 0.5 mm. It was observed that below a mesh size of 1 mm, the void fraction profile of the solid–liquid fluidized bed remained independent of mesh refinement (Fig. 2). Thus, 1 mm was chosen as the grid independent value for all further simulations. The geometry was created in Ansys SpaceClaim R18.1 and exported into Ansys Fluent.

Table 1: Geometrical and particle properties

Bed parameters	Parameter values	Particle properties	Property values
Width of the column (mm)	100	Solid	Glass
Height of the column (mm)	1680	Liquid	Water
Initial bed height (mm)	270	Diameter of the particle (mm)	0.8
Number of grids	1,65,239	Density of solid (kg m^{-3})	1075
Grid size	1 mm × 1 mm	Density of liquid (kg m^{-3})	998.2
Time step	1.5×10^{-4}	Viscosity of liquid (kg m^{-1} s^{-1})	0.001
Max number of iterations per time step	30	Terminal settling velocity (m s^{-1})	0.02

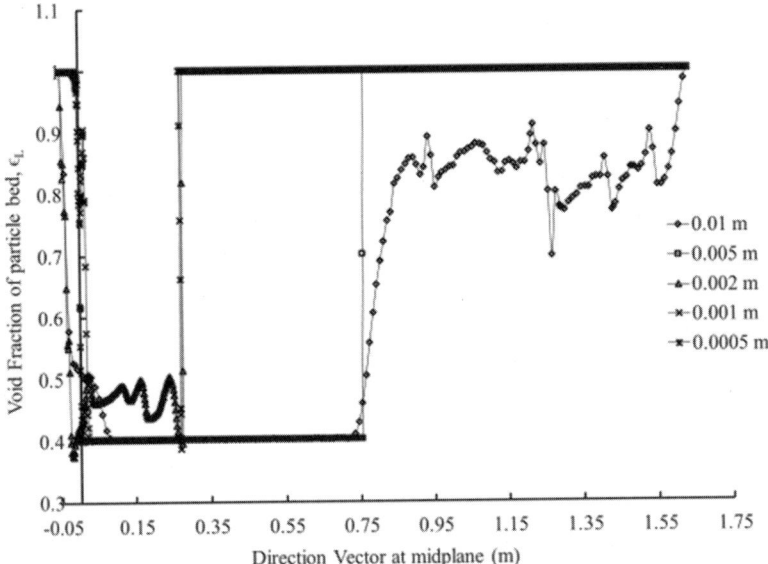

Figure 2: Grid independence study of fluidized bed showing 1 mm as optimal mesh grid size

4. Model formulation

We have utilized an Eulerian multiphase model, which considers the solid and liquid phases as fully interpenetrating continua subject to continuity and momentum equations. The kinetic theory of granular solids is used for closure. The flow is modelled as transient, i.e. time dependence has been considered in the simulation. The mass and momentum balance equations for each phase are as follows:

$$\frac{\partial(\rho_L \epsilon_L)}{\partial t} + \nabla \cdot (\rho_L \epsilon_L \langle u_{Li} \rangle) = 0 \tag{1}$$

$$\frac{\partial(\rho_S \epsilon_S)}{\partial t} + \nabla \cdot (\rho_S \epsilon_S \langle u_{Si} \rangle) = 0 \tag{2}$$

$$\frac{\partial(\rho_L \epsilon_L \langle u_{Li} \rangle)}{\partial t} + \nabla \cdot (\rho_L \epsilon_L \langle u_{Li} \rangle \langle u_{Li} \rangle)$$

$$= -\epsilon_L \nabla \langle p \rangle + \nabla \cdot (\epsilon_L \mu_{eff,L} \nabla \langle u_{Li} \rangle + (\nabla \langle u_{Li} \rangle)^T)) + \rho_L \epsilon_L \vec{g} - K_{LS} \tag{3}$$

$$\frac{\partial(\rho_s \epsilon_s \langle u_{Si} \rangle)}{\partial t} + \nabla \bullet (\rho_s \epsilon_s \langle u_{Si} \rangle \langle u_{Si} \rangle)$$

$$= -\epsilon_s \nabla \langle p \rangle - \nabla \langle p_s \rangle + \nabla \bullet (\epsilon_s \mu_{eff, s} (\nabla \langle u_{Si} \rangle + (\nabla \langle u_{Si} \rangle)^{\mathrm{T}})) + (\epsilon_s \lambda_s)(\nabla$$

$$\bullet \langle u_{Si} \rangle I)) + \rho_s \epsilon_s \vec{g} + K_{LS} \tag{4}$$

where μ_{eff} is the effective viscosity, g is the gravitational acceleration, and K_{LS} is the interphase momentum transfer force. The only force considered for the momentum exchange between the solid and liquid phases is the drag force and the contribution of other forces has been neglected.

The drag coefficient has been modelled using the drag law derived from first principles by Joshi [22] and Pandit and Joshi [23] using energy balance as the starting point. The drag coefficient obtained by them for creeping, transition, and turbulent regimes is as follows:

$Re_\infty < 0.2$

$$C_D = \frac{6}{Re_\infty} \left(\frac{3.6 \epsilon_S}{\epsilon_L^2} + \epsilon_L^2 \right) \tag{5}$$

$0.2 < Re_\infty < 500$

$$C_D = \frac{6}{Re_\infty} \left(\frac{3.6 \epsilon_S}{\epsilon_L^2} + \epsilon_L^2 \right) + \left(C_{D\infty} + \frac{2k}{3V_S^2} \right) \tag{6}$$

$Re_\infty > 500$

$$C_D = C_{D\infty} + \left(\frac{2k}{3V_S^2} \right) \tag{7}$$

The solids flow modelling required an additional energy balance equation to take into account conservation of solid fluctuational energy. The kinetic theory of granular solids was used to introduce a granular temperature Θ in the Eulerian model which considered the solid phase as an interpenetrating continuum. The energy equation accounting for the energy associated with the solid phase is as follows:

$$\frac{\partial \epsilon_s \rho_s \Theta_s}{\partial t} + \frac{(\langle u_{Si} \rangle \partial (\epsilon_s \rho_s \Theta_s)}{(\partial x_j)} = \nabla \langle p_s \rangle \langle u_{Si} \rangle + \nabla \bullet (\mu_s \nabla \langle u_{Si} \rangle \bullet \langle u_{Si} \rangle)$$

$$+ \nabla \bullet (k_{\Theta s} \nabla \Theta_s) - 3K_{LS} \Theta_s - \gamma_{\Theta s} \tag{8}$$

The above equation requires a radial distribution function, g_s, which controls the solids volume fraction, ε_s, as follows:

$$g_s = \left[1 - \left(\frac{\varepsilon_s}{\varepsilon_{s,\,max}} \right)^{1/3} \right]^{-1} \tag{9}$$

Both the radial distribution function, g_s and the solid viscosity, μ_s, are modelled using the expression given by Ding and Gidaspow (1990). The solid pressure, p_s and the bulk viscosity, λ_s, are modelled using the equation given by Lun et al. (1984):

$$p_s = \varepsilon_s \rho_s \Theta_s [1 + 2(1 + e_s) \varepsilon_s g_s] \tag{10}$$

$$\lambda_s = \frac{4}{3} \varepsilon_s \rho_s d_p g_s (1 + e_s) \left(\frac{\Theta_s}{\pi} \right)^{1/2}$$

The flow was taken to be transient and the pressure outlet boundary condition was used. For the inlet, a flat velocity profile was applied with a magnitude of 7×10^{-3} m s^{-1}. No slip was assumed at all the walls for both the solid and liquid phases. The governing equations were discretized using the first order upwind discretization scheme. A time step of 1.5×10^{-4} s was chosen for the simulations. The SIMPLE algorithm was employed for the pressure–velocity coupling. The convergence criterion, or residual limits, were taken as 10^{-4} for all the equations.

5. Results and discussion

Figure 3 shows the solids fraction contour along the bed height (complete bed not shown for sake of clarity). It can be readily inferred that the bed is in the fluidized condition since we started our simulations with a solids fraction of 0.598 in the lower portion of the bed. The contour shows values of the solid fraction varying from this value suggesting that the bed is fluidized. Figure 4 shows the velocity contours for the fluid (water) along the bed height. The figure suggests channelling and internal recirculation in the solids section which is confirmed by the vector plot shown in Fig. 5. Figure 6 shows the velocity of fluid phase plotted along the flow direction suggesting that velocity is maximum in the fluidized region and drops downstream the flow towards the outlet.

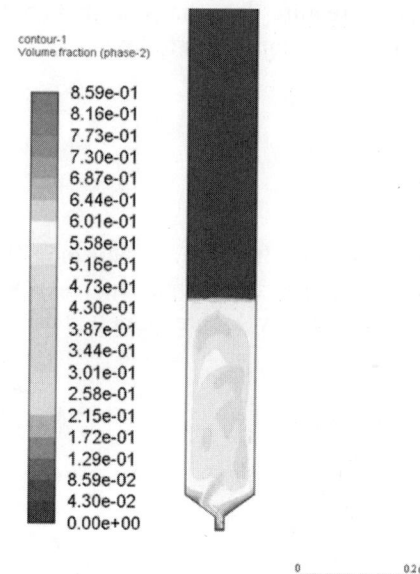

Figure 3: Solids fraction contour along the bed
(complete bed not shown for sake of clarity)

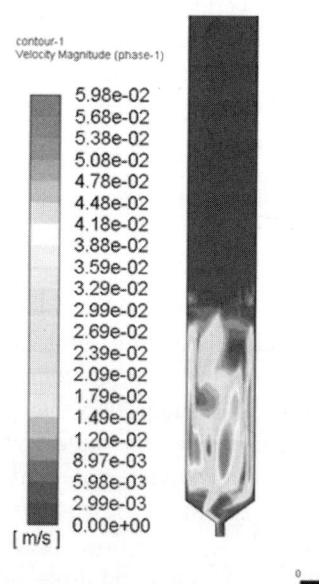

Figure 4: Fluid velocity contour along the bed
(complete bed not shown for sake of clarity)

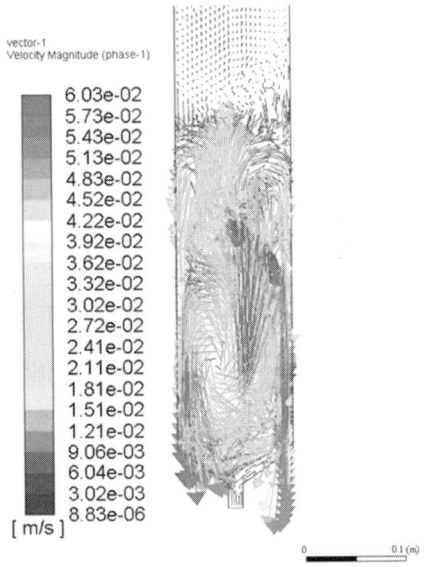

Figure 5: Fluid velocity vectors plotted along the bed
(complete bed not shown for sake of clarity)

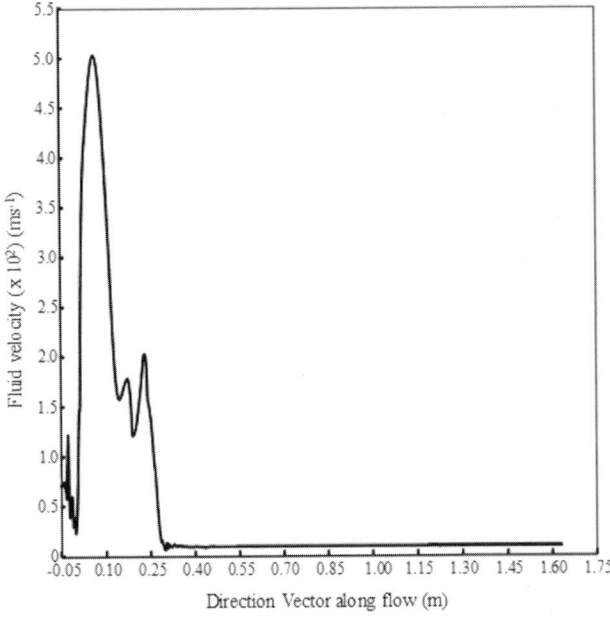

Figure 6: Fluid velocity plotted along the flow direction

6. Conclusion

The basic computational framework for solid–liquid fluidization has been established herein. This will be the basis for the fluidization studies to be carried out using binary, ternary, and quaternary mixture of solids in the work ahead.

7. Notation

C_D: drag coefficient in the presence of other particles based on the total surface area of the particle.

$C_{D\infty}$: drag coefficient of single particle settling in the infinite medium based on the total surface area.

d_p: particle diameter, m.

e_S: coefficient of restitution.

g: gravitational acceleration, m s^{-2}.

g_S: radial distribution function.

I: identity matrix.

$k_{\theta S}$: solid diffusion coefficient.

K_{LS}: solid–liquid momentum exchange force, kg m^{-2} s^{-2}.

p: static pressure, N m^{-2}.

p_S: solid pressure, N m^{-2}.

Re$_\infty$: Reynolds number based on the terminal settling velocity of the particle $(d_p V_\infty \rho_L / \mu_L)$.

t: simulation time, s.

u_L: liquid velocity, m s^{-1}.

u_S: solid velocity, m s^{-1}.

V_L: superficial velocity, m s^{-1}.

V_S: interstitial liquid velocity, V_L / ε_L, m s^{-1}.

$\gamma_{\theta S}$: collisional dissipation of energy, kg m^{-1} s^{-3}.

ε_L: volume fraction of fluid.

ε_S: volume fraction of solid.

Θ_S: granular temperature, m^2 s^{-2}.

λ_S: bulk viscosity of solid, kg m^{-1} s^{-1}.

μ_L: molecular viscosity of fluid, kg m^{-1} s^{-1}.

μ_S: shear viscosity of solid, kg m^{-1} s^{-1}.

ρ_L: liquid density, kg m^{-3}.
ρ_S: solid density, kg m$^{-3.}$

8. References

[1] Das, A. and Sarkar, B., 2018. Advanced gravity concentration of fine particles: A review. *Mineral Processing and Extractive Metallurgy Review*, pp. 1–36.

[2] Nguyentranlam, G. and Galvin, K.P., 2001. Particle classification in the reflux classifier. *Minerals engineering*, *14*(9), pp. 1081–1091.

[3] Ghosh, T., Patil, D., Honaker, R.Q., Damous, M., Boaten, F., Davis, V.L. and Stanley, F., 2012. Performance evaluation and optimization of a full-scale reflux classifier. *Journal of the Coal Preparation Society of America*, *11*(2), pp. 24–33.

[4] Galvin, K.P., 2012. Development of the reflux classifier. *Challenges in Fine Coal Processing, Dewatering, and Disposal*, Englewood, p. 159.

[5] Boycott, A.E., 1920. Sedimentation of blood corpuscles. *Nature*, *104*(2621), p. 532.

[6] Laskovski, D., Duncan, P., Stevenson, P., Zhou, J. and Galvin, K.P., 2006. Segregation of hydraulically suspended particles in inclined channels. *Chemical Engineering Science*, *61*(22), pp. 7269–7278.

[7] Zhou, J., Walton, K., Laskovski, D., Duncan, P. and Galvin, K.P., 2006. Enhanced separation of mineral sands using the Reflux Classifier. *Minerals Engineering*, *19*(15), pp. 1573–1579.

[8] Galvin, K.P., Walton, K. and Zhou, J., 2009. How to elutriate particles according to their density. *Chemical Engineering Science*, *64*(9), pp. 2003–2010.

[9] Galvin, K.P., Callen, A., Spear, S., Walton, K. and Zhou, J., 2010. Gravity separation of coal in the Reflux Classifier: New mechanisms for suppressing the effects of particle size. *International Journal of Coal Preparation and Utilization*, *30*(2–5), pp. 130–144.

[10] Orupold, T., Starr, D. and Kenefick, T., 2014. The lamella high shear rate REFLUX™ classifier. *Journal of the Southern African Institute of Mining and Metallurgy*, *114*(7), pp. 511–518.

[11] Galvin, K.P., Zhou, J., Price, A.J., Agrwal, P. and Iveson, S.M., 2016. Single-stage recovery and concentration of mineral sands using a reflux classifier. *Minerals Engineering*, *93*, pp. 32–40.

[12] Galvin, K.P., Callen, A., Zhou, J. and Doroodchi, E., 2005. Performance of the reflux classifier for gravity separation at full scale. *Minerals Engineering*, *18*(1), pp. 19–24.

[13] Galvin, K.P., Callen, A.M. and Spear, S., 2010. Gravity separation of coarse particles using the Reflux Classifier. *Minerals Engineering*, *23*(4), pp. 339–349.

[14] Galvin, K.P., Doroodchi, E., Callen, A.M., Lambert, N. and Pratten, S.J., 2002. Pilot plant trial of the reflux classifier. *Minerals Engineering*, *15*(1–2), pp. 19–25.

[15] Iveson, S.M., Mason, M. and Galvin, K.P., 2014. Gravity separation and desliming of fine coal: pilot-plant study using reflux classifiers in series. *International Journal of Coal Preparation and Utilization, 34*(5), pp. 239–259.

[16] Roy, S. and Dudukovic, M.P., 2001. Flow mapping and modeling of liquid–solid risers. *Industrial & Engineering Chemistry Research, 40*(23), pp. 5440–5454.

[17] Wen, C.Y. and Yu, Y.H., 1966. Mechanics of fluidization. *The Chemical Engineering Progress Symposium Series, 162*, 100–111.

[18] Doroodchi, E., Galvin, K.P. and Fletcher, D.F., 2005. The influence of inclined plates on expansion behaviour of solid suspensions in a liquid fluidised bed: A computational fluid dynamics study. *Powder Technology, 156*(1), pp. 1–7.

[19] Cornelissen, J.T., Taghipour, F., Escudié, R., Ellis, N. and Grace, J.R., 2007. CFD modelling of a liquid–solid fluidized bed. *Chemical Engineering Science, 62*(22), pp. 6334–6348.

[20] Richardson, J.F. and Zaki, W.N., 1997. Sedimentation and fluidisation: Part I. *Chemical Engineering Research and Design, 75*, pp. S82–S100.

[21] Reddy, R.K. and Joshi, J.B., 2009. CFD modeling of solid–liquid fluidized beds of mono and binary particle mixtures. *Chemical Engineering Science, 64*(16), pp. 3641–3658.

[22] Joshi, J.B., 1983. Solid–liquid fluidised beds: Some design aspects. *Chemical Engineering Research and Design, 61*, pp. 143–61.

[23] Pandit, A.B. and Joshi, J.B., 1998. Pressure drop in fixed, expanded and fluidized beds, packed columns and static mixers: A unified approach. *Reviews in Chemical Engineering, 14*(4–5), pp. 321–371.

Additional Articles

1

Design construction and operation management innovation practice of ten million tons coal preparation plant

Chengxiang Lei, Qingzhou Fang

Huai Mine West Mine Investment Management Co., Ltd. Ordos Inner Mongolia China

Abstract: Through technological innovation of process design, engineering construction management innovation and operation management mode innovation of three 10 million tons coal preparation plants in Ordos area of Huai Mine West Company, the green sustainable development of the coal preparation plant was promoted. The road of technological and economic integration innovation and development of coal preparation plant has been realized. The coal preparation plant has become a coal enterprise with a circular economy and green development.

Keywords: Coal preparation plant; design; operation; innovation; practice

1. Introduction

Huai Mine West Mine Investment Management Co., Ltd.(hereinafter referred to as Huai Mine West Company) has three steam coal preparation plants in Ordos area, which are Selian No. 2 Coal Preparation Plant, Tangjiahui Coal Preparation Plant and Bojianghaizi Coal Preparation Plant respectively. The total design capacity is 35 Mt/a.

The main separation processes of the three coal preparation plants are the combined process of heavy medium shallow slot separator separation, pressure two-product heavy medium cyclone separation and coal slurry water three-stage concentration (FANG, 2015).

2. "6+3" Innovation and practice of coal preparation technology

2.1 Analysis of deep screening technology

(1) Analysis of raw coal screening data

The ash content of raw coal in Selian No. 2 Coal Mine is 23.73%, which is medium ash coal. The dominant particle size is + 50mm, the yield is 23.83%, and the ash content is 25.27%. The content of +13mm lump coal is 53% and the cumulative ash content is 25.09%. The ash content of each particle size of + 13mm is higher than that of the raw coal, and gangue content of lump coal is large. The ash content of lump coal decreases obviously after sorting. The yield of -0.5mm coal slime is 5.02%. But the ash content was 25.31%, which is higher than that of the adjacent granules. Gangue is fragile.

(2) Analysis of the relationship between particle size and calorific value

When the particle size is from 50 mm to 6 mm, the increment of cumulative calorific value of raw coal under the screen before and after sorting decreases from 267.19 kcal/kg to 37.17 kcal/kg. After separation, the ash content of raw coal decreases, but the water content increases. With the decrease of particle size, the specific surface area of coal particles increases exponentially. The greater the degree of sliming is, the greater the increase of water content is, the smaller the increase of calorific value is.

(3) Study on the lower limit of separation particle size

In order to determine the optimum lower limit of separation particle size of raw coal, a series of experiments are carried out. When the lower limit of separation particle size is 6 mm, its economic benefit is higher than that of the lower limit of separation particle size of 0 mm and 13 mm. When the lower limit of separation particle size is 3 mm, the results are as follows. When the classification efficiency is less than 50%, its economic benefit is less than that of the lower limit of separation particle size of 6 mm. When the classification efficiency reaches 50%, its economic benefit is close to that of the lower limit of separation particle size of 6 mm. When the classification efficiency reaches 60% and 70%, the economic benefit improves significantly. Under the existing conditions, 3 mm classification is carried out. When the classification efficiency reaches 60% or more, there is a large technical risk. But 6 mm screening technology is mature (SHI 2014, WANG Yongping et al. 2014, WANG Hong et al. 2018, LIU Yumiao 2016).

2.2 Selection of screening equipment for raw coal

At present, banana screen, post-doctoral screen, and flip-flow screen are commonly used in raw coal screening equipment. The flip-flow screen has the following characteristics: the screen plate makes flip-flow movement, the vibration frequency is 800 times per minute, the amplitude of floating screen frame can be adjusted, and the maximum amplitude can reach 18 mm. The vibration intensity of the material can be adjusted in the range of 10-50 g. The screen hole is not easily blocked. The screening efficiency is higher for difficult screening materials. The classification particle size of dry screening can reach 3 mm. When the water content of raw coal is 24%, the classification screening efficiency of 6 mm is 60-80%, and the classification efficiency is high. The flip-flow screen has a simple structure and small maintenance, which meets the production needs of deep screening in coal preparation plant (LIU Wentong 2015, WANG Haisheng 2014, GONG Sanpeng et al. 2018).

In January 2015, the coal preparation plant of Selian No. 2 Coal Mine was put into trial production. The flip-flow screen is used to classify raw coal of 6 mm. The screening efficiency reaches 70-90% and the screening effect is good. The amount of coal slime produced by filter pressing only accounts for 1-1.5% of the sorting amount. When the total water content of raw coal is 20-25%, the material of 6 mm is screened by flip-flow screen, which is suitable for the coal characteristics in this area.

2.3 Application of three-stage concentration technology

In order to ensure that the coal slime water can be effectively treated, according to the coal characteristics of the raw coal, the coal slime water three-stage concentration process is selected in the process design of Selian No. 2 Coal Preparation Plant.

The three-stage concentration process is to use two thickeners in series. The coal slime water in the production system is preliminarily concentrated by the first stage thickener. The overflow of the first stage thickener serves as the feed of the second stage thickener, and its underflow is sorted and recovered by the coarse slime treatment system. The overflow of the second stage thickener serves as circulating water, and its underflow is recovered by the filter press. The filtrate of the filter press is purified by the third stage thickener. Compared with the single-stage concentration process, the effective sedimentation area of coal slime water in the three-stage concentration process increases. And the process index of coal slime water treatment is improved.

2.4 "6+3" coal preparation process

In 2012, when designing the technological process of Selian No. 2 Coal Preparation Plant, the "6+3" coal preparation process was adopted, that is, lump coal of 200-13 mm is separated by heavy medium shallow slot separator, fine coal of 13-6 mm is separated by heavy medium cyclone, pulverized coal of -6 mm is not separated, and coal slime is recovered by the three-stage concentration process.

"6" indicates that the raw coal system adopts 6 mm deep screening technology, which can reduce the amount of call slime in the system. "3" represents the three-stage concentration process of coal slime water. The process has strong adaptability to coal characteristics and can realize the purpose of coal slime recovery in the plant (WANG Xingyu et al. 2016).

3. Innovation of engineering construction management law

3.1 Design bidding stage

During the design period of the Selian No. 2 Coal Preparation Plant, domestic and foreign experts and directors of coal preparation plants are invited to exchange views. They analyzed and guided the ideas of process design. It ensures that the process design is advanced.

In the project bidding of coal preparation plant of Huainan Mining Group Co., Ltd., the preliminary design of coal preparation plant is used as the project bidding document for the first time in Selian No. 2 Mine. Compared with the feasibility study report as the bidding document, its content is more specific and closed to the reality. And the investment is more controllable.

3.2 Construction stage

(1) Standardize management behavior and formulate reasonable engineering plan

Huai Mine West Company provides technical guidance, service, and supervision to the construction period, quality, investment and safety management of coal preparation plant. The project company is the main body of responsibility for the construction of coal preparation plant and responsible for the construction period, quality, investment and safety of coal preparation plant. Project company organizes preliminary design review and determines process design and main separation equipment. The project company has enriched the personnel of civil construction, electromechanical and safety supervision into the construction management team of a coal preparation plant. It ensures the quality of engineering construction.

A reasonable and rigorous construction plan has been formulated. Through the plan, the general contractor, the construction party, the builder party, and the supervisor can understand the work objectives, task contents, and time nodes. In the process of reviewing the construction plan of coal preparation plant, its rationality, preciseness, and seriousness have been emphasized. It is conducive to the development and connection of work, as well as the management and assessment.

(2) Strengthening construction management and ensuring the qualification of construction quality

Starting from the source, the engineering quality of the coal preparation plant under construction has been supervised. The project company strictly controls every process from the acceptance of incoming raw materials to the construction process and strictly implement it in accordance with the standard requirements. The quality control points are set for the difficulties, key points, and key processes in the construction process. The technical work has been done and the acceptance of concealed works has been strengthened. In the winter construction, a series of measures are taken to ensure the construction quality.

The schedule was refined by inverted schedule and strict assessment criteria have been formulated. Month schedules have been worked out for all the participating projects in the coal preparation plant. Key projects have been broken down into weekly ones and the completion is reported in the form of a weekly report. The reasons for the failure of the project are found out, and the measures for speeding up the progress are formulated. The construction progress is inspected regularly and irregularly by the working staff. Regular engineering meetings are held to coordinate various issues. Construction arrangements and construction projects have been adjusted in time to ensure the completion of the project as scheduled. Within three years, 100 construction meetings were held, which effectively supervised the construction of coal preparation plants and greatly improved the execution capacity of project companies.

(3) Attach importance to important construction nodes and ensure construction safety

In the construction of coal preparation plant, according to the overall work arrangement and the connection of coal preparation plant and mine on production date, the key projects, and completion time are specified in the annual project plan. In the supervision process, the key projects and the construction node period are always emphasized to ensure that the planned objectives are completed on schedule. The construction of coal preparation plant is a general contracting project. There are many unit projects, and some projects have a large workload. At the same time, there are also high-altitude operations. Especially in the case of tight construction period, the construction process is more complex. Safety management runs through the whole construction process of a coal preparation plant. During the construction of the whole coal preparation plant, safety and accident-free have been realized (FANG Qingzhou 2016).

4. Green cycle promotes sustainable development of enterprises

4.1 Comprehensive Utilization of Gangue and Coal Slime

The kaolinite associated with gangue in Tangjiahui Coal Preparation Plant has industrial value. Some enterprises use gangue as a cosmetic filler and advanced ceramics to turn waste into treasure. When the content of kaolinite in gangue is low, it is used for brick-making raw materials and power generation in gangue thermal power plant.

Tangjiahui Coal Preparation Plant adopts thermal drying technology of coal slime. The dried slime is mixed with the blended coal product. The variety of coal slime products has been changed. The pollution of coal slime environment has been eliminated. It has created good social benefits.

4.2 Dust control and comprehensive utilization of water resources

The coal preparation plant has adopted effective dust reduction and dust removal measures, such as increasing the spray to reduce dust in the retransmission point of the raw coal system. Negative pressure water curtain is added to remove dust for large-scale screening equipment. The collected coal slime water can be treated in the coal slime water treatment system of the coal preparation plant. Dust removal sewage is not discharged, which reduces dust pollution in the air.

The production water of coal preparation plant is taken from the clean water after treatment of underground mine sewage, which reduces the water cost of a coal preparation plant. Advanced treatment technology of coal slime water ensures the effective utilization of water resources and the protection of resources and environment.

5. Innovation of coal preparation operation and management model

5.1 New management operation model

Huai Mine West Company has explored a new operation mode. In 2015, Huainan Mining Group Co., Ltd. established Huainan Mine Fenlei Coal Preparation Company. This new type of coal preparation joint venture company operates independently, accounts independently and assumes its own profits and losses. It is an economic entity with independent legal person status. The registered capital of the coal preparation joint venture company is composed of 50% of each party and all of them are financed in monetary form. Through flexible management and job rotation system, the purpose of training staff, training team and retaining excellent staff can be achieved. The problem of surplus labor force of Party A has been effectively solved. Through the brand advantages of Huainan Mining Group Co., Ltd and the management advantages of Fenlei Company, good benefits have been created.

Through the superior technology of both sides, the economic and technological indexes have been formulated comprehensively and scientifically, and the corresponding measures have been taken effectively to implement them. The coal preparation efficiency has been improved. Large-scale and intensive purchasing has been carried out, which reduces the cost of material purchasing, reduces the cost of production, improves the economic benefits of enterprises, and ensures the preservation and appreciation of assets (LEI Chengxiang 2016).

5.2 New operation and management mode has achieved remarkable results

Huai Mine Fenlei Coal Preparation Engineering Technology (Beijing) Co., Ltd. meets the management requirements of "high quality and efficiency" of coal preparation plant and achieves a high level of management.

In terms of production management, mining benefit maximization is the development goal, and the system capability is fully developed. The system process is optimized and the raw coal washing is reasonable. The product structure has been optimized, the product quality has been guaranteed to be stable and the market competitiveness of the product has been improved. The technical inspection system has been improved and coal quality management has been strictly enforced. The quality of commercial coal is stabilized by using an automatic control system. Through strengthening technological innovation, the system is optimized and upgraded.

In terms of safety management, a number of strong safeguard measures have been taken, such as the resident safety officer system, network security officer system, reward and punishment system, etc., and a full-coverage safety management system has been established. At the same time, the safety awareness and quality of employees have been improved, the construction level of safety culture has been improved, and a good situation of "all staff should be safe" has been established. According to the characteristics of spontaneous combustion of coal in Western China, the management of coal bunkers and coal yards has been strengthened. The investment of special funds for safety has been guaranteed.

In terms of electromechanical management, with the equipment spot inspection as the center, it includes the responsibility system of equipment charter, the maintenance management system of disassembly and inspection of electromechanical equipment. The maintenance of electromechanical equipment has been standardized. Overhaul Ledger has been established and updated regularly, spare parts have been managed throughout the life cycle, and preventive and planned maintenance of equipment has been done well. Unified bidding and procurement of materials have been implemented, and the intermediate links have been reduced. Real-time on-line control and unified allocation of inventory materials in various branches are realized by networking software.

In terms of human resources, combined with the human resources advantages of Huainan Mining Group Co., Ltd. and Fenlei Company, human resources sharing has been effectively realized. Employees of enterprises have standardized their actions and operations so as to achieve a high degree of unity in enterprise management. Each employee's working ability of "strict, unified, standardized and efficient" has been realized.

In terms of economic benefits, the leading role of final benefit to coal separation production has been strengthened and the pressure has been scientifically and effectively transmitted to every basic unit of work. Strict production management improves the stability of the system. The consumption of special materials for coal preparation has been strictly controlled and the production cost has been reduced. The efficiency of classification and screening of raw coal is guaranteed. The amount of coal slime brought into the separation system is reduced. Medium consumption and drug consumption are reduced. It promotes the integration of technology and economy.

6. Conclusion

6.1 "6+3" coal preparation technology obtained national invention patent

The "6+3" coal preparation process innovates the coal preparation method, which has important practical and guiding significance for steam coal preparation plants with high ash content, high water content and high degree of sliming.

The statistics of production data show that the yield of coal slime in Selian No. 2 Coal Preparation Plant is 4-5%. Compared with surrounding coal preparation plants, the amount of coal slime produced by it is 2 ~ 3%

higher than that produced by coal preparation plants using +13 mm lump coal separation and -13 mm fine coal bypass technology, but it is 10-15% lower than that produced by coal preparation plants using lump and fine coal full separation technology, which is equivalent to reducing the amount of low-calorific value coal slime by 0.55-1.05 million tons.

6.2 Establishment and operation of ten million tons coal preparation plant

Through management innovation, the engineering quality, efficiency, and investment of Selian No. 2 Coal Preparation Plant have been effectively controlled. After production practice of more than three years, the technical and economic index of the Selian No. 2 Coal Preparation Plant has met the design requirements.

The construction quality of the project is reliable. The total building area of the Selian No. 2 Coal Preparation Plant is 16027 and the total number of equipment is 470.

The effective construction period of Selian No. 2 Coal Preparation Plant is 11 months, which has created a record of the shortest construction period of 10 million tons coal preparation plant in Huainan Mining Group Co., Ltd. and a new record of the shortest construction period of 10 million tons coal preparation plant in Northwest alpine region.

The total investment is 372.8 million yuan, as it also creates a new record of the lowest investment in 10 million ton coal preparation plant of Huainan Mining Group Co., Ltd.

6.3 New operation management mode realizes win-win benefit and market sharing

(1) It has solved the problem that enterprises have a poor ability to resist market risks unilaterally under the current economic situation.

(2) It ensures the reasonable investment of internal coal preparation plant funds, prolongs the service life of the equipment and strongly promotes the safety and quality standardization construction of the coal preparation plant.

(3) The problem of reemployment of surplus personnel of party A after capacity reduction is solved.

(4) The joint venture company of coal preparation has made positive progress in expanding its external market.

7. References

[1] FANG Qingzhou. Technical management practice of coal preparation plants of Huai Mine West Mine Investment Management Co., Ltd.in Ordos[J]. Goal Preparation Technology, 2015, (5):55-59.

[2] FANG Qingzhou. Management practice of engineering construction pattern of Selian No.2 coal mine coal preparation plant[J]. Goal Quality Technology, 2016, (1):65-68,64.

[3] GONG Sanpeng, WANG Xinwen, YU Chi, et al. Study on the motion law of main and floating screen frame of vibrating flip-flow screen with damping[J]. Coal Engineering, 2018, 50(8):126-132.

[4] LEI Chengxiang. A study of the operational mode of the coal preparation plants in Ordos region managed by Huaikuang Western Region Coal Mine Investment & Management Co., Ltd. of Huainan Mining Group[J]. Goal Preparation Technology, 2016, (3):65-67,71.

[5] LIU Wentong. Application of flip-flow screen in deep classification of steam coal[J]. Clean Coal Technology, 2015, (3):18-20,24.

[6] LIU Yumiao. Application of large-sized flip-flop screening machine in cleaning of power coal[C]. Goal Preparation Technology, 2016, (6):30-30,36.

[7] SHI Jianfeng. Study and application of 2SCZ1230 high efficiency dual mass flip-flop screen[J], Goal Preparation Technology, 2014, (6):5-8.

[8] WANG Haisheng. Application of flip -flow screen in Donghuantuo Coal Preparation Plant[J]. Journal of North China Institute of Science and Technology, 2014, (11):60-62.

[9] WANG Hong, CHEN Zhiqiang. Study on shear spring and application of double mass vibrating flip-flop screen[J]. Coal Engineering, 2018, 50(4),149-151,156.

[10] WANG Xingyu, LIU Gang. Application of Fli-flow Screening Machine in Raw Coal Classification-screening of Coal Preparation Plant[J]. Coal Mine Machinery, 2016, 37(5):159-160.

[11] WANG Yongping, ZHANG Xuliang. Application of JFDI-3048 Flip-flop Screen Dry Sieving in Coal Preparation Plant[J]. Value Engineering, 2014, (29):60-61.

2

Improving management level of coal preparation plant applying modern electronic intelligent information technology

Zhang Zhen

Shenhua Ningxia Coal Industry Group, Yinchuan , China

Abstract: Equipment central control system, automatic process system, measurement, and detection monitoring system and equipment protection system have been widely applied in coal preparation. However, the real-time operating parameters, status information, metering test data, video and images acquired from each system are still independent of each other, and they are decentralized with control system and computer information management system of coal preparation plant, leading to the lack of comprehensive analysis of parameters, data, and information, hence, the conclusion and implementation scheme are impossibly given by this system. Through the analysis of equipment central control system and automatic process system as well as monitoring system of measurement and detection of coal preparation plant, the fast improvement of management level of coal preparation plant with modern electronic intelligent information technology is discussed in this paper.

Keywords: Production management of coal preparation; electronic intelligent information technology; remote centralized control; intelligentization

1. Present situation of the production-process system in coal preparation plant

Using automatic control technology, metrology detection monitoring technology, and equipment protection function, the production-process system and equipment of coal preparation plant are concentrated in the central control system.

1.1 Central control system

The central control system of coal preparation plant consists of raw coal system, preparation system, slurry treatment system and loading system, via which the start and stop of all the equipment in the plant is automatically controlled. Much advancement is achieved, such as few workers are needed for monitoring equipment, enhancing the safety of workers, lowering the working intensity of workers, etc. Overall, the central control system is advantageous for improving the coal preparation plant.

1.2 Automation of the production-process system

The automatic production-process system is mainly composed of dense medium system, filter press system and automatic control of single equipment like concentrator, filter press, etc. Due to the imperfection of sensors for ash content, density, concentration, flux, etc., the information transfer is hysteretic and raw coal properties, requirement of products, equipment capacity, production condition and preparation parameters are difficult to be concentrated in one computer platform for analysis to guide preparation process. Hence, up to now, the central control system lacks integrality that the automation system of single equipment is independent of each other, and the intelligent linkage mechanism is not formed.

1.3 Measurement and detection monitoring system and personal protective system for equipment

At present, a large number of techniques for monitoring measurement and detection and protecting equipment are applied in coal preparation plant, including quantity and quality of raw coal and products, electricity consumption, water consumption, fuel consumption, level of coal bunker, liquid level of pool bucket, gate, valve, flap position, important equipment, electric current of motors for 55kw and above, temperature for stator winding and bearing of high voltage motor, important motor, reducer, exciter amplitude and temperature, power cable concentrated cable tray, cable trench cable temperature, dense medium system parameters, flotation feed flow and concentration, methane and carbon monoxide concentration in coal bunker and other position having gas and coal dust, tape machine overload, drawstring, deviation, slipping, blockage protection, important longitudinal tear protection of the tape machine, smoke temperature protection of the important tape machine for transporting dry coal, and scraper protection for the scraper, etc. Industrial TV monitoring equipment was largely used in the coal preparation plant, but all equipment is individual and dispersive. The data of every system is not combined for integrally analyzing the preparation process.

2. Application of modern electronic intelligent information technology

All single equipment and equipment cluster of complex function are combined using the modern electronic intelligent information technology. The parameters of all equipment are collected to establish a big data analysis platform for automatically controlling the preparation process, equipment maintenance, and product loading.

2.1 Fast implementation of central control production equipment using modern electronic intelligence information technology

With the fast development of modern electronic intelligent information technology, it is relatively easy to realize central control of production equipment in coal preparation plant. For achieving equipment control and central control, a large number of flame-retardant control cables are needed to be purchased and installed. Furthermore, a large number of cable bridges and cable conduits are needed, for which a lot of workers are needed, construction period is long, construction path is complex, the construction is difficult, and maintenance workload is large. A technique named wireless frequency hopping spread spectrum communication technology is researched and applied by coal blending center of Shenhua Ningxia Coal Industry Group Co., Ltd, by which all equipment of coal preparation plant is remotely and centrally controlled. This technique is newly applied in coal preparation industry and it subversively changes the traditional technique, thus it is valuable to be popularized and applied.

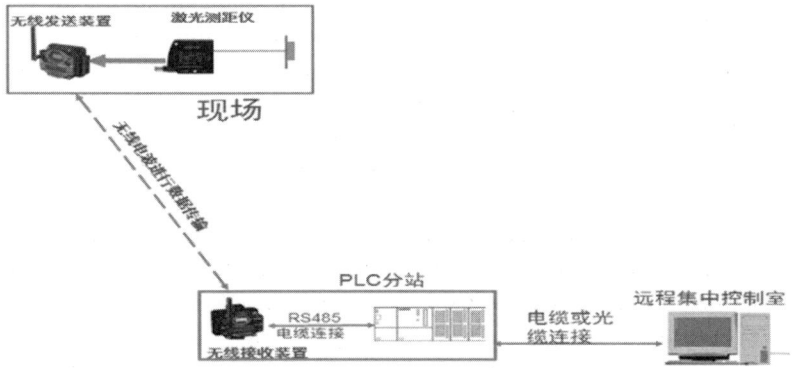

Figure 1: Schematic of wireless frequency hopping spread spectrum communication technology

Fig. 1 shows the schematic of the wireless frequency hopping spread spectrum communication technology. After detecting the displacement change via laser range finder, an analog signal of 0-20 mmA is exported through a data line connecting with the wireless transmitter. The wireless receiver of PLC receives and transfers the signal to the remote control room. On the other hand, the control signal is sent from the remote control room via the wireless receiver, and the wireless transmitter receives the signal to control the displacement of equipment.

Three advantages of wireless frequency-hopping spread spectrum communication technology are discovered: 1) the construction speed is fast, the construction is not difficult, a few workers are needed for the construction, the construction period is short, a small number of cables is needed to be purchased and installed, inspection and maintenance of cables is lowered, 2) strong anti-interference ability contributes to the suitability and application of wireless frequency-hopping spread spectrum communication technology in sites having a large number of frequency converter, motor, wifi, walkie-talkie and other mobile communication devices, and 3) the control of some remote and auxiliary equipment can be easily centrally controlled via the wireless frequency-hopping spread spectrum communication control technology.

2.2 Applying modern electronic intelligent information technology to realize remote and real-time monitor of status and fault diagnosis of large-scale production equipment

With the fast development of modern electronic intelligent information technology, it becomes easy to manage the status, fault diagnosis, maintenance scheme and maintenance plan of large-scale production equipment, such as motors, speed reducers, vibration exciters, etc. Before that, large-scale production equipment is manually inspected, and many disadvantages exist like high labor intensity for the worker, easy negligence for hidden dangers, belated warning for equipment damage, discontinuous and incomplete monitor of data, impossible prediction of potential malfunction, etc. The remote intelligent maintenance system of large-scale electromechanical equipment developed by the China University of Mining Technology has been applied in Chengzhuang Coal Preparation Plant. As shown in Fig. 2, the temperature rise of equipment is detected by the temperature sensor, and the signal is transferred to the background workstation via the wireless frequency hopping spread spectrum communication technology. After data analysis, the equipment failure status and maintenance proposal will be sent to the managers' mobile phones via SMS.

Using the Ann model of the wireless data acquisition system, multipoint temperature and vibration parameters of equipment are collected for establishing typical fault simulation library of large-scale electromechanical equipment in coal preparation plant. The prediction and maintenance technology of fault data is developed for establishing a remote prediction and maintenance platform for large-scale electromechanical equipment in the coal preparation plant, and the maintenance suggestions are sent to the managers in real time via SMS for pre-arranging the maintenance and spare parts preparation, where both the management and maintenance of equipment are intelligent.

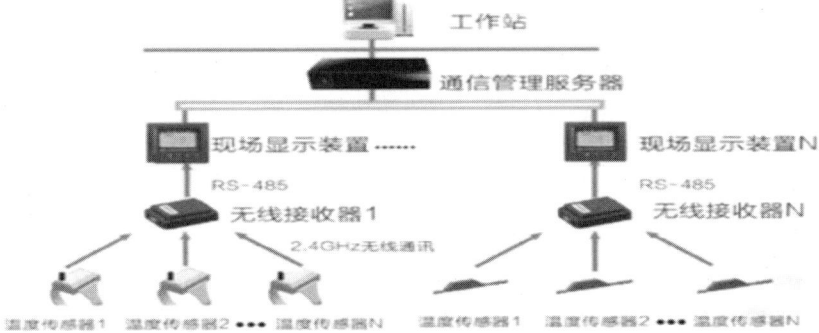

Figure 2: Remote intelligent maintenance system of large-scale electromechanical equipment

2.3 Applying modern electronic intelligent information technology to complete monitoring system for measurement and detection and equipment protection system

The management of the whole production process and automatic management of remote control become easy via the modern electronic intelligent information technology. A series of intelligent solutions should be solved for achieving the intelligent production process of the coal preparation plant, including sampling data analysis, dense medium system, flotation system, slurry treatment, and product sales system.

Numerous labor costs are saved via the application of raw coal automatic blending system in coal preparation plant. Using online ash analyzer in the coal blending system, the properties of coal on the belt are online detected in real time. The instantaneous and average ash content of random time period can be given according to the requirements, which is beneficial for the automatic blending of raw coal and enhancement of product quality. The schematic of the automatic detection of ash content using online ash analyzer for realizing automatic control of density is shown in Fig. 3.

Figure 3: Intelligent density control system based on collection
and analysis of coal ash data

Based on PLC control system, the detection of process parameters, automatic control, and production management of dense medium separation system are concentrated in the dense medium density control system, and the locking relationship of all equipment in the density control system is realized through software programming. The aims that accurate and timely control and adjustment of dense medium density, enhancement of product quality and decline of medium loss per ton coal as well as automatic and intelligentization of the entire production process can be achieved via this system.

3. Conclusion

The modern electronic intelligent information technology has been deeply applied in the coal preparation process. The industrial aim of coal preparation is to achieve intelligentization of coal preparation, sampling, coal analysis, slurry treatment, coal blending, coal sale, equipment management and equipment maintenance combined with big data and artificial intelligence technology as well as the existing automation and digitization of coal preparation plant.

3

Practice of high-frequency dewatering screen for coarse clean coal

Cuixian Cheng, Jianguang Wu, Taiyuan, China

Dongqu Coal Preparation Plant/Xishan Coal Electricity Group Co., Ltd.

Abstract: The dewatering of coarse clean coal is an important part of the production of the coal preparation plant. The dewatering and recovery processing of conventional sieve bend combined with Coarse slime centrifuge often leads to the loss of clean coal due to the wear of the screen surface, the quality of the product is degraded, and it will also affect the separation effect of the flotation column. The high-frequency dewatering screen has a high vibration intensity, the screen surface is not easy to block the hole, the screening efficiency is high, and the processing capacity is large, which is especially suitable for the screening of fine powder materials. Dongqu Coal Preparation Plant now utilizes high-frequency dewatering screen instead of the original dewatering devices. The results show that the clean coal recovery of high-frequency dewatering screen is higher, and the coal ash content can be reduced. There was less coarse clean coal lost in the system, making the flotation column run properly. Also, the economic benefits of the coal plant.

Keywords: Coarse clean coal slurry, dewatering, high-frequency dewatering screen

1. Introduction

Dongqu Coal Preparation Plant is a mine-type coal preparation plant with a washing capacity of 3 million tons/year. A combined process of deliming and pressureless three-product heavy medium cyclone + teeter bed separator (TBS) + flotation is used. The separated process is described as follows: first, the 50-0 mm raw coal entering the separated system is desliming using 1.0 mm desliming screen. The 50-1.0 mm is sorted by the pressureless three-product heavy medium cyclone. Then, -1.0 mm is graded into 1-0.25 mm and -0.25 mm using a classifying cyclone. Clean coal products with ash content below 11% are sorted from 1-0.25 mm using TBS, and enter the fine coal warehouse together with heavy medium cyclone clean coal. Finally, -0.25 mm coal slime is sorted using flotation. Clean coal is recycled using a quick-open filter press and tailing is recovered by concentration + pressure filtration combined process.

2. Problems before the transformation of coarse clean coal dehydration system

After the coarse coal slurry is sorted by TBS, the TBS overflow is concentrated by a concentrated cyclone, and then dehydrated by the curved screen as well as coal slime centrifuge to become a coarse clean coal product. The concentrated cyclone overflow, sifted water of curved screen, and the centrifugate of centrifuge enter the flotation system together. The problems before the system transformation are as follows:

2.1 Problems of concentration cyclone

In actual production, the liquid level of the concentrate barrel is unstable and there often exists an evacuated state, which makes the concentrated cyclone unable to have a stable feed pressure and feed amount. Even if the cyclone overflow is returned to the concentrate tank, the liquid level of the concentrate tank is still uncontrollable, hence the concentrated cyclone has a poor concentration effect.

2.2 Problems of curved screen

In actual production, the dewatering effect of the fixed curved screen with screen seam of 0.5 mm and screen area of 3 m2 is not good, resulting in low centrifuge feed concentration. After the fixed curved screen is used for a period of time, the screen surface is seriously worn, the screen seam is widened, and sifted water contains a large amount of coarse particles, and the product recovery effect is poor. The results of the screening test of the curved screen before the modification are shown in Table 1. It can be seen that the content of +0.5 mm in sifted water is 11.90%, which indicates that a part of the coarse coal particles are lost in the system.

Table 1: Results of screening test of the feed, discharge and undersize materials of the curved screen before the transformation

Size fraction (mm)	Feed		Discharge		Sifted water	
	Yield (%)	Ash content (%)	Yield (%)	Ash content (%)	Yield (%)	Ash content (%)
+1.500	0.85	6.36	1.15	5.51	–	–
1.500-1.000	1.65	5.70	1.20	6.69	–	–
1.000-0.500	44.93	7.46	43.88	7.17	11.90	7.34

Contd...

Contd...

Size fraction (mm)	Feed		Discharge		Sifted water	
	Yield (%)	Ash content (%)	Yield (%)	Ash content (%)	Yield (%)	Ash content (%)
0.500-0.250	33.45	10.62	34.90	9.89	16.83	10.44
0.250-0.125	14.44	37.33	14.99	29.8	38.37	33.56
0.125-0.074	2.61	70.86	2.16	67.02	14.57	69.85
0.074-0.045	1.60	70.72	1.20	66.87	10.80	69.40
-0.045	0.45	44.80	0.50	42.70	7.53	44.58
Total	100.00	15.63	100.00	13.68	100.00	36.54

2.3 Problems of centrifuge

There is a serious problem that the screen seam is seriously scoured by the low concentration material when centrifuge with a sieve basket of 0.5 mm seam is used. In addition, the concentration of the centrifugate is obviously increased, a large amount of coarse coal particles are lost in undersize materials and there is a poor recovery effect when the seam of the sieve basket exceeds 0.6 mm. The results of the screening test of the centrifuge feed, discharge and centrifugate before the transformation are shown in Table 2. It can be seen that the content of +0.5 mm in the centrifugate is 21.95%, which indicates that a part of the coarse coal particles is lost in the system.

Table 2: Results of screening test results of centrifuge feed, discharge and centrifugate before transformation

Size fraction (mm)	Feed		Discharge		Centrifugate	
	Yield (%)	Ash content (%)	Yield (%)	Ash content (%)	Yield (%)	Ash content (%)
+1.500	1.15	5.51	1.10	4.90	–	–
1.500-1.000	1.20	6.69	2.16	5.64	–	–
1.000-0.500	43.88	7.17	46.96	7.76	21.95	8.60
0.500-0.250	34.90	9.89	32.61	10.16	36.99	11.35
0.250-0.125	14.99	29.8	13.35	28.10	23.66	38.88
0.125-0.075	2.16	67.02	1.56	58.88	6.07	68.78
0.075-0.045	1.20	66.87	0.90	55.74	4.66	67.09

Contd...

Contd...

Size fraction (mm)	Feed		Discharge		Centrifugate	
	Yield (%)	Ash content (%)	Yield (%)	Ash content (%)	Yield (%)	Ash content (%)
-0.045	0.50	42.70	1.35	28.79	6.67	40.35
Total	100.00	13.68	100.00	12.69	100.00	25.27

The loss of coarse particles in the curved screen and the centrifuge not only causes the coarse clean coal sorted by TBS not to be effectively recovered, but more importantly, these coarse particles enter the flotation system, which seriously affects separation performance of cyclone micro-bubble flotation column and the short column micro-bubble flotation machine.

3. Transformation plan of coarse clean coal dewatering system

According to the analysis of the above problems, the study of the current advanced coal preparation technology, and the good dewatering effect of our plant using high frequency vibrating screen to recover coarse particles in flotation tailing, hence the transformation plan for the coarse coal dehydration system is as follows: The combined process of concentrate barrel + concentrated cyclone + curved screen is replaced by high frequency coal slime dewatering screen. High-frequency coal slime dewatering screen is a new structure of high-efficiency screen vibrating screening machine. The vibration exciter drives the screen surface to vibrate at high frequency through the electromagnetic but the screen box keeps stationary. The vibration system is designed to work in a resonant state. The vibration frequency of the screen surface is 50Hz, the amplitude is 1~5 mm, and the vibration intensity is 8~10 times of the gravity acceleration, which is 2~3 times of the vibration intensity of the general mechanical vibrating screen. It is characterized by the screen surface that is not easy to block holes, the high screening efficiency, the large processing capacity, convenient and adjustable screen machine angle while this screen is particularly suitable for the screening of fine-grained powder materials. The vibration parameters of the screen machine are controlled by a computer to realize the separate adjustment of the vibration parameters of each vibrator. The high-frequency coal slime dewatering screen selected for transformation is DZSM2436 with a screen area of 8.64 m^2 and a mesh size of 0.1~3 mm. The equipment connection diagrams of the coarse clean coal dewatering system before and after the transformation are shown in Figure 1 and Figure 2, respectively.

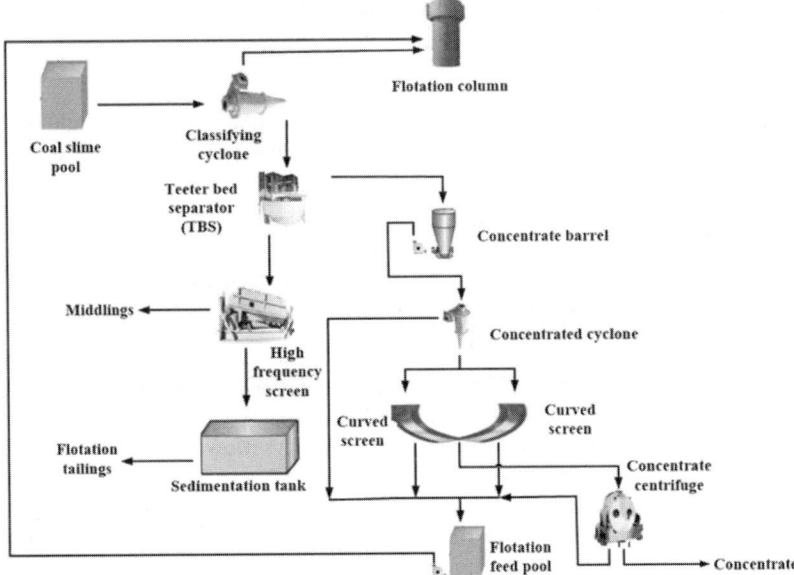

Figure 1: Equipment connection diagrams before transformation of coarse clean coal dewatering system in Dongqu Coal Preparation Plant

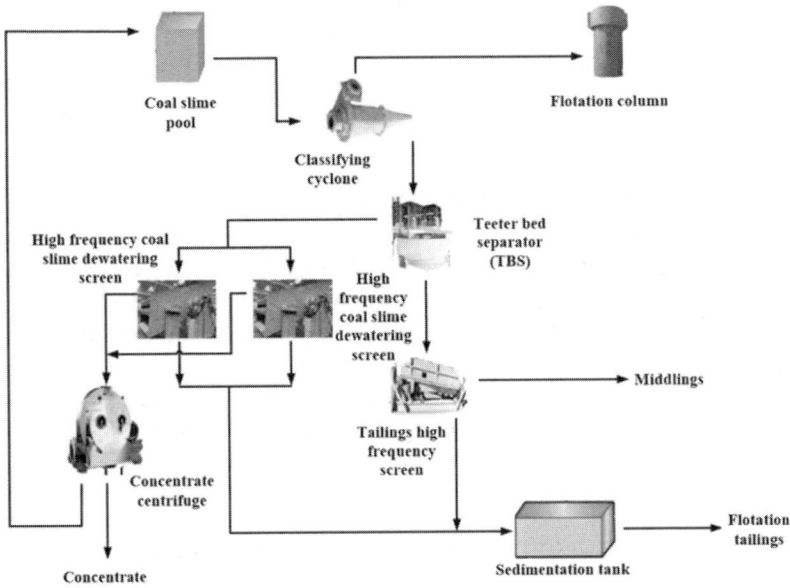

Figure 2: Equipment connection diagrams after transformation of coarse clean coal dewatering system in Dongqu Coal Preparation Plant

4. Results of process transformation

4.1 Significant reduction in coarse particles in the sifted water

The screening test results of the concentrated cyclone overflow and sifted water of the curved screen before the transformation and the sifted water of the high-frequency coal slime dewatering screen after the transformation are shown in Table 3.

Table 3: Screening test results of concentrated cyclone overflow and sifted water of curved screen before transformation and sifted water of high-frequency coal slime dewatering screen after transformation

Size fraction (mm)	After transformation		Before transformation			
	Ssifted water of high-frequency coal slime dewatering screen		Concentrated cyclone overflow		Sifted water of curved screen	
	Yield (%)	Ash content (%)	Yield (%)	Ash content (%)	Yield (%)	Ash content (%)
1.000-0.500	1.76	13.00	8.83	5.46	11.90	7.34
0.500-0.250	4.13	15.31	19.12	7.30	16.83	10.44
0.250-0.125	53.70	40.06	17.51	13.98	38.37	33.56
0.125-0.075	13.59	56.48	7.43	19.13	14.57	69.85
0.075-0.045	5.89	58.20	9.73	18.16	10.80	69.40
-0.045	20.94	40.92	37.38	30.44	7.53	44.58
Total	100.00	42.04	100.00	18.89	100.00	36.54

It can be seen from Table 3 that yield of +0.25 mm coal particles is 5.89% in sifted water from the high-frequency coal slime dewatering screen after the transformation, which indicates the high frequency screen has good dehydration recovery effect. However, the yield of +0.25 mm is 27.95% in the overflow from the concentrated cyclone and is 28.73% in the sifted water from the curved screen before the transformation. This shows that after the transformation, the recovery effect of coarse particles become better and only a very small part of +0.25 mm particles enter the flotation system, which lays a foundation for improving the flotation separation effect.

4.2 Reduction in ash content of coarse clean coal

Before the transformation, the ash content was reduced from 15.28% to 12.66% after the TBS overflow was concentrated by the concentrated cyclone and dewatered by the curved screen as well as the coal slime centrifuge. But after the transformation, the ash content was reduced from 14.00% to 10.00% after the TBS overflow was dewatered by the high-frequency coal slime dewatering screen and the coal slime centrifuge. The ash content of coarse clean coal was decreased after the transformation, which indicates that the high-frequency coal slime vibrating screen has a very good desliming and ash-reducing effect.

4.3 Reduction in moisture content of coarse clean coal

The average processing capacity of the coal slime centrifuge was 17.64 t/h, and the product moisture was 14.3% before the transformation whereas the average processing capacity of the coal slime centrifuge was 22.17 t/h and the product moisture was 12.9% after the transformation. The discharge water from the centrifuge was reduced by 1.4% and the processing volume was increased by 4.53 t/h.

5. Conclusion

(1) The ash content of TBS clean coal was reduced from 14.00% before the transformation to 10.00% after the transformation and the decreasing amplitude of ash content was 28.57%, which was 11.42% higher than the decreasing amplitude of 17.15% before the transformation. Hence, the ash content of TBS clean coal was reduced after the transformation.

(2) The traditional TBS clean coal dewatering process was simplified after the transformation. The moisture content of TBS clean coal was decreased from 14.30% before the transformation to 12.90% after the transformation, decreased by 1.4%, and decreasing amplitude was 9.79%, which was beneficial to reduce the moisture content of the final clean coal.

(3) Concentrate barrels, concentrated cyclones and feed pumps, flotation feed pumps are reduced to use, which saves energy and reduces production costs. The annual electricity cost can be saved by about 2 million yuan.

4

Practice of treatment of refuse discarded by coal heavy medium separation process using a 3 Mt/a coarse and small coal combined dry cleaning system

Ren Shangjin1 Sun He[1] Xia Yucai1 Cheng Jianzhong[2]

[1] *Tangshan Kaiyuan Technology Co., Ltd., Tangshan, China;*
[2] *School of Chemical Engineering & Technology, China University of Mining & Technology, Xuzhou, China*

Abstract: An introduction is made in the paper to the physical property of the refuse produced by the heavy medium separation process for treatment using the combined coarse coal and small coal dry cleaning system, the layout of the main and auxiliary equipment and control system applied, as well as the composite technical and economical result obtained.

Keywords: Coarse coal and small coal combined dry cleaning system; differential dry cleaning machine; Small coal dry cleaning jig; Separating density control

1. Outline

The refuse disposed of by a heavy medium separation plant in China's Shanxi Province has a calorific value of 1816 Kcal/kg and a total sulfur content of 1.57%, and is liable to undergo spontaneous combustion due to its high content of small coal. The refuse material has been stockpiled mixed with sandy soil in the open air for many years. To address this issue, a combined coarse coal and small coal dry cleaning system is specifically developed. The system is comprised of 4 sets of the CFX-12 differential dry cleaning machines and 2 sets of TFX-9 small coal dry cleaning jigs. The overall system is under centralized control with PLC which has density adjustment and control functions. The system with a designed capacity of 3 million tons a year produces 3 final products – clean coal, middlings and reject. The clean coal with a calorific value of 3282 Kcal/kg and a total sulfur of 0.96% is used as fuel for power stations. The middlings product with a calorific value of 1487 Kcal/kg is used as brick-making material while the reject with a calorific value of only 921 Kcal/kg is used as a backfilling material for landscape restoration or land reclamation. The system with a total installed capacity of 2143 kW and an operating cost of 3.41 yuan/t is set up at an initial capital cost of 11.73 million yuan.

2. Analysis of refuse property

The refuse for treatment has a calorific value of 1816Kcal/kg and contains high-density kaolinite, rocks, carbonaceous material, inter-banded coal, and mixed coal with a calorific value of >4000 Kcal/kg. The abovementioned materials differ widely in density (See Table 1 & 2).

Table 1: Screen Analysis of Refuse

Size Fraction (mm)	Mass (kg)	Yield (%)	Calorific value (Kcal/kg)
50 ~ 6	82.4	73.31	1886
6 ~ 0	30.0	26.69	1622
Total	112.4	100.00	1816

Table 2: Float-and-sink Date of the 50-6 mm Refuse

Density fraction (kg/L)	Yield (Percentage in the fraction)	Yield (Percentage in the whole sample)	Calorific value (Kcal/kg)
<1.50	6	4.40	5535
1.50 ~ 1.80	8	5.86	4176
>1.80	86	63.05	1419
Sub-total	100	73.31	1886

As can be seen from Table 2, the <1.80 density fraction with a calorific value of 4778 Kcal/kg and a yield of 10.26% ought to be recovered while the >1.80 faction with a calorific value of 1419 Kcal/kg still contains useful resources that can be recovered through liberation, if possible.

3. Combined coarse coal-small coal dry cleaning system

As shown in Fig 1, the <150 mm refuse feed is first crushed with a sizing crusher. Then the <30 mm size is sent into the 4 sets of CFX12 differential dry cleaning machines to produce 4 primary products – clean coal, middling, reject and fine coal. The coarse clean coal is screened on a slip-flop screen at a size of 10 mm. The >10 mm size after being crushed to less than 10 mm using a ring-hammer crusher is sent into the 2 TFX9 dry cleaning jig together with the flip-flop screen's <10 mm undersize material to produce re-cleaned coal and middling products. The fine coal product of the primary cleaning process and the clean coal obtained through re-cleaning process constitute the final clean coal product while the middling produced through primary and re-cleaning processes constitute the final middling product. The reject of the primary cleaning process is the final reject product.

4. Selection of dry cleaning machine

4.1 Intelligent differential type

As illustrated in Fig. 2, the differential-type dry cleaning machine mainly consists of the differential-type exciter 2 (including the drive arm 1, exciter 2, connecting element 4 and motor 5), separating deck 8, framework 7, hanging-lifting system 10 (including the electric lifting unit and steel-wire rope 6), deck inclination adjustment mechanism 11, air-blast tube 3, dust-collector hood 9 and air chamber 12.

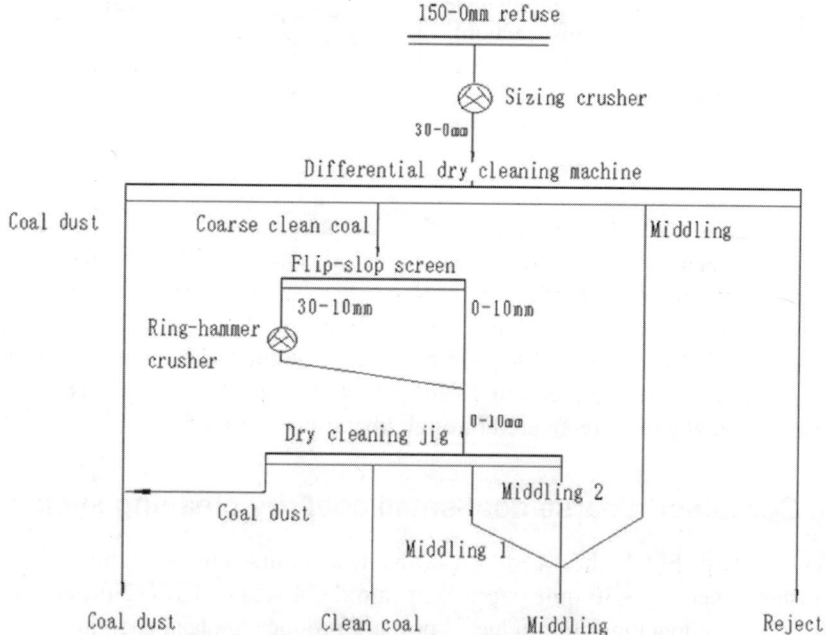

Figure 1: Basic flow-sheet of the combined dry cleaning process

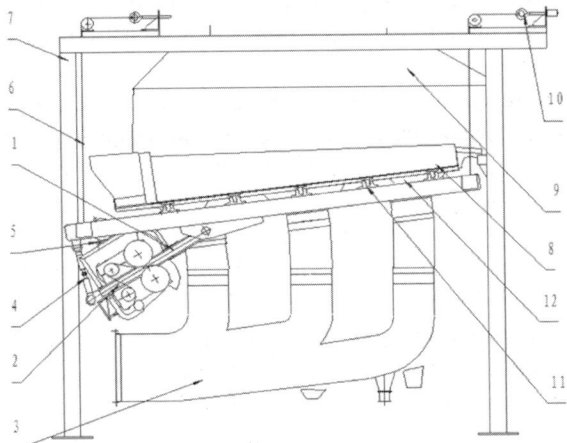

Figure 2: Schematic diagram showing the structure of the differential-type dry cleaning machine

The 4 sets of the CFX-12 differential-type dry cleaning machines used in the system are symmetrically arranged in a manner as shown in Fig. 3. On the top of each separating deck, a dust hood is fitted to form a separate dust collection system. The coarser coal dust passes through the dust collector with the purified air being circulated as blast air in a closed circuit for avoiding the escape of dust. The finer coal dust passes through the bag collector, and only purified air is allowed to be discharged into the atmosphere by the draught fan. The dry cleaning machines are provided with an intelligent density measuring and control system for monitoring the percentage of shale in the feed material, so as to enable the influential factors like feed rate, air blowing volume and vibration frequency of separating deck to be controlled by PLC.

Figure 3: A general view of the CFX-12x4 differential-type dry cleaning machine

Working principle of the intelligent dry cleaning machine: The exciter and the deck are linked up at a certain angle. The deck is brought by the exciter into a differential vibration motion – slower forward motion and faster backward motion. The feed material in the hopper above flows down via the chute and is then uniformly led into the separating deck via the feeder. The air chamber under the deck is connected with the blower through the air tube. The material on the deck becomes gradually stratified under the combined effect of an upward airstream and the mechanical vibration motion of the deck. Because of the obstruction of the riffles on the deck, the lighter particles tend to float gradually upward and is finally discharged into the clean coal chute at the side of the machine while the heavier particles that sink down to the middle and bottom of the material bed are gradually separated along with their motion into middling and reject products, which are discharged into respective chute from the tail of the separating deck. The technical parameters of the intelligent dry cleaning machine are listed in Table 3.

Table 3: Technical Parameters of the CFX Dry Cleaning Machine Set

Separating area, m²	12 x 4
Feed size adaptable, mm	50-0
Capacity, 10000 t/a	300
Surface moisture of feed adaptable, %	<10
Ecart probable Ep, kg/L	0.17
Separating efficiency η, %	<90
Power of main frame, kw	11 x 4
Total installed capacity, kw	340 x 4 (1735)
Concentration of exhaust emission, mg/m³	6.45

It can be seen from Table 4, the intelligent dry cleaning machine is high in capacity, adaptability, separating efficiency yet low in power consumption.

4.2 Intelligent dry cleaning jig

As shown in Fig.4, the jig is mainly comprised of the framework 1, hanging mechanism 2, separating deck 3, leveling device 4, dust hood 5, exciter 6 (class-8 vibrating motor), adjustable air chamber 7, pulsating air supply mechanism 8, discharge mechanism 9 and feeding mechanism 10. The bed body is made up by rectangular box, hand-operated uniform air distribution screen plates. Between the upper and lower screen plates are small partition boxes in which small ball-carrying screen plates are fitted. The small partition boxes are each connected with its corresponding adjustable air chamber. The bed body is also connected with exciters, leveling mechanism 4, material distributor and discharge mechanism 9. The deck body is hanged on the framework 1 via flexible hanging mechanism 2. The feeding mechanism 10 and dust hood 5 are also fixed on the framework 1. The pulsating air supply mechanism 8 is flexibly linked up with the air chamber. The leveling mechanism 4 is fixed inside the deck body for ensuring the even distribution of feed material on deck. All the associated equipment are driven through a frequency converter.

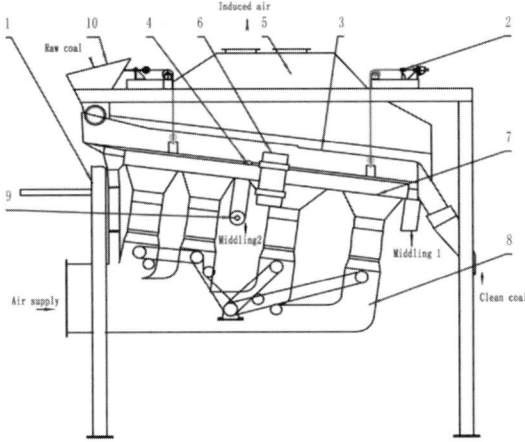

Figure 4: Structural diagram of the dry cleaning jig

The dry cleaning system uses 2 sets of TFX-9 jigs which are symmetrically arranged in a form as shown in Fig. 5. Each jig is an independent machine which consists mainly of the feeding system, air supply system, dust collection system, dust collection and discharge system and product discharge system. Considering that the material treated is small in size and much dust is generated in the cleaning process, and for avoiding blocking of screen plate by the dust in the air chambers beneath the screen plate, an open-circuit pulsating pure air supply system is used for ensuring the air channels are free of obstruction. Because of the use of an open-circuit dust collection system with cyclone and pulsating bag collectors which are series connected with induction fans, the density of gas emission can be reduced to 17 mg/m^3. The dirt content of feed material is monitored by the density measuring and control system and based on the monitored data, the PLC can adjust and control the operating parameters such as the feed rate, air supply volume, deck vibration frequency and longitudinal inclination angle, and discharge rate.

Figure 5: A general view of the TFX-9x2 small coal dry cleaning machine set

Working principle of the jig: Under the combined effect of mechanical vibration of the separating deck and upward pulsating airflow blasted from under the deck, the feed material becomes gradually stratified. The heavy particles gradually sink down to the bottom while the lighter particles gradually go up to the top of the material bed. Under favorable jigging conditions, a well-stratified material bed can be formed. The heavier particles gathering at the bottom of the bed are discharged by the discharge mechanism at the bottom of the bed while the remaining materials continue to go through the separation process under the effect of vibration of the deck. After two times of bottom discharge of heavier particles, the final products of middling 2, middling1 and clean coal can be obtained (see Fig. 4)

The technical parameters of the jigging machine are tabulated in Table 4.

Table 4: Technical Parameters of the Jigging Machine

Feed size, mm	13-0
Feed size adaptable, mm	13-1
Capacity, t/h	≤150×2
Feed moisture adaptable, %	<8
Air volume, $km^3/m^2 \cdot h$	3.9
Air pressure, Pa	2233
Organic efficiency η, %	<90
Ecart probable Ep, kg/ L	0.238
Separating area, m^2	9×2
Power of main body, kW	14.5×2
Total installed capacity, kW (Main & auxiliary equipment included)	180×2（408）
Concentration of emission, mg/m^3	17

As can be seen from Table 4, the two sets of dry cleaning jigs offer a host of merits of being highly adaptable to finer and high-moisture feed, small in Ep value, high in organic efficiency, large in unit separating area, good in separating performance, and remarkably high in capacity which is up to 300 t/h. However, the 2 sets of jigs operate with a very low energy consumption (less than 1/3 of that of coarse coal dry cleaning machines), and the power consumption per ton of coal treated is only 1.08 kWh.

5. Separating performance

5.1 The technical performance data of the differential-type dry cleaning machines is listed in Table 5.

Table 5: Technical Performance Indicators of the Differential-type Dry Cleaning Machine Set

Product	Yield (%)	Total moisture (%)	Calorific value (Kcal/kg)	Total sulfur (%)	Remarks
Clean coal	41.46	3.8	2533	1.18	
Middling	12.78	3.2	1486	1.22	
Reject	40.76	3.9	921	2.30	
Coal dust	5.00	5.3	3340	1.01	Density of emission: 6.45 mg/m^3
Total	100.00		1780	1.63	

It can be seen from Table 5, a very good separation performance can be obtained as evidenced by the calorific values of the products: coarse clean coal – 2533 Kcal/kg, coal dust – 3340 Kcal/kg, middling – 1486 Kcal/kg, and reject – 921 Kcal/kg.

5.2 The technical performance data of the small coal dry cleaning jigs is listed in Table 6.

Table 6: Technical Performance Indicators of the Small Coal Dry Cleaning Jigs

Product	Yield,% (percentage of the fraction)	Yield,% (original fraction)	Total moisture, %	Calorific value, Kcal/kg	Total sulfur, %	Remarks
Clean coal	47.89	19.86	4.5	3279	0.88	
Middling 1	20.75	8.59	3.3	1514	1.08	

Contd...

Contd...

Product	Yield,% (percentage of the fraction)	Yield,% (original fraction)	Total moisture, %	Calorific value, Kcal/kg	Total sulfur, %	Remarks
Middling 2	21.36	8.86	4.3	1462	1.19	
Coal dust	10.00	4.15	5.3	3230	1.3	Density of emission: 17 mg/m³
Total	100.00	41.46		2520	1.03	

As indicated in Table 6, a clean coal with a high calorific value of 3279 Kcal/kg and a total sulfur content as low as 0.88% can be obtained.

5.3 The composite performance of the coarse and small coal integrated dry cleaning system

For the composite performance data of the coarse and small coal integrated dry cleaning system, refer to Table 7.

Table 7: Composite Performance Data

Product		Yield (%)	Calorific value (Kcal/kg)	Total sulfur (%)
Clean coal	Clean coal (jig)	19.86	3279	0.88
	Coal dust(dry cleaning machine)	5.00	3340	1.01
	Coal dust(jig)	4.15	3230	1.3
	Total	29.01	3282	0.96
Middling	Middling 1 (jig)	8.59	1514	1.08
	Middling 2 (jig)	8.86	1462	1.19
	Middling (dry cleaning machine)	12.78	1486	1.22
	Total	30.23	1487	1.17
Reject	Dry cleaning machine	40.76	921	2.3
Raw coal	Total	100.00	1777	1.57

As can be seen from Table 7, the waste material produced by the heavy medium separation process using shallow-trough separators can be successfully processed by the coarse coal and small coal combined dry cleaning system. The clean coal product thus obtained (a yield of 29.01%) with a calorific value of 3282 Kcal/kg and a total sulfur of 0.96% is up to the requirement of power plant for use as fuel (Required standard: calorific value -- > 3000 Kcal/kg; total sulfur -- <1%). The middling product with a yield of 30.23% and a calorific value of 1487Kcal/kg can be used as brick-making material. The rejects with a yield of 40.76% and a calorific value of 921 Kcal/kg can be used as road building material or backfilling material in land reclamation.

6. Economic analysis

Total investment: 11.73 million yuan

    Investment per ton of coal: 4.01 yuan

    Power consumption per ton of coal treated: 3.00 kWh

    Operating cost per ton of refuse treated: 3.41 yuan

    Operating cost per ton of clean coal produced: 11.75 yuan

    For the breakdown of investment and operating costs, refer to Table 8 and Table 9.

Table 8: Breakdown of Capital Cost

No.	Cleaning method	Power (kW)	Price (10000 yuan)
1	Differential dry cleaning	1735	778.65
2	Small coal dry jigging	408	424.45
	Total	2143	1173

Table 9: Breakdown of Operating Cost

No.	Cleaning method	Cost (yuan/t)	Yield (relative to original fraction) (%)
1	Differential dry cleaning	2.85	100.0
2	Small coal dry jigging	1.35	41.46
	Total	3.41	

7. Conclusion

It proves feasible to treat discards of heavy medium separation system by using the coarse and small coal integrated dry cleaning process with intelligent differential-type coarse coal dry cleaning machine and dry cleaning jig to produce multiple products tailored to market needs. The use of the process can yield not only a favorable economic result but also such remarked social benefits as full utilization of coal resources, reduction of the cost incurred by the discharge of pollutants, and reduction of environmental pollution.

8. References

[1] Wang Zurui, The Prospect of the Application of Pneumatic Coal Cleaning in China, [J] Coal Preparation Technology, 1996(2):7-11

[2] Zhou Shaolei, Shan Zhongjian, Deng Xiaoyang et al, China's coal preparation industry – Present status and developing trend, China Coal, 2006.11, 11~14

[3] Wu Shiyu, Developmentt of China's coal preparation over the past 30 years, Coal Processing and Comprehensive Utilization, 2009, 01, 1~4

[4] Li Minghui, A general review of coal preparation and processing over the past 6 decades, Coal Engineering, 2014, 10, 24~29

[5] Ren Shangjin, Sun He, Ren Yandong, Design and Application of the Major Coal Dry Cleaning Machines Used in China, [J] Clean Coal Technology, 2012(5):1-5

[6] Xu Quanheng, Ren Shangjin, Yao Dasuo, Design and Application of the CFX-24x2 Differential Double-deck Pneumatic Cleaning Machine, [J] Coal Processing & Comprehensive Utilization, 2010(2):1-4

[7] Ren Shangjin Sun He,   New-generation Coal Dry Cleaning Equipment – Differential Dry Cleaning Machine, Symposium of the 17th International Coal Preparation Conference, Turkey, 2013

[8] Sun He, Ren Shangjin et al, Application of the TFX small coal dry cleaning jig [J], Coal Processing and Comprehensive Utilization, 2015 (7): 11~13

[9] Ren Shangjin Sun He Chen Jianzhong, Small Coal Dry Cleaning Jig, Symposium of the 18th International Coal Preparation Conference, Russia , 2016

[10] Ren Shangjin, Sun He, Xia Yucai, Chen Jianzhong, The TFX-9 small coal dry cleaning jig [J], Coal Processing and Comprehensive Utilization, 2017

5

Research on the fault diagnosis of coal preparation equipment based on artificial neural networks

PAN Yongtai[1], LI Zekui[1], ZHU Changyong[2], LIU Wenchang[3], LANG Jun[4, 5],
WEI Yinghua[2], YAO Fuqiang[6], LIU Zhen[6]

[1] School of Chemical and Environmental Engineering, Engineering Research Center for Mine
and Municipal Solid Waste Recycling, China University of Mining and Technology(Beijing),
Beijing,China;

[2] Shenhua Ningxia Coal Industry Group Co., Ltd Taixi CPP , Ningxia,China;

[3] Yangquan Coal Industry(Group) Co., Ltd, Shanxi, China;

[4] School of Chemical Engineering and Technology, China University of Mining and Technology ,
Jiangshu,China;

[5] National Engineering Research Center of Coal Preparation and Purification, Jiangshu ,China;

[6] Top Crusher Co., Ltd, Tianjin, China

Abstract: In the process of coal preparation, the abnormal sound signal of equipment often indicates the occurrence of equipment failure, and crusher is one of the equipment with the highest failure rate. This paper takes crusher for an example, the monitoring and diagnosis of the feed fault are studied. In the process of coal crushing, if the impurities like iron or wood are mixed in the feed, it will cause instantaneous and severe impact on the equipment, further the bending deformation of the roller, and even serious accidents. In this paper, the process of crushing single iron, wood and coal in the crusher cavity is regarded as the study subject. Collect the audio signals respectively and calculate the power spectrum by using MATLAB. Through the comparative analysis of the power spectrum curve, differences among signals are determined and feature quantities are extracted. Finally, the artificial neural network model is established according to feature quantities. Through the model verification, the correct recognition rate of iron, wood and coal reaches 80%, 65% and 60% respectively. It preliminarily proves the feasibility of the method and the rationality of feature quantities selection, and provides a novel idea for intelligent monitoring and fault diagnosis of coal preparation equipment.

Keywords: Coal preparation equipment; crusher; sound recognition; Artificial Neural Networks; fault diagnosis

1. Introduction

In the process of coal preparation, the equipment tends to make an abnormal sound when the failure occurs, and the crusher is one of the equipment with the highest failure rate in this process. Generally, the main feeding material of crusher is coal. However, due to the complexity and unpredictability of the composition of the raw ore, the feeding material tends to be mixed with impuritiess like wood or iron occasionally. Especially in the situation that the larger size iron enters the crushing cavity, the iron will jump between the crushing roller since it cannot be occluded effectively by the crushing teeth, which will cause instantaneous and severe impact on the equipment, and further the bending deformation of the roller, and even trigger a huge safety accident. Therefore, it is of great significance to monitor and diagnose the feeding fault of the crusher.

In this paper, mainly focus on the materials of coal, wood and iron as the study subject, firstly collect the audio signals in the process of material crushing, then analysis the power spectrum of each signal and compare the three differences and extract the characteristic. Finally, establish the artificial neural network model, by which it becomes possible to recognize the material types, and provides theoretical guidance for subsequent feeding fault diagnosis research.

2. Apparatus

The signal is the basis of equipment fault diagnosis, in order to obtain the characteristic signal in the crushing process, the corresponding test system should be established. Fig.1 is the layout diagram of the test system, which is composed of the crushing part, the sound test part and the data acquisition part.

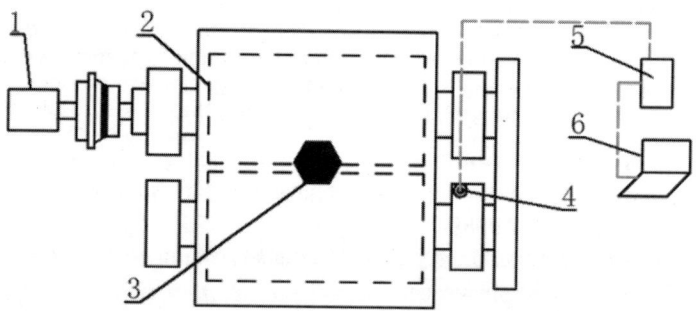

Motor 2-Teeth roller 3-Raw material 4-Sound sensor
5-Data acquisition instrument 6-Computer

Figure 1: Layout diagram of the test system

In the test system, the type of ZKB-II shear crusher is used, with the motor current frequency of 30Hz and motor speed of 864 r/min. The sound signal acquisition device contains an YSY5000 IEPE sound pressure sensor with a frequency response range of 20Hz-20KHz and a YSV8008 data acquisition instrument with a maximum sampling frequency of 51.2KHz.

3. Signal acquisition

Signal acquisition is an extraordinarily important step, which is related to the accuracy of data used in the subsequent processing of the system[1], Fig 2 shows the flow chart of a signal collection in the test process. As a fundamental test, the testing material is arranged with the single particle of coal, wood and iron respectively, and a total of 60 groups of data are collected.

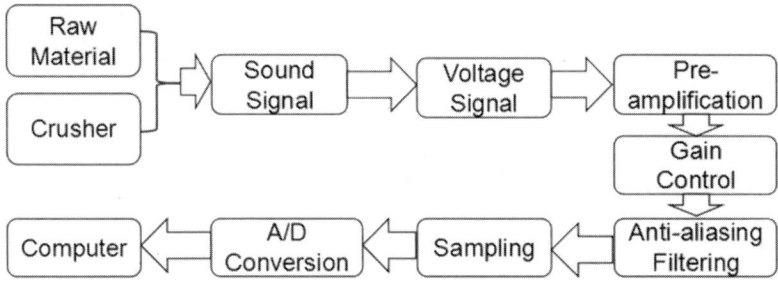

Figure 2: Flow chart of audio signal acquisition

To ensure that the sampling value can reflect the change rule of the waveform, it is required that the sampling frequency of waveform should be high enough. The higher the sampling frequency is, the more samples can be obtained per unit time and the higher the fidelity of the digital audio signal will be. However, the higher the sampling frequency is, the larger the data volume will be and the more storage space will be required [2].In this acquisition process, the sampling frequency is fixed at 10240Hz, which is appropriately lower than the highest sampling rate of the signal. In this way, on the one hand, the length of data to be processed will be as large as possible, and useful information in the data will not be lost; On the other hand, it will also improve the processing speed of data, making the system process faster and more accurate [3].

4. Signal processing and analysis

Matlab is one kind of computer language with powerful function, high efficiency, good interactive numerical calculation and visualization[4]. In this paper, Matlab will be used for the analysis in time domain and frequency domain.

In the time domain, spectrum subtraction is used in the noise reduction of the three audio signals. Spectrum subtraction is a method of speech noise reduction. Suppose the time series of the speech signal is x(n), and the speech signal at the ith frame which is obtained after the processing of windowing and framing is xi(m) and the frame length is N. After FFT transform, the average spectrum value of the leading noise segment and the spectrum value of each frame signal in the frequency domain are calculated, and subtract the average spectrum value of noise from the spectrum value of each frame, then get the spectrum value of noise reduction[5]. FIG.3 shows three audio signals and their noise reduction signals.

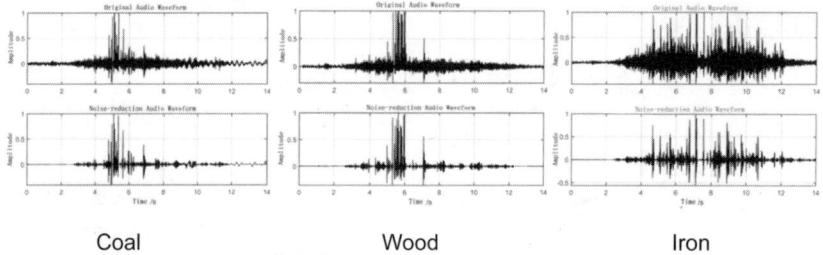

Coal   Wood   Iron

Figure 3: Three audio signals and noise reduction signals

As shown in Fig. 3, there are significant differences between the three audio signals after noise reduction. The audio amplitude and distribution of coal and wood are similar over time, with short-term energy concentration and locally prominent amplitude. The main reason is that coal and wood are brittle materials compared with iron, and they are more likely to be damaged and fracture, in which process the instantaneous crushing energy is more concentrated. In contrast, the distribution of the iron audio is more uniform with a certain periodicity, caused by the periodic impact from the continuous jump of the iron in the crushing chamber since the iron cannot be completely occluded. It can be seen that there are certain differences between the three in terms of time-domain characteristics. However, due to the limitations of time-domain analysis, in-depth analysis will not be conducted here.

In speech signal processing, signal analysis and processing in the frequency domain play an important role. The study of speech signals in the frequency domain makes some features that cannot be shown in the time domain greatly apparent[5]. Power spectrum analysis is one of the key technologies in digital signal processing, and it is an important statistic for describing random signals [2]. In this paper, the power spectrum analysis method in speech signal processing is used to analyze the audio signals of the material crushing in the frequency domain. As shown in Fig. 4, the power spectrum of audio signals of coal, wood, and iron are obtained by using Pwelch function of mean period diagram method.

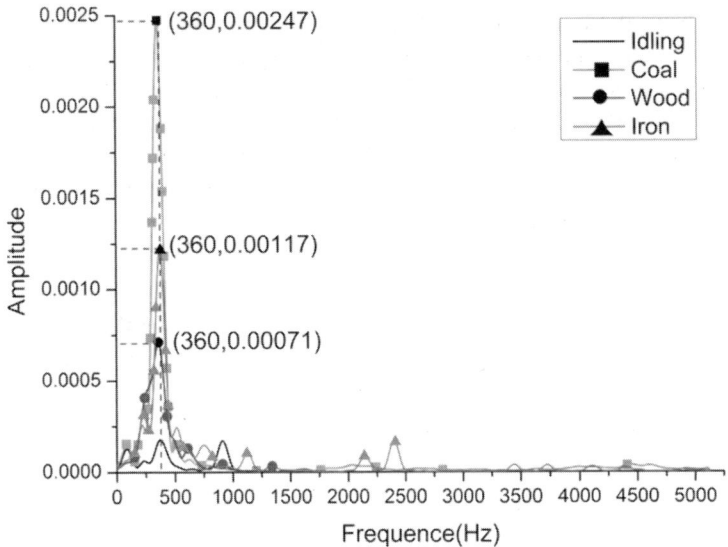

Figure 4: Power spectrum of three audio signals

It can be seen from the comparison in Fig 4 that the power spectrum curves of the three materials are all prominent in the low frequency band (0Hz-1000Hz), indicating that the spectrum component in this frequency band of the three signals is quite rich, and the frequency spectrum energy of the audio signal of the coal is significantly higher than that of iron and wood. In addition, the main frequency peaks of the three are mainly concentrated around 70Hz (1x frequency), 210Hz (3x frequency) and 360Hz (5x frequency), corresponding to the main peak of the no-load spectrum. It can be seen that in the low frequency stage, the frequency of material crushing audio shows up as the odd times of the frequency of the no-load equipment. Though the main frequency of every material is around 360Hz, the amplitude corresponding to this frequency varies greatly.

In the middle frequency band (1000Hz-3000Hz), the power spectrum amplitudes of coal and wood are evenly distributed, while the power spectrum amplitude of iron fluctuates more significantly than those of coal and wood, indicating that the audio frequency of crushing coal and wood is less distributed in this frequency band and lower contribution.

In the high frequency band (above 3000Hz), the spectrum amplitudes of the three are not obvious. In order to find out the difference among the three in this frequency band more accurately, the logarithmic amplitude of the power spectrum is adopted to eliminate the influence of the order of magnitude, and zoom in on the differences between the curves, as shown in Fig 5. As can be seen from the figure, the power spectrum curves of the three in the high frequency band fluctuate distinctively. Among them, the power spectrum curve of wood is similar to the no-load condition, with the smallest fluctuation, which is mostly made up of instantaneous small amplitude fluctuation. The power spectrum curve of iron fluctuates the most violently than the other two, with the continuous frequency peaks. In addition, for the coal, the fluctuation of the power spectrum curve is relatively gentle, with a longer period.

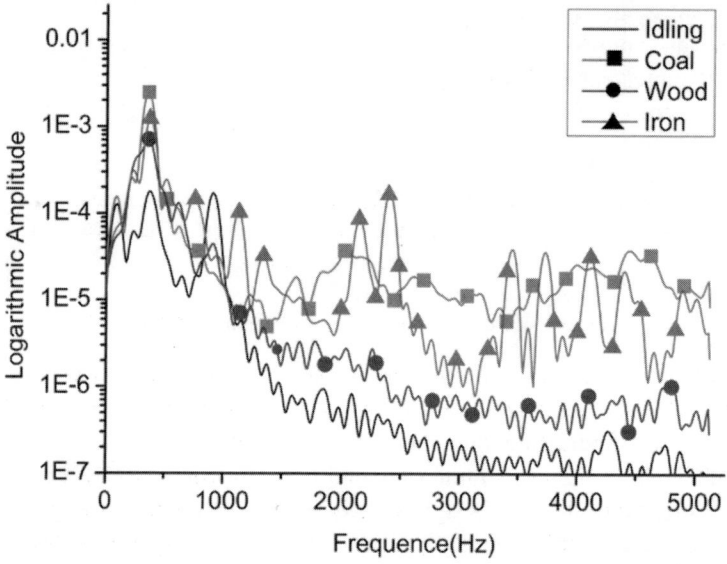

Figure 5: Logarithmic power spectrum of audio signals

According to the power spectrum analysis above, the power spectrum curves of the three materials are different in the low, middle and high frequency band. It is defined that the audio signal of coal crushing is normal signal, while the other two audio signals are fault signals, and their differences are taken as fault characteristics. In this paper, the amplitude corresponding to 360Hz in the low frequency band (0Hz-1000Hz) is taken as the characteristic quantity A, the frequency value corresponding to the wave peak of the amplitude in the middle frequency band (1000Hz-3000Hz) is taken as the characteristic quantity B, and the fluctuation value (standard deviation) of the logarithmic amplitude of the power spectrum of each material in the high frequency band (above 3000Hz) is taken as the characteristic quantity C.

5. Establish recognition model

Artificial Neural Network (ANN) is a nonlinear dynamic model established by simulating human nervous system based on the understanding of human brain structure and operation mechanism [6]. Because of its self-organization, self-learning, associative memory and other functions, it has been widely used in pattern recognition (including speech recognition and image recognition), optimization control and other aspects [7].

According to the fault characteristics above, sixty groups of audio signal data are trained, verified and identified, and an artificial neural network model is established. The distribution of characteristic values of 60 groups of audio signals can be seen in FIG. 6. Compared with the distribution of iron sample data, the distribution of coal and wood sample data is more concentrated and partly intersected, which predicts that the similarity of coal and wood characteristics is relatively high, and the recognition difficulty will be also higher than that of iron samples.

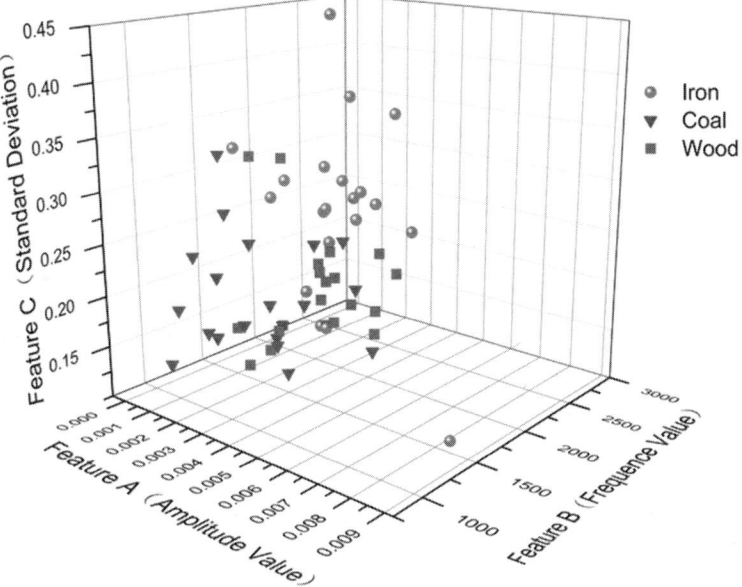

Figure 6: Three kinds of audio signals characteristic value distribution diagram

Generally, the forward network forms of the artificial neural network include BP network, perceptron, linear neural network, etc. BP network greatly improves its computing capacity by using error back-propagation method, and the network can realize the approximation of any continuous function by adjusting the relationship between neurons in the hidden layer[8]. Based on these advantages, in this paper, BP artificial neural network model is selected to recognize the feeding fault.

As shown in Fig. 7 is the model structure diagram that is established, the neural network adopts a 3-layer structure, with 3 nodes in the input layer, 10 nodes in the hidden layer and 3 nodes in the output layer. In addition, 42 groups (70%) of data are randomly used for model training, 9 groups (15%) of data are used for cross-validation, and 9 groups (15%) of data are used for model testing.

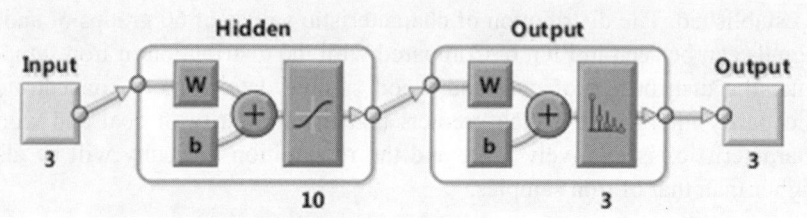

Figure 7: BP neural network structure diagram

6. Model training and result

BP algorithm[9-13] transforms the mapping problem of input and output of the neural network into a nonlinear optimization problem. In this processing, the gradient descent algorithm, the most common optimization method, is used[14]. Iterative operation is used to modify the network weight, by which they mean square deviation between the network output and the expected output can realize minimized[15]. As shown in Fig 8, in the process of network training, the gradient is still in the stage of continuous decline by the 16th iteration, and it is confirmed that the error curve of the sample will not decline for 6 consecutive iterations, also the target error meets the preset requirement, and finally, the network iteration stops.

In the process of iteration, the effective evaluation of network performance is very important. Cross entropy is one of the indicators to evaluate the network training performance. The smaller the value is, the smaller the loss between the model's recognition value and the real value will be, and the higher the similarity will be. In other words, the model generated by this algorithm is the closest to the optimal scheme. As shown in Fig 9, in the iterative process of network training, the cross-entropy values of the training curve, the test curve and the verification curve all keep decreasing, and the cross-entropy value of the test curve is lower than the verification curve value, indicating that the test performance of the network is better than the verification performance, and the test result of the network is effective. In the figure, the iteration number of times corresponding to the minimum entropy value is 10, and this point is the optimal network performance point, and the weight corresponding to this point is used to uniquely determine the parameter values of the network model.

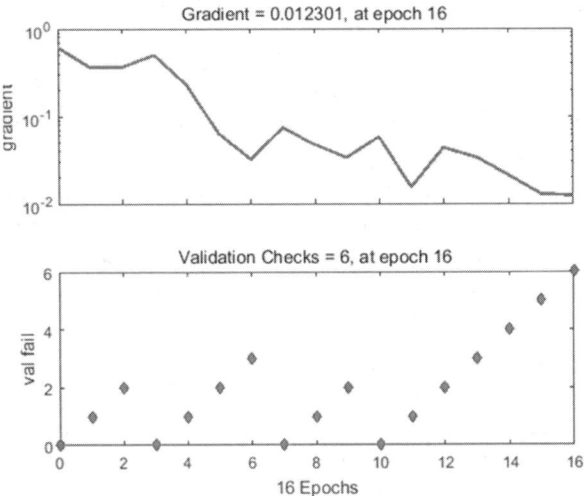

Figure 8: Gradient curve and validation check

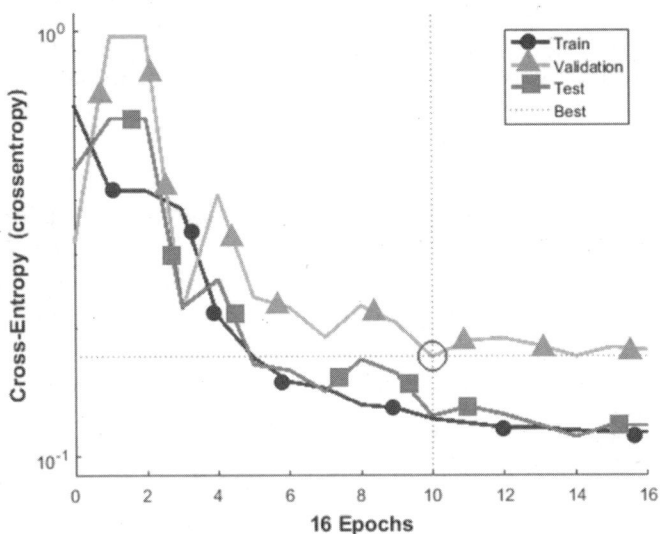

Figure 9: Model training performance diagram

Above, various parameters of the model are determined through multiple iterations of the minimum gradient method, and the accuracy of the model in recognizing sample data is an important criterion for judging if the model is good and reliable. For the analysis of recognition accuracy, the confusion matrix method is commonly used, which is the comprehensive embodiment of algorithm performance and test results in the recognition process through a specific matrix. As shown in Fig. 10 is the confusion matrix of this model. It can be obtained from the figure that the recognition accuracy of iron in the training matrix, verification matrix and test matrix are respectively 84.6%, 50% and 80%. Among the 20 samples, 1 sample is wrongly identified as wood and 3 samples are wrongly identified as coal. However, the identification results of wood and coal are relatively similar, 6 of the wood samples are wrongly identified as coal, while 5 of the coal samples are wrongly identified as wood. Based on the overall confusion matrix, the overall recognition accuracy of the model for iron, wood and coal reaches 80%, 65% and 60%, respectively.

Figure 10: Model confusion matrix diagram

7. Conclusion

(1) To some extent, the result indicates that it is feasible to recognize the iron through the difference of audio signals preliminarily, and the selection of the audio signal characteristic quantity is reasonable. However, due to the high sensitivity of audio signal and poor anti-interference ability to other external factors, the accuracy of iron recognition by audio signal still exists some errors, and other signals such as vibration signal can be added for auxiliary recognition to improve the accuracy and reliability of iron recognition.

(2) The recognition accuracy of wood and coal is obviously lower than that of iron, and misclassification between each other exists, which shows that the similarity of the two kinds of audio signal features is much higher, and the degree of differentiation is much lower. Therefore, it is necessary to study the differences of the audio signals between coal and wood, and optimize the method of the characteristic extracting in order to improve the recognition accuracy for coal and wood.

(3) For other coal preparation equipment, such as vibrating screen and cyclone, the audio recognition method in this paper can also be used to establish the corresponding artificial neural network model to monitor the abnormal sound signal of the equipment in the process of operation. Combined with the big data technology, the fault can be diagnosed online. And then the control system can intelligently adjust the rotation speed, positive and negative rotation, start and stop of the equipment, through which a closed loop control system of intelligent monitoring and diagnosis can be established. The system can not only ensure the safe operation of the equipment to the greatest extent, but also make great significance for the efficient production of the mine.

(4) In addition to the research content of this paper, there are still many factors to be studied in depth to adapt to the practical application in the industrial field, such as the expansion of data collection, the optimization of recognition characteristic quantity, the mixed feeding condition of multi-particles, and the change of material diversity.

8. Reference

[1] Dong Wei. Research on the Application of Feature Extraction and Feature Optimization in Vehicle Acoustic Classification [D]. North University of China. 2010

[2] Liu Bo. Research on vehicle acoustic characteristics analysis and vehicle recognition. [D]. Wuhan University of Technology, 2007

[3] Wei Hongfeng. The parameter model method of picking up the automobile audio signal information. [D]. Northeast Normal University, 2005.

[4] CHEN Jiayan, CHEN Dongjiao, ZHANG Daxiang. Collecting and Processing of Sound Signal with Matlab. [J]. JISUANJI YU XIANDAIHUA, 2005(06) : 91-92+96.

[5] Song Zhiyong. Matlab voice signal analysis and synthesis (Second Edition) [M]. Beijing : Beihang University Press, 2018 : 170, 21-22.

[6] Shen Shiyi.Neural Network System Theory and Its Application [M]. Beijing : Science Press, 2000, 1-20.

[7] Lou Shuntian, Shi Yang. System Analysis and Design Based on MATLAB-Neural Network. [M]. Xian: Xidian University Press, 1998.

[8] Werbos P J. The roots of back propagation [J]. NY : John Wiley & Sons, 1994.

[9] Ku Xiangchen Guo Yuefei Duan Mingde Cao Beibei. Predicting Tool Wear by Vibration Frequency Spectrum J. Machinery Design Manufacture 2017(10):113-116.

[10] Zhang Kaifeng Yuan Huiqun Nie Peng.Prediction of tool wear based on generalized dimensions and optimized BP neural network J. Journal of Northeastern University 2013 : 349, 1292-1295.

[11] Chen Chao Xu Jianlin Huang Jianlong. Tool condition monitoring system based on artificial neural network J.Chinese Journal of Mechanical Engineering 2002 : 388,135-138.

[12] Gao Hongli Xu Mingheng Fu Pan.Tool wear monitoring based on integrated neural networks J. Journal of Southwest Jiaotong University 2005 : 405, 641-653.

[13] Ghosh N. Ravi Y B. Patra A.Estimation of tool wear during CNC millingusing neural network-based sensor fusion J. Mechanical Systems and Signal Processing □ 2007 : 211, 466-479.

[14] Hecht_Nielsen R. Theory of the Back propagation. Proe. IEEE IJCNN, 1989, l : 593-605

[15] Zhang Rui. Research on the Mechanical Fault Diagnosis Technology Based on Artificial Neural Network Theory. [D]. Northeast Forestry University, 2001.

6

Revolution in coal preparation technology – GDRT coal gangue intelligent separation system

Liu Ping, Cao Jialiang

GDRT Mechanical and Electrical Technology (Beijing) Co., Ltd.

Abstract: The widely used coal preparation methods at present are mainly movable sieve jigging, heavy media shallow slot separation, and heavy media vertical wheel separation. Compared with the manual coal preparation method, these coal preparation methods all have disadvantages, such as high investment, complex process, high operation cost and difficulty in later water pollution treatment.

GDRT Coal Gangue Intelligent Separation System. This method makes use of the difference of attenuation of γ rays penetrating the coal gangue of a certain thickness. Through the computer calculation of the signals collected by the sensor, the material is then judged as coal or gangue, and meanwhile, instructions are given to hit the gangue identified, so as to realize the separation of coal and gangue. This system can also be combined with the existing system of a coal preparation plant to conduct pre-separation of gangue from coal in advance, which greatly reduces the cost of coal washing.

Keywords: GDRT; Coal gangue intelligent separation system; Gangue; Separation

1. Principle of GDRT coal gangue intelligent separation system

1.1 Schematic diagram of GDRT coal gangue intelligent separation system

The GDRT coal gangue intelligent separation system (hereinafter referred to as GDRT system) is mainly composed of five parts, namely, the feed system, queuing system, detection identification and control system, and execution system. There are several materials queuing channels on the belt, and each channel runs independently, which improves the operational reliability of the whole machine.

Working principle: After graded sieving of the raw coal, the raw coal lumps enter the feed hopper and are put in sequence under the action of the queuing mechanism. Then they enter several material queuing channels on the belt respectively. Each channel constitutes a separation system composed of an identification system which consists of a set of independent ray sensor, a pneumatic actuator, and a control system. When the material passes through the ray source and the sensor, the sensor amplifies the induced signal and transmits it to the control system. The processor in the control system compares the received signal with the mathematical model of gangue in the system to obtain a weighted value of the density of material passing through the ray. This value is compared with the threshold value set in the system. If the value is higher than the set threshold value, the material is judged as gangue, while if the value is lower than the set threshold value, the material is judged as coal. When the material is judged as gangue, the actuator changes the trajectory of the falling gangue with the high-pressure air injection valve in the process of material falling. If the attenuation per unit thickness of the measured object is less than the comparison threshold value, it is judged as coal, and the air valve does not operate, so that it is naturally thrown into the coal hopper.

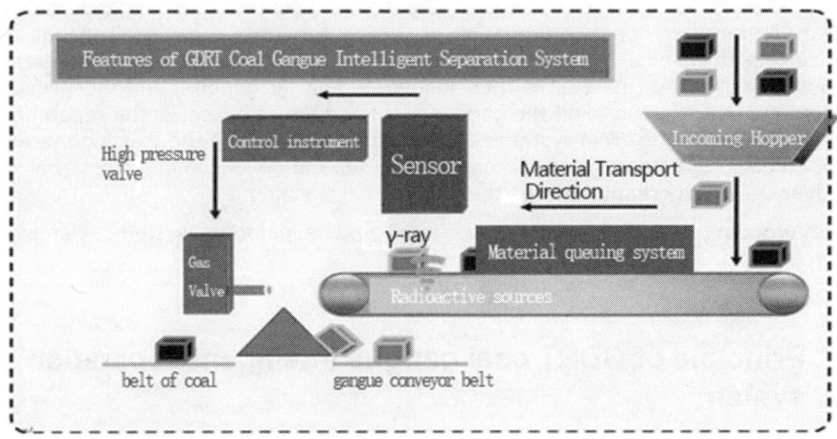

2. Features of GDRT coal gangue intelligent separation system

1. Green, no consumption of water and other media;
2. Low construction investment, small floor space, short construction period and low operation cost;
3. High degree of automation, simple operation and easy maintenance;
4. High separation efficiency, sorting accuracy of gangue over 90%, and the coal content in gangue after separation less than 1%;
5. Applicable to a wide variety of coal, and suitable for gangue removing from all kinds of coal;

6. Strong environmental adaptability, applicable to all regions, and showing obvious advantages in cold or water scarce regions;

3. Analysis of typical application case of GDRT coal gangue intelligent separation system

3.1 Take China guodian corporation pingzhuang liujia mine for example

Pingzhuang Liujia Coal Mine has two sets of GDRT coal gangue intelligent separation system, namely, 30 – 80 mm grain size separation system and 80 – 200 mm grain size separation system respectively. The monitoring data during commissioning acceptance are as follows:

3.2 Acceptance test data of liujia coal mine 30 – 80 separation system

Date	30~80 mm Coal content in gangue after separation	30~80 mm Gangue content in coal after separation	30~80 mm Gangue content in raw coal	30~80 mm Treatment capacity (t/h)
2016-6-30	2.42%	4.97%	30.90%	42t/h
2016-6-30	1.81%	5.41%	30.97%	51t/h
2016-7-5	1.34%	3.62%	33.81%	45t/h
2016-7-5	0.70%	2.58%	22.94%	42t/h
2016-7-27	0.54%	0.26%	85.58%	35t/h
Average Value	1.36%	3.37%	40.84%	

3.3 Acceptance test data of liujia coal mine 80 – 200 separation system

Date	80~200 mm Coal content in gangue after separation	80~200 mm Gangue content in coal after separation	80~200 mm Gangue content in raw coal	80~200 mm Treatment capacity (t/h)
2016-6-29	0.00%	0.22%	31.78%	78t/h
2016-6-29	0.00%	3.45%	29.66%	81t/h
2016-6-30	0.00%	1.54%	26.14%	78t/h
2016-6-30	1.19%	0.23%	34.50%	75t/h

Contd...

Contd...

Date	80~200 mm Coal content in gangue after separation	80~200 mm Gangue content in coal after separation	80~200 mm Gangue content in raw coal	80~200 mm Treatment capacity (t/h)
2016-7-1	0.00%	0.56%	32.91%	80t/h
2016-7-1	0.00%	0.86%	36.92%	77t/h
2016-7-5	1.34%	4.20%	77.74%	89t/h
2016-7-5	0.70%	3.73%	85.58%	87t/h
2016-7-27	0.54%	6.30%	92.84%	84t/h
Average Value	0.42%	2.34%	49.78%	

Based on the above data, the average coal content in gangue after separation with the GDRT coal gangue intelligent separation system in Pingzhuang Coal Mine is 0.42%, and the average gangue content in coal after separation is 2.34%, meeting the high-performance indexes. As mentioned earlier (Conley et al. 2017).

The GDRT coal gangue intelligent separation system has also been applied in the following coal mines, and has achieved great results in coal content in gangue and gangue content in coal:

In April 2015, successfully put into operation in Fuxin Mine Group Baiyinhua Coal Mine (700,000 tons of annual output of raw coal): GDRT-80/200-1.6, 1 set.

In May 2015, successfully put into operation in Rui'an Coal Mine (1,500,000 tons of annual output of raw coal) of China Guodian Corporation Inner Mongolia Pingzhuang Energy Co., Ltd.: GDRT-30/80-1.4, 1 set; GDRT-80/200-1.2, 1 set.

In July 2016, successfully put into operation in Fuxin Mine Group Baiyinhua Coal Mine (700,000 tons of annual output of raw coal): GDRT-80/200-2.0, 1 set.

In March 2017, successfully put into operation in Yuyang Coal Mine (7,000,000 tons of annual output of raw coal) of Chongqing Energy-Songzao C&E Co., Ltd.: GDRT-60/180-1.4, 1 set.

In March 2018, successfully put into operation in Baiyinhua Coal Mine (700,000 tons of annual output of raw coal) of Fuxin Mine Group: GDRT-80/200-2.0, 1 set.

In March 2018, successfully put into operation in Pingmei Group No. 5 Mine (1,800,000 tons of annual output of raw coal) in Pingdingshan, Henan Province: GDRT-80/200-1.2, 1 set; GDRT-30/80-1.2, 1 set.

4. The GDRT coal gangue intelligent separation system has complete independent intellectual property rights, and has been certified as "World Advanced Technology" by China national coal association

4.1 The GDRT coal gangue intelligent separation system is an original product developed with great effort by GDRT Electromechanical Technology (Beijing) Co., Ltd., and all intellectual property rights are owned by the company

1. The GDRT coal gangue intelligent separation technology has obtained two national invention patents, and the patent numbers respectively are:

2. A kind of Coal Gangue Automatic Separation Machine, Patent No.: 2009 1 0107879.9

3. A Queuing Mechanism, Apparatus and System for Solid Materials on the Conveyor Belt Patent No.: 2009 1 0107878.4

   The GDRT coal gangue intelligent separation technology has obtained the intellectual property rights of two national software works, whose copyright registration numbers respectively are:

   GDRT Separation Equipment Group Single Customer Data Receiving and Monitoring Service Software V1.0 Computer Software Copyright Registration Certificate No.: 2015SR188358

4. GDRT Separation Equipment Remote Data Acquisition Terminal Software V1.0 Computer Software Copyright Registration Certificate No.: 2015SR183159

5. GDRT Separation Equipment Group Customers Data Receiving and Monitoring Service Software V1.0 Computer Software Copyright Registration Certificate No.: 2015SR183213

6. GDRT Coal Gangue Intelligent Separation System Detection Software V1.0 Computer Software Copyright Registration Certificate No.: 2010SR019394

7. Control Software of GDRT Coal Gangue Intelligent Separation System with Temperature Correction V1.0 Computer Software Copyright Registration Certificate No.: 2010SR019380

8. GDRT Coal Gangue Intelligent Separation System Control Software V1.0 Computer Software Copyright Registration Certificate No.: 2010SR019381

9. Control Software of GDRT Coal Gangue Intelligent Separation System with Temperature Correction V1.0 Computer Software Copyright Registration Certificate No.: 2010SR019383

10. The GDRT Coal Gangue Intelligent Separation System passed the technical and scientific achievements appraisal of the State Scientific and Technological Commission on April 11, 2014.

11. Expert appraisal opinions and conclusions of "GDRT Coal Gangue Intelligent Separation System" are as follows (with the scanned copies of the expert opinions and appraisal certificates attached at the end of this section)

Entrusted by China National Coal Association, China Coal Processing and Utilization Association organized relevant experts to appraise the "GDRT Coal Gangue Intelligent Separation System" developed by GDRT Electromechanical Technology (Beijing) Co., Ltd. (hereinafter referred to as GDRT) in Beijing on April 11, 2014. After listening to the introduction of "GDRT Coal Gangue Intelligent Separation System" by the enterprise, the experts reviewed the submitted materials and formed the following appraisal opinions after inquiry and discussion:

I. The "GDRT Coal Gangue Intelligent Separation System" is developed with a correct technical route, complete documents and data, and is qualified for technical appraisal.

II. The system uses nuclear physics and computer control technology, and creatively adopts the modified type single photon radiation detection technology with compound sensors and decentralized dynamic line contact mechanical queuing technology. After the identification and separation of 30 – 300 blocks of raw coal, the gangue sorting accuracy reaches 90%, and the coal content in gangue after separation is less than 3%.

III. The system is suitable for gangue separation of various types of raw coal. It has such features: low construction investment, small floor space, short construction period; no consumption of water resources and other media; green, simple operation, convenient maintenance, and low operation cost; strong environmental adaptability.

IV. The system has been used in many coal mines and achieved great economic and social benefits.

The "GDRT Coal Gangue Intelligent Separation System" has obtained two patents for invention, as well as two patents for utility models. Using gamma-ray for dry sorting is the first of its kind in China, and the technology has also reached the international advanced level.

Suggestions: Further improve the processing capacity and expand the range of application.

5. Comparative analysis of several coal preparation processes

Comparative analysis table of several coal preparation processes

Treatment process	Heavy media shallow slot	Jigging	Air separation	X-ray identification technology	γ-ray identification technology
Feed system	20-200	15-150	0-50	Distributing system, without queuing, prone to incorrect hit	Queuing for identification one by one, with an identification accuracy of 100%
Identification principle	Suspension with a density between coal's and gangue's is used as a medium for separation	Stratification is realized according to the density under the action of vertically pulsating water flow	Materials of different densities are projected to different distances under the action of wind	Essentially image identification with the X-ray High requirements for raw materials, and the identification accuracy subject to the influence of water and other factors	The attenuation difference of the γ-ray after penetrating materials only related to the molecular structure, with no requirement for raw materials; point to point identification with high accuracy

Contd...

Contd...

Treatment process	Heavy media shallow slot	Jigging	Air separation	X-ray identification technology	γ-ray identification technology
Identification device	Swirler	Jigging machine	Air dry coal separator	High voltage power supply vacuum tube with high heat generation; use a forced cooling system with a complex structure; short service life and heavy maintenance under continuous high temperature and dust environment	Radiation sources and radiation source sensors, simple structure, unchanged for 20 years, no requirements for environment, and almost zero maintenance
Separation system	Swirler	Jigging machine	Air dry coal separator	Matrix high-pressure nozzle, blowing from the bottom up, materials drifting over the air nozzle, easy to break the air nozzle, upward air nozzle prone to dust jam	High-pressure nozzle set opposite to the belt conveyor headpiece to blow downward, no jam or hit to the air nozzle
Equipment investment 200 t/h	3000	200	300	10 million	6 million
Applicable environments	Not applicable to high cold and water scarce regions	Not applicable to high cold and water scarce regions	Not applicable to high cold and high humid regions	All	All
Applicable coal types	Not applicable to lignite and easy-to-slime coal	Not applicable to lignite and easy-to-slime coal	Not applicable to coal of high water content	Not applicable to coal of high water content or gangue coal	All kinds of coal
Separation effects	Gangue sorting accuracy 70%, coal content in gangue 1%, and gangue content in coal 2%	Gangue sorting accuracy 70%, coal content in gangue 1%, and gangue content in coal 3%	Coal content in gangue 1%, and gangue content in coal 4%	Gangue sorting accuracy about 70% Coal content in gangue about 3% Gangue content in coal about 10%	Gangue sorting accuracy ▢90% Coal content in gangue ≤1% Gangue content in coal ≤ 1 – 3%
Operation cost (RMB)	5	10	4	10~15	1~2
Applicable grain size (mm)	13~150	<80	<50	30~200	30~400

6. Conclusion

After a comprehensive analysis of the principle and structure of the GDRT coal gangue intelligent separation system, and comprehensive analysis and comparison of several common coal preparation processes, the expert opinions in the technical appraisal report of China National Coal Association are used here as the conclusion of this paper:

1. The GDRT coal gangue intelligent separation system is suitable for gangue separation of various types of raw coal lumps. It has such features: low construction investment, small floor space, short construction period; no consumption of water resources and other media; green, simple operation, convenient maintenance, and low operation cost; strong environmental adaptability.

2. The system has been used in many coal mines and achieved great economic and social benefits.

3. The "GDRT Coal Gangue Intelligent Separation System" has obtained two patents for invention, as well as two patents for utility models. Using gamma-ray for dry sorting is first of its kind in China, and the technology has also reached the international advanced level.

Suggestions: Further improve the processing capacity and expand the range of application.

References: GDRT Coal Gangue Intelligent Separation System (Coal processing and comprehensive utilization 2017 Phase III Kong LI)

7

Status and prospect of coking coal washing and processing in China

Niuxi Guo, Hong Wang, Ran Chen, Huaqiong Ding, Wei Zhou, Nengjin Tao

CCTEG Beijing Huayu Engineering Co., Ltd., Beijing, China

Abstract: The general situation of coking coal resource in China is introduced. Based on the analysis of the current situation of coking coal such as the coal preparation process, equipment, auxiliary facilities, operation, management, etc. The application effect of new equipment in coal washing and processing is introduced, and the development trend of coking coal washing and processing is prospected. The continual innovation is proposed to improve the level of coal processing and utilization, which makes coking coal clean production and efficient utilization.

Keywords: Coking coal, washing and processing, current situation, prospect

1. General situation of coking coal in China

1.1 Scare coking coal resource and unbalanced distribution

Based on classification criterions of coal in China, meager lean coal, lean coal, primary coking coal, fat coal, 1/3 coking coal, gas-fat coal, gas coal and 1/2 middle viscous coal in bituminous coal belongs to coking coal. The coal type of coking coal is comprehensive, but the reserves of coking coal are less. The proven resource reserves of coking coal are only 275,800 Mt that takes up 27% of the total coal resource in China. The reserves of economic minable coking coal are only 64,600 Mt that cannot achieve 25% of total reserves of coking coal in China. The resource distribution of coking coal in China is uneven; it is mainly distributed in North China, Southwest, and Northeast China. The main resource of coking coal in China is distributed in Shanxi province, Hebei province, Henan province, Anhui province and Shandong province, the checked coal production capacity of above provinces takes up 57.3% of coking coal output in China (The main coking coal mine fields and their characters in China is shown in Table 1-1). Shanxi province is the big province for coal output, the checked coking coal production capacity is about 1/3 of total checked coking coal production capacity in China, and the main coal type of coking coal is primary coking coal. The coal type structure of coking coal in China is unreasonable, lean coal and merger lean coal take up 15.89%, primary coking coal takes up 23.61%, fat coal and gas-fat coal take up 12.81%, gas coal and 1/3 coking coal take up 45.73%, the resource reserves of main coking coal for fat coal, primary coking coal are less. The coal quality and washability of coking coal in China are poor, difficult and extreme difficult washed coal takes up 62%. Generally, coking coal belongs to middle ash and sulfur coal, the percent of high sulfur coking coal and fat coal exceed 1/3.

1.2 Application and market requirements of coking coal

Coking coal is mainly used for smelting coke. Coke can be classified as metallurgical coke (including blast furnace coke, foundry coke, iron alloy coke, non-ferrous metal metallurgy coke), gasification coke and calcium carbide coke by use. The quality requirements of metallurgical coke are always higher by large capacity blast furnace and intensified operation of the blast furnace, the general requirements should be $Mt \leqq 6\%$, $Ad \leqq 6\%$, $S:0.4\% \sim 0.6\%$. The reactivity of gasification coke should be well and its ash fusion point should be higher. The fixed carbon content of calcium carbide coke should be increased. But in China, high quality coking coal is poor and limited. As the raw material of coke production, the quality of coking coal has a direct impact on heavy industry for steel, metallurgy, chemical industry, non-ferrous metals and machinery manufacturing, it also has important for long-term development of state industry system and nation lives. Thus, reasonable development, utilization and protection of coking coal must be researched.

Table 1: The Main Coking Coal Mine Fields and their Characters in China

Province	Mine Field	County or City	Proven Reserves (100 Mt)	Coal Type	Ash Content of Raw Coal Ad (%)	Sulfur Content of Raw Coal St.d (%)
Shanxi	Liliu	Lin County, Lishi City, Liulin County	203.1	1/3 coking, fat, coking, lean coal	19.01-29.95	0.48-2.92
	Xiangning	Xiangning County, Gu County, Pu County	171.3	Fat, coking, lean coal	19.34-29.49	0.49-5.97
	Xishan	Gujiao City, Jiaocheng County, Qingxu County	185.3	Fat, coking, lean, meager lean coal	19.99-32.09	0.51-2.83
	Huozhou	Hongdong, Lin County, Huozhou City	266.5	1/3 coking, fat, coking, lean, meager lean coal	13.43-32.51	0.35-2.86
	Huodong	Qinyuan, Gu County	91.2	Coking, lean, meager lean coal	12.99-32.33	0.41-2.73
Shandong	Juye	Juye, Liangshan, Yuncheng City, Heze County	64	1/3 coking, fat, gas coal	13.13-15.57	0.54-4.06
	Yanzhou	Yanzhou, Zou County	33	Fat, coking, lean, meager lean coal	12.00-23.96	0.55-3.58

Contd...

Contd...

Province	Mine Field	County or City	Proven Reserves (100 Mt)	Coal Type	Ash Content of Raw Coal Ad (%)	Sulfur Content of Raw Coal St.d (%)
Anhui	Huaibei	Xiao County, Guoyang County, Huaibei, Suzhou, Haozhou City	98.4	Gas, 1/3 coking, fat, coking, lean coal	6.00-39.45	0.10-6.74
Hebei	Handan	Handan City, Xingtai City	53	Fat, coking, lean, meager lean coal	14.50-28.06	0.46-2.51
	Kailuan	Tangshan City	66	Gas, 1/3 coking, fat, primary coking coal	11.85-23.94	0.51-3.68
Henan	Pingdingshan	Pingdingshan, Xucang, Ruzhou, Xiang County, Ruyang County	75	Gas, 1/3 coking, fat, primary coking coal	8.72-35.50	0.24-7.58
Guizhou	Panjiang	Pan County	102	Fat, 1/3 coking, gas, coking, lean coal	18.92-27.73	0.22-3.37
	Shuicheng	Shuicheng County	113	Gas, 1/3 coking, fat, coking, lean coal	15.0-25.0	1.0-4.5
Heilongjiang	Qitaihe	Qitaihe City	11.5	1/3 coking, coking, lean coal	20-30	0.27-0.50
	Jixi	Jixi City, Jidong County, Muling County	25.5	1/3 coking, primary coking coal	17.0-36.0	0.40-0.80
Yunnan	Enhong, Qingyun	Fuyuan County	19.4	1/3 coking, primary coking, lean coal	16.02-26.20	0.19-3.65

1.3 Poor quality trend for coking coal quality, protective development and utilization

Since the 12th Five-Year Plan, the coal industry has entered into a period of fast development in China. Because of long term mining and utilization, low sulfur and good quality coking coal resources are limited. The recovery of the mining area and coking coal output can be significantly improved by coal mining technology application of reject filling to replace coal. Lots of coking coal mines have been mined to deep section, mechanical mining, fully-mechanized coal mining and top caving mining technology has been used, so the ash content, reject content and sulfur content of raw coal are increased. The quality of product coking coal can be improved and steady by enhancing coking coal washing and processing. From 2009 to 2016, the ash content of clean coal for coking coal was about 9.5% that can meet the requirements of the metallurgical industry fully.

Because of limited process technology and equipment, 15% of coking coal is high ash content, high sulfur, primary coking and fat coal that is difficult to wash and use in China. 200 Mt/a of coking coal is preliminarily estimated to use as normal thermal coal. Thus, a lot of valuable coking coal is wasted.

In order to guarantee the future stable supply of coking coal, coking coal resource should be strictly exploited in protection, high precision washing and processing should be developed rapidly, coking coal resource should be fully used and inefficient utilization of poor resource should be limited under the situation of poor, uneven distributed and over exploited coking coal resource. Meanwhile, in order to relieve the shortage of primary coking coal and decrease coking cost in China, optimized coal blending structure technology should be promoted. In 2010, established Classification and Utilization for Scarce and Special Coal Resources, fat coal, primary coking coal and lean coal were classified as scarce coking coal that should be mined in protection and used as prior purpose; scarce coking coal must be washed totally to increase resource utilization. Meanwhile, in the 5th chapter of The 13th Five-Year Plan for Coal, high precision coal washing and processing should be developed rapidly for realizing coal quality upgrading and classification.

2. Status of existing coking coal washing and processing

During Coal Golden Decade (2002 - 2012) in China, coal production and consumption were increased rapidly, as the basis and leading edge of clean coal technology; coal washing was also entered into great development period. Up to 2016, the washed capacity of raw coal was 2,600 Mt, the washed capacity of coking coal washeries was 1,000 Mt, treatment capacity of the single system achieved 6.0 Mt/a. The total washed rate was 68.9% which was the level of developed countries. More than 2000 various types of coal washeries have been built in China and a lot of coal washeries have advanced coal washing technology and equipment in the world. There are 75 oversize coal washeries, their capacities are more than 10.0 Mt/a, designed washed capacity of raw coal exceeds 1,100 Mt that takes up 42% of total washed capacity in China; 11 coal washeries are coking coal washeries, their washed capacity is 145 Mt. For example, Huaibei Linhuan Coal Washery, Huaibei Wobei Coal Washery, Pingdingshan Mine Tianzhuang Coal Washery, Shanxi Huatai Coal Washery, Shandong Longgu Coal Washery, they belong to more than 10.0 Mt/an oversize coking coal washeries (detailed information is shown in Table 2-1). The washed capacity of Huaibei Linhuan Coal Washery is 16.0 Mt/a, which is the largest scale of centralized coking coal washery in Asia.

Above mentioned oversize coking coal washeries were designed by CCTEG Beijing Huayu Engineering Co., Ltd., they have a high capacity, simple and efficient technology process, large scale equipment selection, advanced index control, rational construction investment.

Table 2: The General Situation of Large scale Coal Washeries after 2000

Title of Coal Washery	Capacity (Mt/a)	Coal Type	Technology Process
Huaibei Linhuan Coal Washery	16.0	Fat, primary coking, 1/3 coking, gas coal	50–0.5 mm: non-pressure three-product HM cyclone separation Coarse slime: coal slime HM separation Fine slime: desliming flotation
Huaibei Wobei Coal Washery	12.0	Primary coking, fat, lean, 1/3 coking coal	50–0.5 mm: non-pressure three-product HM cyclone separation Coarse slime: coal slime HM separation (reserved) Fine slime: desliming flotation
Pingdingshan Mine Tianzhuang Coal Washery (Expanding)	10.0	Gas, fat, primary coking, 1/3 coking, lean coal	300–20 mm: inclined wheel separator for primary separation, shallow groove HM separator for secondary separation 20–1 mm: pressure three-product HM cyclone separation 1–0.25 mm coarse slime: TBS separation -0.25mm fine slime: flotation
Shanxi Huatai Coal Washery	10.0	Primary coking, lean coal	200–30 mm: shallow groove HM reject pre-discharge, crushing to –50 mm 50–1 mm: non-pressure three-product HM cyclone separation 1–0.25 mm: TBS separation Fine slime: flotation
Shandong Longgu Coal Washery	10.0	Fat, 1/3 coking, less gas coal	300-50mm: reject discharge by vibrating screen, crushing to -50mm 50–1mm: pressure two-product HM cyclone for primary and secondary separation 1–0.25 mm: TBS separation –0.25 mm fine slime: flotation

2.1 Advanced and improved washing and processing technology, application for new technology and process

Mechanical reject pre-discharge is set for raw coal, which has become the trend. The reject content in raw coal is increased with the promotion of mechanized coal mining operation and underground manless working face in China, so reject hand picking cannot fit for the requirements of development for modern coal washery. HM bath and jigging have to be used as reject removal of washed coal to decrease primary separation system selection load and processing cost of coal washing. For reject sliming, HM bath can be used to primary discharge the reject which decreases the impact on the stability of suspension of primary cyclone process, the operation cost of coal slurry circulating system and medium consumption.

Coal washing and processing has developed in the past several decades, the jigging mode was replaced by HM mode gradually, now production process is advanced and high efficient, settings of process links are mature. At present, -50mm whole size coal is washed after crushing in coking coal washeries. The main technology process of designed coking coal washeries is desliming gravity flow three-product HM cyclone + TBS + flotation, the auxiliary process is non-disliming three-product HM cyclone + slime HM process + flotation. Desliming separation and HM separation process have low medium and energy consumption, less systemic circulation, the low load of medium recycling system and good flexibility. At this stage, the coal washing process is mature, reliable and flexible. The appropriate technological process should be selected as size composition, specific gravity consist, washability and slime content of raw coal.

Desliming Gravity Flow Three-product HM Cyclone + TBS + Flotation

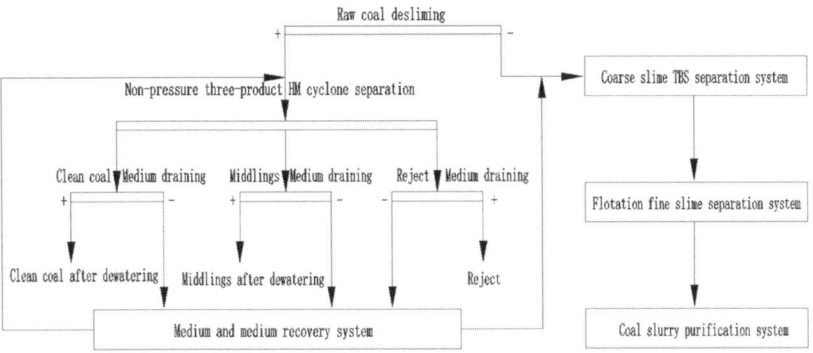

Figure 1: Technological Flow Chart for Desliming

Non-desliming Gravity Flow Three-product HM Cyclone + Coal Slime HM + Flotation

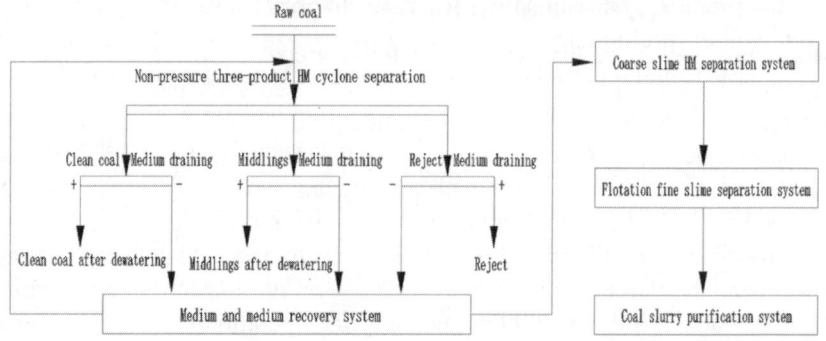

Figure 2: Technological Flow Chart for Non-desliming

In order to recover clean coal for coking as soon as possible, the coarse slime (1.5 – 0.25 mm) separation technology have been widely applied. Coarse slime separation process mainly shows in the equipment selection of slime HM cyclone, TBS and spiral separator. Coarse slime separation technology is selected in accordance with content, washability and separation size range of coarse slime. Coal slime HM cyclone has high energy consumption, complex system, high cost and large capacity. The medium of TBS for coarse slime separation is water, so the TBS has no medium system, its system is simple, power consumption is low and production cost is less. The size range of raw coal separation is finer by coarse slime separation technology, so coking coal can be separated deeply and efficiently.

The recovery rate of scarce coal resource can be increased by middlings rewashing technology process. At present, middlings from separation is almost used as steam coal in China; it is waste for scarce coal type. A part of middlings can be crushed (dissociation between clean coal and gangue) again and rewashed, clean coal can be fully recovered.

In recent years, rewashing technology of flotation tailings has been developed and applied. Generally, rewashing technology of flotation tailings has three types:

(1) traditional one stage of flotation technology process, flotation tailings enters into rewashing flotation machine for flotation;

(2) spiral separation + two stages of flotation technology process, flotation tailings directly enters into classifying cyclone for classification, underflow of classifying cyclone enters into spiral separator for separation, overflow of classifying cyclone enters into rewashing flotation machine for flotation;

(3) water medium cyclone + three stages of flotation technology process, flotation tailings enters into 0.15mm classifying screen after preliminary reject discharge by reject discharge cyclone, +0.15mm materials enter into four stages of water medium cyclone for separation, then clean coal, middlings and tailings can be separated; -0.15mm materials can be separated by three stages of flotation technology process, then clean coal, middlings and tailings can be separated. Hydrocyclone + three stages of flotation technology process has wide application range and good separation effect. In order to increase clean coal recovery and economic benefit, rewashing technology of flotation tailings is used by more coking coal washeries.

In order to make better use of coal resource, coal storage and blending technology has been developed rapidly. On the basis of the requirements of coal quality by the client, different kinds of coal can be blended as a proper ratio to achieve the requirements of the client. In Coal Logistics Development Planning, railway transport capacity will be 3,000 Mt, 11 large scale coal storage and blending bases will be built, 30 coal logistics parks (20 Mt level) will be built, some coal logistics enterprises and coal trade markets will also be developed by 2020. At present, Wuhu Port Coal Storage and Blending Center, Shenhua Ningxia Coal Co., Coal Indirect Liquefaction Project Coal Storage and Transport Devices, Jiangxi Coal Storage Center Jiujiang Coal Storage and Blending Project, Inner Mongolia Yitai Group Huaidong Railway Co. Jungar Banner Bulking Station and Yubei Coal Storage and Blending Base are designed by CCTEG Beijing Huayu Engineering Co., Ltd. The construction capacities of above coal storage and blending bases are more than 10.0 Mt/a, static reserves are above 0.6 Mt. The capacity of Wuhu Port Coal Storage and Blending Center I Phase is 20.0 Mt/a, its static reserves are 1.8 Mt; the capacity of Wuhu Port Coal Storage and Blending Center II Phase is 50.0 Mt/a, its static reserves are 5.0 Mt/a. According to characteristics of feed coal and requirements of product coal by the client, coal blending expert system can supply coal blending schemes, simulation verification, scheme selection and automatic operation. Because of limited dynamic measurement and online testing instruments, coal storage and blending technology is always used for steam coal. Coal blending technology for coking coal can guarantee the coke quality, make better use of coal resource, save good quality coking coal. Now Huaibei Mining Industry Group Linhuan and Wobei Coal Washeries have used the technology.

2.2 Increased large scale and domestic coal washing equipment, extensive use of large scale, high efficient and reliable equipment

Coal washing equipment in China has been passed long term development which has various types and more range of selection. The width of common vibrating screen has been increased to 4.2 – 4.9 m from 3.0 m, flip-flow screen has been introduced in past years. The maximum width of HM bath is 8.0 m. The diameter of small separation HM cyclone has achieved 1.6 m. The diameter of basket for smalls centrifuge has been increased to 1.5 – 1.7 m from 1.4 m. The diameter of TBS has been achieved 3.6 m. Single groove volume of mechanical agitation flotation machine has been achieved 120 m³. The filtration area of full automatic plate type filter press had been exceeded 1050 m². The filtration area of hyperbaric filter has been achieved 200 m². The surface magnetic field intensity of wet magnetic separator has been achieved 180 – 600 mT, recovery ration of magnetic medium has been achieved 99.9%. Chinese coal washing equipment has upsizing, domestication that is high efficient and reliable. Major diameter HM cyclone, large scale coal flotation machine, hyperbaric filter, filter press and pumps are used commonly in more coal Washeries in China.

2.3 Safe, environmental protective, large scale and new type supporting facilities and materials, socialized coal washery operation and maintenance in the world

Besides separation equipment, supporting facilities have been developed for safety, environmental protection and upsizing. The capacity of semi-underground groove bunker can be achieved 0.2 Mt. The total capacity of single closed coal storage have been achieved 0.5 Mt, the diameter of round coal storage can be achieved 120 m, the span of strip type coal storage can be achieved 180 m. The maximum diameter of raw coal and product coal silo is Φ45m, the capacity of single silo is 50,000 t, the silo is completely closed, so it is better for environmental protection and accurate coal blending, this storage mode is always used for coking coal storage.

New type wear-resistant materials are researched and applied. For example, lined ceramics wear-resistant materials are used for cyclone and pipelines; high-chromium alloy is used for overflow parts of slurry pump. Thus, the service life of equipment will be extended and the overall reliability of coal washeries will be improved.

The construction period of coal washeries has been decreased substantially by EPC mode. Modern service industry for coal washery intelligentialize and informatization, hosting operation and maintenance socialization has been promoted and applied. Chinese coal separation process technology and equipment go to the world step by step, the influence of Chinese coal washing and processing has been increased rapidly in the world.

2.4 Existing problems

Most of coking coal washeries are subject to limitation of technology and capital during construction period, so their process system is complex, processing capacity of single system is low (generally less than 2.0 Mt/a), the situation should be improved. Although most of equipment in coal washeries is domestic equipment, key parts of large scale screen and crusher are imported components to guarantee reliability of equipment and technology process. The research, processing and manufacturing level of Chinese equipment has a certain gas with imported equipment that should be improved.

Centralized control and automation level of coal washeries in China have been developed rapidly, automation and informatization are applied in most of coal washeries, for example, unattended operation for receiving coal pit, automatic bunkering for raw coal and product, anti-blocking for raw coal bunker, automation for sewage treatment station, automatic supply for medium, train and truck rapid loading station, etc. But overall automation level of coal washery can be improved step by step, connectivity, automatic diagnosis, unattended intelligentialize and wisdom will be further development direction for coal washeries in China.

3. Development trend and prospect for coking coal washing and processing

To improve social and economic benefits of enterprises, market competitiveness of product, domestic and overseas advanced technology, design concept, high efficient coal washing method, production process and technical equipment should be used. The development tendency of coking coal washing and processing will be energy consumption reduction, resource saving, green mines, clean production and efficient utilization.

3.1 Protective development and increased washing proportion for coking coal resource

Coal is non-renewable primary energy, coking coal should be exploited in protection because of scarcity. Coking coal must be used in cleaning after washing desulfuration and deceasing ash content, the ultimate goal is 100%

washing. The washing proportion of coking coal in developed countries is 80% - 100%, but the requested washing proportion in China should be more than 75% during the 13th Five-Year period. As checking process of coking raw material, coking coal washing and processing should obey the principle of maximum recovery rate under guaranteeing quality of clean coal. Thus, coking coal resource should be exploited in protection, recovery ratio of mine and washing proportion of raw coal should be increased. In addition, in order to protect coking coal resource in China, good quality imported coking coal should be encouraged to import by the government.

3.2 Development direction for large scale, high efficient, simplified, reliable, intelligentized coking coal washing, future development trend for unattended operation intelligentized coal washery

The future development direction of coking coal washing will be "Large scale, high efficient, simplification, reliability, intelligentialize", the management level of new or existing coal washeries should be improved continuously, advanced management concept for "unattended operation and someone inspection" will be the development tendency for future intelligent coal washery.

3.2.1 Technical renewal transformation for old coal washeries

Existing coal washing facilities should be upgraded and reconstructed, technological process should be improved, design and construction mode should be reformed. System management platform for ORACEL ERP should be set to improve operation management level, reduce processing cost and increase economic benefits of enterprises.

3.2.2 Promoting merger for automation and informatization, improving construction level for mechanization, automation, informatization and intelligentialize

The development direction of coal washeries: increase efficiency by downsizing payrolls can be realized by mechanization; operating environment can be improved by automation; management efficiency can be improved by informatization; competitiveness can be improved by intelligentialize. Information technology should be merged into design and management of coal washery to promote digital control, automatic production and remote operational capacity of coal washing equipment.

3.2.3 Using advanced design method and construction mode to improve construction speed and quality of coal washeries

EPC mode is the normal state of coal washery construction, investment and construction mode (BOT, BOOT, etc.) will be the development tendency. Digitization engineering design (BIM) mode and ERP management system for enterprise construction will also be the development tendency for design. During the process of coal washery design, digitization engineering is used for completing intelligent 3D modeling, collision check, reinforcement and steel structure design and realizing 2D drawing, material statement and factory real time roam, etc. to improve design quality. The design method of digitization with modularization can make the standard connection design for non-standard equipment and civil structure, thus most of components can be prefabricated in factory and installed rapidly on site. Financing construction has comprehensive advantages for professional technology service, system management and financing that can save investment and energy consumption as soon as possible. Design is the basis of enterprise establish, integrative ERP information platform can be set for optimizing organization structure, improving management system, formulating the enterprise standard and business process.

3.3 Promoting technology upgrading of coal industry, improving further research for dry separation, coal blending preparation, high sulfur coal washing

In China, the reserves of coking coal resource are less, the types are more, and coal quality is poor, so coking coal should be exploited in protection and then washed and processed for the purpose of clean utilization. In order to simplify process, increase production efficiency, decrease production cost and increase clean coal productivity, the technology process should be confirmed by washability, size composition, specific gravity consist, floatability, sulfur content, occurrence characteristics and reject lithology of raw coal. Meanwhile, coal technology upgrading should be also promoted constantly.

3.3.1 Advancing continuous innovation of coal storage and blending technology, optimizing coal blending process, improving coal blending expert system, setting coal resource allocation system

In China, reserve areas of coking coal are uneven distribution. Meanwhile, coking coal is scarce resource. Because of different requirements for coking coal quality in metallurgy, chemical engineering industry, related data collection should be implemented and resource database for mine field –

district – nation should be established. Thus, overall planning can be made by central enterprises in accordance with database, related national functional department can develop the resource.

3.3.2 Advancing the use proportion for dry coal washing

FGX combined type dry coal washing equipment, TDS intelligent dry separation technology, high efficient material discharge type TBS have been researched and applied in practical dry coal washing projects. TDS intelligent dry separation technology has X-ray intelligent identification method and big data analysis function, coal and reject can be digitized identified by the technology and reject can be discharged by intelligent reject discharge system. This technology has been applied in dry separation links of lump coal (80 – 25mm) of steam coal washeries. But it has not been applied in coking coal washeries because of processing capacity and separation accuracy. Manual reject picking can be replaced by this technology to improve productivity and decrease labor intensity.

3.3.3 The research and development of organic sulfur removal technology of high sulfur coal will be the main research content for long period in future.

Coking coal has high sulfur content and big specific gravity, general desulfurization method can't remove organic sulfur in coal, so the innovation research for microwave with organic auxiliary agent can be used as desulfurization technology. Desulfurization ratio for 0 – 13 mm coking clean coal has been achieved 50% under normal pressure and temperature and organic auxiliary agent, organic sulfur takes up 70%. At present, microwave desulfurization technology has achieved initial effects in laboratory of Shanxi Coking Coal Group.

3.3.4 According to overseas coal washing technology, raw coal can be crushed to –30 mm and then washed wholly

Coal and gangue can be further dissociated, after separation, more coking coal product can be recovered and scarce resource recovery will be increased.

On the basis of electrical reactance frequency spectrum (EIS) sensor technology, innovating and developing flotation reagents on line intelligent feeding system.

3.4 Advancing development and application for advanced technical equipment of large scale coking coal washery

In The 13th Five-Year Plan for Coal, above 10.0 Mt/a advanced coal washing technology should be promoted, related equipment should be researched and applied, the energy and medium consumption, pollutant discharge should be decreased during coal washing process. At present, Tiandi Science and Technology LLC has completed the approval of scientific research project for "key technology and equipment research, demonstration project of above 10.0 Mt/a, large scale, high efficient and intelligent coal washery". So the key technology and equipment research for new type high efficient HM cyclone, high efficient energy saving type HM shallow groove separator, intelligent control technology of HM coal washing system, large scale flip-flow screen has been started to research, the coal washing equipment will be developed along with the direction of upsizing, domestication, new type and high efficiency. Meanwhile, the demonstration project of above 10.0 Mt/a, large scale, high efficient and intelligent coal washery has been started to construct. In addition, super HM cyclone which is researched and manufactured by Tangshan Guohua Scientific and Technical Co. has been used in Shanxi Coking Coal Shuiyu Coal Washery. Single system 6.0 Mt/a HM coking coal washery has been started to design.

4. Conclusion

In China, coking coal is scarce and significant, its security concerns of strategic reserve has been more than simple application. In order to guarantee the development of Chinese iron and steel industry, support Chinese economic society construction and realize the Chinese Dream, it is necessary to implement protective development and utilization for coking coal resource. Coking coal can be produced cleanly and utilized efficiently by improving coal washing technology, equipment, construction and management level to increase economic and social benefits of enterprises.

5. References

[1] Chen Zizhao, Current Situation of Coal Washing and Processing and Coal Quality in China and Prospects for the 13th Five-Year Plan [J], Coal Processing & comprehensive Utilization, 2017 (5): 17-20;

[2] Cheng Hongzhi, Current Situation and Development Trend of Coal Preparation Technology in China [J], Coal Preparation Technology, 2012 (2): 83-87;

[3] Guo Niuxi, Tao Nengjin, Li Minghui. Practices and Prospect on Design of Ultra Large Modernized Coal Preparation Plant [J], Coal Engineering, 2012, 44 (SI): 108-111;

[4] Li Minghui, A Review on 60 Years of Coal Preparation in China [J], Coal Engineering, 2014, 10 (46): 24-29;

[5] Shi Huan, Cheng Hongzhi, Liu Wanchao. Present Status and Development Trend of China's Coal Preparation Technology [J], Coal Science and Technology, 2016 (6): 172-177;

[6] Wang Qinlong, Analysis and Research on Development Prospect of Coal Washing Technology [J], Heilongjiang Science and Technology Information, 2017 (4): 133;

[7] Wu Shiyu, Development of Chinese Coal Preparation in Past 30 Years [J], Coal Processing & Comprehensive Utilization, 2009 (1): 5-9;

[8] National Development and Reform Commission, National Energy Administration. The 13th Five-Year Plan for Coal Industry Development [R], Beijing, 2016;

[9] GB/T 26128-2010 Classification and Utilization of Scarce and Special Coal Resources [S];

[10] GB/T 50359-2016 Coal for Design of Coal Cleaning Engineering [S].

8

Study on dynamic model of secondary optimal cut point in low-quality steam coal jigging

Haibo Yu[1], Xiaoyu Yu[1]

*Fushun Sanyuan Industrial Measurement and Control Technology Research Institute,
Fushun, Liaoning Province, China*

Abstract : In view of the fluctuation of feeding clean coal ash in low-quality steam coal jigging which leads to variation in secondary optimal separation density and affects jigging separation performance, diagram of dynamic model transformation of secondary optimal cut point and new perspectives about the correlation between optimal separation density and optimal ash cut point are presented in this paper. The analysis is combined with a comparison of practical examples to demonstrate insufficiency in previous technology and significant effect of applying the new technology.

Keyword : Steam coal jigging, feeding clean coal ash, optimal cut point, optimal separation density, optimal ash cut point, Automatic Tracking Ash Closed-loop Control (ATACC)

1. Introduction

Low-quality coal makes up about 40% of the coal reserves in China. The uneven washing of low-quality steam coal (Jiongtian et al. 2014) leads to considerable fluctuation of feeding clean coal ash (-1.8g/cm^3), causing variation in secondary optimal cut point of the fixed-sieve jig with simultaneous two-stage reject, which results in considerable fluctuation of cleaned coal quality and coal lost in refuse. Such influence is beyond the control range of the traditional five control factors – jig, air, water, feeding, and discharge (Shiyu et al. 2012) (Ertie 2006) and becomes a key problem for improving quality and increasing efficiency of low-quality steam coal jigging.

The new technology– Automatic Tracking Ash Closed-loop Control (ATACC) (Xiaoyu et al. 2013) (Xiaoyu et al.2016) (Haibo et al. 2013) can solve this problem effectively. On this basis, this paper presents further new research findings including a dynamic model of the secondary optimal cut point under conditions of changing feeding clean coal ash, diagram of dynamic model transformation, the correlation between optimal separation density and optimal ash cut point. Meanwhile, a comparison of practical examples are also provided to demonstrate insufficiency in previous technology and significant effect of applying the new technology.

2. Dynamic model of secondary optimal cut point under changing feeding clean coal ash

2.1 Optimal cut point and a simplified method for aligning optimal cut point

Aiming at simultaneous two-stage refuse discharge of fixed sieve jig, under the condition of cleaned coal without refuse and refuse with minimum coal at the end of the second stage, corresponding two physical quantities are known as the secondary optimal separation density and secondary optimal ash cut point, which are collectively called optimal cut point.

After the stable operation of air, water, feeding, and discharge, observe the cleaned coal on the screen surface during adjusting second-stage discharge quantity (weights are for rough adjustment, and set values are for fine adjustment). When 1-2 pieces of small refuse are occasionally seen on the cleaned coal screen surface within 2 minutes, it aligns the optimal cut point.

2.2 Three measurements/separation density and ash closed-loop control in second stage of a jig

There are three measurements/separation density in the second stage of the jig (Haibo et al. 20022006) (Haibo et al. 2006), i.e. float measuring density, secondary optimal separation density, and secondary actual separation density. Float measuring density is correlated to secondary actual separation density via jigging control system. Based on the deviation between cleaned coal ash and set ash, float measuring density is regulated to allow the secondary actual separation density to track secondary optimal separation density, therefore to enable the cleaned coal ash ≈ set ash. This process is called ash closed-loop control.

2.3 Dynamic model of secondary optimal separation density

2.3.1 Model of aligning optimal cut point under low-ash state

Figure 1 displays an equivalent model of aligning the secondary optimal cut point of a jig when air valves work in the resting phase and the feeding clean coal ash is in a low ash state. Figure 1 shows the following features:

    (a)   From bottom to top, the jig bed contains refuse stratification, misplaced area (where refuse and cleaned coal are not clearly separated) and cleaned coal stratification;

    (b)   When jig discharge sill which is taken as the cut point coincides with the boundary line between cleaned coal and misplaced area, this cut point is the current optimal cut point;

(c) Float measuring density is adjusted to allow the float to fall down on the boundary line between cleaned coal and misplaced area. The discharge control system regulates refuse discharge rate as per the overall thickness of misplaced area and the refuse stratification to maintain a stable position of the optimal cut point as far as possible while ensuring that there are no refuse effluent and refuse discharged with minimum coal.

(d) Optimal separation density reaches high value when feeding clean coal ash is low.

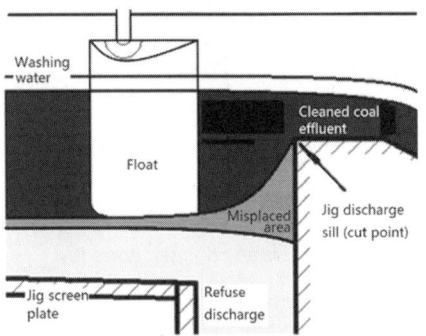   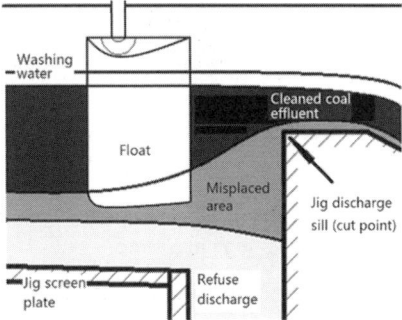

Figure 1: Model of aligning optimal cut point under low-ash state

Figure 2: Change of the cut point model when feeding clean coal ash goes high

2.3.2 Change of the cut point model when feeding clean coal ash goes high

Figure 2 displays the change of the cut point model when feeding clean coal ash goes from low to high. From Figure 2, it can be seen that:

(a) When feeding clean coal ash goes high, cleaned coal density goes up and approaches to refuse density which causes refuse up-mixing, the misplaced area is thickened, optimal separation density becomes low, the high separation density under feeding clean coal low ash is no longer applicable to the operation with low separation density under high ash condition, and part of misplaced area may enter cleaned coal effluent to cause cleaned coal to contain refuse.

(b) Under this situation, the operation staff may observe the quantity of refuse contained in cleaned coal on the screen surface and lower actual separation density by decreasing float measuring density to align new optimal cut point.

(c) With Automatic Tracking Ash Closed-loop Control (ATACC) technology, float measured density may be automatically adjusted as per increase of cleaned coal ash to lower actual separation density.

2.3.3 Model of aligning optimal cut point under high ash state

After adjustment, model of aligning optimal cut point under high ash is shown as Figure 3.

When feeding clean coal ash is at a high state, optimal separation density is at a low state.

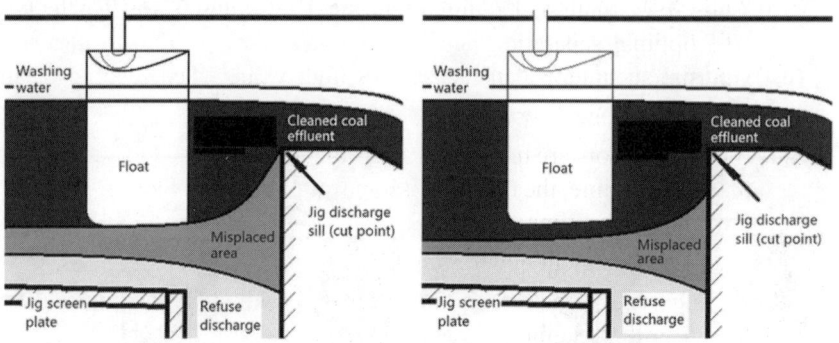

Figure 3: Model of aligning optimal cut point under high ash state

Figure 4: Change of cut point model when feeding clean coal ash goes low.

2.3.4 Change of cut point model when feeding clean coal ash goes low.

Figure 4 indicates that:

(a) When feeding clean coal ash goes from high to low, the misplaced area narrows down, optimal separation density becomes high, and the previous lower separation density under high ash is no longer applicable to the operation with high separation density under low ash state. In this case, if an adjustment is not executed in time, over-discharge and coal lost in refuse may happen.

(b) For manual operation (operation conditions: numerically controlled air valve, automatic feeding, automatic discharge), as the reject extraction chamber is deep and it is difficult for the operators to catch sight of coal lost in refuse immediately, adjustment to aligning new optimal cut point may not be able to be made in time thus causing coal lost in refuse.

(c) With using ATACC technology, float measured density can be automatically adjusted to track the change of cleaned coal ash to raise actual separation density.

2.3.5 Deep analysis on a dynamic model of secondary optimal separation density

The changes of four kinds of optimal separation density dynamic models caused by variation in feeding clean coal ash can be used to compose the state transformation diagram, as shown in Figure 5:

(a) State 1 → State 2. Feeding clean coal ash goes from low to high and optimal separation density goes from high to low and turns into adjustment state.

(b) State 2 → State 3. Both manual adjustments as per the quantity of refuse contained in cleaned coal and ATACC are effective.

(c) State 3 → State 4. Feeding clean coal ash goes from high to low and optimal separation density goes from low to high to turn into adjustment state.

(d) State 4 → State 1:

(1) As the operators are unable to catch sight of coal mixed in, refuse, and make adjustment in time, the transformation from State 4 → State 1 cannot be achieved within a short time.

(2) Manual adjustment under State 4:

It may enter an internal loop under State 4 (as shown by the elliptic loop below State 4 in the diagram) if cleaned coal ash doesn't reach a new high value / optimal separation density doesn't reach a new low value;

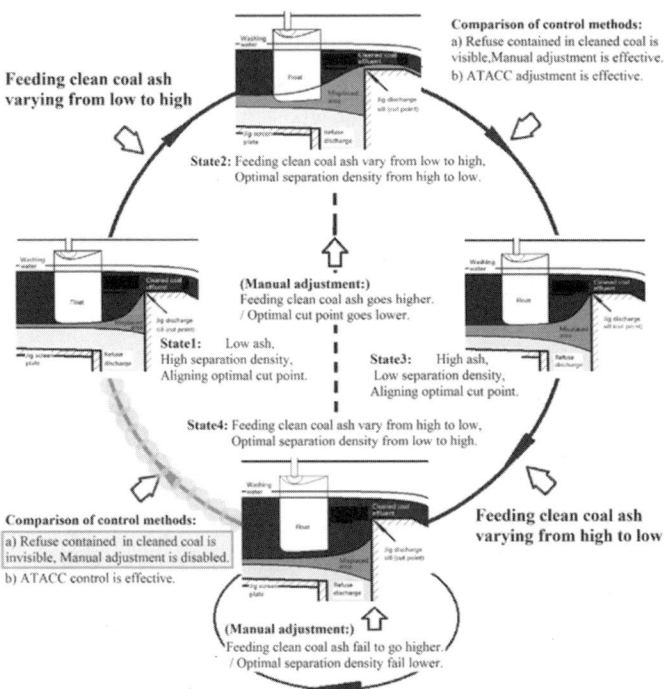

Figure 5: The state transformation diagram of dynamic model about change of optimal separation density caused by variation in feeding clean coal ash

If cleaned coal ash reaches a new high value/optimal separation density reaches a new low value, another round of manual adjustment process (as indicated by the dashed line in the diagram: State 4 → State 2 → State 3 → State 4) will be entered. Thus, a vicious circle will be formed to significantly lower actual separation density, causing to refuse the containing coal under low-ash state, which leads to cleaned coal ash to be too low.

(3) With ATACC control technology, under automatic tracking ash closed-loop control action, ATACC may effectively eliminate defects caused by manual operation to allow normal adjustment cycle from State 1 → State 2 → State 3 → State 4 → State 1 and track the variation tendency of the optimal cut point.

2.3.6 Practical example 1 of theory application

Figure 6 displays a mathematical statistics curve of consecutive 24-hour cleaned coal ash under manual control (numerically controlled air valve, automatic coal feeding, automatic discharge) before ATACC technology is adopted at the coal preparation plant of Daxing Mine. The box indicates the interval of cleaned coal ash index. Due to the above mentioned reasons, the envelope area below ash index shown in Figure 6 takes up to 2/3 of the overall area which means vast ultra-low ash.

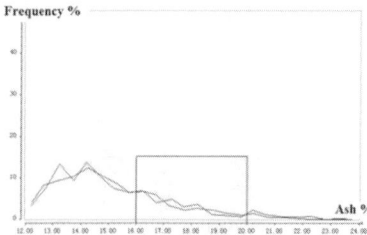

Figure 6: Statistic analysis on cleaned coal ash before ATACC technology is adopted

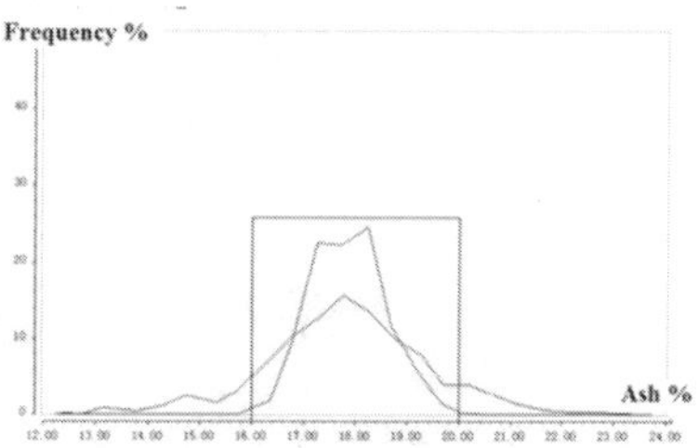

Figure 7: Statistic analysis on cleaned coal ash after ATACC technology is adopted

Figure 7 displays that after ATACC technology is adopted, due to eliminated defects caused by manual operation, the envelope area of cleaned coal ash statistics curve roughly coincides with the ash index interval under conditions of same feeding coal quality and same cleaned coal ash index. This represents the decrease in coal lost in refuse and the significant increase in cleaned coal recovery.

3. Correlation between optimal separation density and optimal ash cut point

After identifying defects of manual adjustment during state transformation of dynamic models about variation in secondary optimal separation density due to change of feeding clean coal ash, it comes to next key problem: How wide is the fluctuation range of secondary optimal separation density of a jig? May it be neglected?

3.1 Correlation between optimal separation density and optimal ash cut point

Each secondary optimal cut point corresponds to two different physical quantities– optimal separation density and optimal ash cut point. In accordance with practical operation, the trend chart of the correlation between secondary optimal separation density and secondary optimal ash cut point is summarized as shown in Figure 8.

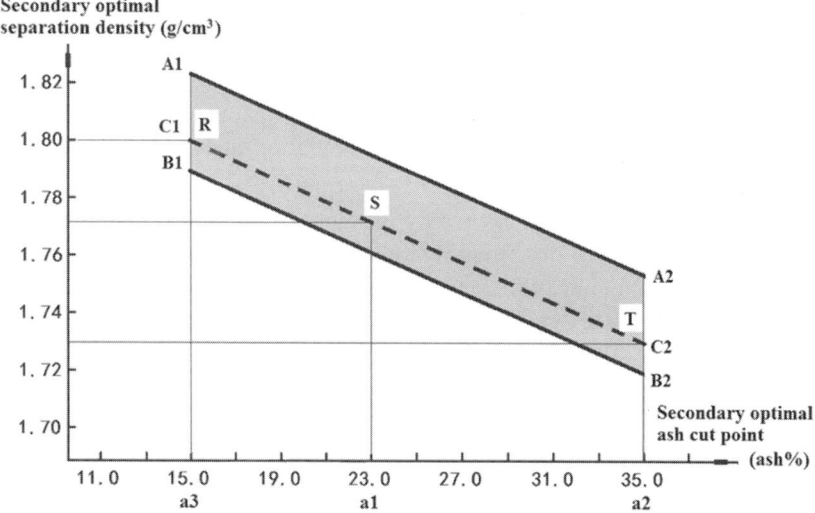

Figure 8: Trend chart of correlation between secondary optimal separation density and secondary optimal ash cut point

Abscissa represents ash at secondary optimal ash cut point and the ordinate represents secondary optimal separation density. Correlation between them is represented by the area formed by line segment A1-A2 and parallel line segment B1-B2; the curve of correlation between secondary optimal separation density and secondary optimal ash cut point is within this area;

profile of this curve is affected by the five control factors (jig, air, water, feeding and discharge) and variation in feeding clean coal quality. However, this does not prevent us from assuming that the correlation curve is C1-C2 to further investigate the correlation between the secondary optimal separation density and the secondary optimal ash cutting point:

When the optimal ash cutting point for feeding clean coal ash into "medium ash" is a1 = 23.0%, the optimal separation density corresponding to point S on the C-C line is 1.77 g/cm³;

When the optimal ash cutting point for feeding clean coal ash into "medium-high ash" is a2=35.0%, the optimal separation density corresponding to T point on the C-Cline is 1.73g/cm³;

When the optimal ash cutting point for feeding clean coal ash into "low ash" is a3 = 15.0%, the optimal separation density corresponding to the R point on the C-C line is 1.80g/cm³.

According to the trend, when secondary optimal ash cut point is relatively low, the corresponding optimal separation density is relatively high; secondary optimal ash cut point goes high while optimal separation density goes low;

In terms of magnitude properties, when secondary ash cut point is between 15.0%~35.0%, corresponding optimal separation density is 1.80g/cm³~1.73g/cm³ which covers the entire separation density range of steam coal jigging.

The changing trend represented by the chart of relationship trend and the order of magnitude is of important reference value during performing ATACC and fine adjustment under site conditions with significant fluctuation range of feeding clean coal ash.

3.2 Practical example 2 of theory application

The Xiaoqing preparation plant is fed with mixture raw coal from No.4 and No.7 coal seams in the pit and coal from heading. Fluctuation range of daily feeding clean coal ash is up to 15.00%~40.00%. Figure 9 displays a mathematical statistics curve of 24-hour cleaned coal ash under manual operation (numerically controlled air valve, automatic coal feeding, automatic discharge). In Figure 9, although the upper limit of statistical cleaned coal ash is up to 30%, as a considerable amount of coal lost in refuse is found during the site inspection, the upper limit of cleaned coal ash index may need to be further increased.

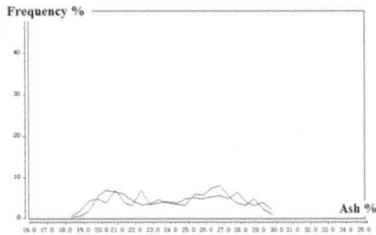

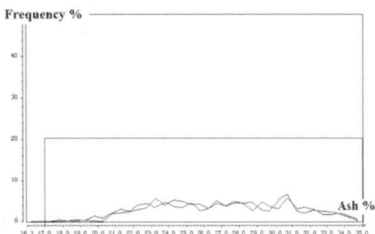

Figure 9: Ash curve of washing lump clean coal in 24-hour manual operation

Figure 10: Ash curve of washing lump clean coal after ATACC technology is adopted

Figure 10 shows the statistical curve of washing block cleaned coal ash within 24 hours after the application of ATACC technology and the expansion of cleaned coal ash control index area to 17.00% ~ 35.00% (see the straight-line box area in figure 10). Compared with Figure 9, it can be seen that the area increased in the cleaned coal ash interval from 30.00% to 35.00% in Figure10 accounts for about 1/4 of the envelope area of the cleaned coal ash statistical curve of that day. At the same time, the area of enveloped by low ash and medium ash moves up obviously. According to the above discussion, it shows that recovery of low, medium and high ash coal has been greatly improved (Jingwen et al. 2011).

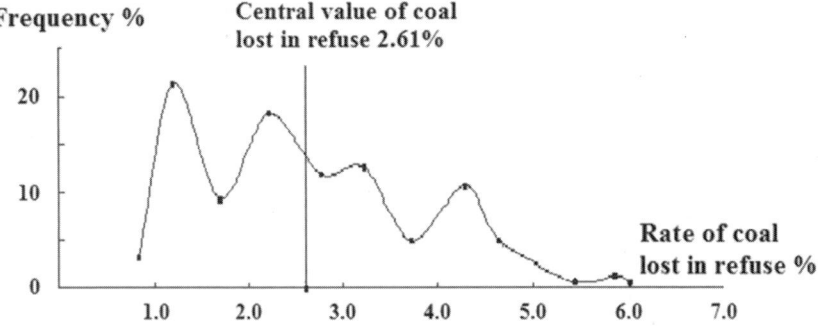

Figure 11: Mathematical statistics curve of coal lost in refuse after application of ATACC technology and relaxation of cleaned coal ash index

Meanwhile, consecutive 15-day inspection is performed on coal lost in refuse in the first and second stages of a jig after ATACC technology is adopted. The results of the statistical analysis are shown in Figure11 which indicates that the central value of coal lost in refuse considerably drops to 2.61%. It further verifies the significant effect provided by efficient jigging under considerable fluctuation of feeding clean coal ash. (Note: This method should be combined with subsequence bunker blending and coal blending to optimize quality.)

4. Supplement

(1) The five control factors (jig, air, water, feeding, and discharge) are the basis for the application of the new ATACC technology (Xiaoyu et al. 2014). The effect of the five control factors exists throughout the discussion about "dynamic model of secondary optimal cut point," mainly reflected in thickness change of the "misplaced area." If the five factors can be well coordinated, the thickness of the misplaced area may be narrowed down and coal lost in refuse is lower after the optimal cut point is aligned. Therefore, the six factors (jig, air, water, feeding, discharge, ATACC) need to be coordinated closely to improve separation effect of low-quality steam coal jigging.

(2) Jigging technology and coordination of control equipment

According to different coal quality conditions and requirements, various processes of washing coal homogenization, bunker blending, and coal blending may be considered.

5. Conclusion

To explore a low cost, improve quality, and increase the efficiency of steam coal jigging control method has been the pursuit of domestic and foreign research staff in coal preparation field. A deep study on the dynamic model of secondary optimal cut point and innovative application of key technology demonstrate that automatic tracking ash closed-loop control (ATACC) is an indispensable and critical factor for low-quality steam coal jigging.

6. References

[1] Ertie Yu, Re-orienting coal jigging technology [J]. Coal preparation technology. 2006 (2) :55-57.

[2] Haibo Yu, Xiaoyu Yu, Control Methods for High-efficiency and Quality Steam Coal Jigging. China ZL2011100928346 [P] 2013-9-18;

[3] Haibo Yu et al. Discussion of Real-time Jig Separation Density Monitoring and Ash Closed-loop Control [J]. Coal Preparation Technology 2002(2):16-17.

[4] Haibo Yu et al. New Technology of Automatic Jigging Control in China [C]. Monographs of the 15th international coal preparation congress. 2006/ Beijing, China: 213-218.

[5] Jiongtian Liu et al. Strategic Research on Coal-quality Improving and Plans of Conveying and Blending [M]. 2014-10.

[6] Jingwen Liu et al. Improvement and Application of Jig Ash Closed-loop Control System in Xiaoqingkuang Coal Preparation Plant [J]. Clean Coal Technology. 2011(3):22-25.

[7] Shiyu Wu et al. A rudimentary Knowledge of Coal Preparation (4th edition)[M]. 2012-11.

[8] Xiaoyu Yu, Haibo Yu，Chunhui Yu. Progress of Steam Coal Clean Efficient Jigging Technology in China[C], XVIII International Coal Preparation congress/28 June-01 July 2016/Saint-Petersburg Russia: 519-524.

[9] Xiaoyu Yu，Haibo Yu, Chunhui Yu. Efficient and Quality Steam Coal Jigging applied in TieFa. China[C], Proceedings of the 17th international coal preparation congress/1-6 OCTOBER 2013/TURKEY: 225-230.

[10] Xiaoyu Yu et al. The Correlation between Jigging Cleaned Coal Ash Automatic Tracking Closed-loop Control & Its Five Main Control Factors [J]. 2014(7):8-11.

9

The development and application of large screens in China

Bo Liu, Shaolei Zhou

Aury (Tianjin) Industrial Technology Co., Ltd, Tianjin, China

Abstract: In the coal industry, the vibrating screen is indispensable equipment. The application of large screens can improve the production efficiency and reduce the production cost; therefore, coal production enterprises have an urgent need for the large-scale and high efficiency screens. However, the traditional design and manufacturing method has characteristics of long working period, low efficiency and high costs, and it cannot satisfy large screen requirements in terms of reliability and high efficient operation. The modern design method and advanced manufacture processes are adopted to produce high reliable and high performance large vibrating screens represents the current trend in industrial development. Using the finite element analysis of modern design method to design the screen structure could effectively optimize the screen structure, avoid resonance, reduce noise and improve screen's reliability significantly. The ABD4885 double deck multi slope screen designed and manufactured by Aury (Tianjin) Industrial Technology Co., Ltd is the largest screen in the world and has successful practical application. This screen has features of highly reliable and highly efficient operation, which has created good economic benefits for the company.

Keywords: Large screens; the screen modern design method; reliability; high efficiency; practical application

In the coal industry, the vibrating screen is indispensable and takes the responsibility in sizing, dewatering, desliming, medium recovery and even separation. The performance and quality of vibrating screens directly impact on coal product and production costs.

With the rapid development of industrial modernization, coal production companies scale becomes bigger and bigger. In order to improve the production efficiency and reduce the production cost, these companies have an urgent need for the large-scale screening equipment and also ask for higher requirements in performance and reliability.

1. Traditional design method

$\sigma = \dfrac{F}{A} \leq \dfrac{\sigma_s}{n}$ The following is a simple example of a traditional calculation method to solve the stress.

The theoretic calculation formula is: [1]

σ   :   the design strength

F   :   the applied load

A   :   the section area

σ_s :   yield strength

n   :   the safety factory

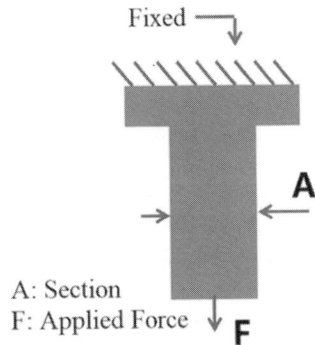

Fixed

A

A: Section
F: Applied Force   F

Figure 1

Because stress concentration will occur at the sharp or rounded angles when various shape especially the complicated parts are under stress. The key and difficulty in the traditional calculation method is the local stress cannot be exactly decided. It's only to make a range for safety factor by experience or experiments. For some crucial parts, they need a large number of experiments and experience to work out the range of the safety factor. It cannot ensure the equipment's safety when the safety factor is too small. However, if the safety factor is excessive, the equipment design will be clumsy, and that is adverse to reduce the cost of manufacture. However it is difficult to decide the safety factor precisely in general case, so it cannot get accurate calculation result by this traditional design method.

For the vibrating screens, traditional design by using the empirical and semi-theoretical method cannot satisfy the screen's requirements in reliability, economical aspect, and adaptability. More advanced design methods need to be developed[2].

2. The modern design method

The finite element analysis method is a modern design method rapidly developed with the development of the computer technology[3]. The FEA application in the design of vibrating screens can precisely calculate the stress in various parts, optimize the screen structure and balance the stress distribution in all parts. The modal analysis carried out by the FEA method could effectively avoid adverse impact by resonance. When necessary, the FEA method can be used for harmonic response analysis to probe the resonant response and check the harm extent of that response.

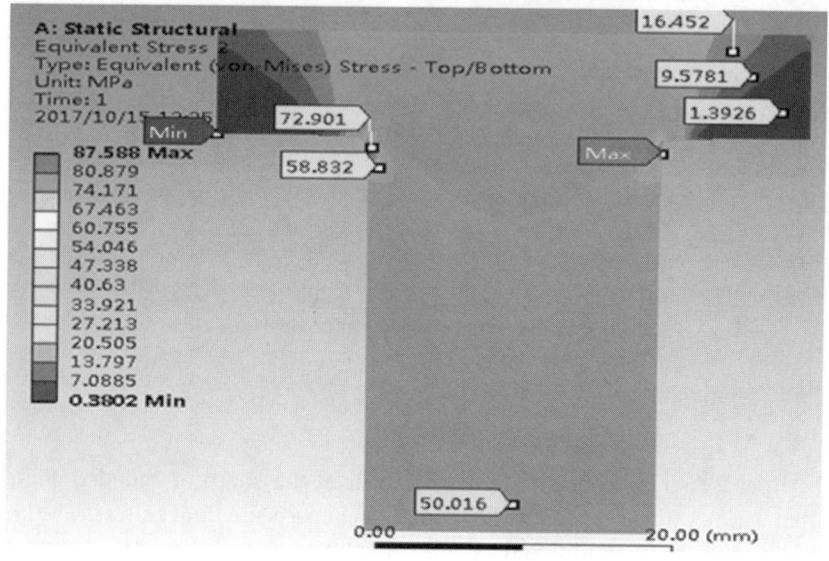

Figure 2

The above case, in which the part in figure 1 is taken as an example to get the analysis result by using static function in ANSYS FEA software.

The stress in different areas can be seen directly from the result picture. Based on this result, the parts can be optimized effectively and rapidly, which cannot be realized by the traditional theoretical calculation.

2.1 The FEA static strength analysis application in the vibrating screen.

2.1.1 Establishment of solid model for the vibrating screen

Screen parts are established by 3D model software and assembled together to be a whole vibrating screen model. The assembly model should be checked for interference to ensure the correct assembly and no interference between parts.

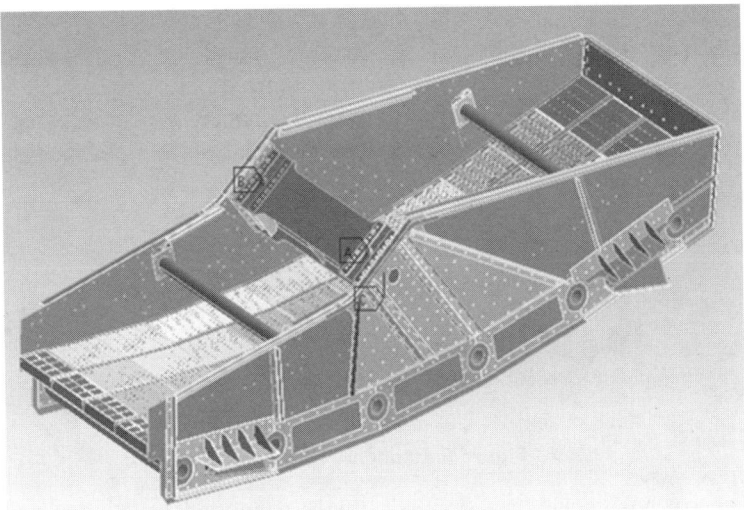

Figure 3: the solid model of the vibrating screen

2.1.2 Establishment of FEA model for vibrating screen

Figure 4: the vibrating screen FEA model

All parts in the vibrating screen are processed by solid element mesh. All rivets and bolts are connected by beam elements, and the exciter is instead of a mass point.

• Impose boundary conditions

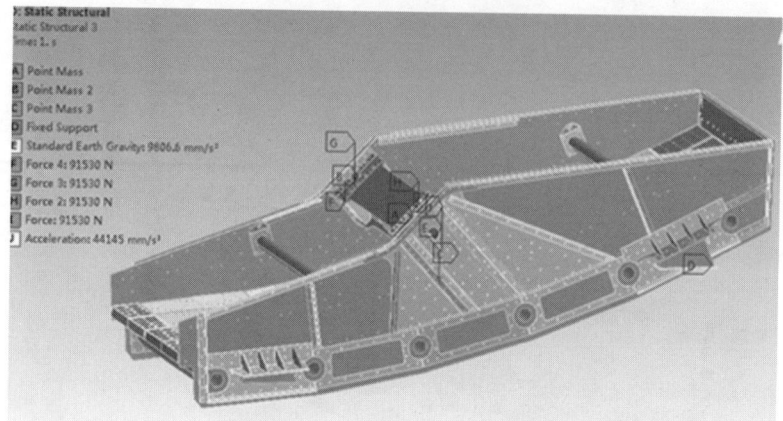

Figure 5: Boundary conditions

The spring brackets are applied simply supported constraints and the exciter is substituted by the mass point for force application. The whole screen is applied an acceleration load in the vibrating direction of the exciter and the gravity load in the vertical direction. The equivalent material load is applied on the screen surface.

• Post-process analysis

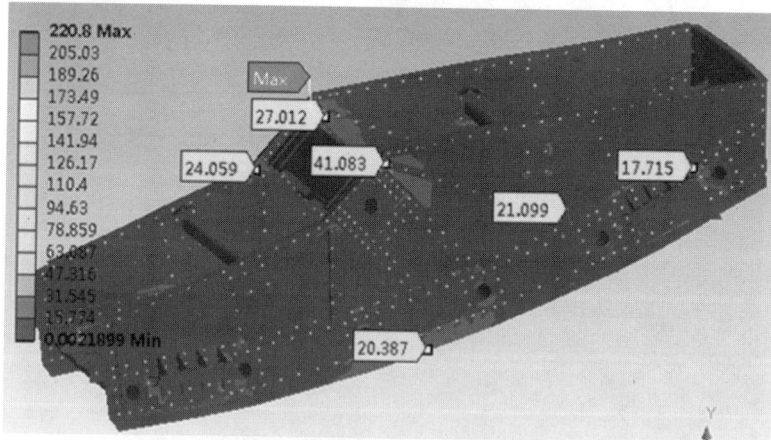

Figure 6: The whole screen stress distribution

Select the stress analysis nephogram and check the position where the maximum stress occurs. Check the stress veracity in stress concentration areas and evaluate whether the stress in each part is able to meet the design requirement.

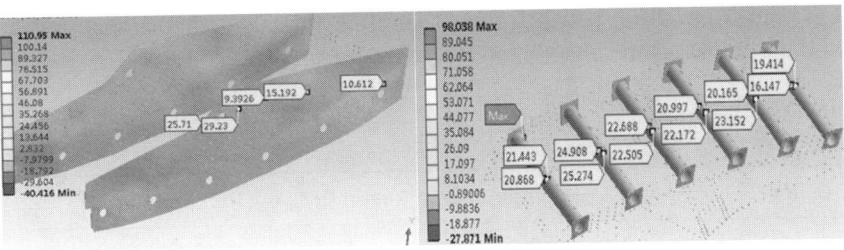

Figure 7: The side plate stress distribution

Figure 8: The cross beam stress distribution

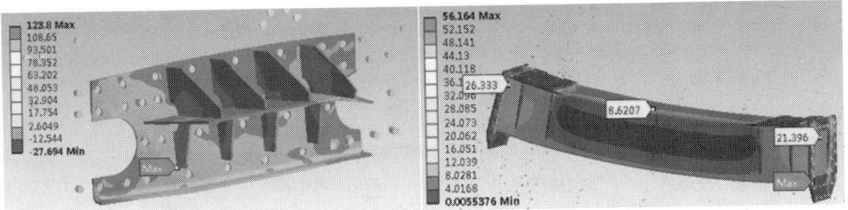

Figure 9: The spring bracket stress distribution

Figure 10: The exciter beam stress distribution

2.2 The FEA modal analysis application in vibrating screens

The modal analysis is an approach to analyze structure dynamic features. The experimental result of the modal analysis should be used to revise the FEA model first. After the FEA model is checked for correctness, the computational modal analysis could be carried out.

2.2.1 The result of modal analysis

The first 15 order modal results are shown in the following pictures, in which the first 6 order modal results are rigid body modal and have no actual meaning.

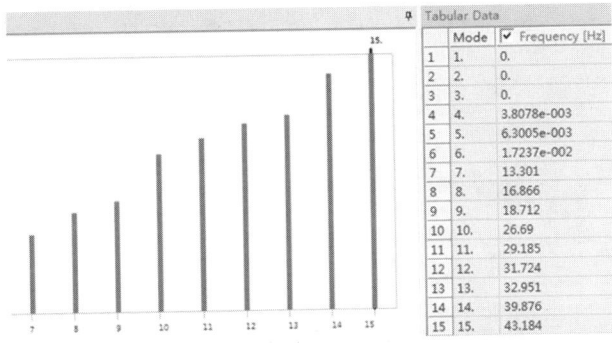

	Mode	Frequency [Hz]
1	1.	0.
2	2.	0.
3	3.	0.
4	4.	3.8078e-003
5	5.	6.3005e-003
6	6.	1.7237e-002
7	7.	13.301
8	8.	16.866
9	9.	18.712
10	10.	26.69
11	11.	29.185
12	12.	31.724
13	13.	32.951
14	14.	39.876
15	15.	43.184

Figure 11: The result of first 15 order modals

The comparison between first 7-12 order modals computational results and the actual test results is shown in table 1.

Table 1: The comparison between first 7-12 order modals computational results and the actual test results

Order	Computational Result (HZ)	Test Result (HZ)
7th Order modal	13.3	13.0
8th Order modal	16.9	16.9
9th Order modal	18.7	18.9
10th Order modal	26.7	26.5
11th Order modal	29.2	28.8
12th Order modal	31.7	31.3

The 7th modal deformation: Swinging deformation along the screen medium section in the horizontal plane

- The 8th modal deformation: Torsional deformation in the vertical plane
- The 9th modal deformation: Torsional deformation in the horizontal plane
- The 10th modal deformation: Swinging deformation along the screen rear end in the horizontal plane
- The 11th modal deformation: Horizontal swinging deformation in lifting beam areas and bottom edges in the center section of screen
- The 12th modal deformation: Horizontal swinging in the front section, swinging deformation in lifting beam area in feed end

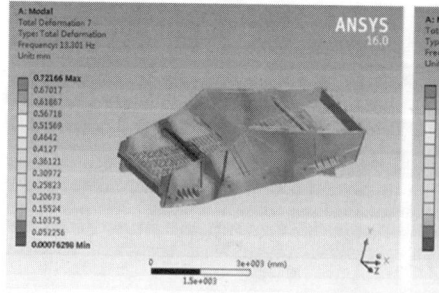

Figure 12: The 7th modal deformation

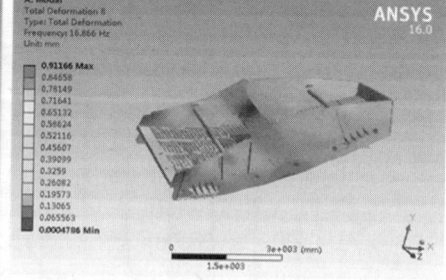

Figure 13: The 8th modal deformation

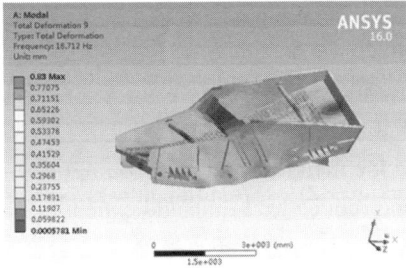

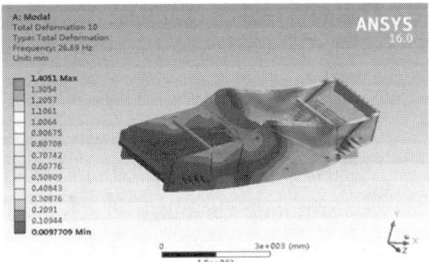

Figure 14: The 9th modal deformation   **Figure 15:** The 10th modal deformation

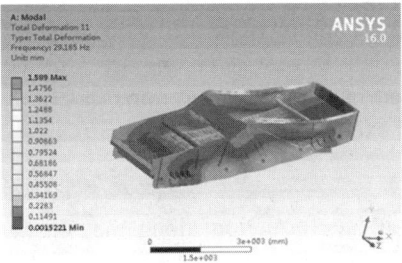

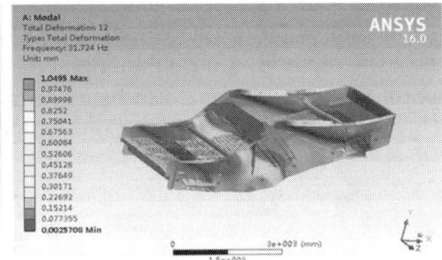

Figure 16: The 11th modal deformation   **Figure 17:** The 12th modal deformation

2.2.2 Modal analysis result evaluation

According to modal the calculation result, the operation frequency could be set at 15HZ, which has at least 10% clearance of the lower 7th and higher 8th natural frequencies. In that, the resonance will not occur. If the screen needs a wider operation frequency, the screen structure should be adjusted in accordance with the results of the 7th and 8th modals. The 7th natural frequency should be reduced and the 8th natural frequency should be increased to achieve a wider operational frequency range[4].

2.3 Summary

The application of the modern finite element analysis method in structural analysis and modal analysis, together with harmonic response analysis and the optimization design, and the adoption of advanced manufacturing processes can produce highly reliable and high-performance vibrating screens. The adoption of FEA of modern design method to design the vibrating screen structure can effectively optimize the screen structure, avoid resonance, reduce noise and significantly improve the reliability, which is theoretically used to take reliable design for any size screen.

The advantages of the FEA method
- Increase design functions and reduce the design cost
- Shorten the design and analysis period
- Improve the product's design reliability
- Adopt optimal design and reduce the material cost or consumption
- Simulate various trial schemes and reduce trial time and expenses

3. The practical application

The ABD4885 double deck multi slope screen that is designed and manufactured by Aury(Tianjin) Industrial Technology Co., Ltd with adopting the FEA method is the vibrating screen with largest screening areas in the world and has succeeded in being applied in Fengjiata coal mine in China. This screen has two exciter beams with 6 exciters for synchronous driving.

Figure 18: Screen photo during test   **Figure 19:** Screen photos on site

Basic information of Fengjiata coal mine project:
- Project name: Fengjiata coal mine of shanxi fengjiata coal mine Co., Ltd
- Location: Fugu county, Shanxi Province, China
- Scale: 10.0Mt/a

Main design parameters of the vibrating screen
- Equipment type: 4.8×8.5m double deck multi slope screen
- Power rate: 2×90kW
- Capacity: 1200t/h
- Feed size: –300mm
- Aperture: 40/25mm

This equipment has been put into use in June of 2016, and all screen parameters have met the design requirement according to the test result. The actual handling capacity of single equipment has reached 1350t/h. The screen is running stable till now and has the features of high reliability and high efficiency, which has created good economic benefits for the end user.

4. Conclusion

(1) Using the finite element analysis in modern design method to design the screen structure could effectively optimize the screen structure, avoid resonance, reduce noise and improve reliability significantly.

(2) Modern design method and advanced manufacture processes are adopted to produce high- reliable and high-performance large vibrating screens, which represents the current trend in industrial development.

(3) The ABD4885 double deck multi slope screen designed and manufactured by Aury (Tianjin) Industrial Technology Co., Ltd is the screen with largest screening areas in the world and has successful practical application. The screen has features of high reliability, high efficient operation and long service life, which has created good economic benefits for end user.

5. References

[1] Ferdinand P.Beer、E.Russell Johnston,Jr、John T.Dewolf、David F.Mazurek Written. TAO Qiu-fan, FAN Qin-shan Translated. Material mechanics[M]. Beijing: Mechanical industry press, 2013.

[2] LIANG Kun-jing, BAI Yong-jun, SHAO Pei-sen. Modern dynamic design method in the vibrating screen development [J]. Coal preparation technology, 2000, (2):12-14.

[3] ZHANG E, MAIMAITIMING Ai-ni. Modern design theory and method[M]. Beijing: Science Press, 2014.

[4] QIAO Chong-quan, WANG Ke-jie. Dynamic analysis and dynamic optimization design of large vibrating screens [J]. Coal mine machinery, 2011, (11):2-5.

10

The road of intelligent coal preparation plant- the design and application of intelligent high efficiency pressure filter

Qingliang Guo[1], Gu Gong[2]

[1]*Jincheng coal industry group, Jincheng, China*
[2]*Sihe coal mine of Jincheng coal industry group, jincheng ,China*

Abstract: Under the background of the industrial 4.0 revolution which is dominated by intelligent manufacturing, the intelligent operation will be the development direction for the coal beneficiation in the future. However, for the filter dewatering, only single machine and set control is automatic. The press filter itself cannot adjust the pressure filtration cycle time according to the property of the coal slime. This cannot meet the development of the modern intelligent coal preparation plant. Here, the author designed a kind of control system for the intelligent operation of press filter. This technological innovation of filter press is a great significance for the realization of the intelligent filter press.

Keywords: Press filter; pressure filtration cycle time; Intelligent; control system; technological innovation.

1. Introduction

The intelligentization of coal preparation plant, which integrates the artificial intelligence, internet of things and other multi-disciplinary knowledge, realizes the intelligent adjustment of process system and parameters of a coal preparation plant. This not only significantly reduce the number of operational personnel and the labor intensity of employees but also stabilize the product quality (Jianhua Huang, 2017). However, the intelligent control of the beneficiation equipment is the key to realize the intelligentization of a coal preparation plant. This mainly includes the intelligent operation and monitoring of equipment.

Filter press, which is an intermittent operation of pressure filtration equipment to separate solids from liquid, is widely used in the chemical industry, metallurgy, coal, food, and other industries. For the coal preparation plant, the filter press is very important to achieve closed cycle washing water and for environmental protection purpose (Zhongkuan Wei et al, 2007).

2. Application status of filter press in coal preparation plant

In a coal preparation plant, the filter press is the main equipment to realize the coal dewatering through pressure filtration. The working process mainly includes compress of filter plate, feeding filling, pressure filtration, and squeezing and discharge process. These processes are mainly controlled by PLC. There is a touch screen combined with a local control button for the PLC and is easy to operate. However, the control of the filter press is just realized the single machine automation operation not connect with other washing process. It needs to check the discharge of pressure filter cake by a human (Dongfang Wu et al, 2017).

In addition, some coal preparation plant with higher technology levels, are only realizing the centralized control of the pressure filter. If there are some changes in the concentration of feeding fineness contain, viscosity or ash content, the working parameters and the filter cycle time can only to be adjusted by the experienced workers. Each cycle time of filter press cannot be intelligently adjustment according to the product indicators by the control system. The monitoring of the operation process of the filter press is mostly supervised by the on-site personnel. the labor cost is high for the production and the reliability is difficult to guarantee. How to effectively solve the above technical problems is the key to achieve intelligent control for filter press.

3. Intelligent operation control system of filter press

In recent years, filter press has developed rapidly in the aspects of filter plate, filter medium, separation efficiency, structure, filtration, process and automation level. With the rapid development of information technology, artificial intelligence, many filter press manufacturers have realized how to liberate the staff from the complex operation, and realize the intelligent operation is the development trend of the filter press. In this paper, the intelligent operation control system of filter press mainly includes three parts: intelligent monitoring of the operation state of filter press, intelligent adjustment and setting of the operation parameters of filter press, and information management platform of filter press equipment.

3.1 Intelligent monitoring of filter press

At present, the whole operation process of the filter press has been controlled with one key automatic operation, but it needs the personnel to check the operation process on the spot, such as: whether there is spraying, whether there is cable pull, chain break and so on. In order to solve these problems, in this design, machine vision technology and sensor wireless early warning are combined in the design of filter presser and the operation of filter presser can significantly reduce labor cost and accident rate.

3.1.1 Intelligent image recognition and linkage cooperation technology

In recent years, with the rapid development of image processing and pattern recognition technology, machine vision technology has been widely used in industrial production(Xiangyang Tang et al,2004). Image intelligent identification and linkage technology has been applied For the scraper equipment, machine vision technology can recognize the scraper chain broken, scraper slanting, debris and automatically shutdown the machine. This technology can be applied to the filter press equipment, used for the intelligent monitoring and early warning of the failure of the filter plate tension, chain breaking, and injection in the filtering stage, etc. in the process of the filter press pull-in, without on-site monitoring of the operation of the filter press personnel.

3.1.2 Wireless monitoring intelligent early warning system

The wireless sensor, non-contact infrared temperature detection, wireless communication technology, and monitoring platform are introduced to the control system of filter press. The working information data of filter press such as feeding pressure, hydraulic system pressure and temperature, bearing temperature of feeding pump, vibration, temperature, vibration, voltage and current of the motor are real-time collected and transformed to the control center. According to the whole information data, the control system can judge the operation state of the filter press and compare with the normal operation data model. The early fault warning, predictive maintenance of equipment can be achieved and significantly reduce the accident rate.

3.2 The intelligent operation control device of filter press

At present, the filter press in China cannot automatically set each cycle of feeding time according to coal quality situation and the dewater effectiveness. The feeding time has to been controlled by an experienced worker. This operation method with hysteresis and instability results in very low efficiency of filter press. In this design, the intelligent adjustment and setting technology are employed.

The pressure filter intelligent control device includes a first stage and a second stage collection bucket, which are respectively equipped with a liquid level probe and an electric valve. The device detects the change of filtrate water flow. According to this change, the PLC control system will automatically control the time needed for filter press filling and filtering. The main composition and working process of the device is shown in figure 1. The filtrate water from the filter press flows into the first-stage collection bucket. In the initial stage, the filtrate water into flow is greater than that outlet flow for the first-stage collection bucket, and the liquid level in the primary collection bucket increases gradually. With the extension of time, the filter cake in the filter chamber of the filter press gradually forms, and the filtrate water discharged from the filter chamber gradually decreases. At the same time, the filtrate water flow into the first-stage collection bucket also decreases and low than the outlet water flow. At this point, the level of liquid in the first-stage collection bucket gradually drops. When the state of the liquid level probe (Fig1-2) changes from touching with filtrate water to not touching, the feed pump will automatically stop working under the control by PLC.

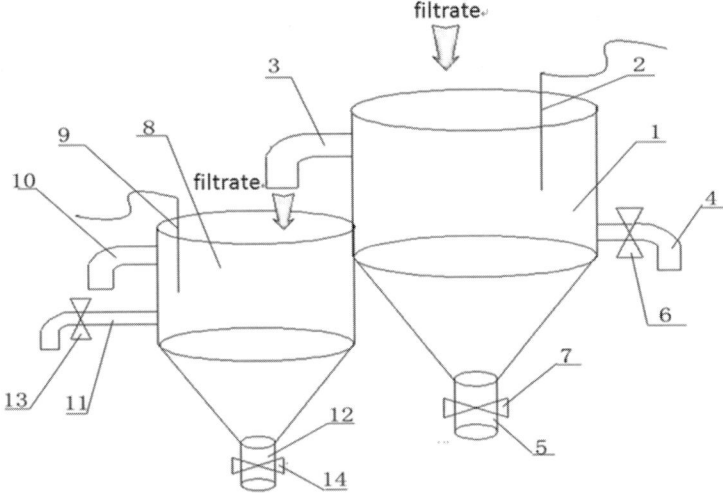

Figure 1: The intelligent operation control device of a filter press

1-first stage tank, 2-first liquid probe, 3-first stage overflow, 4-first outlet, 5-first discharge, 6-first remote control valve, 7-first discharge valve, 8-second stage tank, 9-second liquid probe, 10-second stage overflow, 11-second outlet, 12- second discharge, 13- second remote control valve, 14- second discharge valve

For coal, which is easy to dewatering, the first stage collection bucket is enough for the control of filter press, but for coal, which is hard to dewater, a secondary collection bucket connected with the first one is necessary. The opening of the outlet of the first-stage collection tank can be controlled by the contact time between the second-stage liquid level probe and the filtrate. In the beginning, the inlet flow is bigger than the outlet flow for the second stage tank. The filtrate keeps a high level and the second liquid level probe immersed in the filtrate. With the extension of time, the filter cake in the filter chamber of the filter press gradually forms, and the filtrate water discharged from the filter chamber gradually decreases. At the same time, the filtrate water flow into the second-stage collection bucket from the first one also decreases and low than the outlet water flow. Therefore, the level of filtrate in the second tank is gradually lower and cannot connect with the second probe. At this time, the duration of connect time (CT) between the probe with filtrate will be collected by the PLC and compared with the preset time (ST). If the CT is small than ST, the first stage remote control valve (Fig1-6) will be closed immediately. Most of the filtrate will flow into the second stage tank from the overflow of the first one. The level of filtrate in the second tank will increase and connect again with the second liquid level probe. Then, the first stage control valve will be open again, but the open level is 90% of the former open level. This cyclic process will be over until the CT is bigger than the ST. Then the control of filter press will according to the filtrate in the first stage collection bucket by the PLC.

This device combined with PLC can control automatically the filling and filtering time of filter press according to the change of filtrate water flow rate. This technology not only reduces the staff's labor intensity but also makes the filter press always maintains a high efficiency operation. The control device has been widely used in a coal preparation plant in Shanxi Province.

3.3 Application of weighing sensor in filter press

In recent years, with the development of automation and intelligent technology, online and dynamic weighing sensor measurement technology has been widely applied. The whole frame of the filter press can be mounted on the weighing sensor by means of multi-point support (Fig. 2). This pressure filter wit weighing system is mainly composed of a pressure transmitter and PLC. The filter press filter cake weight can be recorded and the real-time operation parameters of filter press, such as slurry concentration, the average density of the slurry and so on. The design system has been widely applied in the mines in Inner Mongolia.

3.4 Information management platform of filter press

Compared with other equipment, there are many kinds of accessories for the filter press, such as filter plate, press cloth, various valves, pumps, hydraulic system accessories, electrical accessories, and so on. This will generate a large amount of maintenance data information. The construction of an information management platform for the filter press equipment is an important part of the intelligent filter press. Firstly, the filter press equipment file is set up to record the data of all parts inventory and consume, the maintenance cycle of the filter press and affiliated equipment.

Figure 2: Weighing sensor in filter press

This management platform provides a QR code display function. All the data information of the device can be achieved via mobile phones by scanning on the QR code. This can also prompt the technical workers for maintenance and significantly improve the level of enterprise management.

In addition, through the information management platform, operators can control the filter press through mobile intelligent terminals (such as mobile phones or iPads) and can view the real-time monitoring video and the data information of the filter press through wireless monitoring intelligent early warning system.

4. Conclusion

Filter press is the main equipment of the slime water treatment system of coal preparation plant, the intelligent operation of it is the important step to realize intelligent coal preparation plant. In this paper, an intelligent control system of filter press is designed and realize high efficient intelligent operation for filter press in many coal preparation plants. The popularization and application of the system will vigorously promote the technology development of filter press industry and speed up the pace to develop the intelligent coal preparation plant.

5. References

[1] Dongfang Wu, GU Gong. Technical innovation of comincus rapid opening filter and its application performance [C]. Symposium of 2017 China National Coal Preparation Academic Conference.

[2] Jianhua Huang. Preliminary study on the overall concept of intelligent coal preparation process [J]. Coal processing & comprehensive utilization□2017□5□:57-58.

[3] Xiangyang Tang et al., Present Situation and Applications of Machine Vision' s Key Techniques [J]. Journal of Kunming University of Science and Technology (Science and Technology),2004,29(2):36-39.

[4] Zhongkuan Wei,Hailong Li. The present production and application of coal filter press in China [J]. Coal processing & comprehensive utilization□2007□1□:12-13.

Index

Disclaimer